Faraday Division of the Royal Society of **Chemistry**, previously the Faraday Society, founded in 1903 to promote the study of sciences lying between Chemistry, Physics and Biology.

EDITORIAL STAFF

Editor
Philip Earis

Deputy editor
Mary Crichton

Publishing assistant
Rachel Dilworth

Team leader, serials production
Gisela Scott

Technical editor
Helen Lunn

Publisher
Janet Dean

Faraday Discussions (Print ISSN 1359-6640, Electronic ISSN 1364-5498) is published 3 times a year by the Royal Society of Chemistry, Thomas Graham House, Science Park, Milton Road, Cambridge, UK CB4 0WF. Volume 137 ISBN: 0 85404 118 4
ISBN-13: 978 0 85404 118 3

2008 annual subscription price: print+electronic £519, US $1,033; electronic only £467, US $929. Customers in Canada will be subject to a surcharge to cover GST. Customers in the EU subscribing to the electronic version only will be charged VAT. All orders, with cheques made payable to the Royal Society of Chemistry, should be sent to RSC Distribution Services, c/o Portland Customer Services, Commerce Way, Colchester, Essex, UK CO2 8HP.
Tel +44 (0) 1206 226050;
E-mail sales@rscdistribution.org

If you take an institutional subscription to any RSC journal you are entitled to free, site-wide web access to that journal. You can arrange access *via* Internet Protocol (IP) address at www.rsc.org/ip. Customers should make payments by cheque in sterling payable on a UK clearing bank or in US dollars payable on a US clearing bank. Periodicals postage is paid at Rahway, NJ and at additional mailing offices. Airfreight and mailing in the USA by Mercury Airfreight International Ltd., 365 Blair Road, Avenel, NJ 07001, USA.

US Postmaster: send address changes to *Faraday Discussions*, c/o Mercury Airfreight International Ltd., 365 Blair Road, Avenel, NJ 07001. All despatches outside the UK by Consolidated Airfreight.

PRINTED IN THE UK

Faraday Discussions documents a long-established series of *Faraday Discussion* meetings which provide a unique international forum for the exchange of views and newly acquired results in developing areas of physical chemistry, biophysical chemistry and chemical physics.

ORGANISING COMMITTEE, Volume 137

Chair
J P Reid (Bristol, UK)

Editor
P B Davies (Cambridge, UK)

C George (CNRS-Lyon, France)
J A Goree (Iowa, USA)
P H Kaye (Hertfordshire, UK)
D McGloin (St Andrews, UK)

FARADAY STANDING COMMITTEE ON CONFERENCES

Chair
C D Bain (Durham, UK)

K J Edler (Bath, UK)
A J Orr-Ewing (Bristol, UK)
G Jackson (Imperial, UK)
A Rodger (Warwick, UK)

© The Royal Society of Chemistry 2008. Apart from fair dealing for the purposes of research or private study, or criticism or review, as permitted under the Copyright, Designs and Patents Act 1988 and Related Rights Regulations 2003, this publication may only be reproduced, stored or transmitted, in any form or by any means, with the prior permission in writing of the Publishers or in the case of reprographic reproduction in accordance with the terms of licences issued by the Copyright Licensing Agency in the UK. US copyright law applicable to users in the USA. The Royal Society of Chemistry takes reasonable care in the preparation of this publication but does not accept liability for the consequences of any errors or omissions.

Royal Society of Chemistry: Registered Charity No. 207890.

⊚The paper used in this publication meets the requirements of ANSI/NISO Z39.48-1992 (Permanence of Paper).

The Spectroscopy and Dynamics of Microparticles

University of Bristol, UK
2–4 July 2007

FARADAY DISCUSSIONS
Volume 137, 2008

RSC Publishing

The Spectroscopy and Dynamics of Microparticles

Faraday Discussions
www.rsc.org/faraday_d

A General Discussion on The Spectroscopy and Dynamics of Microparticles was held at University of Bristol, UK on 2nd, 3rd and 4th July 2007.

RSC Publishing is a not-for-profit publisher and a division of the Royal Society of Chemistry. Any surplus made is used to support charitable activities aimed at advancing the chemical sciences. Full details are available from www.rsc.org

CONTENTS

ISSN 1359-6640; ISBN 0-85404-118-4
ISBN-13 978-085404-118-3

Cover
See Hinrich Grothe, Heinz Tizek and Ismael K. Ortega, *Faraday Discuss.*, 2008, **137**, 223–234.
Transitions of metastable phases in cloud particles have been detected by electron microscopy, X-ray diffraction and vibrational spectroscopy.

Image reproduced by permission of Dr Hinrich Grothe, from *Faraday Discuss.*, 2008, **137**, 223.

INTRODUCTORY LECTURE

9 **Linear and non-linear spectroscopy of microparticles: Basic principles, new techniques and promising applications**
Richard K. Chang and Yong-Le Pan

PAPERS AND DISCUSSIONS

37 **Identification of biological microparticles using ultrafast depletion spectroscopy**
Francois Courvoisier, Luigi Bonacina, Véronique Boutou, Laurent Guyon, Christophe Bonnet, Benoit Thuillier, Jerome Extermann, Matthias Roth, Herschel Rabitz and Jean-Pierre Wolf

51 **Vibrational exciton coupling in pure and composite sulfur dioxide aerosols**
Ruth Signorell and Martin Jetzki

Enjoyed this discussion?
How about some others...

Faraday Discussions provide a unique discussion forum for original research in physical chemistry, chemical physics and biophysical chemistry.

Previous volumes of interest include:

Chemical Evolution of the Universe
Faraday Discussion 133
This title discusses recent astronomical observations of molecules in various regions and eras of the Universe, and focuses on describing the processes that determine the chemistry.

Atmospheric Chemistry
Faraday Discussion 130
Work is presented in the areas of gas phase spectroscopy, chemical kinetics and photochemistry; atmospheric processes at the air/liquid and air/solid interface; chemical field measurements in the atmosphere and instrument development; development of chemical mechanisms and interpretation of field data through modelling; satellite measurements; and interaction of atmospheric chemistry and climate change.

The Dynamics and Structure of the Liquid-Liquid Interface
Faraday Discussion No. 129
This volume addresses the distinct approaches that have been taken to study liquid-liquid interfaces. The underlying theme is the convergence of the diverse experimental and computational approaches that have been pursued to understand structure, dynamics and transport phenomena associated with liquid-liquid interfaces.

To view the full contents and, where available, purchase an article or a volume visit the website

RSCPublishing   www.rsc.org/faraday_d

Registered Charity Number 207890

65	**MicroParticle photophysics illuminates viral bio-sensing** S. Arnold, R. Ramjit, D. Keng, V. Kolchenko and I. Teraoka
85	**Resonance-based light scattering techniques for investigation of microdroplet processes** Asit K. Ray, Venkat Devarakonda and Zhiqiang Gao
99	**General Discussion**
115	**Charge and charging of nanoparticles in a SiH_4 rf-plasma** W. W. Stoffels, M. Sorokin and J. Remy
127	**Rapid transport of nano-particles having a fractional elementary charge on average in capacitively-coupled rf discharges by amplitude-modulating discharge voltage** Masaharu Shiratani, Kazunori Koga, Shinya Iwashita and Syota Nunomura
139	**Interaction between single dust grains and ions or electrons: laboratory measurements and their consequences for the dust dynamics** Jiří Pavlů, Ivana Richterová, Zdeněk Němeček, Jana Šafránková and Ivo Čermák
157	**Microparticles in plasmas as diagnostic tools and substrates** Gabriele Thieme, Ralf Basner, Ruben Wiese and Holger Kersten
173	**Oxidation of biogenic and water-soluble compounds in aqueous and organic aerosol droplets by ozone: a kinetic and product analysis approach using laser Raman tweezers** Martin D. King, Katherine C. Thompson, Andrew D. Ward, Christian Pfrang and Brian R. Hughes
193	**General Discussion**
205	**The influence of small aerosol particles on the properties of water and ice clouds** T. W. Choularton, K. N. Bower, E. Weingartner, I. Crawford, H. Coe, M. W. Gallagher, M. Flynn, J. Crosier, P. Connolly, A. Targino, M. R. Alfarra, U. Baltensperger, S. Sjogren, B. Verheggen, J. Cozic and M. Gysel
223	**Metastable nitric acid hydrates—possible constituents of polar stratospheric clouds?** Hinrich Grothe, Heinz Tizek and Ismael K. Ortega
235	**Is atmospheric aerosol an acrosol? A look at sources and variability** Ruprecht Jaenicke
245	**Understanding hygroscopic growth and phase transformation of aerosols using single particle Raman spectroscopy in an electrodynamic balance** Alex K. Y. Lee, T. Y. Ling and Chak K. Chan
265	**Deliquescence behaviour of single levitated ternary salt/carboxylic acid/water microdroplets** L. Treuel, S. Schulze, Th. Leisner and R. Zellner
279	**The complex refractive index of atmospheric and model humic-like substances (HULIS) retrieved by a cavity ring down aerosol spectrometer (CRD-AS)** E. Dinar, A. Abo Riziq, C. Spindler, C. Erlick, G. Kiss and Y. Rudich
297	**General Discussion**
319	**Characterization of microparticles with driven optical tweezers** Tiffany A. Wood, G. Seth Roberts, Sarayoot Eaimkhong and Paul Bartlett
335	**Optical manipulation of airborne particles: techniques and applications** David McGloin, Daniel R. Burnham, Michael D. Summers, Daniel Rudd, Neil Dewar and Suman Anand
351	***In situ* comparative measurements of the properties of aerosol droplets of different chemical composition** Jason R. Butler, Laura Mitchem, Kate L. Hanford, Lennart Treuel and Jonathan P. Reid

367	**The spectroscopy and chemical dynamics of microparticles explored using an ultrasonic trap** N. J. Mason, E. A. Drage, S. M. Webb, A. Dawes, R. McPheat and G. Hayes
377	**Using dynamic light scattering to characterize mixed phase single particles levitated in a quasi-electrostatic balance** U. K. Krieger and A. A. Zardini
389	**Elastic light scattering from free sub-micron particles in the soft X-ray regime** H. Bresch, B. Wassermann, B. Langer, C. Graf, R. Flesch, U. Becker, B. Österreicher, T. Leisner and E. Rühl
403	**General Discussion**

CONCLUDING REMARKS

425	**The spectroscopy and dynamics of microparticles** Paul Davidovits

ADDITIONAL INFORMATION

431	**Poster titles**
435	**List of participants**
437	**Index of contributors**

PAPER

Linear and non-linear spectroscopy of microparticles: Basic principles, new techniques and promising applications†

Richard K. Chang* and Yong-Le Pan

Received 9th July 2007, Accepted 26th July 2007
First published as an Advance Article on the web 28th September 2007
DOI: 10.1039/b710441n

In the introduction a brief recollection is made of how one of us (RKC), accidentally, got into this field of linear and nonlinear spectroscopy of a dielectric micro-particle that can be treated as a micro-cavity or a micro-resonator. The basic principles of whispering gallery modes (WGMs) and their relationship with electromagnetic theory are presented. To simplify the mathematics, we only discuss an example from a 2-d case of light illumination perpendicular to the fiber axis. This 2-d example has relevance to semiconductor circular disk lasers, nonlinear optics in torroids, fibers and spheres at the tip of a fiber. The internal and near-field distribution of a WGM are graphically plotted to give the reader a chance to get a physical understanding of the spatial distribution as well as spectral distribution of WGMs. Several new techniques that enable the measurements of: (1) nanometer changes in the cladding diameter over a centimeter length of fiber; (2) some aspects of the morphology of micro-particles by elastic scattering; and (3) biochemical reactions at the interface of liquid media with a sphere at the end of a fiber. A few interesting nonlinear optical experimental results pertaining to stimulated Raman scattering (SRS) are touched upon. We present some preliminary results for promising applications in the area of bioaerosols. These include ambient aerosol characterization and identification with elastic scattering, fluorescence spectroscopy, and other optical and/or biochemical identifiers.

1. Introduction

Traditionally most papers begin with a statement of how important and enabling this field is in technology and its connection with other fields of science. I would like to devote some time to relate to you how one of us (RKC) got involved with this subject of spectroscopy of micro-particles. In 1979, Professor John B. Fenn from the Department of Chemical Engineering at Yale, asked me to help him direct a graduate student named El-Hang Lee in an optics related PhD thesis. At that time I was interested in developing new techniques for combustion diagnostics. The tidal wave created by Surface Enhanced Raman Scattering (SERS) had just reached my laboratory. Knowing that a sphere can model soot, I assigned El-Hang Lee to a topic

Yale University, Department of Applied Physics, P.O. Box 208284, New Haven CT 06520-8284, USA. E-mail: Richard.Chang@yale.edu

† The HTML version of this article has been enhanced with additional colour images.

that involved polystyrene latex (PSL) spheres with an efficient fluorescent dye and adhering to the glass prsim–water interface, that were excited by non-planar waves. This work resulted in a paper called "Angular distribution of fluorescence from liquids and monodispersed spheres by evanescent wave excitation."[1]

Towards the end of a day in 1979, Robert E. Benner, then a post-doctoral fellow in my laboratory and I were having a discussion about the intensity ratio of a single fluorescent tagged sphere and the Raman signal of the H–O stretching band. We borrowed some monodispersed fluorescent tagged PSL spheres from El-Hang Lee and diluted his sample solution many times in order to get only a few PSL spheres per milliliter. When we shined a laser through an optical cell containing the few PSL spheres suspended in water that had drifted into the "sample volume," a broad fluorescent dye profile with many sharp peaks appeared. A sample volume refers to the intersection volume between the focused laser beam and the focal point of the collection lens. It was surprising to see this spectrum with many sharp peaks because our previous measurements involving many monodispersed spheres in a sample volume gave us the typical smooth emission spectrum of a fluorescent dye when excited under the same conditions. We had been expecting that a single PSL sphere would also give us a typical fluorescent dye smooth curve.

Some minutes later, another PSL sphere drifted through the sample volume and the resultant fluorescent curve also had sharp peaks. After the 4th event, we noticed that none of the peaks lined up with each other. Was it conceivable that each PSL sphere had a slightly different radius? With large numbers of PSL spheres, the sharp peaks merged into a broader peak because the average smoothed out the sharp peaks. We immediately attributed these unequally spaced peaks as "random noise." Feeling a little discouraged, we went home.

During the next few days, I vaguely recalled a conference I had attended where Dr Arthur Ashkin had given an invited talk on the optical levitation of a liquid droplet.[2] A droplet was evaporating in the laboratory and whenever the droplet circumference divided by the incident wavelength equalled to some fixed numbers, the scattering would be enhanced and hence the levitational force would be increased. Simultaneously, the increase in the scattered intensity is accompanied by the increase in internal electromagnetic fields that were localized near the rim of the sphere. This enhanced internal field increased the fluorescence. With the help of EM theorist Peter Barber, it became clear to us that within the wide fluorescence spectra there were some wavelengths that were resonant with the PSL spheres. These wavelengths were behaving like microcavities/microresonators with varying high Q values.

Once we understood the microcavity analogy, we realized that the monodispersed bottle of PSL spheres had a distribution of sizes, but mainly peaked at the advertised radius. That was why no two PSL spheres were exactly the same and the deviation from the average smeared out the resonance peaks of the individual microcavity or PSL spheres. It was why El-Hang Lee's experiments with many PSL spheres had smooth fluorescent spectra and Robert Benner's single PSL sphere had many peaks. Once we knew the internal and external electric field distribution for some particular shaped cavity, we could use ray-tracing technique to explain many phenomena that could occur in linear and non-linear spectroscopy.

Now, the 2-dimensional (2-d) microcavity (μ-cavity) enabled the design of several photonic devices needed in the telecommunication field that had very extensive research and development endeavors. Therefore, the topics to be reviewed of light interacting linearly and non-linearly with the various shaped microcavities are too numerous to represent in this introductory paper.

In this paper, we begin with the Fermi Golden Rule and Purcell's statement about the modified density of states for a cavity.[3] We spend sometime highlighting the basics of whispering gallery modes (WGMs) that are unique to highly symmetrical microstructures. These dielectric microstructures can be the basis of a μ-cavity. We want to simplify the mathematics associated with 3-d cavities by emphasizing the 2-d optical cavities, namely with resonances that occur for a disk, a torroid and an equatorial plane of a sphere on the tip of a fiber.

We present examples of elastic scattering of an infinite length dielectric cylinder (approximate model to a fiber) with a beam of light shining perpendicular to the long axis. Then we present current research using elastic scattering by irregularly shaped ambient aerosols when the scattered light is detected from large angles ($\theta \approx 90°$ to $180°$, and $\phi = 0°$ to $360°$) simultaneously. We then move on to inelastic scattering, namely fluorescence from irregularly shaped ambient aerosols where the whole fluorescence spectrum is dispersed and detected by a multi-anode PMT (32-anodes). The topic of lasing in highly symmetrical and slightly perturbed, 2-d cavities is also presented. The emphasis of this paper reveals the authors' current interests, namely μ-cavities and detection/identification of ambient aerosols. The topics of nonlinear optical effects in 2-d semiconductors, glass, polymers and photonic band-gap material require a separate conference by themselves. Suffice to say that some of the key issues still pertaining to nonlinear optics are discussed here. The paper concludes with some conjectures about promising applications.

2. Basic principles of WGMs

2.1 The electromagnetics of WGMs

In an extended sample, the Fermi Golden Rule states that the transition rate is the product of the matrix element, the intensity associated with one photon emission, and the density of states for radiation. Edward Purcell proposed some sixty years ago (Purcell 1946), that the Fermi Golden Rule would be valid for photon emission in a cavity if the density of states is modified.[3] The atomic matrix element in the Fermi Golden Rule is not affected by the size because the particles of interest in this paper are much larger than nanometer size. Hence, there is no quantum confinement effect in μ-particles. For one photon emission in extended media it is just a plane wave. However, for particles in the micrometer size range, the electric field (with wavelength in the visible range) can be greatly modified by the structure of the particle. This EM modification can be grouped into the density of state of the radiation in the Fermi Golden Rule. See Fig. 1.

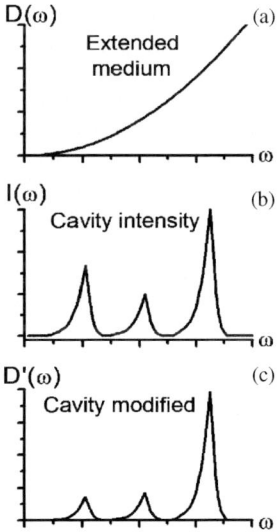

Fig. 1 (a) Density of states for radiation in an extended medium. (b) Scattered intensity as a function of photon frequency. Each peak represents a resonance with the whispering gallery modes (WGMs). (c) The modified density of states due to the presence of the μ-cavity.

In a high symmetry particle, such as a spheroid or a cylindrical fiber, elastic scattering is the important parameter in assessing the effect of morphology on the fluorescence emissions spectra or on the Raman scattering spectra. The size parameter is defined as the particle circumference divided by the effective wavelength, $x = 2\pi a/(\lambda/m)$, where a is the particle radius, λ is the incident wavelength and m is the relative refractive index. In the long wavelength limit, $x < 1$, often referred to as the Debye limit where the electric field within the particle is uniform for plane wave illumination.[4-6] The intermediary range, $x = 1$ to 500, is often referred to as the mesoscopic region. It is an interesting region where all the resonances of the cavity (with or without coupled cavity) may require Fourier transform time domain (FTTD) calculations. In addition, there exists the exact wave formulism such as Lorenz–Mie Theory for perfect spheres, infinite length cylinder and the T-matrix formalism for spheroids. The T-matrix computation becomes computer intensive when $x > 10$. We are in the ray-tracing limit when $x \gg 1$, which oftentimes provides the needed insight into that particular resonance of the cavity.

Much of the physical interpretations related to high symmetry shaped structures in 2-d and 3-d can be reached by using ray-tracing techniques. For asymmetric shaped structures, referred to as asymmetric resonant cavity[7] (ARC), the trajectory of the billiard ball bouncing is equivalent to ray-tracing. During the bounce, the ball obeys Snell's Law. Every bounce off the interface is specified by two numbers: (1) a subtended arc with angle ϕ (relative to the horizontal axis) and (2) the sine of the incident angle ($\sin \chi$). They form the basis of a Poincaré surface of section (SOS), well known in Chaos Theory. The values ϕ and $\sin \chi$ are marked as a dot in the ϕ vs. $\sin \chi$ diagram. The billiard ball is lost or escapes when the incident angle χ is just below the critical angle for total internal reflection $\chi = \sin^{-1}(1/m)$, where m is the relative index of refraction. This is the location and the angle that radiation emerges from the ARC. Thus it is important to image the emission point on the ARC surface while recording the far-field intensity distribution.

For a sphere of any size, the Lorenz–Mie formalism (wave optics) is the most accurate means of calculating the resonant modes of a dielectric sphere.[8] The resonances of a dielectric sphere have been called by many names such as ripple patterns, morphology dependent resonances (MDRs) and lately, whispering gallery modes (WGMs). The acoustical equivalent of the WGMs (in the optical regime) is available to experience at St. Paul's Cathedral in London.

For each angular momentum l mode, there are two distinct types of modes, labeled as TE and TM, where there is no radial electric (magnetic) field component. The boundary conditions for the scattered field just outside the sphere and for the internal field just inside the sphere gives four linear equations for two scattered coefficients (a_l, b_l) and for two interior coefficients (c_l, d_l). The scattered coefficients (a_l, b_l) are defined in eqn (1) and (2). A pole is reached when the real part of any of the 4 coefficients changes sign as x is varied. The poles of these four coefficients are independent of the incident beam regardless whether the beam is Gaussian or plane wave or an evanescent wave supplied by a tapered fiber. The poles of the b_l and c_l are the same and also the poles of a_l and d_l are the same. The poles are complex, where the real part is the resonant frequency of a WGM wave and the imaginary part is the inverse lifetime ($1/\tau$) of the mode.

$$a_l = \frac{J_l(x)J_l'(mx) - mJ_l'(x)J_l(mx)}{H_l^{(1)}(x)J_l'(mx) - mH_l^{(1)'}J_l(mx)} \quad (1)$$

$$b_l = \frac{J_l(mx)J_l(x) - mJ_l(mx)J_l(x)}{H_l^{(1)'}(x)J_l(mx) - mH_l^{(1)}(x)J_l(mx)} \quad (2)$$

The all important quality factor, Q, of a mode is equal to $\omega/\Delta\omega$ and is also equal to $\omega\tau$. In totally transparent particles (with no absorption), the imaginary part of the poles is related to the amount of light leakage associated with WGMs. For a plane wave incident on an infinite flat surface, these evanescent waves carry zero net radiation energy. However, because of the curved surface of a sphere or cylinder, the radiation leakage is already accounted for with the Lorenz–Mie Theory. Whenever there is an abrupt change of refractive index, *e.g.* the corners of a polygon where the evanescent waves are converted into propagating waves, the corners of the polygon appear bright.

The WGMs for a dielectric sphere (with index of refraction n[m]) are specified by the usual notations used in quantum mechanics where the total angular momentum is specified by l[n], the z component of the angular momentum is specified by m[m], the radial quantum number by ν, and the spherical harmonic by $Y_{lm}[Y_{nm}]$. Many EM computational scientists use the notation inside the square bracket. Differences in notation are a source of confusion and often act as a barrier for EM computational scientists to communicate with scientists steeped in quantum mechanics.[9]

The assignment (in terms of mode order ν and angular momentum l) of WGMs observed through an optical spectrograph for a given range wavelength is not a simple unique task. A direct and reliable approach for assigning the angular and the radial quantum numbers of a WGM is to compare the measured WGM peaks with the calculated peaks. In the calculated peaks, the incremental X must be chosen to be commensurate with the linewidth of WGM peaks.

The following three measurements are the easier ones to perform on a single particle: (1) the scattering coefficient Q_{sca}; (2) the extinction coefficient Q_{ex}; and (3) the back-scattered coefficient Q_{back}. The three coefficients relate simply to a_l and b_l in the following way. See eqn (3), (4) and (5).

$$Q_{sca} = 2/X^2 \sum (2l+1)(|a_l|^2 + |b_l|^2) \qquad (3)$$

$$Q_{ext} = 1/X \sum (2l+1)\mathrm{Re}(a_l + b_l) \qquad (4)$$

$$Q_{back} = 1/X^2 \sum (2l+1)(\ 1)^n (a_l - b_l)^2 \qquad (5)$$

where Q_{sca}, Q_{ex} and Q_{back} coefficients are the scattering, extinction and backward cross-sections divided by the cross-sectional area of the particle. Generally, one just measures one of the above three coefficients *versus* wavelength and compares them to the WGMs in experiments such as Raman scattering spectra and/or fluorescence spectra.

2.2 Internal and near-field distribution of a WGM

A comparison between the experimental with the theoretical elastic scattering curve gives us some confidence that they agree reasonably well.[10,11] See Fig. 2.

The experiment for measuring elastic scattering as a function of wavelength is performed with a spectrometer and a tungsten light bulb aligned in the x-direction, the fiber length aligned in the z-direction and the scattered light collection is along the y-axis with an acceptance angle relative to the y-axis. When a broad beam of light is used to illuminate a long section ($\approx$ 2 cm) of a fiber and the detection is limited to the central portion of the illuminated region of that fiber, then that fiber is considered to be a convenient sample to use for scattering experiments. Additionally, that fiber has no end problem, is stationary, needs no levitation, and does not need to be placed on a substrate. The well-accepted 3-d spherical Lorenz–Mie Theory reduces the problem to a single 2-d problem. For a glass sphere formed on a fiber tip, much of its WGM characteristics of the equatorial plane (θ = 90°, a plane perpendicular to the z-axis) can be gleaned from a 2-d problem examination.

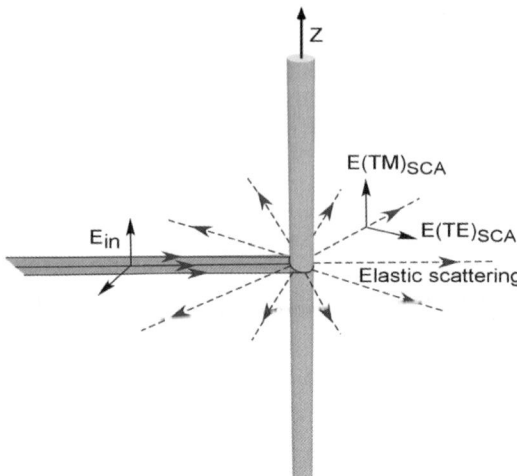

Fig. 2 The experimental geometry for an incoming plane wave to probe the WGMs. Elastic scattering results in all ϕ (azimuthal) directions. The z-axis is reserved to be along the fiber-length axis. Polarization is TE when the electric field is perpendicular to the z-axis. Polarization is TM when the electric field is parallel to the z-axis or the fiber-length axis.

As a function of X (achieved by varying the wavelength), the resonance condition is reached (referred to as ON-resonance) or resonance condition is not reached (referred to as OFF-resonance). With $X = 45.6$, we are in an OFF-resonance condition. Depending on the relative amplitude and phase between the ON-resonance with the non-resonant background of the OFF-resonance, Fano-type lineshapes can be expected to appear in an elastic scattering spectra. See Fig. 3 and 4 for the experimental and calculated scattered intensity profiles as a function of spectrometer wavelength. An ordinary quartz iodine lamp is used as a light source. The scattering angle is at $\phi = 90°$ with a collection angle subtending $14°$. The calculated curves are shown below the experimental curves for TE and TM polarizations. The notations (l, v) of the calculated peaks are l, the mode number and ν, the mode order.

How does the resonance affect the internal field distribution of a fiber as well as the external near-field distribution that starts at the cladding-air interface to a short distance away from the cladding? When $X = 45.6$, it is an OFF-resonance value. The internal field distribution (for TE polarization) in a plane perpendicular to the length of the fiber (defined as the z-axis) is shown in Fig. 5. The intensity outside the cylinder has been deliberately set to zero to bring out the fiber boundary. The peak intensity in the interest region is $7\times$ the incident intensity. This internal "hot spot" can be qualitatively explained by ray optics (red arrows in right column of Fig. 5). The rays near the fiber edge (red arrows in right column of Fig. 5) are refracted by the cylindrically curved illuminated face and then experience total internal refraction (TIR) near the center of the cylindrically curved shadow face. Both the central portion of the illuminated and shadow fiber faces refract the central rays (green arrows) to form an external "hot spot." Similar statements can be made for a spherical particle. The enhanced intensity at the internal and external "hot spots" for spheres exhibits much greater ($10^2\times$) than that for the cylindrical fiber. Using laser-induced breakdown spectroscopy (LIBS) inside or outside of a spherical droplet, we were able to discern for different size droplets and refractive indices, where LIBS first occur either at the inside or outside "hot spot."

When $X = 45.726$, it is an ON-resonance value for a WGM of mode number $l = 53$ and mode order $\nu = 3$. The intensity distribution for this particular WGM is shown in Fig. 6 for TE polarization, again the external field is set to zero, to bring

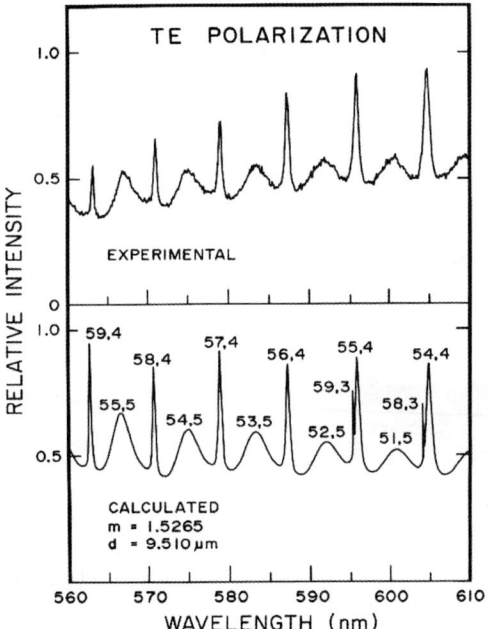

Fig. 3 Comparison between the experimental and calculated elastic scattering profiles at ϕ (TE polarization). The scattering angle is at 90° with a collection angle subtending 14°. The calculated results were for a glass fiber with index of refraction $m = 1.5265$ and $d = 9.510$ μm. The labeled numbers for each peak in the calculated spectrum represent (l, ν) the mode number and the mode order, respectively.[10]

out the fiber–air boundary of the inner and outer fields. Note that for (53.3) WGM, there are 53 sharp peaks going from $\phi = 0°$ to 180° and 3 concentric rings going

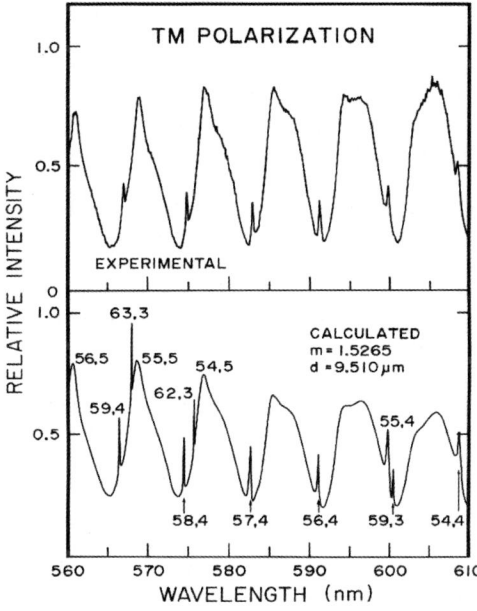

Fig. 4 Same as for Fig. 3, except for TM polarization.[10]

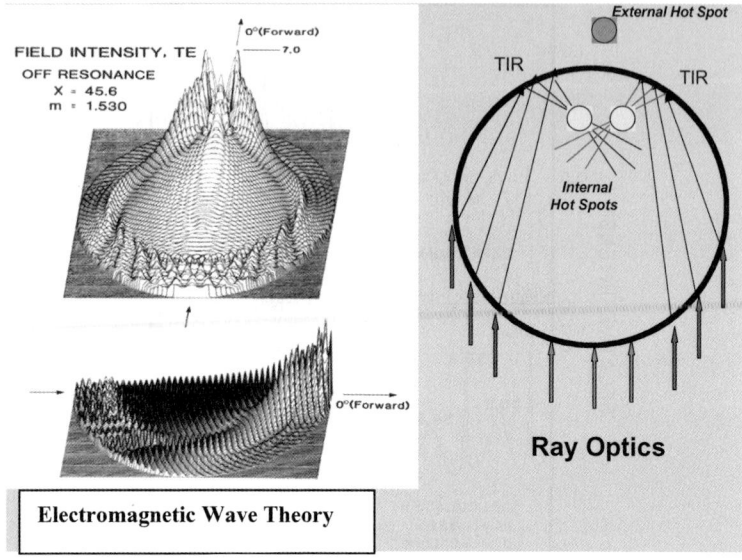

Fig. 5 The internal field intensity of a fiber with $X = 45.6$ and the relative index of refraction $m = 1.530$. The incident plane wave is OFF resonance and has two internal focal areas towards the shadow phase with an intensity increase 7 times (7.0). These two peaks are depicted by ray optics (red arrows in right column) and often referred to as the internal "hot spot." They are usually lumped into one internal "hot spot" region (two yellow circles on right). The other "hot spot" called the "external hot spot," not shown on the left, but shown on the right (green circle), is just outside the shadow phase. It results from the 2-d circular lens with the rays (depicted by the green arrow) near the central portion of the lens (see right column). On the left column, the external field intensity is suppressed to 0 for contrast enhancement purposes and not in the calculation.[11]

from $r = a$ to $r = 0$. Each of the 3 rings with $(2l + 1)$ sharp peaks is uniform in intensity. The ring with the highest intensity has peaks that are 54× the input intensity. Using the coherent sum of the electric field E, this WGM plus all the OFF-resonance modes, the net intensity (EE^*) enhancement is 68.2×. Similar statements could be made for s sphere except the internal "hot spot" could be as much as two orders of magnitude larger than the 2-d case.

Fig. 7 shows the average TE intensity around a circle centered at the z-axis as a function of the radius r. Within the fiber $r < a$, the number of peaks encountered going from $r = a$ to $r = 0$, corresponds to the mode order ν. For mode order ν, there is one peak just beneath the interface for mode order $\nu = 1$, for mode order $\nu = 2$, there are 2 peaks, for mode order $\nu = 3$, there are 3 peaks, *etc.* (Fig. 7(b)–(f)). The Q-value is by far the largest for $\nu = 1$ and decreases as the mode order number increases. By $\nu = 5$, no discernable peaks are found. Peaks can only reside in band $r = a$ to $r = a/m$ where m is the relative index of refraction. For the OFF-resonance case, Fig. 7(a), the contribution of the internal "hot spot" to the ϕ angle averaged for all values of r, there were no discernable peaks.[11]

The peaks are standing wave patterns and exist because of the interference between the clockwise and the counter-clockwise circulating electric fields. The illumination of the left and right portion of the fiber is equal in intensity and phase. Had there been an imbalance in intensity, there would be running waves without sharp peaks, but would have the same radial distribution as for the standing wave case.

Knowing the locations of the WGMs in a 2-d plane (perpendicular to the fiber length) is useful when designing a circular disk laser. For example, the pumping of a WGM mode requires that there be maximum overlap between the WGM and the

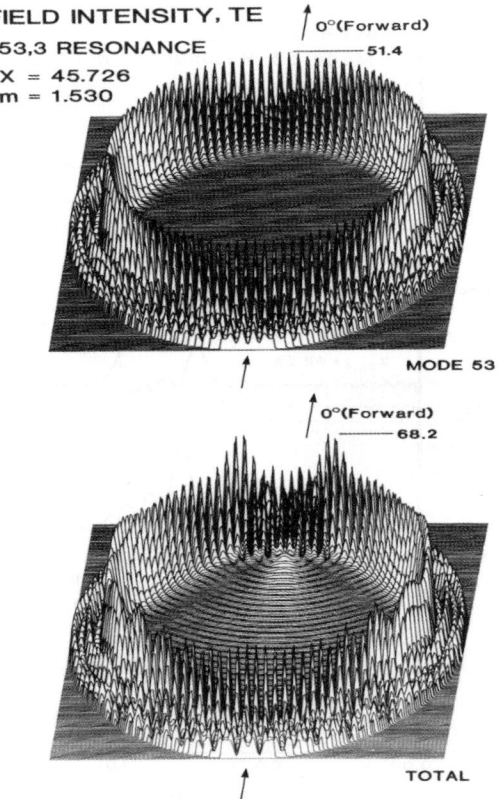

Fig. 6 Same conditions as Fig. 5, except X is tuned to be ON resonance with a WGM having mode number $l = 53$ and mode order $\nu = 3$. The top figure shows the internal distribution just with this WGM alone. For mode number of 53, there are 53 peaks from $\phi = 0°$ to $180°$ and 53 peaks from $180°$ to $360°$. For a mode order of 3, there are 3 rings of 106 peaks each. The maximum intensity for just 53.3 WGMs is $51.4\times$ and for all the modes, it is $68.2\times$.[11]

pumping source. The source may be a thin-ring electrode (for electrical current pumping) or a doughnut-shaped laser beam for optical pumping because of the WGMs, with high Q values, were along the perimeter of the disk.

The near-field spatial distribution outside the fiber, starting from $r = a$, would have been an evanescent wave had the surface been flat. However, because of the curved surface, a certain amount of light is produced tangent to the curved surface. The cavity is certainly "lossy" as opposed to a simpler cavity that is "closed," *i.e.*, there is no loss of radiation.[9] Some of the surface waves get converted to propagating waves. Note that the penetration depth gets shorter as the mode order ν decreases from 5 to 1. For $\nu = 1$, the penetration depth is less than the wavelength, which could be $\lambda = 0.532$ μm. Therefore, the separation between this test fiber and the delivery fiber should be submicron. Thus, this will require e-beam lithography. Research interests of extracting the external pump intensity into this high Q-value mode ($\nu = 1$) having the highest Q value are very intensive.[12] Energy transfer between this mode with the various shaped pump beam and fiber tip grazing the test fiber edge has also been intensively studied. One of the requirements for optimal energy transfer between the source beam or the source tapered fiber technique is to have both waves (tapered fiber and the WGM) phase-matched with good spatial overlap.

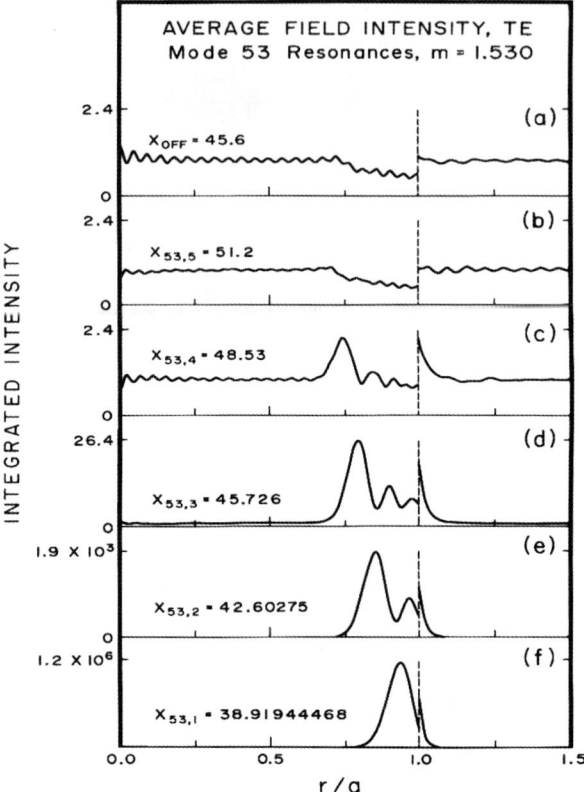

Fig. 7 The average TE intensity around a circle centered at the z-axis as a function of the radius r. (a) OFF-resonance case (b)–(f) modes $\nu = 5$ to $\nu = 1$, in decreasing order.[11]

3. New techniques

3.1 Determination of fiber cladding uniformity along the fiber length

Twenty years later, in 2001, new elastic scattering measurements were made of fiber-cladding diameter using a tunable diode laser (675 nm range with a step resolution of 0.01 nm range).[13] See Fig. 8. The old experiments were performed with a halogen iodide lamp and a spectrometer to specify the wavelength range. The newer experiment used a tunable single-mode laser diode. The aim was to measure the uniformity of WGM-based fiber-cladding diameter, approximately 12.5 µm with nanometer-diameter resolution and simultaneous measurement along a millimeter to centimeter long fiber.

Fig. 9 shows the wavelength of WGM shifts, along the length of the fiber, in a continuous manner. A diode array detector (1024 elements) with the array axis parallel to the z-axis (parallel to the fiber length axis) was used. For each wavelength of the laser diode emission, we record the scattering along the long portion of the fiber. Based upon the WGM wavelength along the fiber, the uniformity of the fiber diameter was measured on a nanometer per millimeter scale. The change in fiber radius (Δa) is related to the wavelength shift of the WGM in the following way: $\Delta a = X (\Delta \lambda)$.

Although having the WGM in the small wavelength range does not uniquely specify the mode order ν or a mode number l, the twists and turns of the WGM along the fiber length indicate how small the deviation is from an ideal fiber, 12.5 µm. Suffice to say, the higher the Q, the higher the resolution of the fiber radius measuring instruments. It appears that the highest Q we have ever measured with

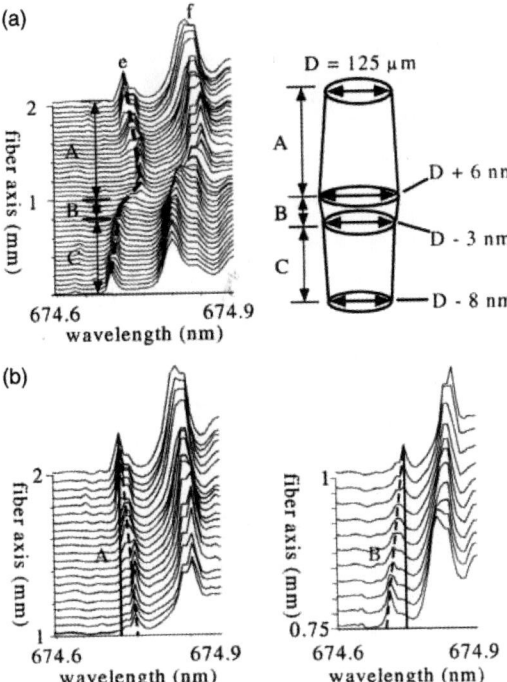

Fig. 8 (a) 3-d plot of the elastic scattering spectra of a fiber measured simultaneously along an ≈2 mm length. Resonances e and f have similar wavelength shifts. Resonance e shifts are denoted A–C. A schematic of the tapered fiber is shown on the right. (b) Expanded view of section A plotted along a 1 mm range compared with that of section B plotted along a 0.25 mm range.[13]

elastic scattering has been limited to $Q = 10^4$. There are several reasons for our failure to observe WGMs with $Q \gg 10^4$. First, the cleanliness of the surface may not be sufficient after removal of the black organic-protective layer. Second, the roughness of the sidewall surface may cause inhomogeneous broadening of the very high Q WGMs. Third, more fundamentally, there may exist inherent limitations of the elastic scattering technique for very high Q WGMs.

This third limitation of elastic scattering to reveal WGMs with $Q > 10^4$ may be a signal-to-noise ratio issue. The maximum value of a_l and b_l scattering coefficients at each pole cannot exceed 1, whereas c_l and d_l coefficients can be much greater than 1. Hence, the total number of scattering photons is proportional to $1 \times \Delta\omega$. Therefore, when one attempts to measure very high Q-value WGMs with a spectrometer, the area under the peak diminishes as $\Delta\omega$ decreases. The prospect of "finding" higher Q-value WGMs in fiber-claddings is good, if a single-mode laser diode with even smaller discrete step jump is used.[4–6]

For $Q \geq 10^6$ it is difficult to extract the linewidth of a peak when the linewidth is comparable to the spectral resolution of the spectrometer or the light source. Time domain measurements of the WGM lifetime τ may be easier to implement, knowing that $Q = \omega\tau$. As an example, if the incident radiation has a photon energy of 2 eV ($\omega = 3 \times 10^{15}$ rad s^{-1}) and $Q = 10^6$, then the WGM lifetime $\tau = 3$ ns, a time response that is readily available for fairly standard photo-detectors.

3.2 Elastic scattering from irregularly shaped particles

We now focus on the topic of elastic scattering from irregularly shaped particles, having just covered the topic of elastic scattering from highly symmetrical particles

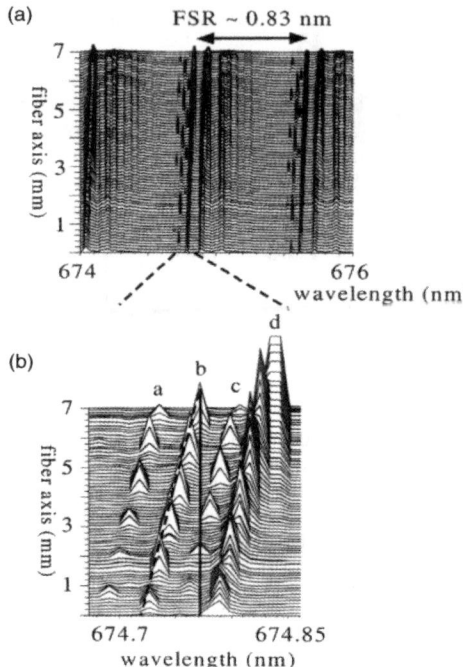

Fig. 9 (a) 3-d plot of the elastic scattering spectra of an optical fiber measured simultaneously along an ≈7 mm length. The WGM wavelength shifts along the entire length indicate that the fiber is tapered. (b) Expanded view of the four resonance modes (denoted a–d) shifted along the fiber with a similar slope. Not all the higher Q modes can be recorded because of the ≈0.01 nm steps in spectra recording. The resonance b shift (shown by the dashed line) suggests a fiber uniformity of ≈1 nm mm^{-1}.[13]

that are invariant to rotation about a given axis, *i.e.*, invariant to the angle ϕ. For lower symmetry particles, it is required that measurements and calculations include as many polar angles θ and azimuthal angles ϕ as possible. The z-axis is usually reserved for the laser propagation direction relative to the conventional 3-d coordinate system for waves emanating from a point source into the 4π stearadians. In the far field, data must be represented as a function of θ and ϕ, where the laser beam direction is always defined to be along the z-axis. We use the acronym TAOS for this Two-Angle (θ and ϕ) Optical Scattering in the 3-d measurement. The technical challenge before us is how to capture simultaneously the elastic scattering from the largest possible angle range, with the highest angular resolution possible in θ and ϕ. Most camera lenses can only capture images from a relatively small portion of 4π.[14] However, an ellipsoidal reflector can capture the scattered image over a much larger angular range. For example, as shown in Fig. 10(c), the ellipsoidal mirror allows the elastic scattered light of an on-the-fly particle to be detected for all ϕ angles between 0° to 360° and θ angles between 0° and 90° [in the forward hemisphere, see Fig. 10(c)] or θ angles between 90° and 180° [in the backward hemisphere, see Fig. 10(b)]. By giving up some ϕ data in the elastic scattering, we can expand the θ-value to include both the forward and backward hemispheres [see Fig. 10(a)]. Because the forward intensity is usually 2–3 orders of magnitude larger than in the backward intensity, this latter configuration demands a larger dynamic range from the detector, especially for the particles with big size parameter, $X > X = 2\Phi a/(\chi/m)$.

Elastic scattering has the largest signal-to-noise ratio of any optical technique. Therefore, the TAOS from single bacteria and/or aggregates of bacteria may be

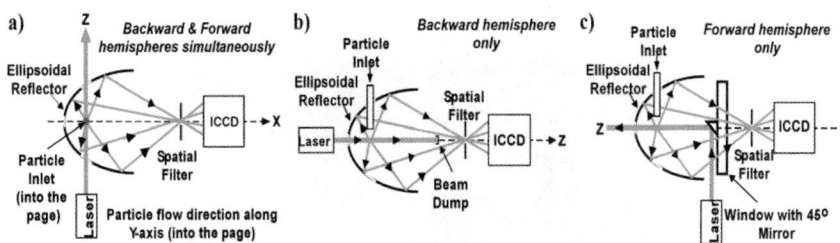

Fig. 10 (a) Top view of the setup used to simultaneously collect the scattered radiation in the TAOS patterns of single aerosols on-the-fly in the forward and backward hemispheres. Fig. 10(b) and (c) Side views of the setup used in previous (*backward or forward hemisphere only*) experiments.[15]

detected with a high-resolution CCD camera. The x–y data stored in the camera memory can be mapped back to θ and ϕ coordinates. To ensure that the mapping is unique, the largest possible range of angles is desirable. The vast amount of TAOS data per laser firing collected by the ellipsoidal mirror enabled us to compare the experimental and calculated TAOS distributions. TAOS distribution is highly sensitive to particle morphology which includes the particle shape, size, absorption, spatial distribution of the index of refraction and the smoothness of the surface. In principle, by measuring the elastic scattering of a particle for all collection angles should enable us to specify, or at least infer, some characteristics of the particle.

However, the inverse problem of taking a camera-taken TAOS image back to a real image of the irregularly shaped particle or cluster is an unrealized dream. Nevertheless, some parts of the TAOS images can be used to characterize and classify some properties of particles. For a sphere, the TAOS technique is very informative in providing the size and the index of refraction, including the real and the imaginary part of the index of refraction, For a finite length dielectric fiber, it can easily distinguish a fiber from a small sphere. Professors Paul Kaye and James Hough of the University of Hertfordshire in the UK are world leaders in this field and together they have made seminal contributions and clever instrument designs that can be deployed in a battlefield. They designed a specially shaped detector with a centered 16-channel array-chip.[16] Additionally, they initiated the technique of Asymmetry factor (Af) that is a measure of the azimuthal variation of light scattered by a particle.

A typical TAOS image of a sphere consisting of concentric rings is shown in Fig. 11[17] for the back hemisphere configuration where the θ value varies from 180° to 90°. For a fixed θ value, the ϕ variation (from 0° to 360°) lies on a circle with radius θ. For spherical particles, the infinite rotation axis can be chosen to be coincident with the direction of the laser beam. For a distorted sphere, we no longer observe a circular ring, but rather a wiggly ring. Fig. 12 (a) shows SEM photos of a roughened surface of a nearly spherical cluster consisting of 13 polystyrene latex (PSL) spheres, Fig. 12(b) the representation of the cluster used in the numerical calculations, and Fig. 12(c) a typical TAOS pattern of PSL cluster measurement in the near-backward scattering direction. Note in Fig. 12(c) the "island-like" structures in the TAOS image.

The spectacular development in computer speed and the enormous increase in computer memory have revived the interest in the seemingly untenable goal of the "inverse" problem. In addition, there have been major achievements by a small number of EM theorists who have mastered several of the approximation techniques, such as the semi-analytical T-matrix, discrete dipole approximation (DDA) and Fourier transform time domain (FTTD) algorithm. The experimentalists are now equipped with high resolution and highly sensitive CCD (single photon detection) camera technology.

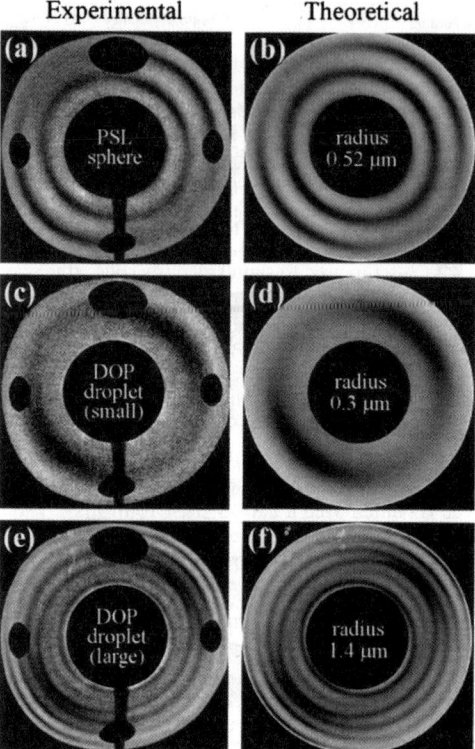

Fig. 11 TAOS patterns of calibration particles (see left column) compared with the Lorenz–Mie calculation (see right column).

Recently, direct comparison of the calculated results with T-matrix and the experimental TAOS images was attempted on a single *Bacillus subtilis* (BG). In modeling a single bacterium, many approximate shapes were considered. For example, we tried ellipsoids, two touching spheres of equal radii, and most successfully, a short cylinder with two hemispheres, one for each end of the cylinder (see the SEM image in Fig. 13).

The TAOS distribution is altered by the orientation of the axi-symmetric axis relative to the laboratory frame axes. In our experiment, the BG is expected to be tumbling at some rate, after the BG exits from a laminar flow nozzle. Because we

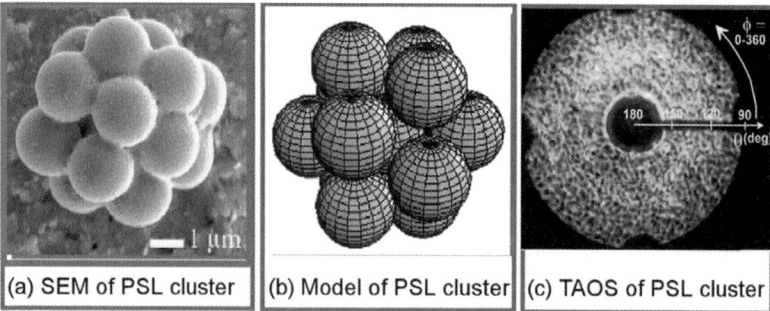

Fig. 12 (a) SEM image of a typical cluster formed by 13 PSL spheres. (b) Representation of the cluster used in the numerical calculations. (c) A typical TAOS pattern of PSL cluster measurement in the near-backward scattering direction.

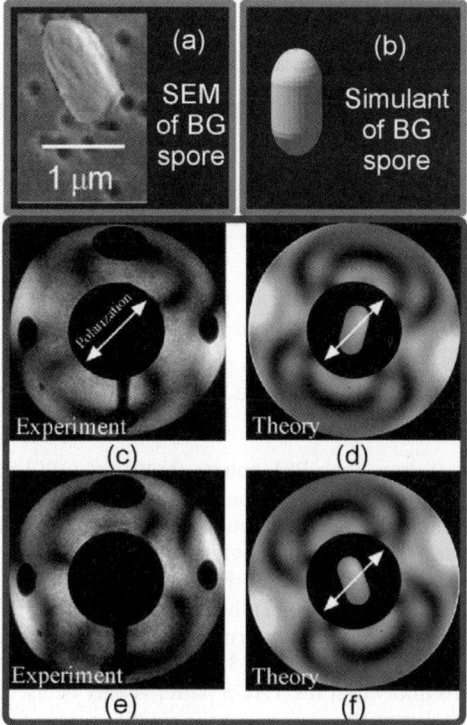

Fig. 13 (a) SEM image of a single BG spore. (b) One of the simulant model for single BG spore. (c) and (e) Typical TAOS patterns of single BG measurement in the near-backward scattering direction. (d) and (f) The corresponding numerical calculated TAOS patterns for single BG spore.

were using a short pulse (≈30 ns) to illuminate the BG, the resulting TAOS distribution is recorded as though the BG was frozen in space in a random orientation. From the many TAOS patterns we have measured for a single BG, only a few patterns have their axisymmetrical axis in a plane perpendicular to the laser beam direction (along the z-axis). The only remaining uncertainty that needs to be determined is the tilt angle of the BG's axisymmetrical axis relative to the laboratory vertical axis. We guessed that when the TAOS image was taken, the axisymmetric axis (in a plane perpendicular to the laser propagation axis) was frozen with a 30° tilt to the vertical axis (see Fig. 14). The computational results were calculated assuming various tilt angles (centered around 30°) and with an axi-symmetrical axis in the plane perpendicular to the z-axis.

The black holes in the experimental TAOS image are drilled holes through the ellipsoidal mirror in order to let the laser beam in and out and to let the particle stream enter and exit. Although these holes totally block the TAOS data in those angular regions, they were essential for the experiment and, in addition, these holes served as excellent spatial markers for the TAOS images.

Fig. 14 shows typical backward hemisphere TAOS patterns from (a) a single 1 μm-diameter PSL sphere, (b) a single BG spore (anthrax simulant), sausagelike, 1 μm in length, 0.5 μm in diameter and (c) a fiber-like riboflavin crystal about 5 μm in length. See the corresponding SEM images at the bottom row. The surface of a spherical PSL particle is smooth, whereas that of a BG spore and riboflavin particle is rough. Furthermore, the refractive indices of the PSL sphere and riboflavin are homogeneous, whereas the refractive index of a BG spore is inhomogeneous. These physical differences are manifested in the distinctive backscattered patterns depicted.

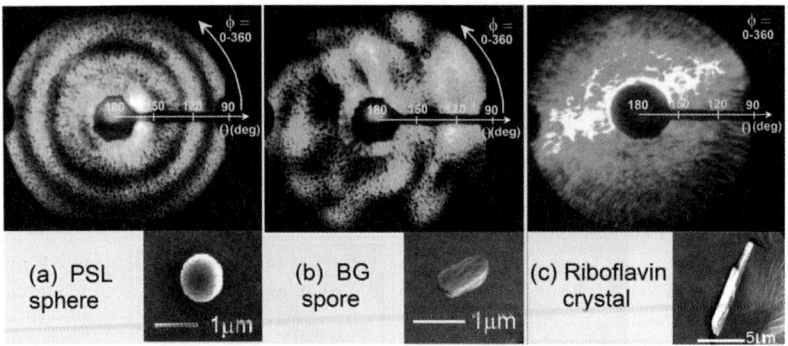

Fig. 14 Typical backward hemisphere TAOS patterns from: (a) a single 1-μm-diameter PSL sphere; (b) a single BG vegetative spore; and (c) a fiber-like riboflavin crystal. A typical SEM image of each type of particle is also shown.

The TAOS patterns from a 1-μm PSL sphere shows the well-known ring structure expected from Mie theory. By contrast, the pattern from a single BG spore is far less symmetrical, with island-like features and even less symmetrical for the fiber-like particles. The patterns from different individual BG spores are not identical because (1) the different BG spores, carried by the gas flow, are at random orientations related to the illuminating beam and (2) the surface irregularity of BG spores varies from one to another. However, all patterns are composed of intensity islands with similar sizes and shapes as shown in Fig. 13. The asymmetries and pattern differences in the TAOS patterns imply that the TAOS patterns may be useful in differentiating and classifying aerosol particles.

3.3 Nonlinear optics with WGM

The cavity enhanced electric field and the roundtrip feedback are distinguishing features of WGMs. The combination of high internal electric fields and high Q factor is the condition for fluorescent dye droplets to act as droplet lasers with optical pumping. Similarly, it is the condition for stimulating Raman scattering that has generated a great deal of research.

During Professor Nicolaas Bloembergen's 70th birthday celebration at Harvard University in 2000, some of his students, myself (RKC) included, gave talks about nonlinear optics.[18] I went through all the nonlinear optical effects that had been predicted by Bloembergen for extended liquid media. All had been seen in droplets.[18–22] There was nothing new about the nonlinear optical mechanisms inside the droplets. However, in almost every case some surprising features occurred that had not been predicted. The main source of these surprises had to do with two effects that did not exist for extended liquid media, but did exist in droplets: (1) the concept of phase-matching of plane waves, applicable in an extended media, had to be formulated into spatial overlap among the WGMs of the various electric fields and (2) the high internal "hot spot" does not have a large spatial overlap with the WGMs, compared to the much larger overlap among the WGMs.

An example of one of the surprises is the observation of many orders of stimulated Raman scattering (SRS) from a droplet.[23,24] A possible reason for this has to do with the differences in the poor spatial overlap between the "hot spot" and the WGMs and with the good spatial overlap among the WGMs themselves.[11] When a plane wave is used to illuminate a droplet, a part of the plane wave gets focused at the internal "hot spot." If the Raman gain at the "hot spot" is sufficient to produce 1st-order SRS that is on a WGM, then it serves as a pump to produce a 2nd-order SRS that is on a different WGM. Both the 1st-order SRS and the 2nd-order SRS are on their respective WGMs. The overlap length of the two SRS's is much longer than that of the "hot spot" and the WGM of the 1st-order SRS. This 1st-order cascading process continues to the nth-order SRS which produces $(n + 1)$th-order SRS. The cascade process only involves the 1st-order Raman scattering process. If the pump is in the blue wavelength region, the total light will appear as a white light source. This cascade process will continue throughout the visible and on to the near-infrared region, where higher order vibrational modes cause optical absorption and stop the SRS cascading process.

Because nonlinear spectroscopy requires the highest electric field possible (up to the sample breakdown limit), femto-second lasers are being used. The tremendous advantage of the femto-second laser pulse over the Q-switched laser pulse (10 ns) is the short-time pulse duration that delivers much less energy, but much greater intensity. Higher order nonlinearities are expected from using a femto-second laser pulse.

In contrast to using the femto-second laser, is the possibility of using cw-lasers, preferably tunable semiconductor laser diodes for very high Q-value microparticles. This provides a possible alternative measuring technique to supplement the femto-second laser. The combination of the WGMs of microparticles with photonic band-gap devices should prove to be a very powerful and versatile tool.

4. Promising applications

4.1 Monitoring morphology of ambient aerosols by elastic scattering

Recently, our laboratory instrument detected TAOS patterns of ambient aerosol particles (in the 1–10 micron range) in the US Army Research Laboratory, Adelphi, MD. Air was drawn (at the rate of 570 L min^{-1}) into the concentrator that was based on the virtual impactor principle, where the smaller particles follow the airflow while the larger particles (with larger inertia) cannot follow the airflow. These larger particles in the 1–10 μm (respiratory) range were channeled into a minor flow exit (≈ 1 L min^{-1}). The minor flow is drawn under slight negative pressure within the surrounding box. The aerosol drawn through an aerodynamic nozzle that forces the particles upon exiting the nozzle to achieve a focused laminar aerosol jet of about 2 cm in length before the aerosol jet becomes chaotic. The aerosol nozzle is positioned approximately 0.5 cm above a triggering focal volume which is defined by the intersecting volume of two laser diodes. The two diode laser beams intersect at 90° and have their emission at 685 nm and 635 nm. When the particle crosses this intersection volume, the scattered light will be at both wavelengths. The scattered light is then detected by the two PMTs with 685 nm and 635 nm interference filters, respectively. The signals from both PMTs go through an AND gate. Only when the AND gate receives a coincidence pulse, would a trigger pulse be issued by the AND gate. A trigger pulse, after some time delay, fires the Nd:YAG laser (532 nm) and turns on the intensified CCD detector that captures the TAOS pattern.

The sample volume for the aerosol is defined as the intersection of the 532 nm laser pulse and one focal point of an ellipsoidal mirror. The 532 nm laser beam is deflected away from the intensified CCD. The particle's scattered light is reflected by the ellipsoidal mirror to its second focal point. The TAOS patterns were collected from $\theta = 75°$ to 135° and the $\phi = 0°$ to 60°. Angles θ and ϕ associated with the scattering can be related directly to the x–y image of the intensified CCD.

By using the laboratory prototype, we were able to record TAOS patterns for approximately 6000 ambient particles captured during an 18-h data gathering run in the Baltimore/Washington area in early October 2004. Fig. 15 shows 20 consecutive TAOS patterns of ambient particles captured during that run. These TAOS patterns can be grossly classified into five different categories: (1) Spherical particles that give concentric rings; (2) Perturbed spherical particles that give imperfect wiggly concentric rings; (3) Particles with complex structure with random patterns; and (4) Fiber-like particles; and (5) Particles with swirl patterns.

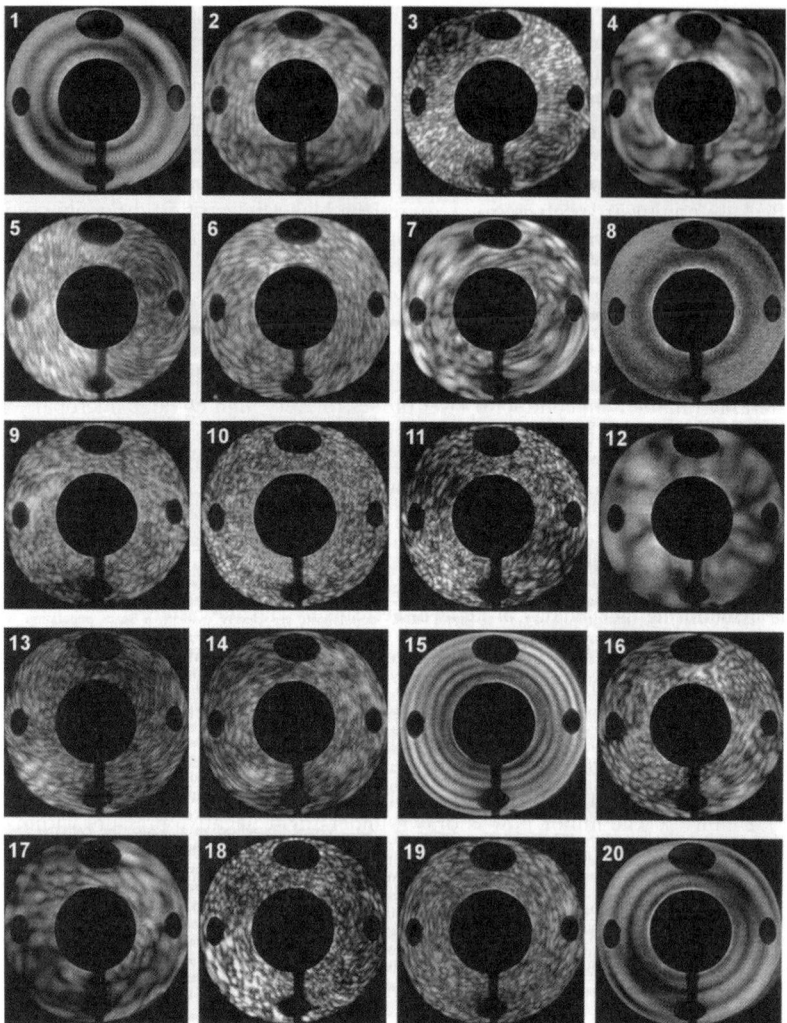

Fig. 15 TAOS patterns in the backward hemisphere recorded consecutively during a data collection run.[17]

Our limited data set gave a surprising result. Specifically, the frequency of appearance of the various TAOS patterns occurred as a function of particle size. The most noteworthy observation is that for nominally 1 μm particles, 65% of the data are spherical or perturbed spheres. The dominance of spherical (or near spherical) in 1 μm particles is consistent for the commonly accepted notion that these spheres are formed in the atmosphere by nucleation and gas-to-particle reaction processes.

For 5 μm particles, only 5% are spheres/perturbed spheres while 71% are complex structures. The complex TAOS patterns suggest that most micron-sized particles or super micron particles appear to be directly injected into the atmosphere. The many "island structures" from the complex TAOS patterns are suggestive of clusters or aggregates of smaller primary particles. The relationship of the island size and numbers are affected by the surface roughness associated with aggregates and clusters. In order to extract more information from the TAOS patterns with complex structure, pattern recognition or neural network techniques are needed to achieve more rigorous classification of the particle's scattering pattern.

To analyze the symmetry of the TAOS patterns, the degree of symmetry is defined as

$$\text{Dsym} = 1 - \sum_{\text{pixel subset}} \text{abs}\left[\frac{I^N(\theta,\phi) - I^N(\theta,\phi+180°)}{2}\right] \qquad (6)$$

where abs stands for absolute value and $I^N(\theta,\phi)$ is the normalized single pixel intensity value at angle (θ, ϕ). The Dsym is similar to the asymetrical factor Af define by Kaye and his group.[16] For a spherical particle, two pixels that have the same scattering angle θ but have ϕ values that are offset by 180° should have the same intensity resulting in a Dsym value of 1. In addition, if the aerosol has a rotation of symmetry axis that is pointed either parallel or perpendicular to the incident polarization, the scattering pattern will also have a 180° rotational symmetry, again resulting in a Dsym value of 1.

Dsym was calculated for the entire data set of ambient particles (5993 patterns). The single PSL spheres (39 patterns), single or small clusters of *Bacillus subtilis* spores (97 patterns), diesel exhaust soot particles (288 patterns), and droplets of dioctyl phthalate (200 patterns) are also included.

As expected, all of the PSL spheres fell between a Dsym value of 0.8 and 0.9. A majority ($\sim 86\%$) of the dioctyl phthalate droplets also fell into this bin, whereas all the droplets had a Dsym value above 0.7. The particulate matter in diesel exhaust was concentrated in the 0.7 to 0.8 bin although there was a considerable number of particles with a Dsym in the 0.8 to 0.9 bin. These results suggest that the particles are quite sphere-like. The Dsym values of single BG spores were fairly uniformly spread over the range from 0.2 to 0.8 with a slight peak in the 0.6 to 0.7 bin. Unlike all the other data sets, the ambient data set appears to have two peaks, one in the bin from 0.6 to 0.7 and another in the bin from 0.8 to 0.9. This implies that the ambient particles can be divided into two categories, a non-spherical class and a spherical class. The ambient data is plotted with a much finer bin scale, as shown in Fig. 16 and a Dsym value of 0.77 is determined to be the threshold level. With this division, 84% of the particles would be classified as non-spherical and 16% of the particles would be classified as spherical.

4.2 Fluorescence spectroscopy of ambient aerosols

We had just reviewed the current status of TAOS, an elastic scattering process. We will now review the latest development of an instrument that measures the fluorescence spectra of ambient aerosols, and then deflect those particles emitting a fluorescence profile similar to a specific preset fluorescence profile. The task of detecting or identifying a few harmful bacteria among mostly harmless bacteria is like "finding a needle in the haystack" and placing the needles in a separate pile for further analyses.

Fluorescence is an inelastic process, with photon energy loss to the vibrational modes of the bioaerosols. For detecting the fluorescence profile of ambient aerosol particles (in the respiratory range of 1–10 μm) the laser induced florescence (LIF) instrument is similar to the TAOS instrument. What is common in both of these instruments is the role of the concentrator and the nozzle that produces a focused aerosol beam. Similar to the TAOS instrument, each aerosol transits through the trigger volume defined by two laser diode beams. The AND gate output triggers a UV-Nd:YAG laser 4th harmonic at 266 nm. The tell-tale signature for a biological particle is the fluorescence from one of the 20 amino acids.

The amino acids fluoresce upon UV excitation with tryptophan being strongest at 359 nm (see Fig. 17). There are additional biologically related fluorescent peaks, in particular the nicotinamide adenine di-nucleotide (NADH), excited by UV radiation at 266 nm (the 4th harmonic of YAG lasers). The 266 nm UV source can also excite

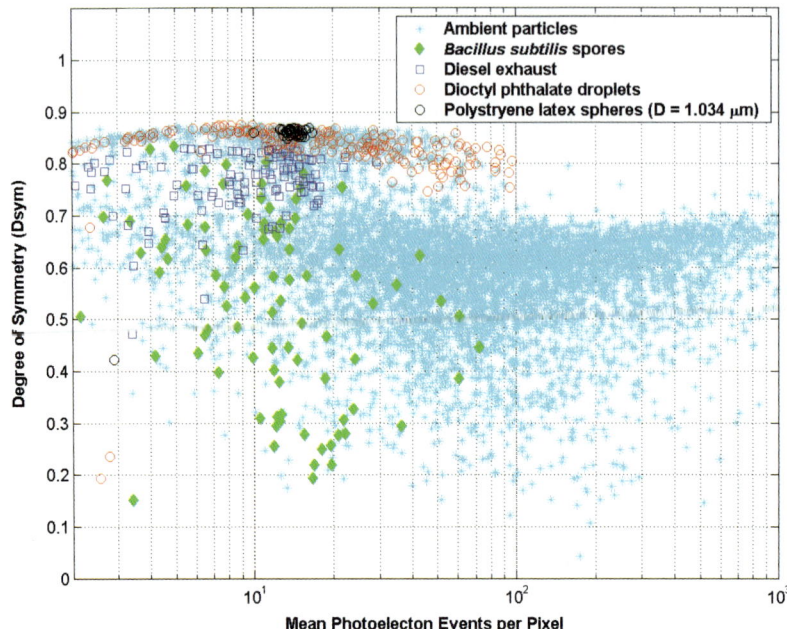

Fig. 16 Degree of Symmetry (Dsym) *versus* the mean number of photoelectron events per pixel for various aerosols.[17]

flavins such as riboflavin. The NADH is a measure of the biological activities within the bioaerosol.[25]

Our instrument can cope with exciting and recording the fluorescence spectra of 90 000 particles s^{-1}. This remarkable rate of quasi real-time data gathering is attributed to a 32-anode photomultiplier (PMT from Hamamatsu) and the fast on-board digital computer (PhotoniQ-OEM from Vertilon). After a 266 nm pulse, the entire LIF spectrum (200 to 700 nm) is dispersed by a reflection/transmission grating and is read by the 32-anode PMT. Sufficient spectral resolution is preserved with the 32-anode PMT, each anode covering about 15 nm. Most groups involved in

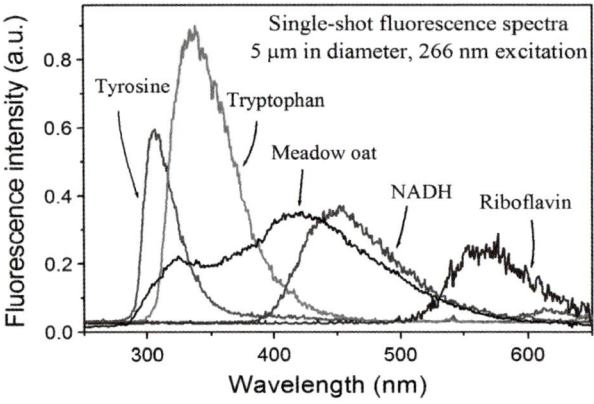

Fig. 17 Single-particle 266-nm-excited fluorescence spectra of common fluorophors found in biological particles and meadow oat pollen particle. Each spectrum is for a nominal 5 μm diameter particle measured with the single particle fluorescence spectrometer. The tyrosine intensities below 300 nm appear to be reduced by the absorbance of the filter (*N*,*N*-dimethylformamide) used to block the 266 nm scattered light.

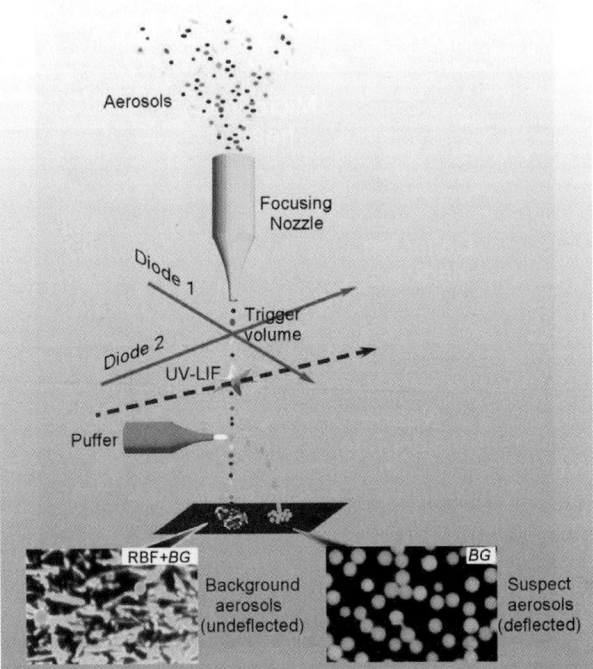

Fig. 18 Schematic of the rapid aerodynamic sorting system. Ambient air is drawn into the system and a laminar air stream is formed from concentrated aerosols. Each individual aerosol particle is illuminated by a single UV laser pulse after it passes through a trigger volume defined by two diode-laser beams. If this flowing aerosol is found to exhibit the same UV-laser-induced fluorescence spectrum as a pathogenic bacteria or virus, this suspect bioaerosol is deflected from the main stream by the puffer for further analysis and pathogen identification. The two bottom insets show tryptophan particles well sorted from a mixture of tryptophan and riboflavin aerosols.[26]

bio-detection use two detection channels; one UV channel (200 to 350 nm) and one visible channel (350 to 700 nm). If the fluorescence of a particle matches pre-determined signature criteria, electronics trigger the air puffer to blast out a puff of turbulent air that then deflects that chosen particle for further analysis. The idea of using an air puffer was suggested by Jean-Pierre Wolf while he was on sabbatical at Yale University in 2000. It was a brilliant but unconventional idea that took us 2–3 years to really implement it. Because of the turbulent puff of air, the chosen particle is deflected towards a broad landing area that may be a substrate (David Schmidt). A re-grouper, also called a particle aerodynamic localizer (PAL), consisting of a cone, can localize the puffed particle into a small area (0.6 mm) on the substrate.[26,27] To improve the collection efficiency of the puffed particles on the substrate, electrostatic means may be employed in conjunction with the puffer and PAL.

The puffer can be activated up to $1000\times$ per second. Therefore, we have the ability to enrich the concentration of the deflected particles that had been chosen based on the similarities of their fluorescence spectra to that of our pre-determined fluorescence spectra. The bottom of Fig. 18 showed us qualitatively that there was a big enrichment factor when trying to separate pure BG from a mixture of RBF and BG. In fact, under microscope counting, the enrichment factor was greater than 10^3 for the separation of *Bacillus subtilis* (BG) from the mixture of riboflavin (RBF) and BG. We had similar results (greater than 10^3 times) for separating BG from the mixture Arizona road dust (ARD) and BG.

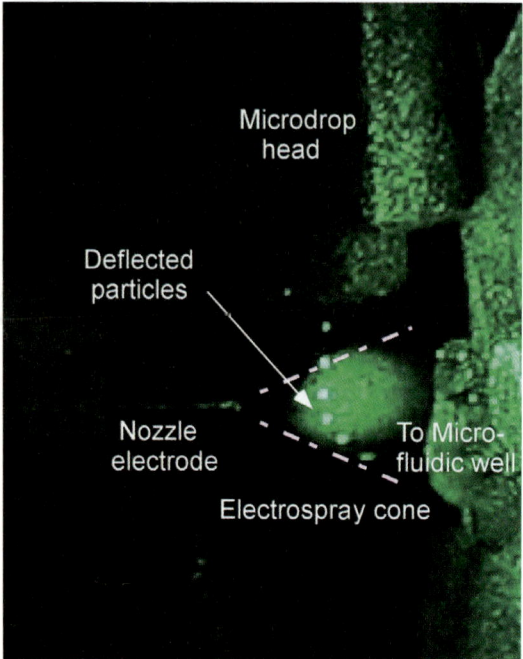

Fig. 19 Liquid droplets (50 micron in diameter) passing through an electrospray acquire a charge and then are attracted to the microfluidic reservoir well by the strong electronic field (New result from Hermes Huang, Yale University, unpublished work).

The collection efficiency of deflected particles onto substrates is dependent upon the type of substrate needed for the identification stage. If Raman scattering is to be used in the identification stage, then a thin aluminium film is deposited on the substrate. For FTIR experiments, the substrate is usually ZnSe with or without a conducting plane. Electrostatic forces are helpful in improving the collection efficiency. Ambient aerosols may be naturally charged (either + or −). However, by making the aerosols traverse a path through the electrospray, we are able to make them all have approximately the same charge (see Fig. 19). These charged particles are susceptible to electrostatic deflection towards the surface of a substrate or a liquid interface. At the electrified interface, the particles are attracted to the liquid surface of the substrate and are prevented from bouncing off. The particles remain there until ready to be used for the second stage intended for identification of the chosen particle. Particularly exciting is collecting these aerosol particles into a reservoir well of a microfluidic cell.

During the time a particle transits the plume of the electrospray, there is a high probability that the particle will collide with one of the many submicron liquid droplets forming the plume. After one collision, the particle acquires a charge of the collided electrospray droplet. The now charged aerosol will repel other submicron droplets and be deflected by an electric field to the opposite electrode.

Our fluorescence instrument (without the puffer and without the electrospray features) has been employed in looking for the fluorescence emission from ambient aerosols in three locations: Adelphi, MD, New Haven, CT, and Las Cruces, NM, with over 124 000 spectra having been collected. We performed hierarchical cluster analysis (HCA) on the 124 000 data spectra. This HCA method has been a useful and unbiased way for us to work with large amounts of data. In this approach, two spectra that are almost similar are combined into a cluster with a spectrum that is the average of the two spectra. Within some special measure of similarity, this combining procedure is repeated until there are no more two similar spectra. One surprising result from these three venues is the 8 to 10 spectra that are common among them.

The HCA spectra from the Maryland data formed eight different line shapes, referred to as clusters. The HCA spectra analyses from the three different locations showed that their corresponding line shapes of clusters 4, 5, 6 and 7 all look very similar (shown in Fig. 20). That was surprising because we expected them to be somewhat different given that one location was a city and the other two were a suburb and a desert-like area, respectively.[28]

We are in the planning stages with Jordan Peccia, a Yale colleague who specializes in the transport of bio-aerosols, to collaborate on the identification of what produced the different fluorescence spectra. Both the puffer deflector and the electrostatic collector will be used to select aerosols that have a good match with the preset fluorescence spectra of one of the clusters common to all three sites (MD, NM and CT).

4.3 Photonics: add/drop filters and coupled 2-d resonators

Two-dimensional microstructures with high symmetry acting as a microcavity have been under intense investigation. One of the most famous devices that make use of WGMs is the add/drop filter.[29] This device shown in Fig. 21 consists of a microdisk with two waveguides (two tapered fibers that were made leaky near the points of contact on the μ-disk) that are submicron separated from the microdisk.

The main waveguide is transporting hundreds of channels and one and only one channel, is on resonance with the WGM of a given microdisk. That one channel will be depleted of its signal strength by transferring its original signal to the secondary waveguide. All the other channels in the main waveguide remain intact upon making one complete turn in the microdisk. This mode of operation is called the "drop mode." Conversely, if one wishes to add a channel to the main waveguide, a channel from the secondary waveguide that is resonant with the WGM of the microdisk, transfers its signal to the main waveguide. This is called the "add mode." Together they form the add/drop filter. Because the coupling between the main and secondary waveguides or fibers occurs nearly at two points on the circular microdisk, attempts are being made to design longer interaction lengths. In so doing, one hopes that the separation or gap between the waveguides and the microdisk would be greater than submicrons so that electron lithography would not be necessary. Hence there exists an intense interest in using various shaped microdisks such as polygons or stadiums as the add/drop filter.

Additionally, we have been interested in extracting more of the internal radiation from a lasing microdisk. Recently, we achieved unidirectional emission with a single beam. We reported what we believe to be the first unidirectional emission with a single beam from a 2 μm high cavity in a microstructure in a 2-d disk with a spiral-shaped cross section[30–34] defined by $r(\phi) = r_0 [1 + \varepsilon(\phi/2\pi)]$, where r_0 is 250 μm and ε is 0.1.

The one-element device is a spiral shown in Fig. 22(a). The asymmetry of this geometric shape lifts the degeneracy of the clockwise (CW) rays and counter-clockwise (CCW) rays. That is, the spiral has a pronounced chirality. The CCW rays, at the notch will escape as a single unidirectional emission beam. We observe that the propagation is normal to the flat edge of the notch when the sidewalls of the microstructure are smooth. The propagation direction is tilted at 30° from the normal when the sidewalls are rough. The dark band around the perimeter on the spiral element is a p-contact ring electrode.

The 2-element device is shown in Fig. 22(b) and the 3-element device is shown in Fig. 22(c). Each element in these 2-element and 3-element devices is electrically isolated from the other elements. For the one-element spiral, the brightest spot occurs at an area perpendicular to the notch. For the 2-element and 3-element

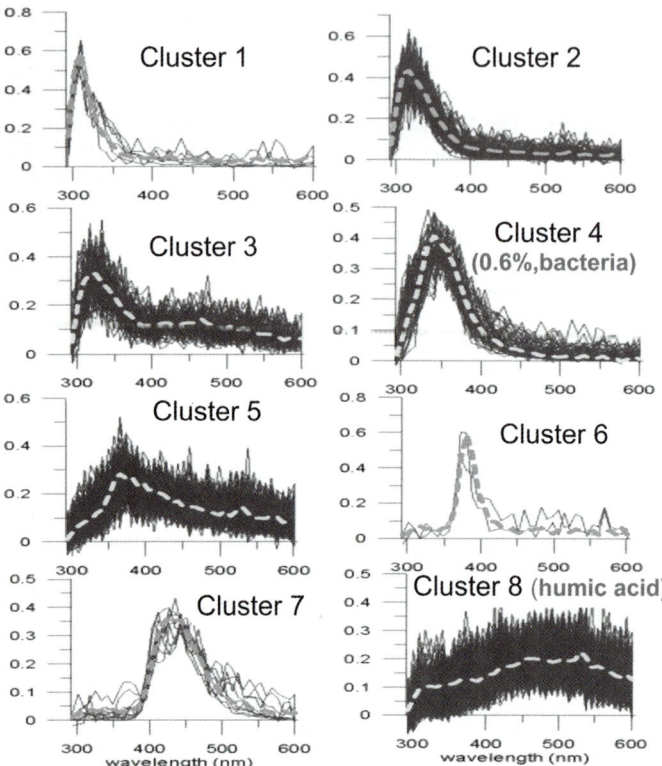

Fig. 20 Template spectra (dotted lines) and normalized clustered particle flu spectra (solid lines) taken Mar. 28, 2003 from 12 : 52 to 13 : 18 at Adelphi, Maryland. A total of 10 000 single particle flu spectra were measured, but only 1127, which exceeded a fluorescence threshold, are included in the analysis. Cluster 4 and 8 are assumed to be caused by bacteria and humic acid materials respectively, by their fluorescence spectra.[28]

devices, the brightest spot occurs at the unattached corner of the flat surface of the semicircle. The measurements are made only from the brightest spot. The output intensity is measured when the pumping current density to an element varies from 0 to some high value, while keeping the current density fixed for the other element(s). We are hoping to see how the changes of one element in the 2-element and 3-element devices, affects the combined unit.

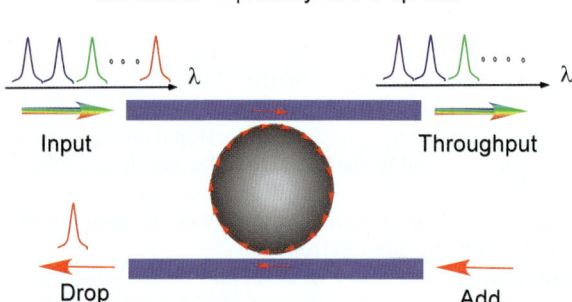

Fig. 21 Add/drop filter where the main waveguide is the lower fiber and the secondary waveguide is the upper fiber.

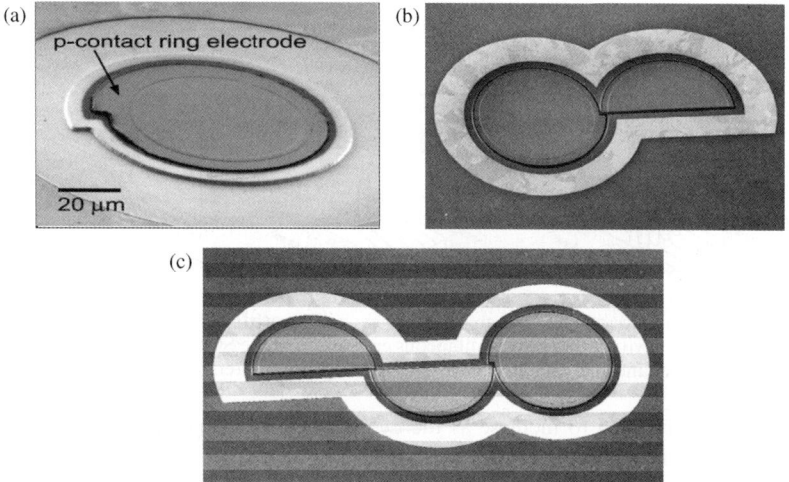

Fig. 22 (a) Optical microscope image of a spiral 2-d microstructure where the electrodes around the rim appear as a dark ring. The rays inside the semiconductor are confined near the interface and can be decomposed into clockwise and counterclockwise rays. (b) Direct coupling of a semicircle with a spiral at the notch. (c) Direct coupling of two semicircles in a line with one semicircle directly coupled with a spiral at the notch (Sp–Sc1–Sc2).

The output intensity as a function of input pumping (either electrical or optical) will grow in three stages: (1) the output intensity varies linearly with input current (or intensity); (2) the output intensity varies superlinearly with input current (or intensity); and (3) saturation sets in when the stimulated rate approaches the pumping rate. The laser threshold is defined to be an inflection point of the superlinear portion of the growth curve of the output intensity.

The lasing threshold of the spiral is determined by the high-Q clockwise circulating stimulated emission and is nearly independent of what gets coupled at the notch. The output emission from the notch is at least 100× stronger than that coming from the curved interface (180° opposite the notch) of the spiral.

For an isolated semicircle, the lasing behavior is similar to a Fabry–Perot laser. The output emission appears at the two flat corners. The reflection at the two corners of the semicircle (about 35% reflectivity) provides the needed feedback for lasing.

By direct coupling of a semicircle edge with a notched edge of a spiral, a two-element device results. The semicircle and the spiral are each pumped by an electrical current that are isolated from each other. However, the counter-clockwise rays cause escapes to leak through the junction as they enter the semicircle. There exists impedance mismatch reflection at the junction. These reflective waves are now converted to clockwise rays that in turn get partially converted back to counter-clockwise rays. The transmitted waves are continuing towards the perimeter of the semicircle and appear as a light source at the opposite corner of the semicircle. The semicircle acts as an active waveguide.

We chose a semicircle microdisk because the mode pattern of the semicircle better approximates that of a spiral. The EM modes of both the semicircle and the spiral can be expanded in terms of cylindrical cavities by two semicircles of different radii. If $r \gg \lambda$, then the semicircles can be replaced by an arc. The arc can be replaced by a linear bar forming a seamless joint with the flat portion of a spiral notch. In this large radius limit, the coupling efficiency for a semicircle directly coupled with a spiral can equal to that of a straight bar coupled to a spiral.

When a semicircle is attached or directly coupled to a notch, the threshold of the spiral seems not to be affected. The higher Q factor of the WGMs of the spiral seems to determine its threshold and seems to be independent of any coupling. The number

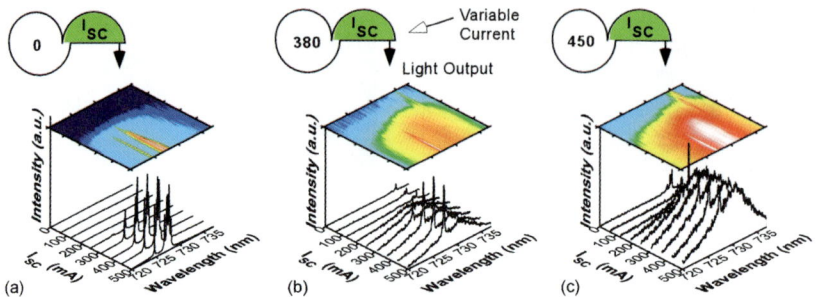

Fig. 23 (a) Dynamics of semicircle acting micro-cavity laser with increasing I_{SC}. (b) Same as (a) except the spiral cavity introduces spontaneous emission into the SC1. (c) Spiral cavity reaches laser current level and the semicircle cavity keeps increasing current.

of modes is observed to increase as the overlap between the notch length and the length of the corner of the semicircle increases.

With a semicircle attached to the spiral, the output of the two-element device depends on the current to the spiral (I_{SP}) and on the current through the semicircle (I_{SC}). Fig. 23 shows several interesting ways that the 2-element device interacts as an unit. The semicircle in the two-element device acts as an absorber with $I_{SC} = 0$ and acts as an amplifier when the current I_{SC} is less than the lasing threshold for a semicircle. When the attached semicircle is in its lasing state, when $I_{SP} = 0$, the increase in I_{SP} from 0 to a current above its threshold, the lasing in the semicircle is quenched and the amplified spontaneous emission (ASE) is observed. That is a demonstration of the overwhelming effect of the spontaneous emission emerging from the spiral on the semicircle and *vice versa*. The ASE of one element depletes the inverted population of the other element. By keeping the results obtained fixing the I_{SP} at 3 different values, $I_{SP} = 0$, 380, and 450 mA. Both 380 mA and 450 mA are above the lasing threshold.

The cascading of a second semicircle of equal radius with a semicircle of a 2-element device offers many more possibilities. One noteworthy feature is the current control of the middle element that has the ability to switch wavelengths between two discrete values. Again, each device is current regulated by a current power supply. Each element is electrically isolated from the others. The middle element, semicircle #1 (SC-1), controls the output wavelength characteristics at the far corner of semicircle #2 (SC-2) which is directly coupled to SC-1 (see Fig. 22).

Summary

The aim of this paper is to introduce the foundation onto which the reader can build his/her own creative research topics. We have given a glimpse of the many facets of the effect of WGMs on linear and nonlinear spectroscopy. The internal and near-field distribution of a WGM determines where thin electrodes should be placed around the perimeter of the microstructure and how light escapes from the microstructure. The advances in semiconductor lasers and high resolution CCD cameras have enabled a series of new diagnostic techniques such as the determination of fiber-cladding uniformity along the fiber length and large angle elastic scattering to provide some morphological information. Although we have only presented three promising applications, two of which are related to ambient aerosols, there are hundreds more promising applications in this field and some of the very best ones are being presented in this *Faraday Discuss.* meeting. The third promising application is in the photonics field and the telecommunication field, such as add/drop filters and oscillators/amplifiers with 2-coupled or 3-coupled lasers. There will be many more new promising applications with respect to nonlinear optics combined with WGMs. The tremendous advantage of the femto-second laser pulse over the Q-switched laser pulse (10 ns) is the short-duration pulse that delivers much less energy, but much greater intensity. With respect to the femto-second laser pulse even higher order nonlinearities are expected. The future of linear and nonlinear spectroscopy of microparticles looks very bright as several allied fields such as photonic crystals, nanoparticles, biomedical techniques, atmospheric chemistry begin to complement each other. Nothing can match serendipity.

Acknowledgements

RKC particularly appreciated the early interaction he had with Peter Barber and Steve C. Hill. They taught him a great deal about electromagnetic interactions. Throughout much of the reported works here, the close collaboration with Steve C. Hill, Ronald Pinnick and Jean-Pierre Wolf was very much appreciated by the authors. The list of RKC's former and present students and former post-doctoral fellows is too long to list, but they all know well how very much RKC depended on them and appreciated them. This article is a review of all their collaborative efforts. Much appreciation goes to the following agencies for their partial support of the work reported here: US Army Research Laboratory; US Army Research Office; US Air Force of Scientific Research; US Air Force Research Laboratory; DARPA-SUVOS, DARPA-SSBA; RAAD; and DOE.

References

1. E. H. Lee, R. E. Benner, J. B. Fenn and R. K. Chang, *Appl. Opt.*, 1978, **17**, 1980.
2. A. Ashkin and J. M. Dziedzic, *Phys. Rev. Lett.*, 1977, **38**, 1351–1354.
3. E. M. Purcell, *Phys. Rev.*, 1946, **69**, 681.
4. S. C. Hill and R. E. Benner, *J. Opt. Soc. Am. B*, 1986, **B3**, 1509.
5. P. W. Barber and S. C. Hill, *Light Scattering by Particles Computational Methods*, World Scientific, Singapore, 1990.
6. C. F. Bohren and D. R. Huffman, *Absorption and Scattering of Light by Small Particles*, John Wiley & Sons, New York, 1983.
7. J. U. Nockel, A. D. Stone and R. K. Chang, *Opt. Lett.*, 1994, **19**, 1693.
8. S. C. Hill and R. E. Benner, *Optical Effects Associated with Small Particles*, eds. P. W. Barber and R. K. Chang, World Scientific, Singapore, 1988, Note: Much of Section 2.1 comes from this most useful reference on EM theory and WGMs.
9. F. S. C. Ching, P. T. Leung and K. Young, in *Optical Processes in Microcavities*, eds. R. K. Chang and A. J. Campillo, World Scientific, Singapore, 1996, p. 1.
10. J. F. Owen, P. W. Barber, B. J. Messinger and R. K. Chang, *Opt. Lett.*, 1981, **6**, 272.
11. J. F. Owen, R. K. Chang and P. W. Barber, *Opt. Lett.*, 1981, **6**, 540.
12. K. J. Vahala, *Nature*, 2003, **424**, 839–846.
13. A. W. Poon, R. K. Chang and D. Q. Chowdhury, *Opt. Lett.*, 2001, **26**, 1867.
14. S. Holler, Y. L. Pan, R. K. Chang, S. Hill, R. G. Pinnick and J. R. Bottiger, *Opt. Lett.*, 1998, **23**, 1489.
15. G. E. Fernandes, Y. L. Pan and R. K. Chang, *Opt. Lett.*, 2006, **31**, 3034–3036.
16. P. H. Kaye, *Meas. Sci. Technol.*, 1998, **9**, 141–149.
17. K. B. Aptowicz, R. G. Pinnick, S. C. Hill, Y. L. Pan and R. K. Chang, *J. Geophys. Res., [Atmos.]*, 2006, **111**, D12212.
18. R. K. Chang, "Some of Bloembergen's Nonlinear Optical Effects Revisted in Single Micrometer-Sized droplets", in *Resonances*, eds. M. D. Levenson, E. Mazur, P. S. Pershan and Y. L. Shen, World Scientific, Singapore, 1990, p. 200.
19. S. C. Hill and R. K. Chang, *Studies in Classical and Quantum Nonlinear Optics*, ed. O. Keller, Nova Science Publisher, New York, 1995, p. 171.
20. R. Symes, R. M. Sayer and J. P. Reid, *Phys. Chem. Chem. Phys.*, 2004, **6**, 474–487.
21. J. L. Cheung, J. M. Hartings and R. K. Chang, Nonlinear Optics of Microdroplets Illuminated by Picosecond Laser Pulses, in *Handbook of Optical Properties II*, ed. R. E. Hummel and P. Wibmann, CRC Press, New York, 1997.
22. M. Fields, J. Popp and R. K. Chang, Nonlinear Optics in Microspheres, in *Progress in Optics 44*, ed. E. Wolf, Elsevier Science B. V., Amsterdam, 2001, p. 1.
23. S. X. Qian, J. B. Snow, H. M. Tzeng and R. K. Chang, *Science*, 1986, **231**, 486.
24. S. X. Qian and R. K. Chang, *Phys. Rev. Lett.*, 1986, **56**, 926.
25. S. C. Hill, R. G. Pinnick, S. Niles, Y. L. Pan, S. Holler, R. K. Chang, J. R. Bottiger, B. T. Chen, C. S. Orr and G. Feather, *Field Anal. Chem. Technol.*, 1999, **3**, 221.

26 Y. L. Pan, V. Boutou, J. R. Bottiger, S. S. Zhang, J. P. Wolf and R. K. Chang, *Aerosol Sci. Technol.*, 2004, **38**, 598.
27 M. Frain, D. P. Schmidt, Y. L. Pan and R. K. Chang, *Aerosol Sci. Technol.*, 2006, **40**, 218–225.
28 R. G. Pinnick, S. C. Hill, Y. L. Pan and R. K. Chang, *Atmos. Environ.*, 2004, **38**, 1657.
29 K. Vahala, *Optical Microcavities*, World Scientific, Singapore, 2004.
30 G. D. Chern, H. E. Tureci, A. D. Stone, R. K. Chang, M. Kneissl and N. M. Johnson, *Appl. Phys. Lett.*, 2003, **83**, 1710–1712.
31 M. Kneissl, M. Teepe, N. Miyashita, N. M. Johnson, G. D. Chern and R. K. Chang, *Appl. Phys. Lett.*, 2004, **84**, 2485–2487.
32 N. Tsujimoto, T. Takashima, T. Nakao, K. Masuyama, A. Fujii and M. Ozaki, *J. Phys. D: Appl. Phys.*, 2007, **40**, 1669–1672.
33 T. Y. Kwon, S. Y. Lee, M. S. Kurdog S. Rim, C. M. Kim and Y. J. Park, *Opt. Lett.*, 2006, **31**, 1250–1252.
34 J. Y. Lee and A. W. Poon, *Proceedings of the IEEE 8th International Conference on Transparent Optical Networks*, IEEE, New Jersey, 2006, pp. 62–65.

PAPER www.rsc.org/faraday_d | Faraday Discussions

Identification of biological microparticles using ultrafast depletion spectroscopy

Francois Courvoisier,[a] Luigi Bonacina,[a] Véronique Boutou,[b] Laurent Guyon,[b] Christophe Bonnet,[b] Benoit Thuillier,[b] Jerome Extermann,[a] Matthias Roth,[c] Herschel Rabitz[c] and Jean-Pierre Wolf*[ab]

Received 19th October 2006, Accepted 8th December 2006
First published as an Advance Article on the web 21st August 2007
DOI: 10.1039/b615221j

We show how an ultrafast pump–pump excitation induces strong fluorescence depletion in biological samples, such as bacteria-containing droplets, in contrast with fluorescent interferents, such as polycyclic aromatic compounds, despite similar spectroscopic properties. Application to the optical remote discrimination of biotic *versus* non-biotic particles is proposed. Further improvement is required to allow the discrimination of one pathogenic among other non-pathogenic micro-organisms. This improved selectivity may be reached with optimal coherent control experiments, as discussed in the paper.

1. Introduction

Rapid detection and identification of pathogenic aerosols from potential bio-terrorism release and epidemic spreads are urgent safety issues. In order to efficiently protect populations, bioaerosol detection devices have to be *very fast* (typically minutes) and *very selective* to discriminate pathogenic from non-pathogenic particles and minimize false alarm rates. This very difficult task initiated recent major research and development efforts.

Biochemical identification procedures such as Polymerase Chain Reaction (PCR),[1–5] antibiotic resistance determination,[6,7] or matrices of biochemical micro-sensors[8,9] are selective, but slow (at least some hours). On the other hand, optical techniques provide information in "real time" but until now lack in specificity. In particular, several optical systems, based on fluorescence[11–14] and/or elastic scattering[15,16] have been developed to distinguish bio- from non-bioaerosols. The most advanced experiments address *individual* aerosol particles, within which fluorescence is spectrally analyzed.[11,17,18] The major drawback of these instruments is, however, frequent false alarms triggered by other organic aerosols, such as diesel particles or cigarette smoke.[17,19] Fig. 1 shows, as an example, the similitude in the fluorescence spectra of diesel fuel, soot particles, the amino acid tryptophan (Trp) and *Bacillus subtilis*, which is a biosimulant for *Bacillus anthracis*. As shown in the figure, the major contribution in the UV-Vis fluorescence (around 340 nm) in biological

---

[a] GAP-Biophotonics, University of Geneva, 20, rue de l'Ecole de Médecine, 1211, Geneva 4, Switzerland. E-mail: jean-pierre.wolf@physics.unige.ch; Fax: +41 22 379 39 80; Tel: +41 22 379 65 94
[b] LASIM (UMR 5579), Université Claude Bernard Lyon 1, 43, Bd du 11 Novembre 1918, 69622, Villeurbanne, France
[c] Department of Chemistry, Frick Laboratory, Princeton University, Princeton, NJ USA

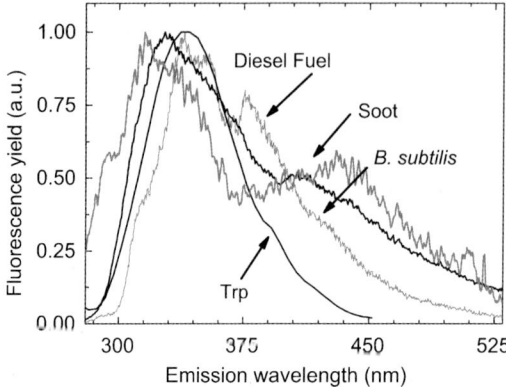

Fig. 1 Comparison of the one-photon (at 270 nm) excited fluorescence spectra of different organic compounds. Despite small shifts in the emission bands, an identification based on the one-photon excited fluorescence spectroscopy is impossible. Data for *Bacillus subtilis* are shown with the courtesy of S. C. Hill (ARL, Adelphi, MD, USA). Reprinted from ref. 32 © 2006, with permission from Elsevier.

particles is due to the amino acid tryptophan. The longer wavelength tail of the fluorescence is attributed to the emission of nicotinamide adenine dinucleotide (NADH; around 450 nm) and flavins (riboflavin, flavin mononucleotide FMN and flavin adenine dinucleotide FAD around 560 nm).[17] Due to the interference with Polycyclic Aromatic Hydrocarbons (PAHs) containing soot and diesel, the identification of bioaerosols in a background of traffic related particles (typical of urban conditions) is therefore extremely difficult. It is even more elusive to expect discrimination of different types of bacteria.[20] An interesting approach was recently reported,[10] which combines optical and biochemical analyses. A fluorescence/scattering device was used to sort "on-line" bioaerosols from other particles, which could then be subsequently chemically analyzed *in situ*.

Optical techniques are also attractive as they can provide information remotely. The Lidar (Light detection and ranging) technique[21] allows for mapping aerosols in 3D over several kilometres, similar to an optical Radar. Lidars are able to detect the release and spread of potentially harmful plumes (such as pathogen releases from terrorists or legionella from cooling towers) at large distance and thus allow taking measures in time for protecting populations or identifying sources. So far, Lidar detection of bioaerosols has been demonstrated either using elastic scattering[21] or UV-LIF.[21,22] However, the distinction between bio- and non-bioaerosols was either impossible (elastic scattering only) or unsatisfactory for LIF-Lidars (interference with pollens and organic particles like traffic-related soot or PAHs).

To overcome these difficulties, there is an interest in exciting the fluorescence with ultrashort laser pulses in order to access specific molecular dynamical features. Recent experiments using coherent control and multiphoton ultrafast spectroscopy have shown the ability to discriminate between molecular species that have similar one-photon absorption and emission spectra.[23,24] Two-Photon Excited Fluorescence (2PEF) and pulse shaping techniques should allow for selective enhancement of the fluorescence of one molecule *versus* another that has similar spectra. Optimal Dynamic Discrimination (ODD)[25] of similar molecular agent provides the basis for generating optimal signals for detection.

In this paper, we demonstrate that femtosecond pump–probe spectroscopy allows for distinguishing biological microparticles from PAH-containing ones. More precisely, we could distinguish amino acids (Trp) and flavins (riboflavin RbF, FMN and FAD) from PAHs (naphthalene) and diesel fuel in the liquid phase using a "Pump–Pump Depletion" (PPD) technique. We also applied the technique to live bacteria and individual bioaerosol particles, for which PPD showed even higher selectivity than for molecules in the liquid phase.

2. Experimental

PPD uses a sequence of two femtosecond pulses (one UV-Vis and the other at 810 nm), separated by a variable time delay Δt. The fluorescence is recorded as a function of Δt, which reflects the molecular dynamics within the intermediate state (see below). The experiments (Fig. 2) use a Kerr lens mode-locked Ti : sapphire oscillator and a chirped pulse amplifier that delivers 120 fs pulses at 810 nm. The output is frequency doubled and tripled in two consecutive β-Barium Borate (BBO) crystals so that the pulses at 270 nm for Trp, naphthalene and diesel (405 nm for Rbf, FMN and FAD) and at 810 nm are synchronously emitted. After splitting both pulses with a dichroic mirror, the second (near infrared) pump pulse is delayed using a stepper motorized delay line, with a resolution of 16 fs. Both beams are then recombined, carefully spatially superimposed and focused onto a 1 mm length quartz flow cell, leading to intensities up to $I_{270} = 4 \times 10^5$ W cm^{-2} ($I_{405} = 9 \times 10^8$ W cm^{-2}) and $I_{810} = 2 \times 10^{11}$ W cm^{-2}. The fluorescence is dispersed by a Jobin Yvon Y10 spectrometer (2 nm resolution) or filtered with BandPass (BP) (10 nm) filters centred at 340 nm close to the maximum intensity of Trp and PAH fluorescence bands (or at 520 nm for flavins) and detected by PhotoMultiplier Tubes (PMT). Trp or flavins are solvated in water, whereas naphthalene is dissolved in cyclohexane.

The concentrations used are typically 0.2 g L^{-1} for Trp, 0.3 g L^{-1} for naphthalene, 0.1 g L^{-1} for Rbf and from 0.1 to 1 g L^{-1} for FMN and FAD. We checked that our results do not depend on the different solution concentrations. In the further experiments, solutions of bacteria (*Escherichia coli*, *Bacillus subtilis*, *Enterococcus fæcalis*) have been prepared by dissolving lyophilized cells and spores in water (Strain W ATCC 9637 for *Escherichia coli*, ATCC 6633 for *Bacillus subtilis*) or in a nutrient buffered (Symbioflor for *Enterococcus fæcalis*) solution at concentrations of 10^7 to 10^9 bacteria cm^{-3}. Diesel fuel has also been used to simulate the complex mixture of PAHs contained by organic particles emitted by diesel engines. A circulator ensures that solutions in the excitation volume are renewed at each laser pulse.

For the experiments performed in air, an aerosol generator based on a piezo-electric-driven nozzle is used. This allows producing the particles "on demand" and therefore to synchronize the generator with the incoming laser pulses, so that each droplet is irradiated by a single laser pulse pair (pump–probe type of experiment). The size of the droplets can be varied from 20 to 60 μm by tuning the appropriate voltage pulse on the piezoelectric element. The aerosol stream speed is about 1 m s^{-1}. The particle diameter was stable over hours within a few percent (measured by

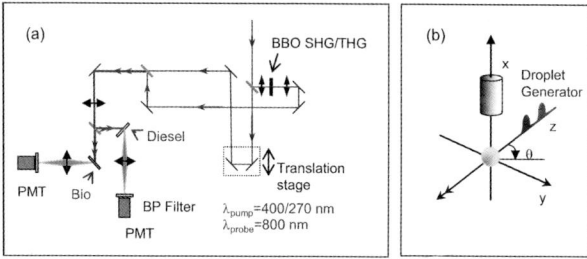

Fig. 2 Experimental set-up of the PPD experiments. (a) Optical layout used for the experiments in liquids. For the experiments in air, bioaerosols are generated with a droplet generator (b), which replaces the flow cell in (a). The droplet generator is synchronized with the femtosecond pulse pair.

forward Mie scattering with a Helium–Neon laser). The laser is slightly focused with a 1 m focal length lens, so that the beam waist remains 5–10 times larger than the droplet diameter. A (synchronized) pulsed LED and a CCD camera permanently monitor the droplet position and the laser waist.

3. Pump–probe spectroscopy to identify bacteria and biomolecules in liquids

As mentioned above, a major drawback inherent in LIF instruments is the lack of selectivity because UV-Vis fluorescence is incapable of discriminating different molecules with similar absorption and fluorescence signatures. While mineral and carbon black particles do not fluoresce significantly, aromatics and polycyclic aromatic hydrocarbons from organic particles and diesel soot strongly interfere with biological fluorophors such as amino acids.[17,19] The similarity between the spectral signatures of PAH and biological molecules under UV-Vis excitation lies in the fact that similar π-electrons from carbonic rings are involved. Therefore, PAHs (such as naphtalene) exhibit absorption and emission bands similar to those of amino acids like Tyrosine or Trp. Some shifts are present because of differences in specific bonds and the number of aromatic rings, but the broad, featureless nature of the bands renders them almost indistinguishable (Fig. 1).

In order to discriminate these fluorescing molecules, we applied a novel femtosecond PPD concept (Fig. 3). It is based on the time-resolved observation of the competition between Excited State Absorption (ESA) into higher lying excited states and fluorescence into the ground state. This approach makes use of two physical processes beyond that available in the usual linear fluorescence spectroscopy: (1) the dynamics in the intermediate pumped state (S_1) and (2) the coupling efficiency to higher lying excited states (S_n).

As sketched in Fig. 3, a first femtosecond pump pulse (at 270 nm for Trp and PAHs, 405 nm for flavins), resonant with the first absorption band of the fluorophores, coherently excites them from the ground state S_0 to a set of vibronic levels $S_1\{\nu'\}$. The vibronic excitation relaxes by internal energy redistribution to lower $\{\nu\}$ modes. Fluorescence relaxation to the ground state occurs within a lifetime of several nanoseconds. Meanwhile, a second 810 nm femtosecond "repump" pulse is used to transfer part of the $S_1\{\nu\}$ population to higher lying electronic states S_n. The depletion of the S_1 population under investigation depends on both the molecular

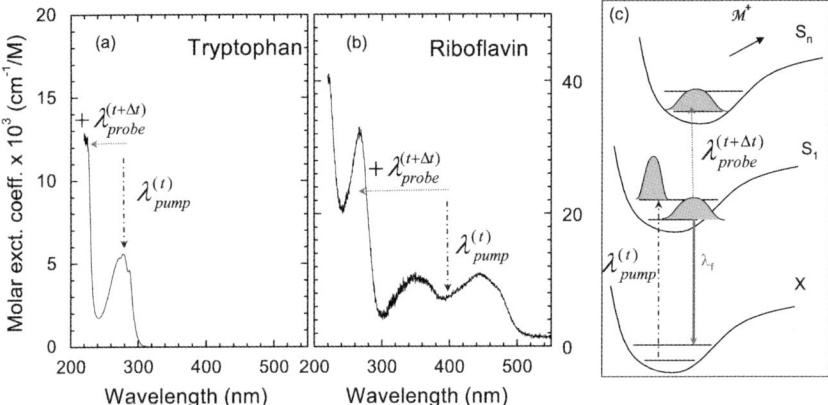

Fig. 3 Absorption spectra of Tryptophan (a) and Riboflavin (b). (c) PPD scheme in Trp, flavins and polycyclic aromatics. The pump pulse brings the molecules in their first excited state S_1. The S_1 population (and therefore the fluorescence) is depleted by the second pump pulse. Reprinted from ref. 32 © 2006, with permission from Elsevier.

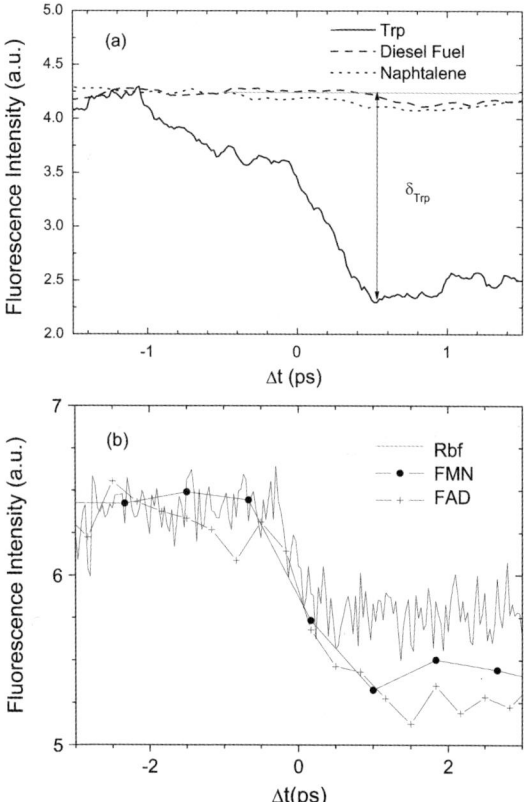

Fig. 4 (a) PPD experiment on Trp and PAHs, demonstrating discrimination capability between the amino acid and other aromatic molecules. (b) Similar results obtained in flavins.

dynamics in this intermediate state and the transition probability to S_n. The relaxation from the intermediate excited state may be associated with different processes, including charge transfer, conformational relaxation[26,27] and intersystem crossing with repulsive $\pi\sigma^*$ states.[28,29]

S_n states are both autoionizing[29] and relaxing radiationlessly[30] into S_0. By varying the temporal delay Δt between the UV-Vis and the IR pulses, the dynamics of the internal energy redistribution within the intermediate excited potential hypersurface S_1 is explored. The S_1 population and the fluorescence signal is therefore depleted as a function of Δt. As different species have distinct S_1 hypersurfaces, discriminating signals can be obtained.

Fig. 4(a) shows the PPD dynamics of S_1 in Trp as compared to diesel fuel and naphtalene in cyclohexane, one of the most abundant fluorescing PAHs in diesel. While depletion reaches as much as 50% in Trp, diesel fuel and naphtalene appear almost unaffected (within a few percent), at least on these timescales.[31] The depletion factor δ is defined as $\delta = (P_{undepleted} - P_{depleted})/P_{undepleted}$ (where P is the fluorescence power). This remarkable difference allows for efficient discrimination between Trp and organic species, although they exhibit very similar linear excitation/fluorescence spectra (Fig. 1).

Two reasons might be invoked to understand this difference: (1) the intermediate state dynamics are predominantly influenced by the NH- and CO-groups of the amino acid backbone and (2) the ionization efficiency is lower for the PAHs. Further electronic structure calculations are required to better understand the process, especially on the higher lying S_n potential surfaces.

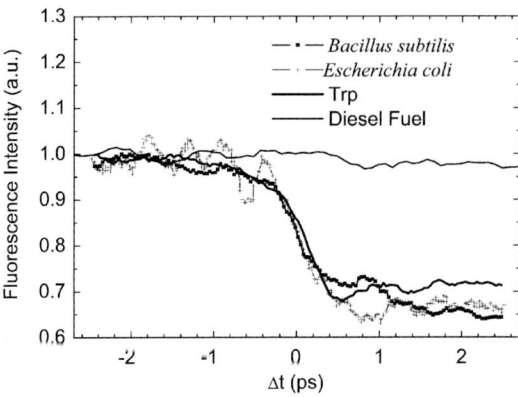

Fig. 5 Discrimination between bacteria and diesel fuel using PPD ultrafast spectroscopy. Reprinted from ref. 32 © 2006, with permission from Elsevier.

Fluorescence depletion has also been obtained for RbF, FMN, and FAD (Fig. 4b). However, the depletion in this case is only about 15% (with a maximum intensity of 5×10^{11} W cm^{-2} at 810 nm). We then repeated the experiment but exciting the flavins at 270 nm, as for Trp. Flavins indeed absorb in both spectral regions (Fig. 3). The second excitation step by the 810 nm pulse then brings the molecules in the excited states around 200 nm. The fluorescence depletion observed with the 270 nm excitation reached 35%. This is an indication that the branching ratio to autoionization in the 270 nm band is much lower than around 200 nm. To get the depletion while exciting at 405 nm, two photons at 810 nm probably have to be used. This is confirmed by intensity dependencies, which show a quadratic dependence in I_{810} in this latter case. A model based on rate equations was recently developed in order to quantitatively explain the PPD behaviour in Trp and RbF.[32]

In order to more closely approach the application of detecting and discriminating bioagents from organic particles, we applied PPD spectroscopy to live bacteria (λ_1 = 270 nm and λ_2 = 810 nm), such as *Escherichia coli*, *Enterococcus* and *Bacillus subtilis*. Artefacts due to preparation methods have been discarded by using a variety of samples, *i.e.* lyophilized cells and spores, suspended either in pure or in biologically-buffered water (*i.e.* typically 10^7–10^9 bacteria cm^{-3}). The bacteria-containing solutions replaced the Trp- or flavin-containing solutions of the formerly described experiment. The observed pump-probe depletion results are remarkably robust (Fig. 5), with similar depletion values for all the considered bacteria (results for *Enterococcus*, not shown in the figure, are identical), although the Trp micro-environment within the bacteria proteins is very different from water.

On the other hand, the very similar depletion behaviour for all bacteria and Trp also shows the limitations of PPD spectroscopy in the present configuration. Biomolecules can be distinguished from other aromatics but PPD is unable to discriminate two different bacteria in solution. We are currently exploring new experimental ODD approaches to reach this goal (see Section 5).

4. Multi-Photon Excited Fluorescence (MPEF) and PPD in aerosol microparticles

Femtosecond lasers provide high intensity pulses with low energy, which allows inducing non-linear processes in particles without deformation due to electrostrictive and thermal expansion effects.

The most prominent feature of non-linear processes in aerosol particles is strong localization of the emitting molecules within the particle, and subsequent backward enhancement of the emitted light.[33,34] This unexpected behavior is extremely attractive for remote detection schemes, such as Lidar applications. Localization is achieved by the non-linear processes themselves, which typically involve the nth power of the internal intensity $I^n(\mathbf{r})$ ($\mathbf{r}$ for position inside the particle). The backward enhancement can be qualitatively understood by the reciprocity (or "time reversal") principle: re-emission from regions with high $I^n(\mathbf{r})$ tends to return toward the illuminating source by essentially retracing the direction of the incident beam that gave rise to the focal points. This backward enhancement has been observed for both spherical and non spherical[35] microparticles.

More precisely, we investigated, both theoretically and experimentally, incoherent multiphoton processes involving $n = 1$ to 5 photons.[34] For $n = 1, 2, 3$ (at 800 nm incident wavelength), MPEF occurs in bioaerosols because of the absorptions of amino acids (tryptophan, tyrosin), NADH, and flavins. The strongly anisotropic MPEF emission was demonstrated on individual water microdroplets containing tryptophan, riboflavin, or other synthetic fluorophors in ethanol.[33–35] The experiment was performed with the aerosol source described in Section 2. MPEF angular distribution for the one- (400 nm), two- (800 nm) and three-photon (1, 2 µm) excitation show that the fluorescence emission is maximum in the direction towards the exciting source. The directionality of the emission is dependent on the increase of n, because the excitation process involves the nth power of the intensity $I^n(\mathbf{r})$. The ratio $R_f = P(180°)/P(90°)$ increases from 1.8 to 9 when n changes from 1 to 3 (P is the emitted light power). For 3PEF, fluorescence from aerosol microparticles is therefore mainly emitted backwards, which is ideal for Lidar experiments.

At higher intensities, significant ionization occurs in water itself, involving $n = 5$ photons. The growth of the plasma is also a non-linear function of $I^n(\mathbf{r})$. We showed that both localization and backward enhancement strongly increases with the order n of the multiphoton process, exceeding $R_f = P(180°)/P(90°) = 35$ for $n = 5$.[36] Notice that the light emitted by the plasma has the potential of providing information about the aerosol composition, as recently demonstrated in LIBS (Laser Induced Breakdown Spectroscopy) experiments on bacteria.[37,38]

This unique backward emission behavior allowed us to demonstrate the first MPEF Lidar detection of biological aerosols[39,40] using the "Teramobile" system. The Teramobile (www.teramobile.org) is the first femtosecond-terawatt laser-based Lidar, and was developed by a French–German consortium, formed by the Universities of Iena, Berlin, Lyon, and the Ecole Polytechnique (Palaiseau).

The bioaerosol particles, consisting of 1 µm-size water droplets containing 0.03 g l^{-1} of Riboflavin were generated at a distance of 50 m from the Teramobile system. Riboflavin was excited by two photons at 800 nm and emitted a broad fluorescence around 540 nm. The broad fluorescence signature was clearly observed from the particle cloud (typically 10^4 p cm^{-3}), with a range resolution of a few meters.

Primarily, MPEF-Lidar is advantageous as compared to linear LIF-Lidar for the following reasons: (1) MPEF is enhanced in the backward direction and (2) the transmission of the atmosphere is much higher for longer wavelengths. For example, if we consider the detection of Trp with 3-PEF, the transmission of the atmosphere is typically 0.6 km^{-1} at 270 nm, whereas it is 3×10^{-3} km^{-1} at 810 nm (for a clear atmosphere, depending on the background ozone concentration). This compensates the lower 3-PEF cross-section compared to the 1-PEF cross-section at distances larger than a couple of kilometres. The most attractive feature of MPEF is, however, the possibility of using pump-probe techniques, as described in Section 3, in order to discriminate bioaerosols from background interferents, such as traffic-related soot or PAHs.

In order to get closer to the real application, we then performed ultrafast PPD spectroscopy in bioaerosols and, in particular, water microdroplets that contain Trp and/or FMN (Fig. 2). The droplet radius was about 25 µm, which is larger than the

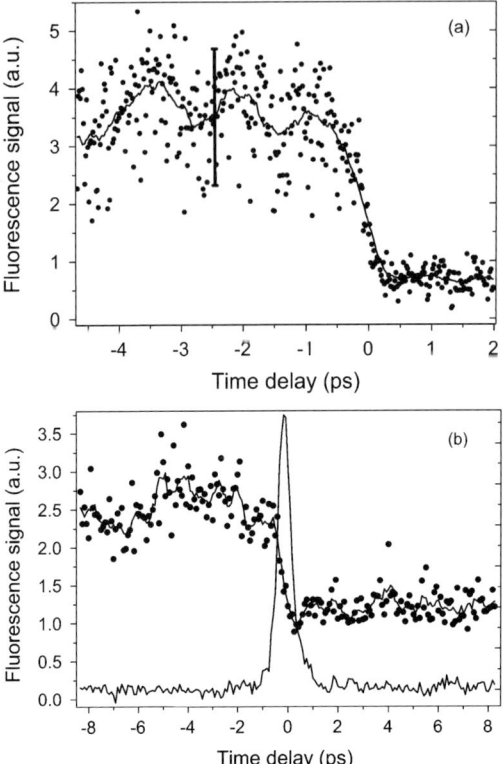

Fig. 6 PPD spectroscopy in bioaerosols: (a) Trp-containing microdroplets, (b) FMN-containing microdroplets. Depletion is as much as 80% for Trp and 60% for FMN. The peak in (b) shows the cross-correlation of both laser pulses, which indicates the time resolution of the experiment.

size of single bacteria (1 µm) or even bacteria clusters (typically 10 µm), but still constitute an acceptable model.

The laser intensities at 810 nm and 270 nm for Trp (405 nm for FMN) that excited the microparticles were similar to the intensities used for the experiments in liquids.

The most impressive result of these experiments is the very high PPD efficiency as compared to depletion ratios in liquids. The depletion factor δ reaches 80% for Trp droplets and 60% for FMN droplets (to be compared to 50% and 15% in liquids, respectively). Some tentative explanations could be invoked for this unexpectedly high efficiency, but the definitive reason is not clear yet: (1) The spatial overlap between pump and probe pulses might be enhanced by the shape of the droplet; (2) the spherical shape induces hot spots inside the droplet where intensities are up to 100 times higher than the incident one,[33,34] but the total hot spot volume is rather small; (3) there might be some surface effects (orientation of molecules on the surface) that would enhance the two photon absorption.

Notice that these time-resolved pump–probe experiments could also potentially be used up to the size of the microparticle. Time-resolved two-photon absorption (and fluorescence) has indeed been used in droplets to observe the trajectories of the light bullets within the particle.[41,42] Each time the femtosecond pulses crossed within the droplet, the fluorescence signal was enhanced. The time delay between two fluorescence maxima corresponds to the roundtrip time $t_R = 2\pi am/c$ and thus to the radius of the droplet a (m is the refraction index = 1.33). In our case, as we measure fluorescence depletion, the second pulse (at 810 nm) could lead to further depletion after a roundtrip. Unfortunately, this was not observed in our experiments (Fig. 6, with

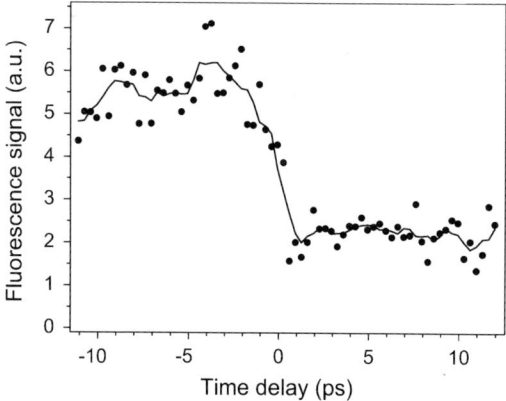

Fig. 7 PPD spectroscopy in 20 μm droplets containing about 100 *E. coli* bacteria.

$t_R = 0.7$ ps), because of the limited signal-to-noise ratio (the intensity of the pulse after 1 roundtrip is reduced by at least one order of magnitude, see ref. 42 and 43).

We finally applied this technique to 20 μm water droplets containing typically 100 live bacteria (*Escherichia coli*). As shown in Fig. 7, the depletion factor δ is again greatly enhanced as compared to bacteria in bulk water: 60% depletion in the microdroplet and 20% in solution ($\lambda_1 = 270$ nm, $\lambda_2 = 810$ nm).

This experiment is also interesting for field applications, as bacteria and viruses are efficiently transmitted by droplets of saliva (coughing, breathing, speaking, *etc.*).

We propose to use the unique discrimination capability of PPD as a basis for a novel selective bioaerosol detection technique that avoids interference from background (traffic-related) organic particles in air. For instance, let us consider a mixture of N_B bacteria and N_D diesel particles. The excitation shall consist of a PPD sequence (270 nm and 810 nm pulses). The fluorescence power P emitted by the mixture shall be measured as the second laser (at 810 nm) is alternately switched on and off (denoted as P_{on} and P_{off}). Without the second laser pulse, the fluorescence cross-sections are $\sigma_D(I_{810} = 0) = \sigma_D$ and $\sigma_B(I_{810} = 0) = \sigma_B$, while with the probe laser the cross-sections will be reduced by a factor R_B ($=0.2$) and R_D ($=0.98$) for the biological and the diesel particles, respectively. This differential procedure allows for determining the concentration of the two types of particles:

$$N_B = \frac{R_D P_{off} - P_{on}}{I_{270}\sigma_B(R_D - R_B)} \text{ and } N_D = \frac{R_B P_{off} - P_{on}}{I_{270}\sigma_D(R_B - R_D)}$$

The method's performance strongly depends on the difference $R_B - R_D$ between the two species to be discriminated, which is large in our case. Notice that in order to precisely quantify the concentration of the diesel particles, quantitative knowledge of the cross-sections is required, which might be difficult to determine because of the variety of possible organic particles.

The PPD-technique will be especially attractive for active remote sensing techniques such as Lidar, where the lack of discrimination between bioaerosols and transportation related organics is currently most acute. The implementation of the PPD-technique in the Teramobile system is therefore presently under consideration.

5. Towards a coherent identification of bacteria in air

PPD spectroscopy is an attractive technique for discriminating bioaerosols from other organics. However, it is unable to discriminate one type of biological aerosol from another (Fig. 5). A possible reason is that the averaged dynamics in the excited states of the bio-fluorophors (embedded in complex proteins) are quite similar in all living organisms. A natural approach is therefore to extend the PPD technique to coherent control, where the amplitude and phase of every spectral component of the exciting pulses are shaped in order to best-fit the potential hypersurfaces. The method is then extremely sensitive to the details of the potential hypersurfaces, which provides unprecedented selectivity.[23,24] For this, a large number of parameters (corresponding to the amplitude and phase of each spectral component within the exciting laser pulse(s)) has to be controlled. This "pulse shaping" technique is usually performed by introducing a liquid crystal array in the Fourier plane between two gratings (4f arrangement). In 1992, the concept of "optimal control" was introduced,[44,45] in which a feedback loop optimizes the laser pulse characteristics to reach most efficiently the desired target. Excellent results have been obtained using coherent control schemes in atomic and molecular systems (mostly in the gas phase).[46]

A major technical limitation in quantum control experiments is the relatively narrow spectral region available (often around 800 nm) and the limited available bandwidth. This is a major drawback for large molecules exhibiting broad and featureless absorption bands. We therefore intend to develop a spectrally-broad laser source with a full control of the laser parameters. This will allow for the first time to coherently populate or depopulate a broad energy domain of the electronically excited intermediate state. This is the basis of the "Optical Dynamic Discrimination" (ODD) technique.[25] The physical processes that we intend to use to get the selectivity are extensions of the PPD-technique, namely shaped PPD, intrapulse pump–dump with detection of the fluorescence and stimulated emission, and CARS (Coherent Antistokes Raman Scattering).

In order to adapt the pulses to the very broad absorption bands of the biological fluorophors, and to the related short coherence time, spectral broadening of the laser pulses by filamentation in a gas cell may be used. This technique was recently successfully applied to generate pulses at 800 nm as short as 5 fs, by recompressing the broadened pulse with chirped mirrors.[47] Filaments[47,48] arise in the non-linear propagation of ultrashort, high-power laser pulses in transparent media. They result from a dynamic balance between Kerr-self-focusing and defocusing by a self-induced plasma. These spatio-temporal solitonic structures are capable of generating an extraordinary broad supercontinuum by Self-Phase Modulation (SPM) and Four Wave Mixing (FWM), spanning from the UV to the IR.[48] First broadening experiments at 400 nm using filamentation have been conducted in order to obtain a broadband laser source for flavins (and NADH) excitation. Filamentation was produced in an Ar gas-filled cell (7 bars) of 100 cm length. While the incoming laser bandwidth was only 3 nm (0.5 to 1 mJ, 150 fs), it reaches 15 to 20 nm (FWHM) after filamentation, as shown in Fig. 8.

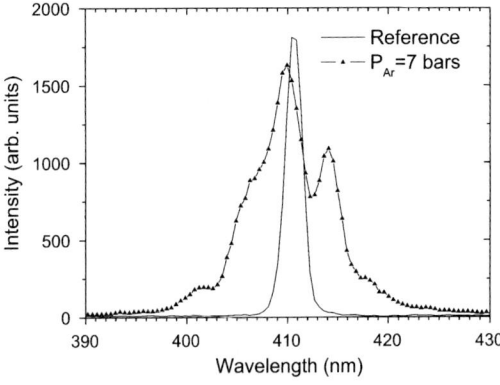

Fig. 8 400 nm laser pulse broadening by filamentation in Ar.

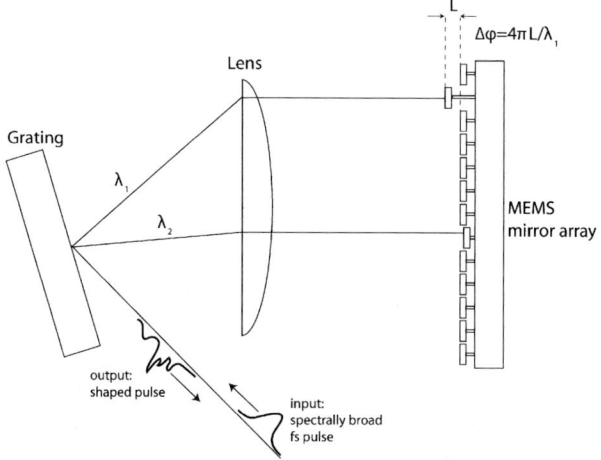

Fig. 9 Set-up for the shaping of broadband fs pulses.

In a later phase, the same approach will be applied in the UV, in order to coherently excite the amino acids, such as Trp.

At present, pulse shaping techniques are mainly based on Liquid Crystal Display (LCD) arrays and acousto-optic modulators, but their transmission is rather limited in the UV-Vis domain. We therefore intend to adopt as an active element a novel and versatile generation of MicroElectroMechanical Systems (MEMS) represented by an array of movable micro mirrors. Recently, this technology has been applied for phase-only shaping of femtosecond pulses at 400 nm.[49] Within this approach, shaping of the UV-Vis pulses is carried out in the experimental arrangement shown in Fig. 9, which acts as a 4-f zero dispersion compressor, with the mirror array placed in the Fourier plane. The pulses are first dispersed by a reflection grating; successively, a cylindrical lens focuses the spectrum on the mirror array. The phase modulation is accomplished by varying the overall optical path length of the different components, acting on the pistons of the mirror elements. Note that the unique spectral range of the apparatus is represented by the maximum travel of the mirror pistons, which limits the maximum phase shift at a given wavelength $\Delta\Phi(\lambda)$.

The new device that we are currently designing should allow both phase and amplitude shaping of pulses in a wide spectral region, ranging from 250 to 900 nm. An additional degree of freedom will be added to the mirrors, in that the intensity attenuation will be performed by tilting the mirrors in the vertical plane.

These developments should allow evaluating the ultimate capabilities of ODD for discriminating bioaerosols from other organics, bacteria from pollen, and bacteria among each other.

Conclusion

Femtosecond spectroscopy opens new ways for the optical detection and identification of microorganisms in water and in air. Its unique capability of distinguishing molecules that exhibit almost identical absorption and fluorescence signatures is a key feature for identifying bacteria in a background of urban aerosols. The technique can also be applied for the remote detection of microorganisms in air, if a non-linear Lidar-based configuration is used, as for the Teramobile system. A more difficult task will be the distinction of one bacteria species from another, and in particular the identification of pathogen from non-pathogen bioaerosols. A possible way to reach this difficult goal might be coherent control and ODD. Significant technical improvements are currently developed, which should provide a definitive answer on the potential of coherent excitation schemes. Another option could be provided by LIBS, as the wealth of emission lines under femtosecond excitation might allow the targeting of some biological process that is characteristic from one type of bacteria.

Acknowledgements

The authors acknowledge the members of the groups of L. Woeste at the FU Berlin, R. Sauerbrey at the University Jena, and A. Mysyrowicz at the ENSTA (Palaiseau) (members of the Teramobile consortium). We also wish to thank J. Kasparian, J. Yu, E. Salmon, and G. Méjean at the University Lyon 1, and M. Moret and P. Bejot at the University Geneva. The backward enhancement measurements on aerosols were performed in collaboration with the groups of R. K. Chang (Yale University) and S. C. Hill (US Army Research Laboratories). For these studies, RKC and JPW acknowledge NATO support SST-CLG977928. J. P. Wolf gratefully acknowledges the support of the Swiss National Foundation for Research (Grant number 200021-111688, project "CIBA") and the Schmidheiny Foundation. H. Rabitz acknowledges support from ARO-MURI and NSF grants.

References

1. F. C. Tenover and J. K. Rasheed, in *Manual of Clinical Microbiology*, ed. P. R. Murray, E. J. Baron and M. A. Pfaller, American Society for Microbiology, Washington, DC, 1999.
2. J. Ho, *Anal. Chim. Acta*, 2002, **457**, 125.
3. P. Belgrader *et al.*, *Science*, 1999, **284**, 449.
4. S. I. Makino, H. I. Cheun and M. Wateral *et al.*, *Appl. Microbiol.*, 2001, **33**, 237.
5. B. Beatty, S. Mai and J. Squire, in *FISH: A practical approach*, Oxford University Press, Oxford, 2002.
6. F. Pourahmadi, *Clin. Chem.*, 2000, **46**, 1151.
7. B. K. De, *Emerg. Infect. Dis.*, 2002, **8**, 1060.
8. C. Hagleitner, *Nature*, 2001, **414**, 293.
9. P. Francois, M. Bento and P. Vaudaux *et al.*, *J. Microsc. Methods*, 2003, **55**, 755.
10. Y. Pan, V. Boutou and J. R. Bottiger *et al.*, *Aerosol Sci. Technol.*, 2004, **38**, 598.
11. Y. L. Pan, J. Hartings and R. G. Pinnick *et al.*, *Aerosol Sci. Technol.*, 2003, **37**, 627.
12. F. L. Reyes, T. H. Jeys and N. R. Newbury *et al.*, *Field Anal. Chem. Technol.*, 1999, **3**, 240.
13. J. D. Eversole, W. K. Cary and C. S. Scotto, *Field Anal. Chem. Technol.*, 2001, **5**, 205.
14. G. A. Luoma, P. P. Cherrier and L. A. Retfalvi, *Field Anal. Chem. Technol.*, 1999, **3**, 260.
15. Y. L. Pan, K. B. Aptowicz and R. K. Chang *et al.*, *Opt. Lett.*, 2003, **28**, 589.
16. P. Kaye, E. Hirst and T. J. Wang, *Appl. Opt.*, 1997, **36**, 6149.
17. S. C. Hill, R. P. Pinnick and S. Niles *et al.*, *Field Anal. Chem. Technol.*, 1999, **5**, 221.
18. Y. L. Pan, P. Cobler and S. Rhodes *et al.*, *Rev. Sci. Instrum.*, 2001, **72**, 1831.
19. R. G. Pinnick, S. C. Hill and Y. L. Pan *et al.*, *Atmos. Environ.*, 2004, **38**, 1657.
20. Y. S. Cheng, E. B. Barr and B. J. Fan *et al.*, *Aerosol Sci. Technol.*, 1999, **31**, 409.
21. C. Weitkamp, *Lidar*, Springer-Verlag, New York, 2005.
22. F. Immler, D. Engelbart and O. Schrems, *Atmos. Chem. Phys. Discuss.*, 2004, **4**, 5831.
23. T. Brixner, P. Niklaus and G. Gerber, *Nature*, 2001, **414**, 57.
24. T. Brixner, B. Kiefer and G. Gerber, *J. Chem. Phys.*, 2003, **118**(8).
25. B. Li, H. Rabitz and J. P. Wolf, *J. Chem. Phys.*, 2005, **122**, 154103–1-8.
26. J. T. Vivian and P. R. Callis, *Biophys. J.*, 2001, **80**, 2093.
27. P. R. Callis and J. T. Vivian, *Chem. Phys. Lett.*, 2003, **369**, 409.
28. C. Dedonder-Lardeux, C. Jouvet and S. Perun *et al.*, *Phys. Chem. Chem. Phys.*, 2003, **5**, 5118.
29. H. B. Steen, *J. Chem. Phys.*, 1974, **61**(10), 3997.
30. Y. Iketaki, T. Watanabe and S.-i. Ishiuchi *et al.*, *Chem. Phys. Lett.*, 2003, **372**, 773.
31. F. Courvoisier, V. Boutou and V. Wood *et al.*, *Appl. Phys. Lett.*, 2005, **87**(6), 063901.
32. F. Courvoisier, V. Boutou, L. Guyon, M. Roth, H. Rabitz and J. P. Wolf, *J. Photochem. Photobiol.*, 2006, **A180**, 300.
33. S. C. Hill, V. Boutou and J. Yu *et al.*, *Phys. Rev. Lett.*, 2000, **85**(1), 54.
34. V. Boutou, C. Favre and S. C. Hill *et al.*, *Appl. Phys. B*, 2002, **75**, 145.
35. Y. Pan, S. C. Hill and J. P. Wolf *et al.*, *Appl. Opt.*, 2002, **41**(15), 2994.
36. C. Favre, V. Boutou and S. C. Hill *et al.*, *Phys. Rev. Lett.*, 2002, **89**(3), 035002.
37. M. Baudelet, L. Guyon and J. Yu *et al.*, *Appl. Phys. Lett.*, 2006, **88**, 053901.
38. M. Baudelet, L. Guyon and J. Yu *et al.*, *J. Appl. Physics*, 2006, **99**, 084701.

39 J. Kasparian, M. Rodriguez and G. Méjean *et al.*, *Science*, 2003, **301**(5629), 61.
40 G. Méjean, J. Kasparian and J. Yu *et al.*, *Appl. Phys. B*, 2004, **78**(5), 535.
41 J. P. Wolf, Y. Pan and S. Holler *et al.*, *Phys. Rev. A*, 2001, **64**, 023808.
42 L. Méès, J. P. Wolf and G. Gouesbet *et al.*, *Opt. Commun.*, 2002, **208**(371–375), 2002.
43 L. Méès, G. Gouesbet and G. Gréhan, *Opt. Commun.*, 2001, **199**, 33.
44 R. Judson and H. Rabitz, *Phys. Rev. Lett.*, 1992, **68**, 1500.
45 S. Warren, H. Rabitz and M. Daleh, *Science*, 1993, **259**, 1581.
46 M. Dantus and V. Lozovoy, *Chem. Rev.*, 2004, **104**, 1813.
47 C. P. Hauri, W. Kornelis and F. W. Helbing *et al.*, *Appl. Phys. B*, 2004, **79**, 673.
48 L. Bergé, S. Skupin and U. Peschel *et al.*, *Phys. Rev. E*, 2005, **71**, 016602.
49 M. Hacker, G. Stobrawa and R. Sauerbrey *et al.*, *Appl. Phys. B*, 2003, **76**, 711.

Vibrational exciton coupling in pure and composite sulfur dioxide aerosols

Ruth Signorell* and Martin Jetzki

Received 9th January 2007, Accepted 26th January 2007
First published as an Advance Article on the web 13th July 2007
DOI: 10.1039/b700111h

Icy aerosol particles determine weather processes in planetary atmospheres and are discussed as sites of chemical reactions in interstellar dust. One of the interesting ice systems in this context are pure and composite SO_2 particles. Vibrational exciton coupling is exploited in the present study to understand the properties of SO_2 aerosol particles from their infrared spectroscopic signatures. The condensation to pure SO_2 ice particles leads to a partially amorphous structure of the particles, whose infrared spectra do not show pronounced shape effects. The simultaneous condensation of SO_2 and CO_2 gas results in aerosol particles with a core–shell structure. Icy SO_2 forms the core, whereas CO_2 remains in the shell. This structure is mainly determined by the different thermodynamic properties of the two substances and the less favorable intermolecular interactions between SO_2 and CO_2 compared with the pure substances. The example of mixed sulfur dioxide/ammonia particles demonstrates that nanosized aerosol particles are ideal nanoreactors. They facilitate an easy homogeneous mixture of the reactants providing an interface-free reaction environment.

1. Introduction

Icy particles consisting of simple molecules, such as sulfur dioxide, play a fundamental role as aerosols in planetary atmospheres and as components of interstellar dust. Through phase transitions, mass transfer processes, chemical and photochemical reactions they critically influence the energy balance and composition of planetary atmospheres. Sulfur dioxide ice exists on Jupiter's satellites Io and Europa (see ref. 1–4 and references therein). On Io, SO_2 constitutes a major component on the surface of the planet and in its atmosphere. At temperatures on Io between 65 K and 135 K, SO_2 may be present as icy aerosols and surface frost particles. Icy grain mantles in interstellar dust, which contain SO_2, are discussed as a sulfur reservoir and as sites for chemical reactions.[1,5] An important example for the latter is the condensed phase reaction of SO_2 with NH_3.

The goal of the present contribution is to exploit vibrational dynamics to characterize the properties of pure and composite SO_2 aerosol particles. For this purpose, we combine particle formation in a collisional cooling cell with rapid–scan infrared spectroscopy *in situ* (see ref. 6 and 7 and references therein). The focus of the present study lies on intrinsic particle properties such as particle shape, size, architecture, and structure. As demonstrated in previous studies

Department of Chemistry, University of British Columbia, 2036 Main Mall, Vancouver, BC, Canada V6T 1Z1. E-mail: signorell@chem.ubc.ca; Fax: +1 604 822-2847

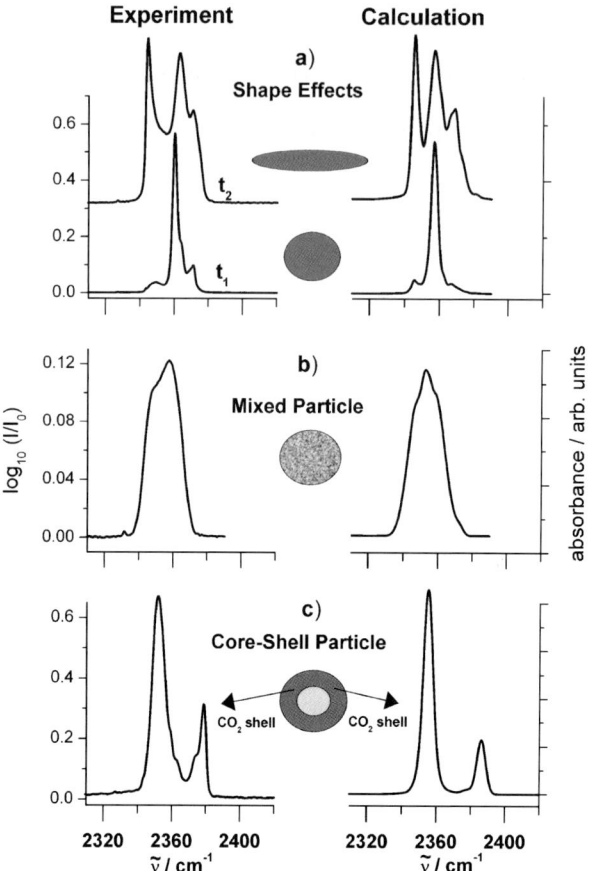

Fig. 1 Infrared spectra of CO_2 aerosol particles in the region of the antisymmetric stretching vibration of CO_2. Left: Experimental spectra. Right: Exciton calculations. As parameters for the exciton calculations for CO_2, we used $\tilde{\nu}_M = 2355$ cm^{-1} and $\delta\mu = 0.32$ D[6]. (a) Spectra for different particle shapes. Lower trace: For globular particles. Upper trace: For elongated particles. Globular particles are formed directly after the particle formation (t_1) and elongated particles are formed after several minutes (t_2). (b) Spectrum of a statistically mixed particle with a CO_2 fraction of 50%. (c) Spectrum of a core–shell particle with CO_2 in the shell. $\tilde{\nu}$ is the wavenumber in cm^{-1}.

(see ref. 6 and 7) particle properties manifest themselves particularly strongly in intense infrared absorption bands, for which transition dipole coupling is the dominant interaction. For SO_2 ice particles, the symmetric and antisymmetric stretching vibrations fulfil this requirement. When transition dipole coupling is the dominant interaction, the band structures in the infrared spectra can be understood and predicted directly from molecular properties. The corresponding quantum mechanical model is called the vibrational exciton model.[7–9] Fig. 1 shows previous results from our research group for pure and composite CO_2 ice particles. These examples demonstrate how particle properties determine the structure of the strong antisymmetric stretching vibration through exciton coupling. The experimental spectra are depicted on the left and the prediction from exciton calculations are shown on the right. Panel (a) illustrates the influence of the particle shape. A crystalline globular particle (lower trace) leads to a narrow band with one pronounced maximum. Crystalline elongated particles (upper trace) are characterized by a much broader band exhibiting several pronounced maxima. If CO_2 is statistically mixed with another substance (mixing ratio 1 : 1) one finds a single broad band as depicted in panel (b). The structure of this band is largely independent of the particle shape. Composite particles do not have to form statistical mixtures. Another possibility is the formation of core–shell particles. Panel (c) shows the spectrum of a CO_2 shell of a core–shell particle. It is characterized by two separate absorption bands. For all these different types of particles, the exciton calculations clearly reflect all significant spectral features found in the experimental spectra. It is important to note in this context that these calculations are true predictions and do not involve any fit whatsoever to the experimental spectra.

The vibrational dynamics of pure sulfur dioxide aerosol particles formed in a collisional cooling cell is discussed in Section 4. Spectroscopic observations of Io have revealed that CO_2 might be an additional component of SO_2 ice on this moon.[1,10] This was our motivation for the investigations of mixed SO_2/CO_2 particles, the results of which are summarized in Section 5. As mentioned above, the reaction of SO_2 with NH_3 is not only important in interstellar chemistry but can also influence processes in aerosol particles in the Earth's atmosphere. Moreover, it has been discussed as a way to remove sulfur dioxide emitted from stationary sources.[11] In Section 6, we report our first results obtained for this reaction in icy aerosols.

2. Experimental methods

Pure and composite SO_2 particles were generated by collisional cooling. A detailed description of the collisional cooling cells used for the present study is given in ref. 6 and 7 and references therein. The aerosol particles are formed by introducing a warm sample gas into a cold bath gas. This leads to supersaturation of the sample gas and thus to particle formation. The particle formation is influenced by the pressures and the temperatures of the sample gas and of the bath gas. As bath gas, we used helium at pressures between 10 and 1000 mbar and temperature between 5 and 80 K. The pure SO_2 particles were formed by introducing a single gas pulse through a magnetic valve into the cold cell. The same method was also used to generate statistically mixed SO_2/CO_2 particles, but instead of a pure SO_2 sample gas we used in this case a premixed SO_2/CO_2 sample gas. Typical opening times of the magnetic valve lay at several hundred ms. The SO_2/CO_2 core–shell particles were prepared with a twin-gas–pulse inlet.[6,12] The two different sample gases are introduced into the cell through two spatially separated inlet tubes, which are connected to two independent magnetic valves. Core–shell aerosol particles are formed when the opening times of the two valves are shifted in time. This time delay typically amounted to 500–1000 ms. The size of the particles investigated here lies around 50 nm (10^6 molecules per particle) and the particle density amounts to about 10^6 particles cm^{-3} (see ref. 13 and 14 for more details). This size region is particularly well suited to study shape effects, mixing effects, and the influence of the particles' architecture on mid-infrared spectra. On the one hand these particles are already too large to exhibit prominent surface effects and on the other hand they are still small enough so that contributions from elastic scattering can be neglected.[7]

The infrared spectra were recorded *in situ* with a Bruker IFS 66 v/S rapid–scan Fourier Transform InfraRed (FTIR) spectrometer. The rapid–scan option allows us to observe time–dependent processes with a time resolution of 30 ms at a spectral resolution of 2 cm^{-1}. The cooling cells are equipped with White optics with an optical path lengths of about 16 m. All spectra were recorded using a mid–infrared Globar light source and a KBr beam splitter.

SO_2 (99.98%), CO_2 (99.995%), and NH_3 (99.98%) were purchased from Messer–Griesheim. Helium gas (99.999%) from Messer–Griesheim was further purified to eliminate water traces.

3. Modelling

Modelling plays a key role for the understanding of the infrared spectroscopic signatures of such complex systems. Our goal is to explain the features observed on a true molecular level since this forms the basis for a comprehensive understanding of aerosol properties. With the many degrees of freedom of these systems (10^7), however, this is a difficult task. Standard quantum chemical methods, for example, are not suitable here. We have demonstrated in previous contributions[6,7,9,14–17] that for ice particles consisting of small molecules like NH_3, CO_2, etc. the band structures of vibrational bands with strong molecular transition dipoles (transition dipoles > 0.1–0.2 D) are determined by transition dipole coupling between all molecules in a particle. This quantum mechanical model is called the vibrational exciton model and has been used previously to model infrared spectra of bulk CO_2 ice and small CO_2 aggregates.[8,18–21] We have extended this model to treat large particles and composite particles and we have demonstrated its suitability by comparison with experimental results and by comparison with quantum chemical calculations for small clusters. The main components of this model can be summarized as follows. The vibrational Hamiltonian has the form

$$\hat{H} = \hat{H}_0 + \hat{H}_D \qquad (1)$$

with

$$\hat{H}_D = \sum_{i<j} -\frac{1}{4\pi\varepsilon_0} \vec{\mu}_i \cdot \frac{3(\vec{\mu}_j \cdot \vec{r}_{ij}) \cdot \vec{r}_{ij} - (\vec{r}_{ij} \cdot \vec{r}_{ij}) \cdot \vec{\mu}_j}{r_{ij}^5}. \qquad (2)$$

$\hat{H}_0$ is the sum over the vibrational Hamiltonians of the uncoupled molecules in a particle. $\hat{H}_D$ includes all pairwise dipole–dipole interactions between the molecules in the aggregate. $\vec{r}_{ij}$ is the distance between the centres of mass of the molecules labeled i and j and $\vec{\mu}$ is the dipole moment operator. Up to first order terms in the vibrational coordinates, the model contains only two parameters: the transition wave number of the uncoupled molecule $\tilde{v}_M$ and the transition dipole of the uncoupled molecule $\delta\mu = \langle 0|\mu|1\rangle$. Both quantities can be extracted from gas phase measurements or from *ab initio* calculations on smaller clusters. The diagonalization of $\hat{H}$ leads to the vibrational eigenfunctions and eigenvalues of the particles. The infrared spectra of the particles are then calculated from these quantities.[9]

In the case of SO_2 the antisymmetric stretching vibration ν_3 at 1335 cm^{-1} and the symmetric stretching vibration ν_1 at 1149 cm^{-1} both have a strong molecular transition dipoles ($\delta\mu = 0.24$ D and $\delta\mu = 0.09$ D, respectively[22]). In addition, next nearest neighbour molecules in the particles are separated by small distances ($r_{ij} < 4.2$ Å[23]). This means that all conditions are fulfilled for exciton coupling to be the dominant interaction for the two stretching bands.[9,16] The best proof, however, comes from the comparison of exciton calculations with quantum chemical calculations for small clusters, as demonstrated in more detail for CO_2 and N_2O clusters in ref. 9. Fig. 2 shows a comparison between exciton calculations and calculations using B3LYP/6-31+G*[24] for the ν_3 band for three small crystalline SO_2 clusters. The same behavior was also found for the ν_1 band, so we do not show these results here. The crystal structure was taken from ref. 23. As input parameters for the exciton calculations, *i.e.* the transition wavenumber $\tilde{v}_M$ and the transition dipoles $\delta\mu$ of the uncoupled molecules, we used the values obtained from B3LYP/6-31+G* for an isolated SO_2 molecule. The comparison in Fig. 2 clearly demonstrates that exciton coupling is the main interaction determining the band structure in the infrared spectra. We will thus use this model in the following to predict infrared spectra of large aerosol particles.

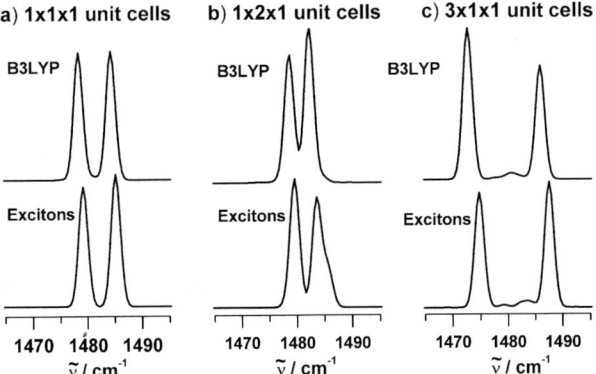

Fig. 2 Comparison of quantum chemical calculations using B3LYP/6-31+G* (upper traces) with exciton calculations (lower traces) for the ν_3 band of three different small crystalline SO_2 clusters. As input parameters for the exciton calculations, we used $\tilde{\nu}_M = 1481$ cm^{-1} and $\delta\mu = 0.25$ D. $\tilde{\nu}$ is the wavenumber in cm^{-1}.

4. Pure SO_2 aerosol particles

Fig. 3a shows particle infrared spectra in the region of the antisymmetric stretching vibration ν_3 (left) and the symmetric stretching vibration ν_1 (right). One spectrum was recorded directly after the particle formation ($t = 0$ s) and the other one about 8 min later. After that time the spectra did not change any further with increasing time. Our particle spectra in Fig. 3 look different compared with bulk spectra reported in the literature.[2,3,5,25–33] This deviation is also reflected in the different band positions found for the particles, which are listed in Table 1 together with corresponding values for the bulk. This difference is not surprising if the band structure is determined by exciton coupling. In this case, the particle shape and architecture have a pronounced influence on the band shape. In the case of SO_2,

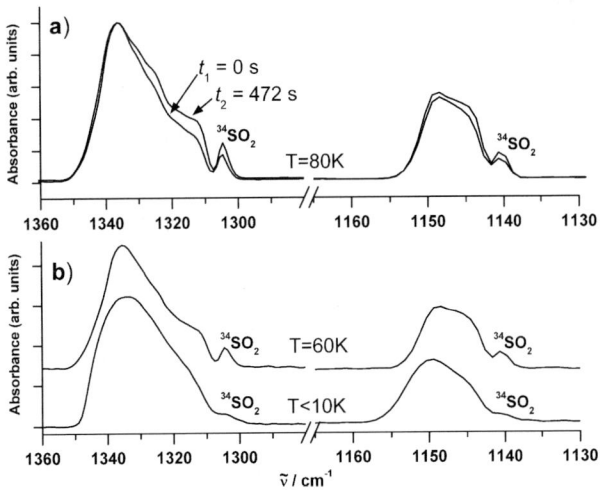

Fig. 3 Experimental SO_2 particle infrared spectra in the region of the antisymmetric stretching vibration ν_3 (left) and the symmetric stretching vibration ν_1 (right). (a) Spectra recorded as a function of time at a bath gas temperature of 80 K. (b) Spectra recorded directly after the particle formation ($t = 0$ s) at bath gas temperatures of 60 K (upper trace) and below 10 K (lower trace). $\tilde{\nu}$ is the wavenumber in cm^{-1}.

Table 1 Fundamental transitions in cm^{-1} in infrared spectra of SO$_2$ particles and bulk SO$_2$

Ref.	Particles	Bulk	Band
Fig. 3(a)	1145 sh		ν_1
	1149		ν_1
	1312 sh		ν_3
	1324 sh		ν_3
	1330 sh		ν_3
	1335		ν_3
25		1142	$\nu_1 - \nu''$
		1147	ν_1
		1308	ν_3
		1316	$\nu_3 - \nu''$
		1330	ν_3
26		1144	ν_1
		1310	ν_3
		1322	ν_3
		1334	$\nu_3 + \nu_L$
27 [a]		1144	$\nu_1(A_1)$
		1148	$\nu_1(A_2)$
		1312	
		1324	$\nu_1 + R_x$
		1341	$\nu_3(B_1$ and $B_2)$
28 [b]		1147	$\nu_1(A_1)$
		1313	$\nu_3(B_1)$
		1327	$\nu_3(B_2)$
29 [c]		1135 α	ν_1
		1304 α	ν_3
		1143 β	ν_1
		1297 β	ν_3
		1119 γ	ν_1
		1259 γ	ν_3

[a] Raman spectrum. [b] At 20 K. [c] Adsorbed on CsCl films. sh: shoulder. ν', ν'': torsional modes. ν_L and R_x: lattice vibrations. α, β, γ: different phases. A_1, A_2, B_1, B_2: symmetry of the vibration.

however, it is worth mentioning that the agreement between different bulk spectra is not perfect either and that over the years the assignment of the bulk spectra has changed. As indicated in Fig. 3, the absorption bands of the isotopomer ^{34}SO$_2$ appear at the low wavenumber sides of the main peaks. Since the two isotopomers lie so close together, their coupling through exciton interaction cannot be neglected and is thus included in the calculations (see below).

Our expectation was that SO$_2$ particles, like particles of NH$_3$, CO$_2$, N$_2$O, SF$_6$, CHF$_3$, C$_2$H$_2$, and phenanthrene,[6,14,34–36] show pronounced shape effects for strong mid-infrared bands (see, for example, shape effects in infrared spectra of CO$_2$ particles in Fig. 1(a)). As demonstrated in previous investigations,[7] there are two experimental parameters that allow us to modify the shape of particles formed in collisional cooling cells. This is the bath gas pressure in the cell and the time after the particle formation has set in. The general observation is that at low bath gas pressure elongated particles are formed, whereas at high bath gas pressures globular particles dominate. Similarly, directly after the particle formation, globular particles dominate, but with increasing time elongated particles become more and more important. For SO$_2$ particles, however, we found that none of these parameters had a strong influence on the band structure. This is illustrated in Fig. 3(a) by the two spectra recorded at different times after particle formation was completed. The difference between these two spectra is much smaller than the difference observed for CO_2 particles with different shapes in Fig. 1(a).

There are three possible explanations for this observation: the particle shapes in the case of SO_2 could be more or less independent of the experimental conditions, or the infrared bands of SO_2 are not very sensitive to the shape, or both. As a first step we tested whether or not SO_2 particle infrared bands are shape-sensitive. For this purpose, we have performed exciton calculations for many different particle shapes under the assumption that the particles are crystalline. As a result we found that the simulations for different particle shapes differ strongly, similar to the behavior shown for CO_2 particles in Fig. 1(a). In other words, the spectra of crystalline SO_2 particles clearly exhibit pronounced shape sensitivity. If we form crystalline SO_2 particles in our cell we must therefore conclude that the distribution of their shapes does not evolve in time. This immediately raises the question whether we form crystalline particles at all. The results of an exciton calculation for the ν_3 band of a typical mixture of elongated SO_2 particles with a crystalline structure is shown in Fig. 4(a). The mixture (see figure caption for details) corresponds to the particle shapes reported as typical for macroscopic crystalline particles.[2] The comparison with the experimental spectrum in Fig. 3(a) (left side) shows almost no agreement. Obviously, the narrow absorptions in the calculated spectrum cannot explain the broad band observed experimentally. We have also performed exciton calculations for crystalline particles with many other shapes. The result was always the same: crystalline particles exhibit absorptions which are too narrow to explain the experiment. We thus concluded that we do not produce perfect crystalline particles in our cooling cells.

The next logical step to follow was the investigation of particles with non-perfect crystalline structures or even amorphous structures. A recent investigation of bulk SO_2[3] reports the formation of amorphous SO_2 at temperatures below 30 K. The infrared spectrum of this amorphous state is characterized by structureless broad asymmetric bands. Another characteristic of this state is the fact that the absorption bands of the isotopomer $^{34}SO_2$ can no longer be distinguished as separate bands, but appear as shoulders close to the absorption of the main isotopomers. The particle spectra in Fig. 3(a) do not obviously show all the characteristics of this amorphous state. The main absorption bands still exhibit some structure and the $^{34}SO_2$ isotopomer bands are well separated from the main absorption band. Obviously, these particles are not completely amorphous. This statement is also supported by the particle spectra shown in Fig. 3(b) recorded at temperatures below 80 K. The 60 K spectrum in panel (b) still looks the same as the 80 K spectrum in panel (a). Below 10 K, however, the spectrum exhibits the characteristics of the amorphous state. The bands are less structured and the $^{34}SO_2$ absorption is no longer clearly separated. We thus conclude that amorphous particles are formed in our cell below 10 K.

From the above results, it is clear that Fig. 3(a) neither shows the spectra of completely crystalline particles nor of completely amorphous particles, but possibly of particles with a partially amorphous structure. To clarify this point we have performed exciton calculations for such partially amorphous structures. In a first step, we allowed the molecules to rotate by a certain angle out of their crystalline position around their centres of mass. The angle was randomly varied between 0° (crystal) and a certain maximum value. As expected, this leads to a strong broadening of the absorption bands. For a maximum rotation angle around 40°, we obtained the same broad absorption bands as in the experimental spectra in Fig. 3(a). In a second step, we also allowed shifts in the centres of mass out of the crystalline position. Again, the shift was randomly chosen between 0 Å (crystal) and some maximum value. Even maximum shifts of up to 1 Å, which are physically rather implausible,[23] produce only very minor spectral broadening. Thus, compared to the effect of rotating the molecules, the effect of shifting them can be neglected.

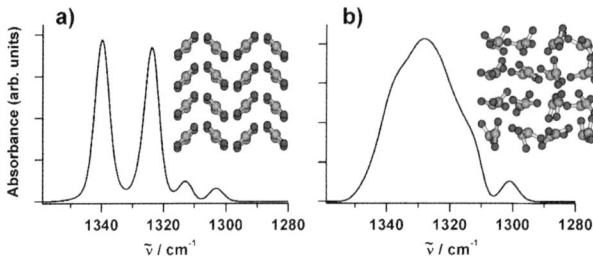

Fig. 4 (a) Exciton calculation for the ν_3 band for a mixture of elongated SO_2 particles with a crystalline structure. For the calculation we assumed a 6 : 1 : 1 mixture of three particle shapes with an axis ratio of 1 : 1 · 4, 1 : 4 : 1, and 4 : 1 : 1, respectively. The crystal structure was taken from ref. 23. As parameters for the exciton calculations, we used $\tilde{\nu}_M = 1325.5$ cm^{-1} and $\delta\mu = 0.24$ D for $^{32}SO_2$ and $\tilde{\nu}_M = 1305.0$ cm^{-1} and $\delta\mu = 0.24$ D for $^{34}SO_2$.[22,32] (b) Exciton calculation for the ν_3 band for the same mixture of elongated SO_2 particles as in panel (a) but with a partially amorphous structure. $\tilde{\nu}$ is the wavenumber in cm^{-1}.

Fig. 4(b) shows an exciton calculation for the same particle mixture as in panel (a), but for a partially crystalline structure. The structure was created by random rotations of the molecules by angles between 0° and 40°. The agreement of the ν_3 absorption band of these partially amorphous particles with the experimental absorption band on the left hand side in Fig. 3(a) is much better than the agreement with the calculated absorption band of crystalline particles shown in Fig. 4(a). It is also important to note that, as in the experiment, the band of the isotopomer $^{34}SO_2$ is still clearly separated from the main absorption in the spectrum of the partially amorphous particles. Again, this is an important feature to distinguish from completely amorphous particles. We found the same results also for the ν_2 vibration, so we do not show these results here.

Although the agreement between calculation in Fig. 4(b) and experiment in Fig. 3(a) is not completely perfect, the above findings clearly show that between about 60 and 90 K we form partially amorphous particles in our cooling cell. This partially amorphous state is surprisingly stable. For example, it is not influenced by the bath gas pressure, which determines the cooling rate during the particle formation. It was also not possible to crystallize the particles by increasing the temperature by several degrees. There are several reasons why experiment and calculation do not agree perfectly. Although exciton coupling clearly determines the main spectral features (see Fig. 2), it cannot explain minor details in the spectra that arise from other less important intra- and intermolecular interactions acting in the particles. It is important to note here that for particles consisting of tens of thousands of molecules it is more than satisfactory to have a quantum mechanical model that is able to explain the dominant features in the spectra. The simple way we construct the partially amorphous particles and the simple ensemble of particles we have chosen for the simulations in Fig. 3 are two other important points that contribute to the deviations between the calculated and the experimental spectrum. By varying these two parameters, we could certainly achieve much better agreement between simulation and experiment. Apart from increasing the aesthetic value, however, this would not have any physical meaning. We have clearly demonstrated that the main features in the experimental spectrum can be predicted with simple, physically meaningful assumptions. In our opinion this is more important than perfecting the agreement by fitting minor details.

As a last point, we had to clarify how shape-sensitive the spectra of partially amorphous particles are. For this, we performed again many exciton calculations for different particle shapes. As a result we found that the spectra of partially amorphous particles are much less sensitive to the particle shape than the spectra of crystalline particles. For this reason we cannot draw the same firm conclusions about the particle shapes from the spectrum of SO_2 particles as was possible in the case of CO_2, depicted in Fig. 1 (a). The only firm statement we can make is that the behavior of the two experimental spectra in Fig. 3(a) would not be consistent with a very drastic change of the particle shape, such as an increase of the axis ratio by a factor of ten. In other words, the temporal behavior of the spectra in Fig. 3(a) implies that the particle shape changes only slightly with increasing time after particle formation. A completely stationary distribution of particle shapes would of course also be consistent with the above findings. From our experience with other ice particles,[7] however, this would appear rather implausible.

5. SO_2/CO_2 aerosol particles

In planetary atmospheres and interstellar ice grains, it is often composite ices that are of interest. A common ice component that appears together with SO_2 is CO_2. On Io, for example, CO_2 molecules have been identified, albeit tentatively.[1,10] Our first goal was to find out what happens if SO_2 and CO_2 are co-condensed in our cooling cell. To this end we prepared a premixed gas sample ($SO_2 : CO_2 = 1 : 1$) that was then introduced into the cold cell. In accordance with previous observations, *e.g.* in the case of CO_2/NH_3 mixtures,[14] we expected the formation of statistically mixed aerosol particles through the rapid cooling in our cell.

Fig. 5(a) shows the experimental spectra of SO_2/CO_2 aerosol particles that were generated from such a premixed gas sample. Compared with the spectrum of pure CO_2 particles (lower traces in Fig. 1(a)) the band of the antisymmetric stretching vibration of CO_2 in Fig. 5(a) is much broader and shows a pronounced shoulder on the high-wavenumber side. The SO_2 band is unchanged compared with the spectrum of pure SO_2 particles (Fig. 3(a)). This behavior is not what we had expected for statistically mixed particles. Our expectation had been to find two broad unstructured bands (as for statistically mixed particles in Fig. 1(b)). To support this expectation we performed exciton calculations for statistically mixed SO_2/CO_2, which indeed show two broad structureless bands for the CO_2 band as well as for the SO_2 band, as illustrated for the antisymmetric stretching vibration of CO_2 in Fig. 6(a). This disagreement between calculation and experiment is a clear hint that we do not form statistically mixed particles from premixed gas samples.

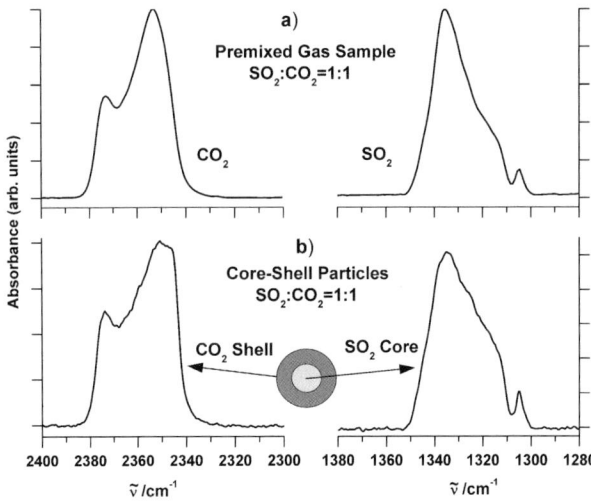

Fig. 5 Experimental infrared spectra of SO_2/CO_2 particles. The ratio of $SO_2:CO_2$ amounts to 1 : 1. Left: Region of the antisymmetric stretching vibration of CO_2. Right: Region of the antisymmetric stretching vibration of SO_2. (a) Particles formed from a premixed SO_2/CO_2 gas sample. (b) SO_2–CO_2 core–shell particles. $\tilde{\nu}$ is the wavenumber in cm^{-1}.

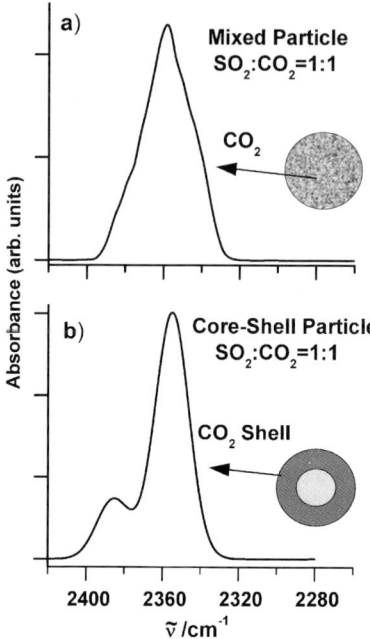

Fig. 6 Exciton calculations for SO_2/CO_2 particles in the region of the antisymmetric stretching vibration of CO_2. The ratio of $SO_2 : CO_2$ amounts to 1 : 1. (a) Statistically mixed SO_2/CO_2 particles. (b) SO_2–CO_2 core–shell particles. As parameters for the exciton calculations for CO_2, we used $\tilde{\nu}_M = 2355$ cm^{-1} and $\delta\mu = 0.32$ D.[6] $\tilde{\nu}$ is the wavenumber in cm^{-1}.

If the two substances do not form statistical mixtures what other type of particles could be generated? Solid CO_2 and SO_2 have very different sublimation (195 K) and boiling points (263 K), respectively. In addition, compared with the pure substances, the intermolecular interactions of sulfur dioxide and carbon dioxide are expected to be less favorable. SO_2 can bind through strong dipole coupling to other SO_2 molecules. Since CO_2 has no dipole this interaction is lost when binding to a CO_2 molecule. For these reasons another realistic case seems to be the formation of core–shell particles from premixed gas samples with SO_2 in the core and CO_2 in the shell. Since the boiling point of SO_2 is so much higher than the sublimation point of CO_2, it is plausible that SO_2 condenses first and CO_2 then forms a layer on the SO_2 particles as soon as the temperature is low enough.

We have tested this hypothesis by generating core–shell particles in a controlled manner.[6] For this purpose we used the twin-gas-pulse inlet described in Section 2 and two different pure sample gas mixtures, one consisting of CO_2 and the other consisting of SO_2 gas (in He). With a first gas pulse, we introduced the SO_2 gas into the cell. After the SO_2 particle formation was complete (about 500–1000 ms) we introduced the CO_2 gas with a second gas pulse. In this way, the CO_2 condenses onto the SO_2 particles and forms a shell. The spectrum of these core–shell particles is depicted in Fig. 5(b). The agreement with the spectrum obtained from a premixed gas sample in panel (a) strongly supports our hypothesis that core–shell particles are formed from premixed SO_2/CO_2 gas samples. Another clear hint that this is indeed the case comes from exciton calculations for core–shell particles. The exciton calculation for the antisymmetric stretching vibration of the CO_2 shell of SO_2–CO_2 core–shell particles is depicted in Fig. 6(b). It shows a double band, which is highly characteristic for a shell structure (see also Fig. 1(c) and ref 6). This is exactly the feature found in the experimental spectra in Fig. 5. The fact that the agreement between calculation and experiment is not perfect can again be traced back to the simplicity of the model, as already mentioned above for pure SO_2 particles. One important point is that we modelled the CO_2 shell as perfectly crystalline. The boundary region between CO_2 and SO_2, however, will not have this perfect structure. It will rather consist of a somewhat distorted crystalline structure.

The results obtained here show that co-condensation of SO_2 and CO gas does not lead to statistically mixed particles. From the spectral features it is found that SO_2 condenses first and afterwards CO_2 forms a layer on the SO_2 particles. In contrast to the behavior of N_2O/CO_2 core–shell particles,[7] the SO_2–CO_2 core–shell particles are stable after the particle formation. For N_2O/CO_2 we had observed that the core–shell particles turn into statistically mixed particles within two minutes. This mixing proceeds *via* diffusion within the particles, or *via* evaporation and recondensation, or both. Due to the less favorable intermolecular interactions compared with the pure substances, diffusion is suppressed in the SO_2–CO_2 core–shell particles. Evaporation and recondensation is also very unlikely to happen at 80 K because of the high boiling point of SO_2.

6. Nanoreactors: reaction of SO_2 with NH_3

Another important ice mixture is SO_2 with NH_3. SO_2 and ammonia have very similar melting points (201 K and 195 K, respectively) and boiling points (263 K and 240 K, respectively). Both molecules have strong dipoles (1.63 D and 1.42 D, respectively) so that we expected to form statistically mixed aerosol particles from premixed gas samples. Fig. 7(a) shows the particle spectrum obtained from a premixed gas sample containing equal amounts of sulfur dioxide and ammonia. This spectrum was a surprise for us. For a statistical mixture we had expected to find the same absorption bands as for pure NH_3 and SO_2 particles (Fig. 7(c) and (d), respectively), broadened compared with the spectra of the pure substances, as observed for statistically mixed CO_2 particles in Fig. 1(b). The spectrum in Fig. 7(a), however, clearly shows additional bands compared with the spectra of the pure particles. These additional bands appear around 710 cm^{-1}, 1050 cm^{-1}, 1465 cm^{-1}, and as a very broad band around 3000 cm^{-1}. It is very unlikely that the broad band around 3000 cm^{-1} corresponds to the NH-stretching bands of ammonia in a statistically mixed hydrogen-bonded SO_2/NH_3 particle. The broadening and shift compared with the pure hydrogen-bonded ammonia particles (trace (c)) can hardly be explained simply by mixing with SO_2.

The only explanation of these observations is a chemical reaction taking place in the particles in the cold cell (80 K). How can we obtain more experimental evidence for hypothesis? If sulfur dioxide reacts with ammonia obviously not all SO_2 is used up in the reaction of a 1 : 1 mixture, as can be seen by the remaining absorptions labelled "SO_2" in Fig. 7(a). These remaining SO_2 bands should thus disappear in the presence of an excess of ammonia in the reaction mixture. Fig. 7(b) shows the spectrum that results from a premixed gas sample with an SO_2 to NH_3 ratio of 1 : 4. In this spectrum, all bands arising from pure SO_2 have indeed disappeared, *i.e.* all SO_2 has now reacted. Since ammonia was in excess there is, apart from the reaction products, also pure ammonia left, as indicated by the absorptions labelled "NH_3" in trace (b). To ascertain that what we see is not a simple gas phase reaction in the sample mixture (very unlikely due to the low concentrations we use[37]) we have also made a series of experiments in which the two gases were introduced into the cell separately but at the same time. The results were the same as for the premixed samples, which proves that no gas phase reaction happened in the premixed samples.

In a second step, we formed core–shell particles with sulfur dioxide in the core and ammonia in the shell and *vice versa* to study whether this influences the reaction velocity. The results were unambiguous. No differences compared with the statistically mixed particles could be observed. Obviously, the reaction of SO_2 and NH_3 occurs immediately upon particle formation, independently of whether the substances were statistically mixed initially or whether core–shell particles were formed.

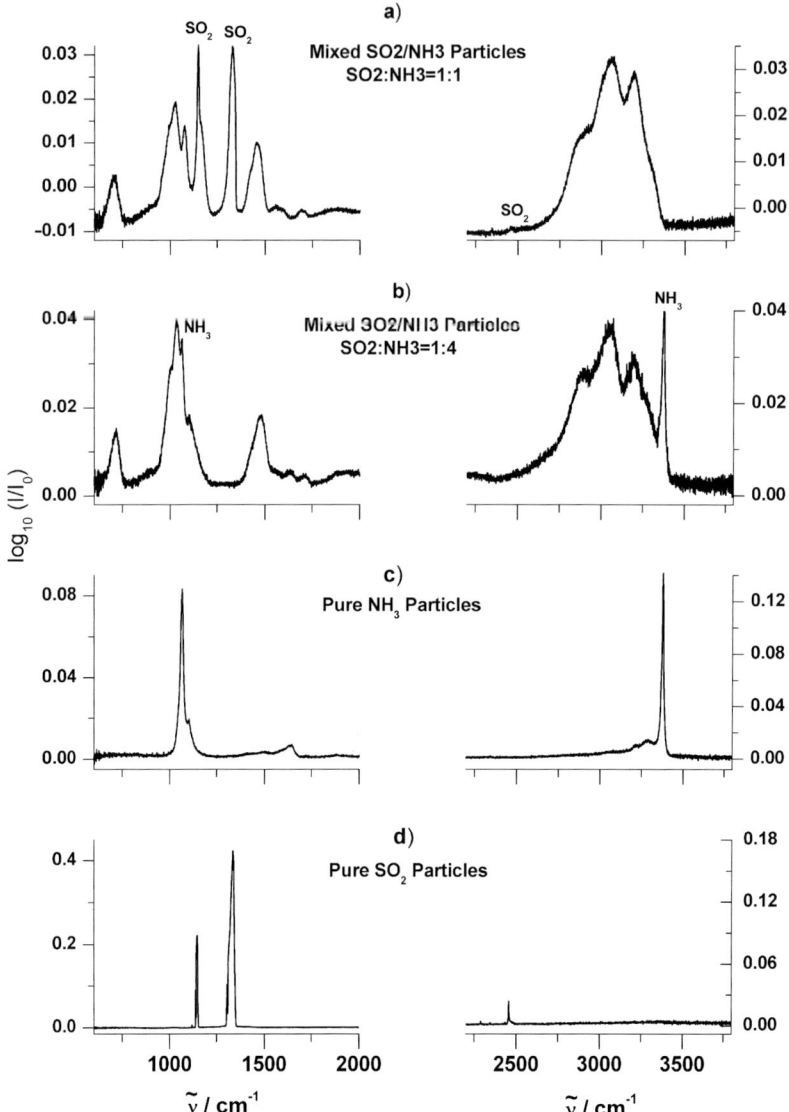

Fig. 7 (a) and (b) Particle infrared spectra of two different SO_2/NH_3 mixtures. (c) and (d) Infrared spectra of pure NH_3 and SO_2 particles, respectively. The particles were formed at a bath gas temperature of 80 K. $\tilde{\nu}$ is the wavenumber in cm^{-1}.

Ref. 38 and 39 report the reaction of SO_2 and NH_3 if the two substances are co-condensed on a window at temperatures down to 10 K. These investigations clearly reveal that, in the condensed state, the two substances react with negligible activation energy. This is consistent with what we observe for our aerosols particles in the cooling cell.

We propose the following scenarios for the reactions in the particles. The statistically mixed particles start to react as soon as condensation to the liquid state (around 240 K) takes place. The reaction is thus a liquid phase reaction in this case. The situation is somewhat different for the core–shell particles. Here a solid core (80 K) is formed first. The second substance than coats this core. We expect that the heat of reaction,[37] and to a lesser extent the heat of condensation, melt the interface between the two substances so that also in this case the reaction takes place in the liquid rather than in the solid phase. It is important to note here that the particle size only lies around several ten nanometres. On a macroscopic scale, the core–shell particles have still to be considered as homogeneous mixtures.

Ref. 38 and 39 have studied the reaction of sulfur dioxide and ammonia for various systems, concentrations, and over a time span of one day up to years. The reaction products strongly depend on these parameters and are generally very complex and could be analyzed only partially. Our observations correspond most likely to what previous investigations call the initial products.[37–39] Among these are the adducts $NH_3 \cdot SO_2(s)$ and $(NH_3)_2 \cdot SO_2(s)$, the structure of which is unclear, as well as $(HNSO)_x$. So far, however, we have not yet unambiguously assigned the bands observed in our particle spectra. The identity of the reaction products thus remains unclear at this stage. We hope to get additional hints from high level *ab initio* calculations.

7. Conclusions and outlook

Our infrared spectroscopic investigations of pure SO_2 aerosols reveal a partially amorphous structure of the particles at temperatures down to 20 K and a completely amorphous structure at even lower temperatures. As a consequence the particle shape does not lead to pronounced band structures for the stretching bands, although vibrational exciton coupling is strong for these vibrations. We plan to investigate SO_2 particles at higher temperatures as well. This might lead to crystalline particles.

Vibrational exciton coupling was exploited to analyze the properties of mixed sulfur dioxide/carbon dioxide particles. The co-condensation of SO_2 and CO_2 leads to a core–shell structure with sulfur dioxide in the core and carbon dioxide in the shell. This architecture is determined by the different thermodynamic properties of the two substances and the intermolecular interactions acting. It will be interesting to apply this method to other mixed systems of astrophysical interest such as mixtures of sulfur dioxide with methanol.[5] As demonstrated in ref. 16, the vibrational exciton model also allows us to calculate optical constants for particulate systems. Corresponding data for pure and mixed SO_2 particles will be valuable for comparison with experimental data and thus as reference data for remote sensing.

The reaction of sulfur dioxide with ammonia demonstrates the suitability of collisional cooling to study condensed phase reactions taking place under the cold conditions in planets and interstellar dust. Compared with bulk measurements it has two important advantages. The reaction partners can instantaneously be mixed homogeneously and the reaction in the aerosol particles is not influenced by any side effects arising from sample holders. Rapid-scan *in situ* infrared spectroscopy allows us to study such processes with a time resolution of 10^{-5} s. Due to diffusion of the aerosol particles in our cell, the study of reactions in aerosol particles as demonstrated is, however, limited to a total observation time of one to two hours.

The condensed phase reaction of SO_2 and NH_3 studied in the present contribution is also important in aerosol particles in the Earth's atmosphere, where both substances can condense, adsorb on, or be absorbed by aerosol particles. To better understand this reaction we plan to extend our experimental investigations to broader concentration and broader temperatures ranges (up to 240 K and down to 4 K). In addition, we hope to obtain further insight into the assignment of the reaction products from high level *ab initio* calculations. Finally we plan to extend the approach presented here to other condensed phase reactions such as the reaction of ammonia with water, which is important on Io and Europa.[1,3,40]

Acknowledgements

This project was financially supported by the Natural Sciences and Engineering Research Council of Canada, by the Canada Foundation for Innovation, and by the

Deutsche Forschungsgemeinschaft. We thank Roman Ueberschaer for help with some of the measurements.

References

1. S. A. Sandford and L. J. Allamandola, *Icarus*, 1993, **106**, 478–488.
2. D. B. Nash and B. H. Betts, *Icarus*, 1995, **117**, 402–419.
3. L. Schriver-Mazzuoli, H. Chaabouni and A. Schriver, *J. Mol. Struct.*, 2003, **644**, 151–164.
4. J. R. Spencer, E. Lellouch, M. J. Richter, M. A. López-Valverde, K. L. Jessup, T. K. Greathouse and J.-M. Flaud, *Icarus*, 2005, **176**, 283–304.
5. A. C. A. Boogert, W. A. Schutte, F. P. Helmich, A. G. G. M. Tielens and D. H. Wooden, *Astron. Astrophys.*, 1997, **317**, 929–941.
6. R. Signorell, M. Jetzki, M. K. Kunzmann and R. Ueberschaer, *J. Phys. Chem. A*, 2006, **110**, 2890–2897.
7. G. Firanescu, D. Hermsdorf, R. Ueberschaer and R. Signorell, *Phys. Chem. Chem. Phys.*, 2006, **8**, 4149–4165.
8. R. M. Hexter, *J. Chem. Phys.*, 1960, **33**, 1833–1841.
9. R. Signorell, *J. Chem. Phys.*, 2003, **118**, 2707–2715.
10. B. Schmitt, C. de Bergh, E. Lellouch, J.-P. Maillard, A. Barbe and S. Douté, *Icarus*, 1994, **111**, 79–105.
11. H. Chang and Y. Choi, *Aerosol Sci. Technol.*, 2000, **32**, 268–283.
12. M. K. Kunzmann, PhD Thesis, Cuvillier Verlag, Göttingen, 2002.
13. M. K. Kunzmann, R. Signorell, M. Taraschewski and S. Bauerecker, *Phys. Chem. Chem. Phys.*, 2001, **3**, 3742–3749.
14. M. Jetzki, A. Bonnamy and R. Signorell, *J. Chem. Phys.*, 2004, **120**, 11775–11784.
15. R. Signorell and M. K. Kunzmann, *Chem. Phys. Lett.*, 2003, **371**, 260–266.
16. A. Bonnamy, M. Jetzki and R. Signorell, *Chem. Phys. Lett.*, 2003, **382**, 547–552.
17. G. Firanescu, D. Luckhaus and R. Signorell, *J. Chem. Phys.*, 2006, **125**, 144501.
18. G. Cardini, V. Schettino and M. L. Klein, *J. Chem. Phys.*, 1989, **90**, 4441–4449.
19. R. Disselkamp and G. E. Ewing, *J. Chem. Soc., Faraday Trans.*, 1990, **86**, 2369–2373.
20. D. J. Wales and G. E. Ewing, *J. Chem. Soc., Faraday Trans.*, 1992, **88**, 1359–1367.
21. M. A. Ovchinnikov and C. A. Wight, *J. Chem. Phys.*, 1994, **100**, 972–977.
22. K. Kim and W. T. King, *J. Chem. Phys.*, 1984, **80**, 969–973.
23. B. Post, R. S. Schwartz and I. Fankuchen, *Acta Crystallogr.*, 1952, **5**, 372–374.
24. M. J. Frisch, G. W. Trucks, H. B. Schlegel, G. E. Scuseria, M. A. Robb, J. R. Cheeseman, V. G. Zakrzewski, J. A. Montgomery, Jr, R. E. Stratmann, J. C. Burant, S. Dap-prich, J. M. Millam, A. D. Daniels, K. N. Kudin, M. C. Strain, O. Farkas, J. Tomasi, V. Barone, M. Cossi, R. Cammi, B. Mennucci, C. Pomelli, C. Adamo, S. Cliord, J. Ochter-ski, G. A. Petersson, P. Y. Ayala, Q. Cui, K. Morokuma, D. K. Malick, A. D. Rabuck, K. Raghavachari, J. B. Foresman, J. Cioslowski, J. V. Ortiz, A. G. Baboul, B. B. Ste-fanov, G. Liu, A. Liashenko, P. Piskorz, I. Komaromi, R. Gomperts, R. L. Martin, D. J. Fox, T. Keith, M. A. Al-Laham, C. Y. Peng, A. Nanayakkara, C. Gonzalez, M. Challa-combe, P. M. W. Gill, B. Johnson, W. Chen, M. W. Wong, J. L. Andres, C. Gonzalez, M. Head-Gordon, E. S. Replogle and J. A. Pople, *GAUSSIAN 98 (Revision A.7)*, 1998.
25. R. N. Wiener and E. R. Nixon, *J. Chem. Phys.*, 1956, **25**, 175.
26. P. A. Giguère and M. Falk, *Can. J. Chem.*, 1956, **34**, 1833–1835.
27. A. Anderson and R. Savoie, *Can. J. Chem.*, 1965, **43**, 2271–2278.
28. A. Anderson and M. C. W. Campbell, *J. Chem. Phys.*, 1977, **67**, 4300–4302.
29. I. Hussla and J. Heidberg, *J. Electron Spectrosc. Relat. Phenom.*, 1986, **39**, 213–222.
30. M. H. Brooker, *J. Mol. Struct.*, 1984, **112**, 221–232.
31. R. K. Khanna, G. Zhao, M. J. Ospina and J. C. Pearl, *Spectrochim. Acta, Part A*, 1988, **44**, 581–586.
32. A. Tafi, P. Procacci, E. Castellucci and P. R. Salvi, *Chem. Phys.*, 1991, **150**, 205–217.
33. Y. Song, Z. Liu, H. Mao, R. J. Hemley and D. R. Herschbach, *J. Chem. Phys.*, 2005, **122**, 174511.
34. D. Hermsdorf, A. Bonnamy, M. A. Suhm and R. Signorell, *Phys. Chem. Chem. Phys.*, 2004, **6**, 4652–4657.
35. Ó. Sigurbjörnsson, G. Firanescu and R. Signorell, in preparation, 2007.
36. Ó. Sigurbjörnsson, G. Firanescu, C. R. Viteri and R. Signorell, in preparation, 2007.
37. R. Landreth, R. G. de Pena and J. Heicklen, *J. Phys. Chem.*, 1974, **78**, 1378–1380.
38. B. Meyer, B. Mulliken and H. Weeks, *Phosphorus, Sulfur Relat. Elem.*, 1980, **8**, 291–300.
39. B. Meyer, B. Mulliken and H. Weeks, *Phosphorus, Sulfur Relat. Elem.*, 1980, **8**, 281–290.
40. L. Schriver-Mazzuoli, A. Schriver and H. Chaabouni, *Can. J. Phys.*, 2003, **81**, 301–309.

PAPER

MicroParticle photophysics illuminates viral bio-sensing

S. Arnold,* R. Ramjit, D. Keng, V. Kolchenko and I. Teraoka

Received 26th February 2007, Accepted 11th April 2007
First published as an Advance Article on the web 17th July 2007
DOI: 10.1039/b702920a

The authors present an approach for specific and rapid unlabeled detection of a virus by using a microsphere-based whispering gallery mode sensor that transduces the interaction of a whole virus with an anchored antibody. They show theoretically that this sensor can detect a single virion below the mass of HIV. A micro-fluidic device is presented that enables the discrimination between viruses of similar size and shape.

Introduction

None of civilization's socio-political catastrophes (*e.g.* world wars) have caused an equivalent destructive effect on the world's population as biological pandemics.[1] Exponentially growing pathogens are difficult to contain and eliminate unless they can be detected early on. Some years ago, one of us (S.A.) reflected on this problem as a friend was dying from a viral infection. His friend's diagnosis came too late; real-time methods for testing for the virus were not available. A decision was made to direct the MicroParticle PhotoPhysics Lab toward finding a solution. This paper represents its first expose.

Our approach is to sense bio-particles using the high sensitivity afforded by the perturbation that an adsorbed molecule has on high Q ($> 10^7$) optical resonances of a microparticle.[2] In particular, bio-particle adsorption will be sensed from the associated shift in resonance frequency.[3,4] We are interested in the optical spectroscopy of microparticles, but not in the spectroscopy of the nanoscopic bio-particles (*e.g.* protein, DNA, virus). Although optical spectroscopy can be particularly useful for small molecules, bio-particles such as protein are difficult to distinguish by optical spectroscopy, since they are considerably larger and share common vibrational and electronic states, as a consequence of being made up of the same 20 amino acids. The same can be said for distinguishing between strands of DNA, since they have four nucleotides in common. Biology is a great teacher in this respect. Through all the eons of evolution, biology has not taken a spectroscopic approach; there is no light in our bodies. Instead nature evolves bio-nano-probes that specifically grab onto protein, DNA and foreign invaders through physio-chemical interactions. Our approach is to use these bio-nano-probes as surface-bound recognition elements and the microparticle to transduce (report) the interaction.

We are not interested in labeling the bio-particle with markers (*e.g.* fluorescent tags). Labels can structurally and functionally interfere with the assay, may not be specific and block our goal of real-time detection by involving an additional step.[5]

MicroParticle PhotoPhysics Lab (MP^3L), Polytechnic University, 6 MetroTech CenterBrooklyn, NY 11201, USA. Web: http://www.poly.edu/microparticle. E-mail: sarnold935@aol.com; Fax: +1 718 260 3139; Tel: +1 917 568 6549

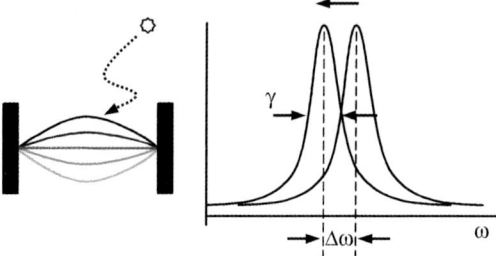

Fig. 1 The frequency response of an oscillator before and after a perturbation.

There are other unlabeled biosensors such as the Surface Plasmon Resonance (SPR) device, however it has limited sensitivity, which has not substantially improved from its inception.[6,7] Our ultimate goal is to demonstrate the unlabeled sensing of a single bio-particle.

We are also not interested in identifying a virus from a multi-step analysis of geonomic information within its protein coat. Instead, we will seek to identify the whole virus by tranducing the immobilization that takes place when a coat protein on its surface interacts with a complementary antibody anchored to the microparticle surface.

Resonant sensors-general considerations

Each and every oscillator, whether a mass on a spring, a violin string, the thorax of a cricket or a Fabry–Perot cavity has the common property of resonance. If they are driven by a harmonic source, the square of the oscillator's amplitude $|A|^2$ (*i.e.*, energy) will demonstrate a Lorentzian-shaped frequency response with maximum at ω_r and linewidth γ (Fig. 1). At the same time they can be sensitive to perturbations. Allowing dust to fall on a violin string causes its tone to be reduced by $|\Delta\omega|$. One can imagine putting bio-nano-probes on the violin string in order to detect specific dust (*e.g.* anthrax spore). However, one is apt to find in a world awash with noise that the frequency shift may not be sufficient for real time detection. The principal difficulty is in measuring a frequency shift much smaller than the linewidth (Fig. 1).

We will characterize this competition by a "measurement acuity factor", $F = |\Delta\omega|_{min}/\gamma$ where $|\Delta\omega|_{min}$ is the smallest measurable frequency shift. Clearly, smaller F is better, but more difficult to achieve. One thing is certain: for a given F the minimum frequency shift that can be measured is proportional to the line width, so that the fractional minimum shift $|\Delta\omega|_{min}/\omega_r = F\gamma/\omega_r$. Whereas F and ω_r are controlled principally by the non-dissipative physics of the oscillator and the bandwidth of the detection system, γ is principally controlled by dissipation. To reduce the minimum measurable shift, one can reduce dissipation. By convention we will represent the linewidth-to-frequency ratio by $1/Q$, where Q is the so-called Quality factor; $\gamma/\omega_r = 1/Q$. With this definition

$$|\Delta\omega|_{min}/\omega_r = F/Q. \qquad (1)$$

The larger Q, the smaller the dissipation and the smaller the dust particle that can be detected. Since a mechanical system like a violin string agitates the fluid around it, it is a far from an ideal sensor; dissipation is assured. By the same token, a Fabry–Perot cavity with metalized mirrors has dissipation due to Ohmic losses on reflection. The least dissipative reflection in this regard is Total Internal Reflection (TIR), in which light propagating in a medium with refractive index n_1 is reflected at a sufficient angle θ from a medium with a lower refractive index n_2. TIR is, in principle, without loss. Light that bounces off the interior surface of a sphere

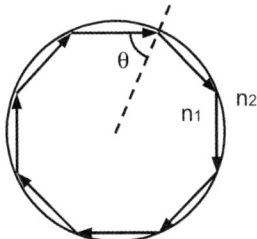

Fig. 2 WGM within a dielectric microsphere.

while executing a polygonal orbit has this appeal. Such an orbit [Fig. 2, with $\theta > \sin^{-1}(n_2/n_1)$] is known as a Whispering Gallery Mode (WGM).

The WGM in Fig. 2 has no apparent loss. This implies an infinite Q. However, it is well known that Q has limits. The largest Q for a WGM mode currently measured in a silica microsphere is $\sim 10^{10}$.[8] Clearly, Fig. 2 is misleading, light does not have stationary states in a dielectric. Photons orbiting in the polygon are only partially trapped. They can "tunnel" into free space modes. Before providing a more complete description of a WGM, we will attempt to obtain a heuristic estimate for the shift of its resonant frequency due to a perturbing layer.

WGM layer perturbation: heuristic approach

There is always a wave-particle duality associated with the photon as there is for the electron. Instead of the description in Fig. 2 of a particle bouncing against the interior of the microsphere, we now turn to a wave description for estimating the sensitivity of the microsphere resonance frequency to adsorption. Fig. 3(a) shows this point of view. It represents the mode by a wave that circumnavigates near the surface of a sphere of radius R and returns in phase.

This picture, which is analogous to a Bohr–de Broglie atom, may be appropriately called a Photonic Atom.[9] By this analogy, the angular momentum of the photon is characterized by a quantum number l that is equal to the number of wavelengths in the orbit. In what follows, this wave description will be used to obtain an estimate for the minimum thickness of an adsorbed layer that is required to produce a measurable shift of the mode's frequency. Later, this quantum analog will be expanded to determine the fields and probability densities associated with these photonic modes.

Let us suppose that a material having the same dielectric constant as the microsphere adsorbs on the sphere's surface to a thickness t (Fig. 3(b)). This layer causes all lengths in Fig. 3(b) to be scaled up, while l for the given mode remains invariant. As a result the wavelength increases so that its fractional increase is the

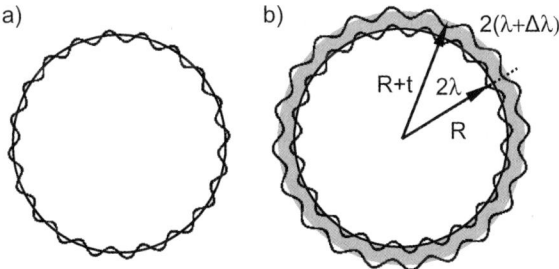

Fig. 3 (a) Photonic Atom Mode; (b) Anticipated wavelength change caused by the addition of a spherically-symmetric layer.

same as the fractional increase in radius; $\Delta\lambda/\lambda = t/R$. Since frequency and wavelength are inversely related,

$$|\Delta\omega|/\omega_r = t/R. \quad (2)$$

The smallest measurable thickness is found by combining eqn (1) and (2),

$$t_{min} = F\,R/Q. \quad (3)$$

For a conservative estimate for t_{min}, we take the measurement acuity $F = 1$, $R = 100$ µm and $Q = 10^6$, for which our minimum thickness is 0.1 nm (the size of a hydrogen atom). With a greater effort we have found experimentally that F can be reduced to 1/50, corresponding, for our example, to a minimum detectable thickness $t = 0.002$ nm. This thickness is approximately one hundredth the size of a hydrogen atom! Although too small to be physical, it certainly shows the promise of the microsphere as an adsorption sensor.

Our analysis thus far is strictly heuristic. Although it only applies to a spherically-symmetric layer adsorbed on a homogeneous sphere and does not allow the layer to have a different refractive index from the microsphere, it provides a great deal of guidance. Perhaps the most important rule, for our homogeneous spherical substrate, is that the smallest amount of adsorbate is detected by minimizing the ratio R/Q.

Before we can address the problem of single bio-particle sensitivity, we must first discuss a theory for the effect of local dielectric perturbations on microsphere resonances.

This requires an understanding for the fields associated with the resonances. We will approach this subject through a quantum analog using a pseudo-potential approach. Following the discussion of the theory for perturbation of microsphere resonances by bio-particles, we will present our virus experiments and show specific real time detection of a virus from the frequency shift of a microcavity resonance, for the first time.

WGM field: A quantum analog approach

Ultra-high Q WGM resonances were first seen in the microparticle area in optical levitation measurements in air.[10] Their interpretation came through Mie theory, which includes both incident and scattered fields.[11] An easier approach, pioneered by Nussenzvieg,[12] utilizes a quantum analog and may be best described as a Photonic Atom (PA) model.[9] Here, we briefly describe the model for completeness. For anyone who has studied the quantum mechanics of hydrogen this approach should be familiar. More importantly, it is concise and particularly physical.

In the PA particle description, the photon is seen as a particle that orbits by bouncing off the interior wall of the dielectric microsphere (Fig. 4a). This "quantum billiard"[13] possesses quantized orbital angular momentum l just as the electron in a Bohr atom.

The PA wave description of a Transverse Electric (TE) mode is constructed by inserting an electric field of the form

$$\mathbf{E} = A\hat{L}\,[\psi_r(r)\,Y_{l,m}/r] \quad (4)$$

into the electromagnetic vector Helmholtz equation, where $\hat{L}$ is a dimensionless angular momentum operator, $\hat{L} = -i\mathbf{r} \times \nabla$, $Y_{l,m}$ is a spherical harmonic and A is a constant. As a result the radial wavefunction ψ_r is found to follow a one dimensional Schrödinger equation

$$\frac{d^2\psi_r}{dr^2} + (E_{\text{eff}} - V_{\text{eff}})\psi_r = 0, \quad (5)$$

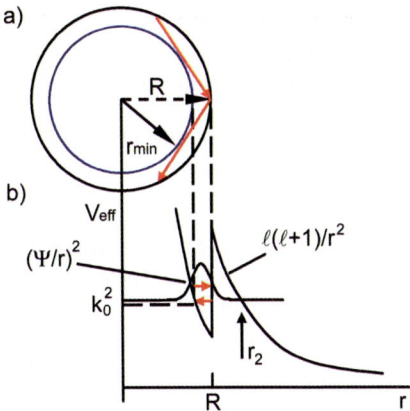

Fig. 4 The inter-relationship between the WGM model and the Photonic Atom Model for a microsphere having a uniform refractive index.

where the effective energy $E_{\text{eff}} = k_0^2$, the square of the free space wave vector and the effective potential

$$V_{\text{eff}} = k_0^2[1 - n(r)^2] + l(l+1)/r^2. \qquad (6)$$

The first term in this potential describes dielectric confinement for a radially variable refractive index $n(r)$, while the second represents centrifugal repulsion. Fig. 4 connects up the particle (4a) and wave (4b) points of view for a homogeneous sphere.

The analog particle with effective energy k_0^2 is caught in a potential "pocket" in V_{eff} as it oscillates radially between the classical turning points at r_{min} and R. However, since the particle is a quantum analog it can tunnel through the barrier extending from R to r_2. Within the barrier the probability density falls. This nearly-exponential fall between $r = R$ and $r = r_2$ is known in electromagnetic theory as the region of the evanescent field (the analog probability density is proportional to the square modulus of the electromagnetic field, $\mathbf{E^* \cdot E}$). Any probability density left at the end of the barrier leads to energy loss in the form of an outward radiating spherical wave. Consequently, unlike the electron in the hydrogen atom, confinement in a dielectric Photonic Atom cannot be complete; the intrinsic Quality factor is finite.

The probability density illustrated for the mode in Fig. 4b has one peak in the potential pocket and is known as a first order mode ($\nu = 1$). At higher energies, modes may form with more interior peaks, corresponding to higher order numbers. In all, four attributes are required to describe a mode: radial order number ν, angular momentum quantum number l, z-component of angular momentum m and polarization P (TE or TM). As in hydrogen, m can have integer values between $-l$ and l. A generalized mode will be labeled as $P_{l,m}^\nu$. For a sphere having a uniform refractive index, the radial wavefunctions ψ_r in the interior and beyond are so called Riccati–Bessel functions;

$$\psi_r = r\, j_l(n_s k_0 r) \text{ for } r \leq R, \qquad (7)$$

$$\psi_r = [j_l(n_s k_0 R)/h_l(n_m k_0 R)]\, r h_l(n_m k_0 r) \text{ for } r \geq R, \qquad (8)$$

where j_l and h_l are spherical Bessel and spherical Hankel functions and n_s and n_m are the refractive indices in the microsphere and the surrounding medium. Mode energies are found, as in quantum mechanics, by matching the logarithmic derivatives of the Ψ_r functions on either side of the interfacial boundary.

Local dielectric perturbations of WGMs

Molecules that approach the surface of a microsphere interact with a WGM as they enter the evanescent field. This oscillating field polarizes molecules, and as a consequence causes a frequency shift of the mode.

Fields due to photons are polarized due to the photon's spin. If a microsphere is in a single photon resonant state of energy $\hbar\omega$, it will have an associated semi-classical field $E(r, t) = Re[E_0(r)e^{i\omega t}]$. With a bio-particle outside the sphere at position r_j, an interaction will occur; the bio-particle will be polarized and develop an oscillating dipole moment in excess of the displaced solvent, $\Delta p(t) = Re[\Delta p_0 e^{i\omega t}]$. The time-averaged energy required to polarize the bio-particle serves as the perturbation that shifts the photon energy of the resonant state by

$$\hbar\Delta\omega = -\frac{1}{2}\langle \Delta p(t) \cdot E_0(r_j, t)\rangle_t = -\frac{1}{4}Re[E_0(r_j)^* \cdot \underline{\underline{\Delta\alpha}}_j \cdot E_0(r_j)] \quad (9)$$

where $\underline{\underline{\Delta\alpha}}$ is the bio-particle's excess polarizability tensor. By dividing this energy shift by the energy of the mode

$$\hbar\omega_r = (1/2)\int \varepsilon(r) E_0(r)^* \cdot E_0(r) dV \quad (10)$$

we arrive at a useful expression for the shift associated with a single bio-particle interaction,

$$\left(\frac{\Delta\omega}{\omega_r}\right)_j = -\frac{Re[E_0(r_j)^* \cdot \underline{\underline{\Delta\alpha}}_j \cdot E_0(r_j)]}{2\int \varepsilon(r) E_0(r)^* \cdot E_0(r) dV} \quad (11)$$

where $\varepsilon(r)$ is the dielectric function within the microparticle and in its surroundings (absent the bio-particle). Eqn (11) provides a much more general result than had been presented previously.[14] By simply switching from a real to imaginary operator in eqn (11), one may also obtain an expression for the change in the linewidth $\Delta\gamma$ of the resonant mode due to molecular absorption,

$$\left(\frac{\Delta\gamma}{\omega_r}\right)_j = \frac{Im[E_0(r_j)^* \cdot \underline{\underline{\Delta\alpha}}_j \cdot E_0(r_j)]}{2\int \varepsilon(r) E_0(r)^* \cdot E_0(r) dV} \quad (12)$$

For our experiments, Δ can be assumed to have no imaginary part, since experiments will be carried out at low enough energies to avoid absorption by protein or DNA. In this way, our measurements can be used to obtain added information about these bio-particles.

So long as one uses photon energies well below excited electronic states, water-soluble proteins have dielectric properties that deviate less than 1% from one to another. In addition, they also share very similar mass densities. On this basis the trace of the polarizability tensor is proportional to the volume of the protein,[15] and to the mass. This allows the Whispering Gallery Mode Biosensor (WGMB) to enjoy a distinct advantage. The resonance shift contributed by a protein contains molecular weight and size information. This is distinct from sensing schemes that use labels or detection involving protein charge.[16,17] Another distinct advantage is contained within the form of the interaction.

Since the shift is proportional to $E_0(r_j)^* \cdot \underline{\underline{\Delta\alpha}}_j \cdot E_0(r_j)$, a non-spherical protein molecule when adsorbed on a surface will show a different shift for a field polarized perpendicular to the interface (TM polarization) than for the parallel case (TE polarization). This approach has recently been used for determining the orientation of the protein Bovine Serum Albumin (BSA) on a silica microsphere in biological buffer.[18]

Many viruses that infect our cells are nearly spherical. For example HIV, HPV and HSV are icosahedral. This is fortunate since the excess polarizability tensor for such a shape is essentially diagonal with identical elements (*i.e.* isotropic) and the basic interaction between the field and the bio-particle at r_j in eqn (11) may be written as

$$E_0(r_j)^* \cdot \Delta\underline{\underline{\alpha}}_j \cdot E_0(r_j) = \Delta\alpha_j E_0(r_j)^* \cdot E_0(r_j) \quad (13)$$

On this basis, the shift due to a single bio-particle perturbation is

$$\left(\frac{\Delta\omega}{\omega_r}\right)_j = -\frac{\Delta\alpha_j E_0(r_j) \cdot E_0(r_j)^*}{2 \int \varepsilon(r) E_0(r)^* \cdot E_0(r) dV}. \quad (14)$$

The shift caused by a large number of bio-particles is most easily computed by assuming that the field at a particular bio-particle is not influenced by the contributions from its neighbors,[19] so that eqn (14) can be summed over a number density $\rho(r)$,

$$\left(\frac{\Delta\omega}{\omega_r}\right) = -\frac{\int \rho(r)\Delta\alpha(r) E_0(r) \cdot E_0(r)^* dV}{2 \int \varepsilon(r) E_0(r)^* \cdot E_0(r) dV}. \quad (15)$$

By taking a more condensed view (*i.e.*, $\rho(r) \gg \lambda^{-3}$), the factor $\rho(r) \Delta \alpha(r) E_0(r)$ within the integrand in eqn (15) may be replaced by $\Delta\varepsilon(r) E_{in}(r)$, where $\Delta\varepsilon(r)$ and $E_{in}(r)$ are the excess permittivity and the "true" local field at r respectively. With this Maxwellian approach, a truly continuum equation for the frequency shift perturbation evolves

$$\left(\frac{\Delta\omega}{\omega_r}\right) = -\frac{\int \Delta\varepsilon(r) E_{in}(r) \cdot E_0(r)^* dV}{2 \int \varepsilon(r) E_0(r)^* \cdot E_0(r) dV}. \quad (16)$$

This is a familiar result obtained from traditional perturbation theory for dielectric particles within metallic cavities[20] and near dielectric cavities.[21] Our approach of starting from a molecular property is somewhat non-traditional, however, it leaves us with several useful results for perturbations that are discrete (eqn (11)) or continuous (eqn (15)).

Eqn (15) may be applied to analyte molecules nearby in solution as well as those adsorbed. The former, which is normally referred to as refractive index sensing, is of little interest to our current investigation.

There are alternative ways to utilize the above theory in dealing with adsorbed virus. If the adsorbed virus particle is much smaller than the evanescent field length then the field in its vicinity may be considered to be uniform, and one may expect eqn (14) to apply with $|r_j| = R$. However, if the virus particle is comparable to or larger than the evanescent field length then one must consider it to be in a non-uniform field. A model for its excess dielectric form may then be input for $\Delta\varepsilon(r)$ in eqn (16) and the numerator evaluated.

Although the numerators in eqn (11), (14), (15) and (16) depend on the specific virus being adsorbed and the evanescent field characteristics, the mode energy integral within the denominators in all of these equations are the same and may be simply estimated. In what follows, we will evaluate the shift due to a single bio-particle with radius $a \ll \lambda$ adsorbed on a sphere resonating in an equatorial TE mode; $TE_{l,l}^r$. This mode is of particular interest since it is possible to excite it selectively using an optical fiber.

For a high Q resonance, the mode energy is dominated by the interior energy.[22] Consequently, we will approximate the denominator of eqn (14) by limiting the integration only to the interior. From eqn (4) and (7), the field in the interior within an equatorial mode $E_0 = A_{in} j_l (n_s k_0 r) \hat{L} Y_{l,l}$, and consequently eqn (14) becomes

$$\left(\frac{\Delta\omega}{\omega_r}\right)_j \approx -\frac{\Delta\alpha_j [j_l(n_s k_0 R)]^2 |\hat{L} Y_{l,l}|^2}{2\varepsilon_s \int [j_l(n_s k_0 r)]^2 r^2 dr \int |\hat{L} Y_{l,l}|^2 \sin(\theta) d\theta d\phi} \quad (17)$$

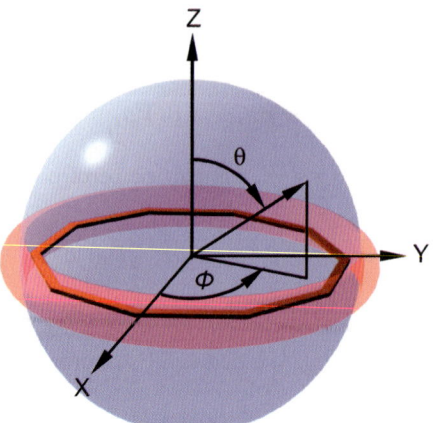

Fig. 5 Depiction of the intensity associated with an equatorial mode ($l = m$).

for $l \gg 1$, $|\hat{L}Y_{ll}|^2 \propto |Y_{ll}|^2$.[23] In addition, the spherical harmonic is normalized with respect to solid angle, $\oint |Y_{ll}|^2 d\Omega = 1$, which reduces eqn (17) to

$$\left(\frac{\Delta\omega}{\omega_r}\right)_j \approx -\frac{\Delta\alpha_j [j_l(n_s k_0 R)]^2 |Y_{ll}|^2}{2\varepsilon_s \int [j_l(n_s k_0 r)]^2 r^2 dr}. \quad (18)$$

On resonance, the integral in the denominator of eqn (18) may be asymptotically ($2\pi R/\lambda \gg 1$) related to the surface value of j_l^2 through

$$\int_0^R [j_l(n_s k_0 r)]^2 r^2 dr \cong \frac{R^3}{2} [j_l(n_s k_0 R)]^2 \frac{n_s^2 - n_m^2}{n_s^2}. \quad (19)$$

where n_m is the relative permittivity of the surrounding medium.[24] By combining eqn (19) with eqn (18), the shift due to a single bio-particle appears,

$$\left(\frac{\Delta\omega}{\omega_r}\right)_j \approx -\frac{\Delta\alpha_j |Y_{ll}(\hat{r}_j)|^2}{\varepsilon_0 R^3 (n_s^2 - n_m^2)}. \quad (20)$$

Where ε_0 is the permittivity of free space ($\varepsilon_0 = \varepsilon_s/n_s^2$), and the unit position vector $\hat{r}_j$ is used to denote the angular orientation at which the spherical harmonic is evaluated. This result does not depend on the resonance order, but possesses a strong dependence on polar angle and microsphere radius. In particular, the spherical harmonic in eqn (20) has a square modulus that peaks at ($\theta = 90°$), producing a "band" of sensitivity around the equator (Fig. 5). With eqn (20) in hand, it is now possible to estimate the Limit Of Detection (LOD).

We will first concentrate on this equatorial region, since particles adhering to this region produce the largest possible shift. For a small number of virus particles N, we will assume that each particle produces an equivalent shift. To estimate the smallest detectable number N_{LOD} we allow the accumulated shift $|\Delta\omega|$ using eqn (20) to be equal to $|\Delta\omega|_{min}$ in eqn (1);

$$N_{LOD} \approx \frac{R^3 (n_s^2 - n_m^2) F}{(\Delta\alpha/\varepsilon_0)|Y_{ll}(\pi/2)|^2 Q}. \quad (21)$$

N_{LOD} is controlled by a number of factors. Since purified silica is readily available (*i.e.*, optical fiber) and easily shaped, we choose silica for the microsphere. We assume the virus to be blood borne, so the medium will be essentially aqueous. F is fixed by the measurement system and fluctuation theory, however, we find that a 10 Hz bandwidth can be achieved with $F = 1/50$, and we will use this number in

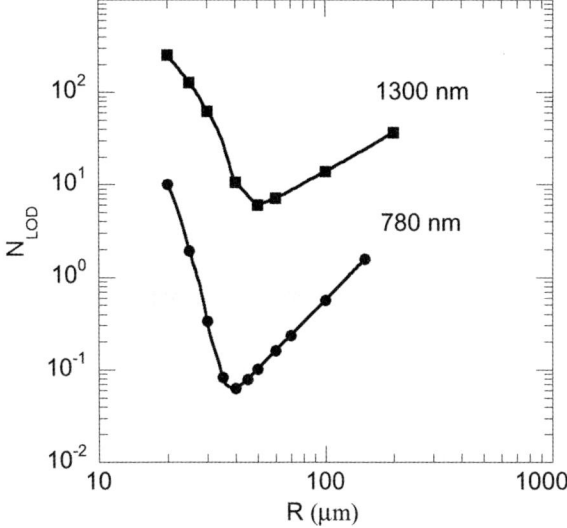

Fig. 6 Smallest theoretical detectable number of HIV virus particles adsorbed on the equatorial rim of a homogeneous silica WGMB as a function of microsphere radius R for $F = 1/50$ and wavelength near 1300 nm or 780 nm.

subsequent calculations. Microsphere size plays a critical role. It explicitly affects the R^3 factor in the numerator and implicitly affects the $|Y_{ll}(\pi/2)|^2 Q$ factor in the denominator. At a fixed wavelength, the spherical harmonic factor increases with l and therefore increases with R. For large l, $|Y_{ll}(\pi/2)|^2 \sim R^{1/2}$, so $R^3/|Y_{ll}(\pi/2)|^2 \sim R^{5/2}$. As for Q, the story is more complicated. Q is a measure of the rate at which a mode's energy decays (i.e. $\gamma = \omega/Q$). The paths for decay are numerous, including intrinsic loss rates due to tunneling (ω/Q_{int}), absorption (ω/Q_{abs}), Raman scattering (ω/Q_{Raman}) and Rayleigh scattering ($\omega/Q_{Raleigh}$, including surface roughness); (ω/Q) = $\omega/Q_{int} + \omega/Q_{abs} + \omega/Q_{Raman} + \omega/Q_{Raleigh}$. Surface scattering is expected to limit the ultimate Q for a large silica sphere in vacuum ($\sim 10^{12}$).[25] For our experiments in aqueous buffer solution, Q for spheres with radii from 20 μm to 200 μm is limited to values orders of magnitude below the ultimate vacuum value, due to loss of dielectric contrast (i.e., increased tunneling) and aqueous absorption;

$$\frac{1}{Q} \approx \frac{1}{Q_{int}} + \frac{1}{Q_{abs}} \quad (22)$$

As R increases above 20 μm, both tunneling and absorption decrease, causing $1/Q$ to decrease faster than $R^{-5/2}$. As a result, N_{LOD} decreases with increasing R. But tunneling decreases much more rapidly than absorption, causing it to have a minority influence as the size further increases. This causes N_{LOD} to begin to increase. Fig. 6 shows a theoretical plot for N_{LOD} for HIV virus particles as a function of the silica sphere radius for two wavelength regions, 1300 nm and 780 nm. The substantially smaller N_{LOD} values in the 780 nm region are due to considerably smaller absorption by water at this wavelength. At either wavelength, we predict very high sensitivity, with N_{LOD} dipping below one tenth at 780 nm for a microsphere approximately 40 μm in radius.

Fig. 6 was constructed from an estimate of the HIV polarizability. Although the excess polarizability of HIV has not been measured, it is not difficult to estimate $\Delta\alpha$ from the virus' mass. Viruses are principally composed of protein and protein have the convenient property of raising the refractive index of an aqueous solution by nearly the same amount for the same mass concentration; the differential refractive index of protein in an aqueous buffer $dn/dc \approx 0.18$ cm^3 g^{-1} (visible).[26] Tobacco

Mosaic Virus (TMV) is one of the few examples for which dn/dc has been measured and the correspondence with protein verified (λ = 488 nm).[27] Although dispersion is projected to lower dn/dc by 0.006 between 780 nm and 1300 nm,[28] we have neglected this difference and used the visible value for our calculations. The excess polarizability is related to dn/dc and the molecular mass m through

$$\frac{\Delta\alpha}{\varepsilon_0} = 2n_m \frac{dn}{dc} m. \tag{23}$$

HIV is a fusion virus with a lipid bilayer surrounding its capsid. From the density and average size of the viral particle (110 nm), we estimate its mass to be 8×10^{-16} gm and $\Delta\alpha/\varepsilon_0$, from eqn (23), to be 4×10^{-4} μm^3.

Although we will capture virus particles from a fluid flow, we are currently not at the point of having them deposit only at the micro-sphere equator. Instead, the entire spherical surface will be functionalized. Here, the more relevant question is the theoretical limit of detection associated with surface mass density, $\sigma_{m,Lod}$, which is derived by adding individual frequency shifts (eqn 20) from bio-particles placed at random positions on the surface and setting this overall shift to the minimum measurable shift (eqn (1)). Before calculating $\sigma_{m,Lod}$ we will first obtain the shift due to a uniform layer of bio-particles.

The frequency shift due to a random layer of bioparticles is found by summing eqn (20) over random position coordinates on the microsphere surface. This discrete sum may be made continuous by defining a number of adsorbates per unit solid angle dN/d$\Omega = \sigma 4\pi R^2/4\pi$ and integrating over solid angle,

$$\frac{|\Delta\omega|}{\omega_r} = \frac{\Delta\alpha\,\sigma}{\varepsilon_0(n_s^2 - n_m^2)R}. \tag{24}$$

The result has the same $1/R$ dependence that we heuristically obtained in eqn (2). The two equations may be further related by describing the bioparticles as having a refractive index n_{bp} and forming a layer of thickness t in which the protein volume fraction is f. With an effective medium approach $\Delta\alpha\,\sigma/\varepsilon_0 = f(n_{bp}^2 - n_m^2)t$, and consequently

$$\frac{|\Delta\omega|}{\omega_r} = \frac{f(n_{bp}^2 - n_m^2)}{(n_s^2 - n_m^2)} \frac{t}{R}. \tag{25}$$

A uniform silica layer on silica corresponds to $f = 1$ and $n_{bp} = n_s$ for which eqn (25) agrees with eqn (2). We can now calculate the limit of detection for surface mass density by setting $\sigma = \sigma_{m,Lod}/m$ and in eqn (24). The results for 760 nm and 1300 nm are shown in Fig. 7.

It should be noted that we have indicated the typical baseline sensitivity levels for other label-free sensing techniques in Fig. 7. Interestingly, the sensitivity for the WGMB at 1300 nm for a relatively large microsphere (200 μm) is potentially better than the typical baseline sensitivity for the Quartz Crystal Microbalance (QCM), SPR,[7] or Micro-Cantilevers (MC),[29] and the minimum on the 780 nm curve beats the best of these by more than two orders of magnitude.

In what follows, we will describe our experimental approach and present our results on specific virus sensing.

Experimental approach

There are two essential parts to our sensor design: the WGM resonator and the driver that stimulates it.

The WGM resonator is formed from silica by rotating the end of a bare telecommunication fiber in an oxygen–propane micro-flame. The end of the glass softens and flows into a spheroidal shape. Our spheroids are oblate with an eccentricity ~6% and equatorial radii between 75 and 200 μm.

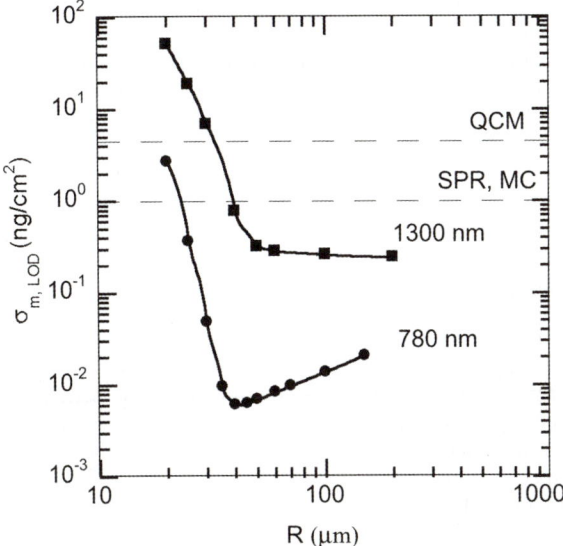

Fig. 7 Smallest theoretical detectable uniform mass density on a silica WGMB as a function of microsphere radius R for $F = 1/50$ and wavelength near 1300 nm or 780 nm.

The use of spheroids may come as a surprise, since the theory in the last section was developed for spheres. However, our theory applies equally well to modes with orbits near the equator of a spheroid so long as the perturbation does not change the eccentricity.

The resonator is driven by evanescently coupling it at its equator to a tapered optical fiber wave guide. Excitation of a micro-spheroid mode is signalled by a dip in the transmission through the fiber.[30] The fiber is flame tapered adiabatically from a diameter of 125 μm to a skirt width of ∼2 μm using an asymmetric pulling device,[31] and directed perpendicular to the spheroid's axis of symmetry (z, Fig. 8a). In this configuration, energy is most efficiently coupled into the equatorial mode ($m = 1$). Other modes with orbits tilted from the equatorial plane have $|m|$ values less than l, and as a consequence of the spheroidal shape differ in circumference and resonance frequency (Fig. 8b).

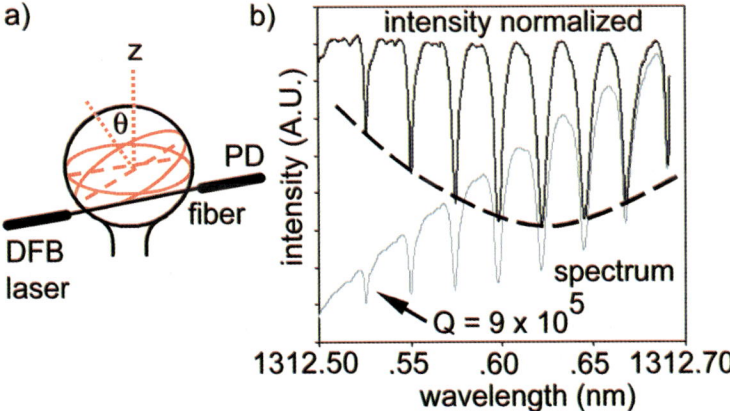

Fig. 8 (a) The optical configuration. (b) A recorded transmission spectrum in water for a silica spheroid having a 200 μm equatorial radius. Adjacent resonances differ by $|\Delta m| = 1$.

The fiber transmission spectrum taken in water in Fig. 8b has a complete range comparable to the resolution of a good fluorescence spectrometer (0.2 nm), however, for our experiments the resolution is much higher. It is ultimately controlled by the linewidth of a Distributed Feedback Laser (DFB, <0.00001 nm) operating near 1312 nm. This DFB laser is tuned directly from its power supply, by driving the laser with a saw tooth current source. As the forward current increases, both the output power and wavelength of the laser increase. The lower fiber transmission spectrum in Fig. 8b reflects this tuning approach with the resonant dips dropping from a saw tooth backbone. The upper curve is constructed by normalizing to constant intensity. Adjacent resonances differ by $|\Delta m| = 1$.

The perturbation in the frequency/wavelength of a resonance in a spectrum such as Fig. 8b is tracked by the movement of a resonance dip by the use of a five point parabolic fit and frequency/wavelength positions are recorded every 200 ms as a "dip trace". For a resonance having a $Q \sim 10^6$, the fluctuations in the base line of the dip trace have an rms value as small as 0.00002 nm for a time period of 5 s.

Two fluidic configurations were chosen. For non-specific binding experiments the sphere and fiber are immersed in a 1 cm^3 volume containing a stirred aqueous buffer solution. For specific binding experiments it is necessary to look at both adsorption and desorption rates. For this purpose, a much smaller micro-fluidic cell with a motorized fluid exchange system was developed. This device will be described in more detail along with our specific binding experiments.

We will next focus on surface preparation for non-specific adsorption experiments and attempt to detect virus particles in this way. Following this, we will turn to the preparation of a surface for specific adsorption and demonstrate specific detection though the discrimination between similar viruses.

Surface modificatoin: non-specific sensing

Our interest is in functionalizing a silica surface for either non-specific or specific sensing. In all cases reported in this paper, we start by cleaning the surface in an oxygen plasma. In the presence of background humidity, silanol groups (Si–OH) form at the surface. Our surfaces will be functionalized by reacting the protruding hydroxyl groups covalently with the methoxy/ethoxy components of linking compounds known as silanes.

The silane compound chosen for non-specific sensing is one that leaves the silica surface with an opposite charge to the net charge on the adsorbate. Most proteins acquire a negative charge at a biological pH (~ 7.4). To produce a positively charged surface we react the protruding hydroxyl groups with 3-aminopropyltriethoxysilane (APTES), since its amino terminus is protanated to NH_3^+ as a consequence of the pK_a of NH_2 (~ 8). APTES is deposited from the vapor phase in a partial vacuum.[32] After bringing the vacuum chamber to atmospheric pressure, the microspheres are transferred to an oven at 120 °C, where the silane rug is cured (*i.e.* cross-linked). The chemical progression is shown in Fig. 9.

Non-specific sensing of virus

For a preliminary test of our sensor, we non-specifically adsorbed BSA on its surface. At sufficient concentration (>100 nM) it produced a Langmuir-like saturation with a frequency shift consistent with eqn (24), based on the molecular weight of BSA (66.4 × 10^3 Da) and a maximum surface density built up by random sequential adsorption.[33] Although BSA is relatively small in comparison with a virus particle, similar experiments using carboxylated polystyrene beads with diameters of 100 nm (mass $\sim 3.5 \times 10^8$ Da) gave good agreement with eqn (24).

For virus adsorption, we chose a virus that kills *E. coli* but is friendly to humans as a model system for both non-specific and specific sensing. The RNA virus known as MS2 is icosahedral in shape, as is HIV, although it has a far smaller size with a

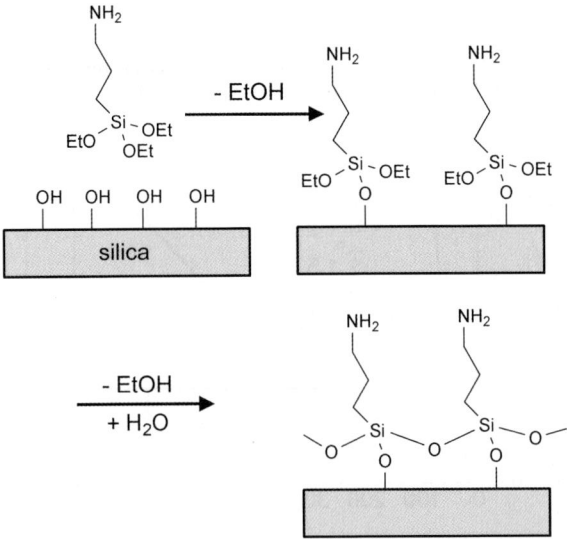

Fig. 9 Amino-silanization using APTES on cleaned microsphere support.

diameter of only 23.6 nm and a mass of 3.6×10^6 Da. MS2 phage has a pI of 3.9,[34] making it suitable at pH 7.4 to take on a negative charge.

Virus stock was obtained from American Type Culture Collection (ATCC, Manassas, VA). MS2 phage was propagated in *E. coli* K91 host and incubated on LB plates (courtesy of N.L. Goddard, Harvard University). After filtration, the virus was resuspended in PBS and stored at 4 °C.

Our experiment is begun by injecting 30 μL of MS2 solution [3×10^9 pfu (plaque forming units)] into 1 mL of PBS buffer (pH = 7.4) in the sample cell of our non-specific sensing system. As a micro-stirring bar homogenizes the solution to a concentration of 5 pM, all resonance dips shift toward longer wavelengths. The dip trace represented in Fig. 10 by plotting the normalized shift $R\Delta\lambda/\lambda$, shows the adsorption of virus in real time and is indicative of monolayer formation. We can track the density of adsorbed virus by inverting eqn (24),

$$\sigma = R \frac{\Delta\lambda}{\lambda} \frac{(n_s^2 - n_m^2)}{\Delta\alpha/\varepsilon_0}. \tag{26}$$

With appropriate values for the refractive indices of silica ($n_s = 1.452$) and the buffer ($n_s = 1.32$), and by calculating $\Delta\alpha/\varepsilon_0$ from eqn (23) (3.7×10^{-18} cm^3), we find that the surface density associated with the equilibrium shift in Fig. 10 is $\sigma_e = 2.6 \times 10^{10}$ cm^{-2}. This layer is far from compact. In our previous experiments with protein, surface densities approached the maximum density for random sequential adsorption, $\sigma_{rsa,max}$.[33] This density is 55% of the inverse "foot print" area ($\sigma_{rsa,max} = 0.55$ $\sigma_{fp} = 0.55 /(\pi a^2) = 1.2 \times 10^{11}$ cm^{-2}). The reason for the small surface density at equilibrium in Fig. 11 lies in the balance between adsorption and desorption rates. Although there was not enough incubated virus to make runs at higher concentrations, measurements made at lower concentration reveal that the equilibrium wavelength shift is in proportion to concentration [v] up to 5 pM (insert, Fig. 10). This may be understood from Langmuir's isotherm. Here, the fraction of maximum surface coverage at equilibrium $\theta_e = k_a [v]/(k_a [v] + k_d)$, where k_a is the adsorption rate constant and k_d is the desorption rate. For our equilibrium shift to be proportional to [v], $k_a[v]$ must be considerably less than k_d, so that the Langmuir isotherm has the approximate form $\theta_e = k_a[v]/k_d$. By making the reasonable

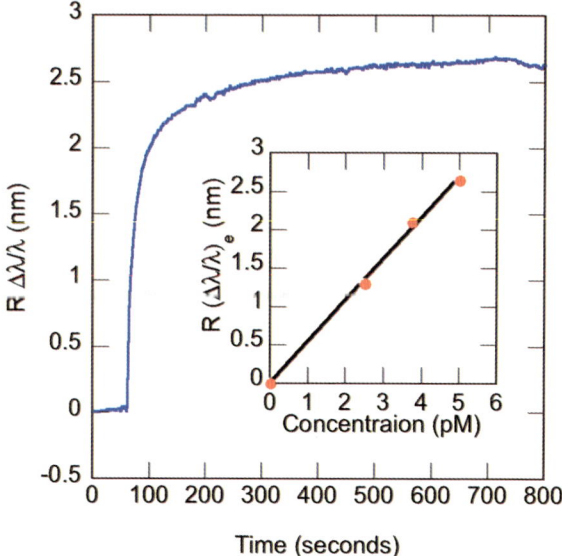

Fig. 10 Dip trace for non-specific MS2 adsorption for a 5 pM concentration. The insert shows the equilibrium shift isotherm below 5 pM.

assumption that $\theta_e \simeq \sigma_e/\sigma_{rsa,max}$, we estimate an upper limit for the equilibrium constant $K_e = k_a/k_d \sim 10^{11}$ M^{-1}.

So far, we have shown that we can detect the binding of virus particles to a microsphere from the shift in wavelength of a whispering gallery mode and extract the associated equilibrium constant. Our ultimate goal, specific sensing, will require the ability to follow the desorption process. This requires a microfluidic system and the design of a surface that can be specific to MS2.

Before discussing our microfluidic system, it should be pointed out that the mass density sensitivity associated with the rms baseline signal in Fig. 10 is consistent with our calculation of $\sigma_{m,LOD}$ in Fig. 7 for $R = 200$ µm; 0.2 ng cm^{-2}.

Micro-fluidic system for specific sensing

Specific sensing will be tested through the difference in desorption rates associated with different virus on a surface functionalized with antibody. This can reasonably be done by utilizing a microfluidic flow system that incorporates the microsphere and a tapered optical fiber.

Our specific sensing system is shown in Fig. 11. It consists of a tapered fiber bound by UV adhesive to a glass slide as a foundation. Below this foundation is a thermoelectric stage and above it is a polymer cap with multiple entry points. This transparent cap, formed by moulding poly-dimethyl siloxane (PDMS) in a micromachined master, is accesible to fluids, the microsphere, laser excitation and an optical detection port. The cross section of the microfluidic channel is 1.5 mm × 2.5 mm and has a total volume of 100 µL.

A fluidic system was designed to handle the various solutions used in our experiments. This system was supplied by two syringe pumps that are directed by a custom LabVIEW driver program. Each pump is fitted with a 25 mL syringe having 24000 steps of total travel (1 µL per step). The step rate is controlled by the driver that delivers liquid over a range from 1 step per 20 d to 100 steps per 1 s. A typical experiment is run at 38 µL min^{-1}. One pump is dedicated to the buffer used for washing in between each sample injection. The other is the sample pump that pushes the contents of a 700 µL teflon sample loop into the fluidic channel. The loop

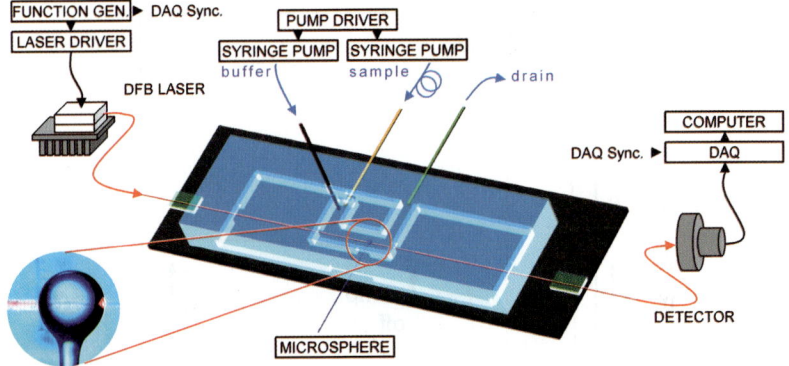

Fig. 11 Microfluidic system for sensing.

is washed with buffer between each sample injection, ensuring that the syringe is free of cross-sample contamination.

Surface modification: specific virus detection

Anti-MS2 (in PBS, pH 7.4, Tetracore) antibodies were covalently linked to the microsphere using amine–carboxyl coupling chemistry (Fig. 12). Briefly, spheres were cleaned by oxygen plasma for 4 min. Immediately after, spheres were immersed in 2% 3-(Triethoxysilyl)propylsuccinic anhydride in ethanol/acetic acid for 2 min then rinsed with ethanol to yield anhydride groups on the silica. Spheres were dried at 115 °C for 20 min. After this, a sphere was coupled to fiber inside the flowcell and slightly basic buffer (pH 8.3) was introduced into the cell, which hydrolyzes the anhydride to create a carboxyl rug across the sphere's surface (Fig. 13).

The sphere was allowed to incubate in the flow cell for 45 min. The rest of the assembly is followed by the dip trace in Fig. 13. Next, activation of carboxyl groups by EDC/NHS provided for a semi-stable amine, highly-reactive ester surface. 200 µL of 0.4 M 1-ethyl-3-(3-dimethylaminopropyl)-carbodiimide (EDC) was mixed with

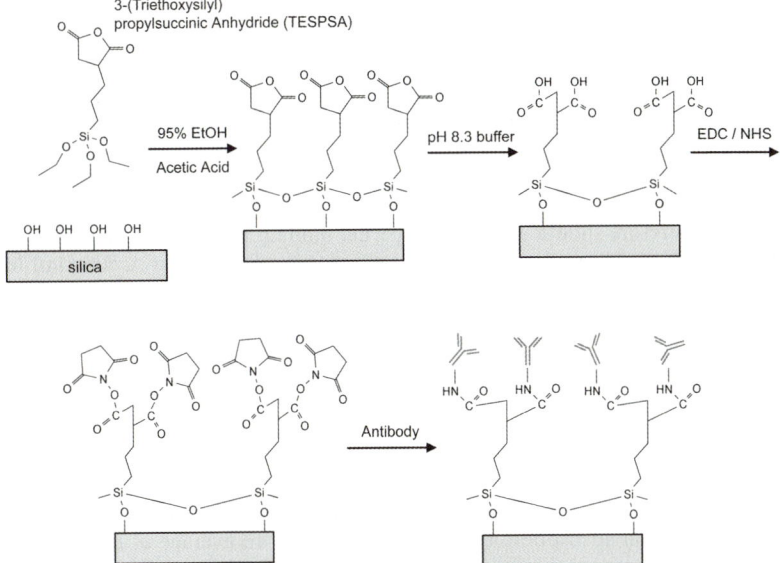

Fig. 12 Specific surface chemistry.

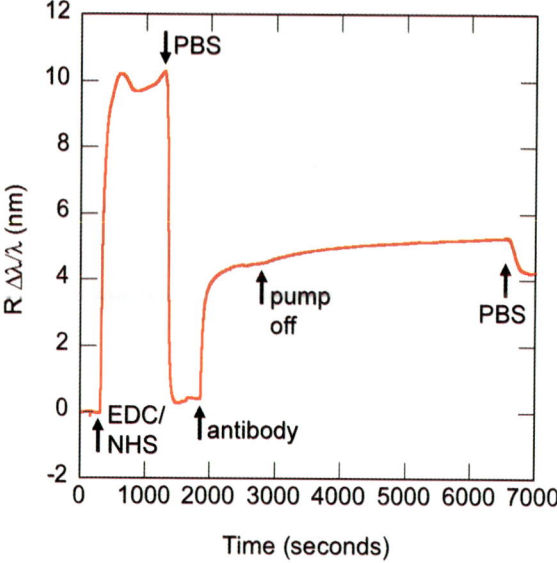

Fig. 13 Dip trace following the preparation of a silica microsphere with antibody.

200 μL 0.1 M N-hydroxysuccinimide (NHS) and injected into the flow channel. NHS esters react with amine groups on the antibody to form covalent links. Under constant flow, the surface was allowed to react for 5 minutes. After rinsing with PBS, MS-2 antibodies were injected at a concentration of 620 nM and allowed to incubate for 60 min with flow off. From the overall shift of 5 nm and the molecular weight of the antibody (155 000 Da), the surface density was estimated from eqn (26) to be 1.1×10^{12} cm^{-2}, about 5 antibodies within an MS2 footprint. Unbound antibodies were then washed away and the surface was blocked with 1 M ethanolamine (pH 8.5). Remaining antibodies are covalently bound to the surface. After a thorough rinse, surfaces were ready for specific virus detection. All steps involving flow were performed at a flow rate of 38 μL min^{-1}. The setup and microsphere surface are now ready for virus specificity detection.

Specific virus detection

In order to discriminate between different viruses, we designed a cycling scheme that included another *E. coli* virus, Phix174, as a negative control. Phix174 is a DNA virus of icosahedral shape and has a mass and diameter of 6.2×10^6 Da[35] and 28.4 nm, respectively. Its coat proteins contain epitopes different from the MS2 virus, and therefore should not exhibit specific binding to our anchored anti-MS2.

The experiment proceeds as follows. At a flow rate of 38 μl min^{-1}, 500 μl of MS2 phage at a final concentration of 5 pM was flowed through the microfluidic channel and then the pump was turned off. MS2 was allowed to bind to the surface for 5 min. We observed that all resonant dips shifted to longer wavelengths, indicating interaction between the modified surface and virus (top trace, Fig. 14). Just after, PBS was flowed into the channel for 10 minutes. We observed a resonance frequency shift towards smaller wavelength, indicating that weakly-bound virus was present on our surface. However the signal did not return to the value of pre-injected MS2. The residual surface bound MS2 after this wash was calculated (eqn (26)) to be $\sigma_v = 9 \times 10^9$ cm^{-2}. Once again, this surface density is far less than the maximum density for random sequential adsorption, even though there are an average of five antibodies within an MS2 footprint. Under the supposition that few antibodies are in a position and orientation to be viable, a decision was made to lower the concentration and

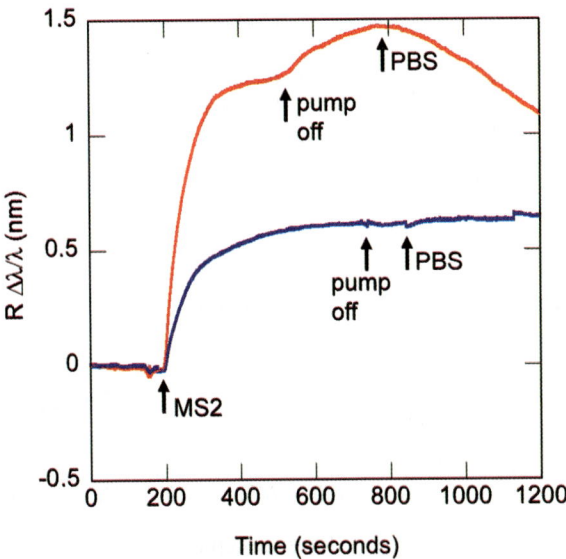

Fig. 14 MS2 virus experiment at 5 pM (upper) and after regeneration at 2.5 pM (lower).

repeat the experiment. This required an antibody-preserving regeneration step, which will be discussed later. The lower trace in Fig. 14 shows this second experiment on the same microsphere, at half the previous MS2 concentration. It should be noted that although the buffer wash caused virus to be removed at the higher concentration run, at 2.5 pM there was no evidence of desorption. We conclude that our supposition is correct; only a small fraction of the antibodies are viable for specific adsorption.

The very useful regeneration step previously noted deserves some attention at this point. It is accomplished by introducing 10 mM glycine into the flow channel (pH 2.0) and allowing the microsphere to bathe in this solution for 5 min. The glycine solution competitively disrupts the electrostatic portion of the antibody–antigen interaction.[36] During this time, the residual signal associated with specifically bound virus is eliminated without eliminating the covalently bound antibody. To complete the regeneration step, PSB is flowed into the channel for ∼10 min. At this point, the microsphere is ready for reuse with another virus (*e.g.* the control Phix174).

With the same microsphere in place, we next tested our sensor against Phix174 (Fig. 15). Phix174 was injected into the cell at 5 pM and allowed to bind as above. After injection, the frequency shifted toward longer wavelength, reaching equilibrium before the pump was turned off. About 100 seconds later PBS was pumped into the channel and the wavelength dropped toward the baseline, indicating that Phix174 non-specifically binds to our surface. The MS2 trace from Fig. 14 is reproduced in Fig. 15 for comparison. This figure clearly shows that the WGMB technique can discriminate between the non-specific binding by Phix174 and specific binding by MS2. The experiments were repeated several times with similar results.

Conclusions

By using perturbation ideas from microparticle photophysics, combined with biochemical surface functionalization and microfluidics, we have found a means for specifically sensing a virus. The axial symmetry associated with a slightly perturbed sphere allows for simple analytical equations that connect up wavelength shifts with the polarizability and surface density of the adsorbed virus. By using

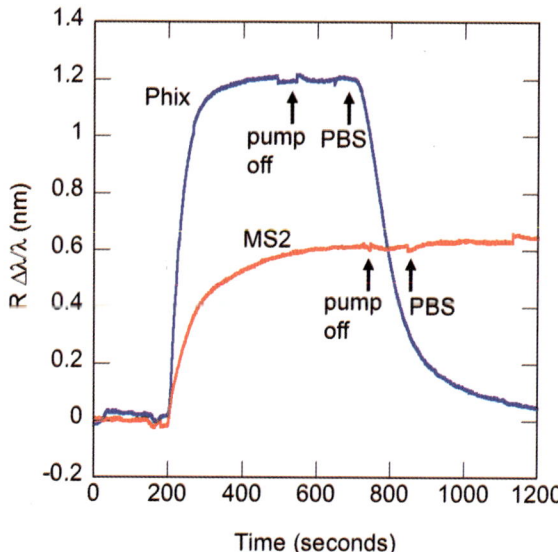

Fig. 15 Comparison between Phix174 and MS2 experiments on the same sphere.

chemical regeneration, our microparticle-inspired real time sensing approach allows us to follow virtually all aspects of the surface modification and viral sensing on the same sphere *in situ*.

Although our calculations looked in part at the question of single virion sensing, our experiments concentrated on the demonstration of specific detection using higher surface densities. This is not a limitation of baseline noise. Rather, the problem involves not being able currently to spatially localize adsorption solely to the equator of the sphere. We are currently designing a photolythographic technique that will allow us to place antibodies only along the equator. Once implemented, a shorter wavelength DFB laser will be employed in order to test our single virion calculations in Fig. 7. We expect the dip trace to contain a randomly-spaced staircase. Based on the agreement between our surface saturation measurements and theory (see the section titled "Non-specific sensing of virus"), the height of each step should reveal the molecular weight of the adsorbate.

We have merely scratched the surface in our use of the WGMB technique. Within a flow cell there can be a multitude of resonant microcavities each functionalized with different antibodies for different viruses. One can do this while multiplexing both the fiber and fluid channel. Other cross configurations will surely be imagined.

Acknowledgements

This research was supported by NSF through the Division of Bioengineering and Environmental Systems, Grant No. 0522668. We thank N. L. Goddard, Harvard University, for incubating the MS2 virus used in our experiments.

References

1. P. W. Ewald, *"Mastering Disease"*, in *The Next Fifty Years*, ed. John Brockman, Vintage Books, 2002, pp. 289–301.
2. A. Serpengüzel, S. Arnold and G. Griffel, *Opt. Lett.*, 1995, **20**, 654–656.
3. S. Arnold, *Am. Sci.*, 2001, **89**, 414–421.
4. F. Vollmer, D. Braun, A. Libchaber, M. Khoshsima, I. Teraoka and S. Arnold, *Appl. Phys. Lett.*, 2002, **80**, 4057–4059.
5. R. Rich and D. Myszka, *J. Mol. Recognit.*, 2005, **18**, 1–39.

6   U. Jonsson, L. Fagerstam, B. Ivarsson, B. Johnsson, R. Karlsson, K. Lundh, S. Lofas, B. Persson, H. Roos and I. Ronnberg, *BioTechniques*, 1991, **11**, 620.
7   R. Karlsson and R. Stahlberg, *Anal. Biochem.*, 1995, **228**, 274–280.
8   M. L. Gorodetsky, A. A. Savchenkov and V. S. Ilchenko, *Opt. Lett.*, 1996, **21**, 453–455.
9   S. Arnold, J. Camunale, W. B. Whitten, J. M. Ramsey and K. A. Fuller, *J. Opt. Soc. Am. B*, 1992, **9**, 819–824.
10 A. Ashkin and J. M. Dziedzic, *Phys. Rev. Lett.*, 1997, **38**, 1351.
11 P. Chylek, J. T. Kiehl and M. K. W. Ko, *Phys. Rev.*, 1978, **A18**, 2229.
12 H. M. Nussenzveig, *Comments At. Mol. Phys.*, 1989, **23**, 175.
13 H. G. L. Schwefel, H. E. Tureci, A. Douglas Stone and R. K. Chang, in *Optical Processes in Microcavities*, "Progress in Asymmetric Resonant Cavities: Using shape as a design parameter in dielectric microcavity laser", ed. R. K. Chang and A. J. Campillo, World Scientific, Singapore, 1996.
14 S. Arnold, M. Khoshsima, I. Teraoka, S. Holler and F. Vollmer, *Opt. Lett.*, 2003, **28**, 272–274.
15 F. Frolich, *Theory of Dielectrics*, Oxford University Press, London, 1958, p. 28.
16 W. U. Wang, C. Chen, K. Lin, Y. Fang and C. M. Lieber, *Proc. Natl. Acad. Sci. U. S. A.*, 2005, **102**, 3208.
17 F. Patolsky, G. Zheng, O. Hayden, M. Lakadamyali, X. Zhuang and C. M. Lieber, *Proc. Natl. Acad. Sci. U. S. A.*, 2004, **101**, 14017.
18 M. Noto, D. Keng, I. Teraoka and S. Arnold, *Biophys. J.*, 2007, **92**, 4466–4472.
19 I. Teraoka and S. Arnold, *J. Appl. Phys.*, 2007, **101**, 0235051.
20 R. P. Harrington, *Time-Harmonic Electromagnetic Fields*, Wiley-IEEE Press, 2001.
21 I. Teraoka and S. Arnold, *J. Opt. Soc. Am. B*, 2006, **23**, 1381–1389.
22 D. Q. Chowdhury, S. C. Hill and M. M. Mazumder, *IEEE J. Quantum Electron.*, 1993, **29**, 2553.
23 J. D. Jackson, *Classical Electrodynamics*, Wiley, New York, 1962, 753.
24 M. Khoshsima, *"Perturbation of Whispering Gallery Modes in Microspheres by Protein Adsorption: Theory and Experiment"*, PhD thesis, Polytechnic University, New York, 2004.
25 M. L. Gorodetsky, A. D. Pryamikov and V. S. Ilchenko, *J. Opt. Soc. Am. B*, 2000, **17**, 1051–1057.
26 S. H. Armstrong, Jr, M. J. E. Budka, K. C. Morrison and M. Hasson, *J. Am. Chem. Soc.*, 1947, **69**, 1747.
27 H. J. Coles, B. R. Jennings and V. J. Morris, *Phys. Med. Biol.*, 1975, **20**, 310.
28 C. E. Alupoaei, J. A. Olivares and L. H. Garcia-Rubio, *Biosens. Bioelectron.*, 2004, **19**, 893.
29 T. P. Burg and S. R. Manalis, *Appl. Phys. Lett.*, 2003, **83**, 2698.
30 G. Griffel, S. Arnold and A. Serpenguzel, *Opt. Lett.*, 1996, **21**, 695–697.
31 D. Keng, *"Asymmetric fiber puller for biosensor experiments"*, in preparation, 2007.
32 K. H. Choi, J. P. Bouroin, S. Auvray, D. Esteve, G. S. Duesberg, S. Roth and M. Burghard, *Surf. Sci.*, 2000, **462**, 195–200.
33 M. Noto, M. Khoshsima, D. Keng, I. Teraoka, V. Kolchenko and S. Arnold, *Appl. Phys. Lett.*, 2005, **87**, 223901.
34 L. R. Overby, G. H. Barlow, R. H. Doi, M. Jacob and S. Spiegelman, *J. Mol. Biol.*, 1966, **91**, 442–448.
35 A. B. Burgess, *Proc. Natl. Acad. Sci. U. S. A.*, 1969, **64**, 613–617.
36 D. Ilchmann, D. Helbig, H. Gohler, M. Stopsack, H. J. Thiele and W. Hubl, *J. Clin. Chem. Clin. Biochem.*, 1990, **28**, 677–681.

Resonance-based light scattering techniques for investigation of microdroplet processes

Asit K. Ray,[*a] Venkat Devarakonda[b] and Zhiqiang Gao[a]

Received 9th February 2007, Accepted 12th March 2007
First published as an Advance Article on the web 30th July 2007
DOI: 10.1039/b702122d

An elastic or a Raman scattering intensity *versus* size parameter spectrum from a droplet shows a series of resonances. Each resonance contains a unique relation between the size and the refractive index of the scattering droplet, and the resonances from homogeneous droplets behave significantly differently than the resonances from layered droplets. These characteristics can be used to analyze observed resonances, and to determine the size and the refractive index of a homogeneous droplet or the inner and outer radii and the core and shell refractive indices of a layered droplet. We show that many microdroplet processes can be studied by applying resonance-based light scattering techniques to single droplets suspended in an electrodynamic balance and to highly monodisperse droplets in linear arrays. This paper focuses on deciphering internal composition distributions in microdroplets that develop due to fast physical or chemical processes. Experiments were conducted on linear streams of droplets that were generated by a modified vibrating orifice aerosol generator from a solution of non volatile dibutyl phthalate dissolved in volatile freon. The residence time of the droplets in the gas phase was altered by varying the distance between the droplet generator and a laser beam that illuminates the droplets. The variation of the frequency of the droplet generator causes the droplet size to change in a prescribed manner, and thus, the scattering intensity as a function of the frequency shows a series of resonances due to the variation of the size. We have analyzed the resonance peak frequencies to obtain the size and the composition distribution inside the droplets as functions of time. During evaporation, the resonances of non-uniform composition droplets shift differently from those of uniform composition droplets. Specifically, lower order resonances from non-uniform droplets shift significantly more than higher order resonances. This is the basis for the determination of the size along with the composition distribution from the observed resonances. The experimental data show that various resonances observed in the Raman scattering spectra shift differently with time, as predicted by the theory. The size and composition distribution results obtained from the analysis of resonances show behavior that is expected from a droplet evaporation model.

[a] *Chemical and Materials Engineering, University of Kentucky, Lexington KY 40506, USA. E-mail: akray@engr.uky.edu*
[b] *BlazeTech Corp., Cambridge MA 02141, USA E-mail: venkat@blazetech.com*

1. Introduction

Phenomena associated with microdroplets play critical roles in the climate, the generation of tailored micro- and nano-particles from aerosol-based processes, and in many industrial processes. Our understanding of microdroplet processes is in a nascent stage, and many practical applications involve trial and errors. Processes in a microdroplet depend on the size and are transient in nature (*i.e.*, the behavior at any given instant depends on the history). In an aerosol, polydispersity and interparticle separation distances play major roles. It is difficult to measure time-dependent size and composition distributions of an assemblage of particles. Experiments involving initial and final states provide little phenomenological information. In view of this situation, a number of experimental approaches can be taken to obtain a fundamental understanding of microdroplet processes. Relatively slow processes can be studied through experiments on single droplets that are suspended in controlled environments, while the fast processes can be examined through identical droplets that traverse identical paths in time-invariant gas phases. The latter approach requires highly monodisperse droplets and provides a unique space-time relation; thus, the variation of a droplet with time can be obtained by examining droplets at various locations in the path. In both cases, resonance-based light scattering techniques can be used for *in situ* characterizations of the droplets. A number of sharp intensity peaks that are called resonances appear in an elastic or a Raman scattering intensity *versus* size parameter spectrum from a droplet. Each resonance provides a unique relation between the size and the refractive index, and the relations for two distinct resonances differ sufficiently enough that the size and the composition of the scattering droplet can be accurately determined by establishing the correct identities of the observed resonances. In this paper, we show that physical and chemical dynamics of microdroplets can be studied by applying resonance-based light scattering techniques to single droplets suspended in an electrodynamic balance and to highly monodisperse droplets in a linear array. We provide various examples that include: determination of thermodynamic and kinetic parameters from unsteady state evaporation and growth of multicomponent droplets; stability limits of charged droplets; collection efficiencies of single charged droplets; and adsorption of vapor in microdroplets. The main objective of this paper is, however, to demonstrate that the internal composition distribution in a microdroplet can deduced from various order resonances that appear in Raman scattering.

2. Experimental

Experiments were conducted either on single droplets that were suspended in electrodynamic balances or on linear streams of highly monodisperse droplets that were generated by a modified Vibrating Orifice Aerosol Generator (VOAG). Detailed descriptions of the experimental systems based on electrodynamic balances can be found elsewhere. In one type of system, single microdroplets were suspended in a gas stream having a precisely controlled composition of vapor.[1–7] In the other type of system, single droplets were suspended inside a modified thermal diffusion cloud chamber.[7–11] By placing a pool of liquid at the bottom plate and by maintaining a temperature differential between the top and the bottom plates of the diffusion cloud chamber, a well-defined vapor concentration field (from unsaturated to supersaturated) as a function of position was created. In these experimental systems, a droplet is illuminated by a fixed wavelength (*e.g.*, He–Ne or Ar–ion) or a tunable ring dye laser beam. Two PhotoMultipliers Tubes (PMTs) are used to detect elastic scattering light in the planes parallel and perpendicular to the plane of polarization of the incident beam. This scheme allows detection of Transverse Electric (TE) and Transverse Magnetic (TM) mode resonances seperately by the two PMTs.

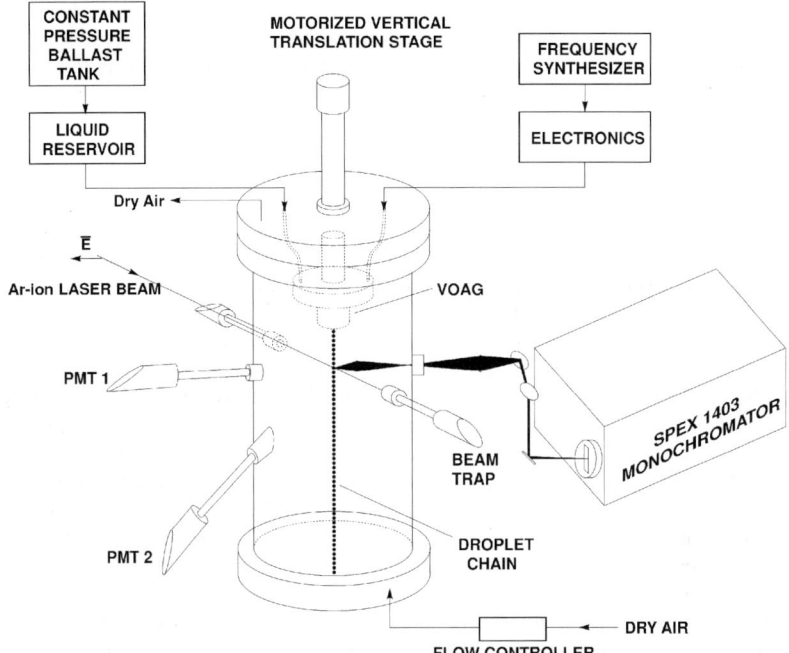

Fig. 1 Experimental system based on linear stream of monodisperse droplets generated by VOAG.

Fig. 1 shows a schematic of the experimental system based on a linear array of droplets that are generated by a modified VOAG.[12–14] The details of the VOAG and its operation can be found elsewhere;[12,15,16] here, we restate the salient features of the experimental system. A desired solution from a reservoir is forced through a vibrating orifice under a constant gas pressure from a ballast tank and the resulting cylindrical liquid jet emerging out of the orifice breaks into monodisperse droplets by the vibration of the orifice. When the solution flows at a volumetric rate of Q, through the orifice vibrating at a frequency f, the droplet radius is given by

$$a = (3Q/4\pi f)^{1/3}. \tag{1}$$

To obtain highly monodisperse droplets, the orifice is driven by a square wave with a frequency stability on the order of 1 part in 10^6, while the pressure difference driving the solution through the orifice is kept within 1 part in 10^5. We have shown[12] that short term fluctuations in the droplet size can be kept within 1 part in 10^5. The droplet generation system is housed inside an environmentally controlled cylindrical chamber through which a gas stream of desired composition flows past the droplets falling down the axis. Each droplet experiences the same history and the axial distance from the generation point is uniquely related to the residence time of the droplets in the chamber environment. Therefore, by examining droplets at various distances from the orifice, the variation of a droplet with time can be deciphered. The droplets are illuminated by a polarized laser beam at a fixed position of the chamber. The orifice assembly is mounted on a motorized vertical translation stage, and can be positioned accurately to any distance, from 1 to 25 mm from the laser beam. The residence time of the droplets at a given orifice position is determined from the diffraction fringe spacing,[12] and can be varied from 10 μs to 10 ms by changing the distance between the droplet generator and the laser beam. Since the residence time is independent of the real time, the system is suitable for studying fast (<0.1 ms) processes. Two PMTs are used to detect elastically scattered light, while a double monochromator records inelastically (*e.g.*, Raman) scattered light. The elastic and Raman scattering data are used to characterize droplets at various residence times. In the following section, we provide a brief review of the light scattering theory and applications of resonance-based light scattering techniques for characterization of homogeneous and layered droplets.

3. Light scattering

Scattering intensity data are analyzed using Lorenz–Mie theory[17–19] to obtain the size and the refractive index of a droplet. When a homogeneous spherical particle of radius a, intercepts a laser beam of wavelength λ, the intensity of light scattered by the particle at a point depends on the refractive index of the particle relative to the surrounding medium, m, and the size parameter, which is defined as $x = 2\pi a/\lambda$. A scattering intensity *versus* size parameter spectrum shows a series of sharp peaks that are often referred to as the Morphology-Dependent Resonances (MDRs). Resonances occur when electromagnetic waves are trapped by almost total internal reflections as they propagate around a narrow annulus near the surface of the particle, and at a resonance that the intensity of light inside the particle can enhance dramatically. This gives rise to intensity peaks in elastic and inelastic scatterings. Mathematically, the resonances arise from the peaks of the scattered field expansion coefficients of TM and TE modes. Each scattering coefficient exhibits an infinite number of peaks. A resonance associated with a peak is identified by a set of three MDR parameters (n, ℓ, μ), where n and μ stand for the mode number and polarization of the associated scattering coefficient (*i.e.*, $\mu = 1$ or 2 for TM or TE mode) and ℓ denotes the order number of the peak (*i.e.*, $\ell = 1$ for the first peak).[7,20] The position (*i.e.*, x value) of a resonance of given MDR parameters, as well as its width (*i.e.*, FWHM, $\Delta x_{1/2}$), is uniquely related to the refractive index m, of the scattering sphere. The characteristics of a resonance are dictated by the MDR parameters associated with it. For example, for a given change in m, the x value of a lower order resonance shifts more than that of a higher order resonance. Moreover, the width of a resonance of a given mode number n, increases as the order number ℓ, increases, while for a given ℓ, the width decreases as n increases. The discussion so far applies only to the elastic scattering. The theory of inelastic scattering by a sphere is not developed enough to be computationally amenable.[21,22] The elastic scattering theory can, however, predict the positions of input and output resonances that appear in the inelastic scattering when the wavelengths of the incident light and the inelastic emission satisfy, respectively, the resonance conditions.

Widths of resonances vary widely; some resonances are extremely narrow. For example, a resonance with MDR parameters (300, 1, 2) has a width of $\Delta x_{1/2} = 3.9 \times 10^{-44}$ at $m = 1.45$, which corresponds to $\Delta a_{1/2} \approx 3.9 \times 10^{-38}$ Å in terms of the size in the visible spectrum. Due to the presence of slight shape distortions and surface roughness on the order of molecular dimensions, extremely sharp (*i.e.*, low order) resonances cannot be detected in the elastic scattering, but can be observed in Raman scattering. Typically, the sharpest resonance detectable in the elastic scattering has FWHM $\Delta x_{1/2} \geq 1 \times 10^{-4}$, while in Raman scattering a first order resonance can be observed.[12,23] In addition, higher order broad resonances become indistinguishable because they overlap with one another. To detect resonances, one needs a scattering intensity spectrum obtained by varying the size parameter, x. Such a spectrum can be obtained by varying: (i) the wavelength of the incident light (wavelength-dependent spectra), (ii) the droplet size through evaporation or growth processes (time-dependent spectra), and (iii) the size through the modulation of the droplet generation frequency (frequency-dependent spectra). By identifying the resonances observed in a spectrum, the size and the refractive index of the scattering droplet can be determined with high accuracies.[20,24,25] For example, Huckaby *et al.*[25] have analyzed resonances in elastic scattering intensity *versus* incident wavelength spectra to obtain the size and the refractive index of a droplet with accuracies of 3 parts in 10^5. We have developed a number of automated procedures for high precision determination of the size and the refractive index from time and frequency-dependent spectra,[6,7,12,25,26] and used these procedures to determine thermodynamic parameters from steady and unsteady state evaporation and growth of multicomponent droplets.

3.1 Applications of resonance based techniques

3.1.1 Resonances of wavelength-dependent spectra. Among the three types of scattering spectra, the wavelength-dependent spectra are only useful for droplets with time invariant, or slowly varying, size. They are, however, particularly suitable for detecting small size and/or refractive index changes occurring in a microdroplet due to its interactions with the surrounding environment. The size parameter value of a resonance of given MDR parameters is unique for a given refractive index. Therefore, when a particle changes size by Δa, from its initial radius a_0, without any change in the refractive index, the wavelength of a resonance appearing initially at λ_0, shifts according to the following relation

$$\frac{\Delta a}{a_0} = \frac{\Delta \lambda}{\lambda_0}. \tag{2}$$

Dhariwal et al.[3] have used this method to study the collection efficiency of a single charged droplet suspended in a stream of charged sub-micron particles. They detected resonances in elastic scattering from an electrodynamically suspended dioctyl phthalate (DOP) droplet by scanning a ring dye laser, and observed the positions of the same resonances after an exposure of the suspended DOP droplet to a stream of charged DOP aerosol. Since the deposition did not alter the refractive index of the host droplet, they observed that the wavelengths of all the resonances shifted by the same amount, and determined nanometre-level size changes due to the deposition using eqn (2). It should be noted that size changes as small as 1 part in 10^6 (i.e., ~ 2 Å for a 20 μm particle) can be determined from the shift of a resonance.[26] The position of each resonance is uniquely related to the refractive index, and the relation is dictated by the MDR parameters of the resonance. When the size and the refractive index of a droplet change simultaneously, the observed shifts of two resonances are, in principle, sufficient to determine the size and refractive index changes. Ray and Huckaby[2] used this principle to detect size and refractive index changes of single DOP droplets due to the absorption of sparingly soluble water vapor.

The scattering intensity at a point from a concentrically-layered sphere depends on the core and outer radii as well as on the core and shell refractive indices.[27] Resonances of a layered droplet behave differently from those of a homogeneous droplet. For small size and/or refractive index changes, both TM and TE mode resonances observable in elastic scattering shift almost identically as long as a droplet remains homogeneous, but when a layer forms on a homogeneous droplet TM and TE mode resonances shift substantially differently. In addition, when the core and outer radii and/or the refractive indices of the core and shell phases change, TM and TE mode resonances also shift differently. The shifts of TM and TE mode resonances provide a basis for an unambiguous discrimination between small size changes of a droplet due to the absorption and adsorption of vapor molecules, and also for detecting minute changes occurring in a layered droplet. These theoretical findings have been substantiated with experimental data.[2,4] Huckaby and Ray[4] had shown that when a Santovac droplet was exposed to saturated Fomblin vapor, wavelengths of two resonance peaks (i.e., one TE and another TM) shifted by an identical amount. The identical shifts of TE and TM mode resonances indicate that the droplet remained homogeneous, and a few nm size change occurred due to the absorption of Fomblin molecules. Furthermore, when the droplet was exposed to Fomblin vapor at a supersaturated level, the TM mode resonance peak shifted significantly more than the TE mode resonance. The significant difference in the peak shifts provides an evidence of the formation of a layer on the droplet due to the adsorption of Fomblin molecules, and they determined the adsorbed layer thickness from the shifts.

3.1.2 Resonances of time-dependent spectra. A scattering intensity *versus* time spectrum from a droplet shows a series of resonances (*i.e.*, peaks) as the droplet undergoes changes in the size and/or composition as a result of some physical or chemical processes. We have developed an automated procedure for analysis of such a spectrum, utilizing the times at which resonances appear in the spectrum. Tu and Ray[7] have described the procedure in detail. Briefly, the size and refractive index of the droplet at each resonance appearance time are determined by aligning the observed resonances with those theoretical resonances that minimize the difference (*i.e.*, alignment error) between the observed and the calculated appearance times. The procedure can be applied to a broad range of problems, and requires either the relations with unknown constants for the variations of the size and the refractive index, or the identities (*i.e.*, MDR parameters) of the observed resonances. The relations may be obtained by modeling the processes that control the variations, and can be algebraic or implicit in the forms of differential equations. The alignment procedure provides the unknown constants (*e.g.*, diffusion coefficient, vapor pressure, activity coefficients) of the relations. Using experimental data from single droplets undergoing dynamic changes under a variety of conditions, we have demonstrated that the size and the refractive index can be determined with relative errors of less than 1×10^{-4}. The procedure has been used to determine activity coefficients from resonances observed in the scattering spectra from single binary droplets during unsteady state evaporation and growth,[8,11] and the parameters of activity coefficient models from evaporation of multicomponent droplets.[28] Because of the high sensitivity, the procedure can be used to study microdroplet phenomena that involve small size changes. For example, Li *et al.*[9] have examined minute mass losses (*i.e.*, size changes) from charged droplets that occur at charge instability induced breakups.

3.2 Variable composition droplets

The above review of prior studies shows that resonance-based light scattering techniques can be used to determine the sizes and compositions of homogeneous and layered droplets with high precision. Many physical and chemical processes, such as rapid evaporation and chemical reactions, induce significant concentration gradients in microdroplets, and rates of many processes, such as nucleation and reactions in microdroplets, depend on the local composition. To understand such processes, one needs to know the composition distribution in a droplet. Despite the availability of the theory to describe scattering by a radially inhomogeneous droplet,[29,30] no technique currently exists for the characterization of droplets with composition gradients. Resonances of a variable composition droplet exhibit behaviors that are between those of uniform composition and layered droplets. When a homogeneous droplet of uniform composition undergoes size and refractive index changes while maintaining uniformity in composition, a TM mode resonance shifts nearly identically to that of a TE mode; thus, the separation distance between the two resonances remains nearly the same. When the size change is accompanied by the formation of a layer, the shift of a TM mode resonance differs substantially from the shift of a TE mode resonance,[2,4] as discussed before. If the size change induces a concentration gradient in the droplet, the separation distances between various order resonances change significantly. Thus, the separation distances between resonances provide a basis for distinguishing uniform, variable concentration and layered droplets unambiguously.

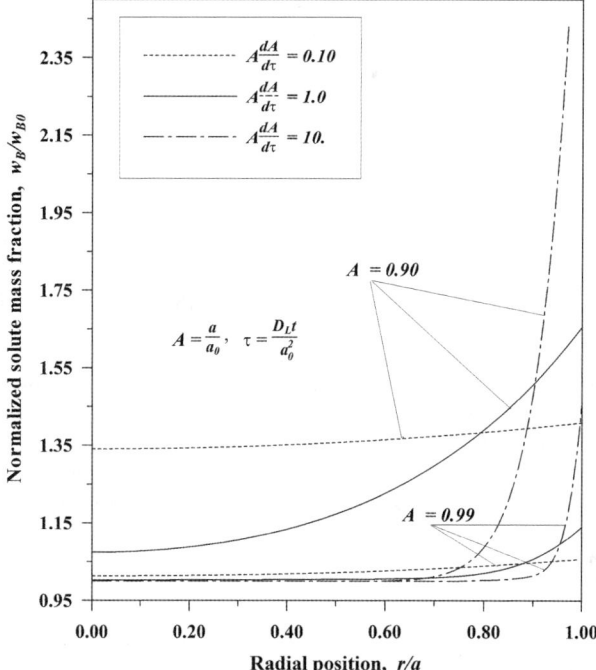

Fig. 2 Solute composition inside a microdroplet as a function of dimensionless radial position for various dimensionless evaporation rates.

To understand the foundation of the method for determining the radial concentration distribution inside a droplet, we consider droplets containing a non-volatile solute B dissolved in a highly volatile solvent A. To examine solute concentration distributions inside such evaporating droplets, we solved the continuity equations that describe the droplet phase diffusion and the conservation of mass along with moving boundary conditions. Fig. 2 shows calculated composition distributions of solute B for varying dimensionless evaporation rates of a droplet that initially has a radius a_0, and contains uniformly distributed solute at a mass fraction w_{B0}. The results show that, at a given dimensionless size (i.e., $A = a/a_0$), the non-uniformity in the solute concentration distribution becomes more pronounced as the dimensionless evaporation rate, $A(dA/d\tau)$, increases, while for a given evaporation rate the non-uniformity in the composition distribution increases as the droplet size decreases, as a result of evaporation of the solvent. To examine the effect of the radial composition distributions on the position of a resonance, we have calculated the positions of various order resonances for the results shown in Fig. 2, using refractive index values of $m_A = 1.358$ and $m_B = 1.493$, for the solvent and the solute, respectively. In addition, we have used an ideal solution assumption for the mixture, and a volume fraction-weighted average value for the mixture refractive index.[2,5] Fig. 3 shows the positions (i.e., size parameter values) of TE and TM mode resonances of various orders for three different situations: (a) for the initial uniform composition droplet (i.e., $A = 1$, $w_{B0} = 0.14278$, and solute volume fraction, $v_B = 0.20$), (b) when the droplet has evaporated at a rate of $A(dA/d\tau) = 5.0$, to a dimensionless size of $A = 0.95$, with the composition distribution shown in Fig. 2, and (c) when the droplet evaporates slowly (i.e., $A(dA/d\tau) \leq 0.1$) to a size $A = 0.95$, maintaining a near uniform composition distribution, as shown in Fig. 2. Comparing the resonance positions for situations (a) and (c), we find that all the resonances shift almost by the same size parameter value when the composition remains uniform. For situation (b), however, various order resonances shift differently from

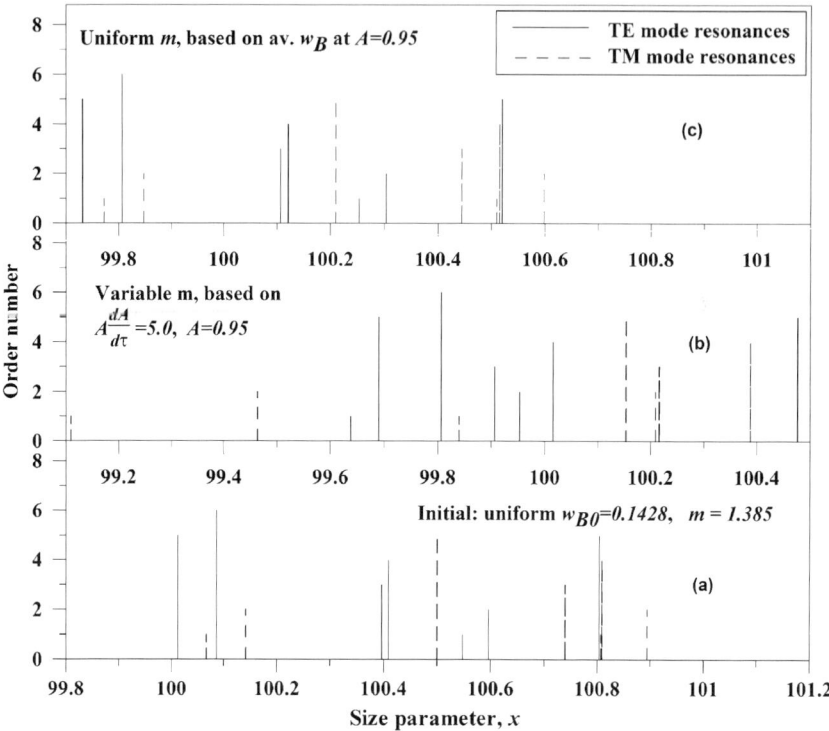

Fig. 3 Comparison between initial and final positions of various order resonances from a droplet after a 5% size reduction. For a uniform distribution (a), all the resonances shift nearly the same, while for a non-uniform distribution (b), lower order resonances shift more than higher order ones.

their initial positions at $A = 1$. Specifically, the lower order resonances shift significantly more than the higher order resonances. To further illustrate this finding, in Fig. 4 we have plotted the shifts of 1st and 5th order TE and TM mode resonances from their initial positions as functions of the dimensionless size for an evaporation rate of $A(dA/d\tau) = 5.0$, along with the shifts for the situation where the composition distribution remains uniform during the entire evaporation process. The results show that when the composition remains uniform, all the resonances shift almost identically, that is, the separation distances between the resonances remain nearly the same, as shown in Fig. 4. For the faster evaporating droplet, however, 1st order resonances shift significantly more than the 5th order resonances and the separation distance between the 1st and 5th order resonances increases as the droplet size decreases. This is expected, since the difference between the solute mass fractions at the center and at the surface of the droplet increases as the droplet shrinks to smaller size under a faster evaporation rate, as shown in Fig. 2. Unlike a layered droplet, the results in Fig. 4 show that TE and TM mode resonances of the same order shift identically when the refractive index varies radially in a homogenous droplet. The calculations presented here provide a basis for the determination of the radial concentration profile from the separation distances between the low and the high order resonances that can be detected in Raman scattering.

In the following section, we describe the procedure with data from binary droplets of non-volatile dibutyl phthalate (DBP) dissolved in highly volatile freon. The data were obtained using the modified VOAG-based experimental system shown in Fig. 1. Linear streams of highly monodisperse microdroplets were generated from solutions containing 10 vol% and 20 vol% of DBP in freon. The droplet generator

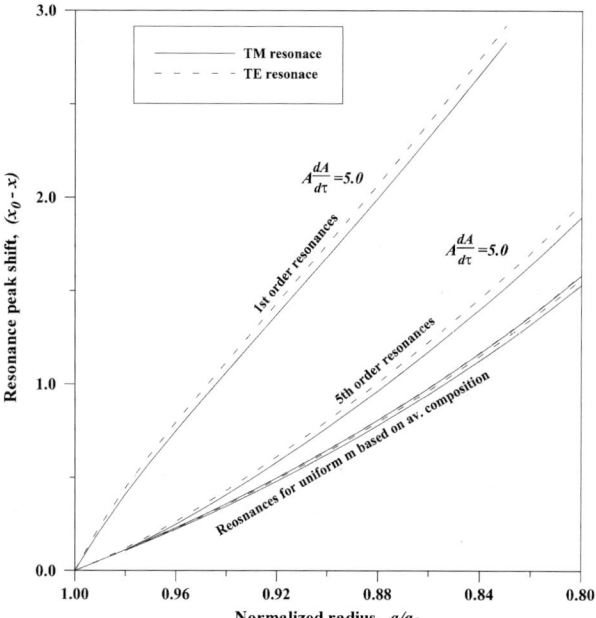

Fig. 4 Shifts of various order resonance peaks from their initial positions as functions of droplet radius. The results for fast evaporation and slow rates are compared.

was moved to different positions, and at each position the droplet generator was scanned over a frequency range of 50 kHz to 100 kHz. The elastic light scattering intensities (PMT 1 and PMT 2) and Raman scattering intensity at a wavenumber shift of $\Delta\nu = 652$ cm^{-1} were recorded as functions of the droplet generation frequency. The Raman shift of 652 cm^{-1} corresponds to a peak associated with freon molecules.

4. Results and discussion

Fig. 5 shows typical Raman scattering intensity *versus* frequency spectra at various distances from the orifice. The residence time at each position is indicated in the figure. We chose the distance $z = 2$ mm as the reference position, and a few of the resonances observed in the spectrum at that position are labeled as A, B, C, ... Their positions (*i.e.*, frequencies) at various distances from the orifice are indicated in the figure. The data show that the frequency at which a particular resonance is observed at the reference position shifts to a lower frequency as the distance from the orifice (*i.e.*, residence time) increases, due to the decrease in the droplet radius by evaporation. A closer examination reveals that some of the resonances shift more than the other resonances in terms of the frequency. To demonstrate the differences in the frequency shifts among the observed resonances, in Fig. 6 we have plotted the shifts of five different observed resonances with respect to their appearances at the reference position. The experimental data in Fig. 6 show that the difference between the frequency shifts of any two resonances increases as the distance from the orifice increases, and large differences (as high as 6000 Hz) among the frequency shifts of the observed resonances indicate the presence of concentration gradients in the droplets. Also, the increasing differences with increasing residence time suggest that the composition variation in a droplet becomes more pronounced as the residence time increases, as suggested by the theoretical analysis. We now describe the procedure used for determining the size and the radial concentration profile inside the droplets from the observed resonances.

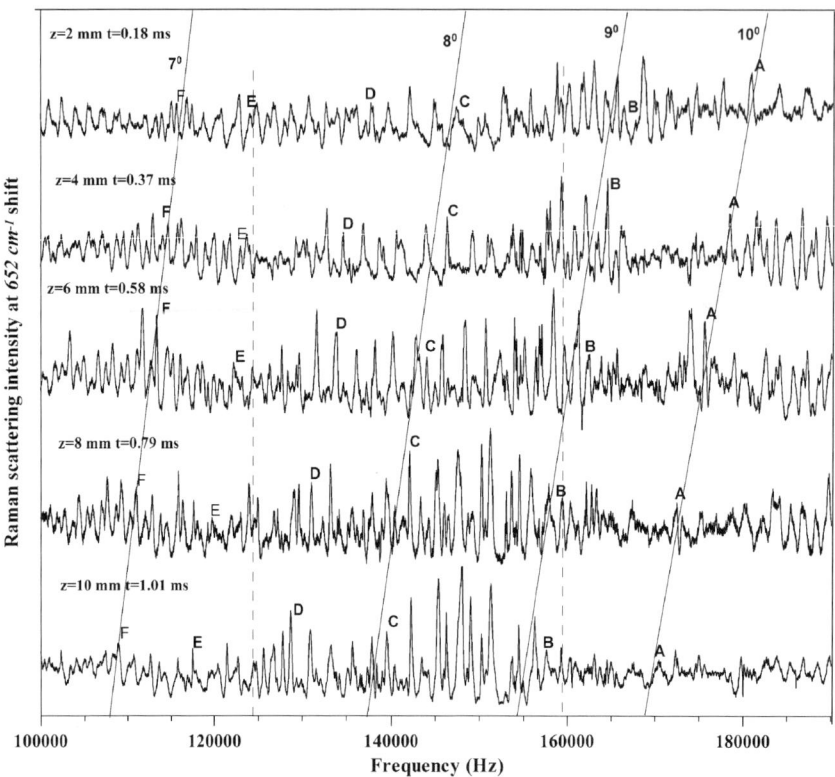

Fig. 5 Experimental Raman scattered intensity *vs.* frequency spectra at various distances from the orifice. Resonance peaks shift to lower frequencies as the residence time increases because of droplet evaporation.

An experimental Raman scattering intensity *versus* droplet generation frequency spectrum provides the frequencies at which resonances appear. From the observed resonances, we selected N sharp resonances and recorded their appearance frequencies. Therefore, the experimental data at a given generator position provide the frequencies $\{f_1, f_2, \ldots, f_N\}$ corresponding to the resonances, and in addition, the droplet size varies with the frequency according to eqn (1). Since the liquid flow rate through the orifice was maintained constant during an experiment, the following relation applies for the variation of the droplet size parameter

$$x = cf^{-1/3} \qquad (3)$$

where c is a constant. At a given generator position, we assumed that the refractive index variation can be described by a polynomial of the form:

$$m = b_0 + b_1\left(\frac{r}{a}\right) + b_2\left(\frac{r}{a}\right)^2 + b_3\left(\frac{r}{a}\right)^3 + b_4\left(\frac{r}{a}\right)^4 \qquad (4)$$

The problem now involves the determination of six unknown quantities, parameters c, b_0, b_1, b_2, b_3 and b_4, from the appearance frequencies of the resonances in the Raman scattering spectrum at that position. To analyze the observed resonant frequencies for assumed values of the six parameters, we first established mode and order numbers of all the theoretical resonances that can possibly appear in the experimental Raman spectrum. Using the procedure described by Ray and Bhanti,[31] we then calculated the size parameter values of these resonances for the refractive index profile given by eqn (4). A subset of N calculated resonances with size

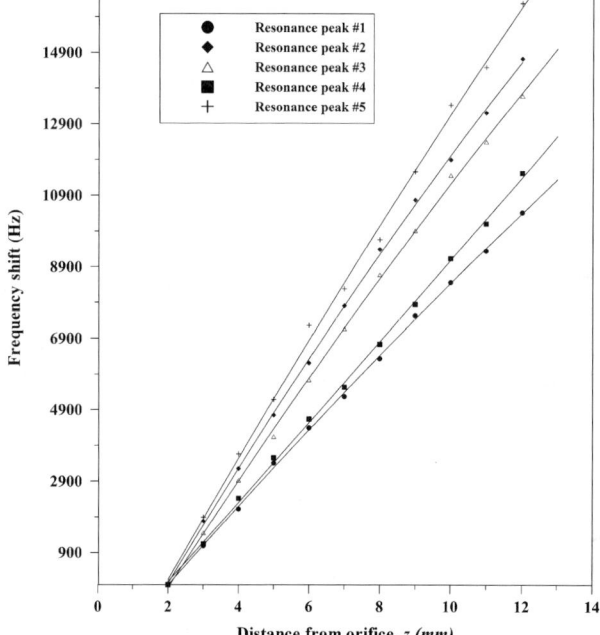

Fig. 6 Frequency shifts of various resonance peaks observed in Raman intensity *vs.* frequency spectra. Large differences (as high as 6000 Hz) among frequency shifts of resonances indicate the presence of concentration gradients in the droplets.

parameter values $\{x_1, x_2, \ldots, x_N\}$ was chosen for alignments with the observed resonances at frequencies $\{f_1, f_2, \ldots, f_N\}$, and we estimated the error in the peak alignments using the following relation:

$$S(c, b_0, b_1, b_2, b_3, b_4) = \sum_{i=1}^{Npt} (x_i - cf_i^{-1/3})^2 \qquad (5)$$

For an assumed set of values for the six parameters, we considered all possible theoretical peaks for alignment with an observed peak. By varying the parameter values we determined all the minima in a 7-dimensional plot of S-c-b_0-b_1-b_2-b_3-b_4, and on the basis of the global minimum in the alignment error we established the best estimates (*i.e.*, optimum values) of the six parameters.

Fig. 7 shows the refractive index profiles obtained by applying the alignment procedure to resonances observed in Raman scattering from droplets that were generated from solutions containing 10 vol% DBP in freon. It should be noted that the continuous curve for the refractive index profile at each of the residence times shown in Fig. 7 was calculated from eqn (4), using the optimum values of b_0, b_1, b_2, b_3 and b_4 obtained by applying the alignment procedure to the resonances observed at the corresponding residence time. The results show that the refractive index varies significantly even at small residence times. The refractive index decreases from the maximum value at the surface to the minimum at the center, with the sharpest gradient at the surface. This is expected for the solute composition profiles shown in Fig. 2. As the residence time increases, the refractive index at the center rises from its initial value because of the buildup of DBP. The alignment procedure also provides the optimum value of c at each position (*i.e.*, residence time) of the droplet generator. From the values of c, we constructed droplet size *versus* time data using eqn (3) and (1). Fig. 8 shows size *versus* time results obtained from droplets of various initial compositions. Since the droplet size varies with the frequency, we have chosen an

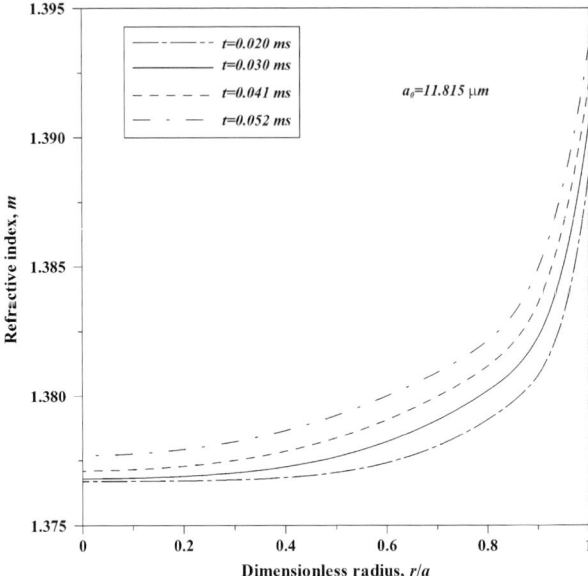

Fig. 7 Radial distributions of refractive index at various times determined from resonances observed from droplets containing 10% DBP in freon.

initial size of $a_0 = 11.8$ μm for presenting the results on the basis of a common reference point. As expected, the results show that the size of pure freon droplets decreases faster than the size of solution droplets. For solution droplets, the results show that droplets containing 10% DBP evaporate faster than the droplets containing 20% DBP, initially. It should be noted here that we can, in principle, compute the composition of DBP as a function of the radial position from a refractive index profile. Because of the rapid evaporation of the droplets involved in the present experiments, temperature gradients exist in the droplets, and the temperature distributions affect the refractive index profiles. To obtain an accurate composition distribution, a correction for the temperature distribution must be incorporated on the basis of the dependence of the individual refractive indices on the temperature.

5. Conclusions

Prior studies have shown that various resonance-based light scattering techniques can be used to determine radii and compositions of homogeneous and layered droplets with high accuracies, and these techniques have been applied to study various processes associated with microdroplets. We have presented a technique for the determination of the size and the radial concentration profile in a microdroplet that is undergoing rapid changes due to some physical or chemical processes. The resonances from a variable composition droplet exhibit behaviors that are distinct from those from uniform composition and layered droplets. When a uniform composition, homogeneous droplet undergoes size and refractive index changes while maintaining uniformity in composition, the separation distances between the resonances remain nearly the same. When the size change is accompanied by the formation of a layer, the shift of a TM mode resonance differs substantially from a TE mode resonance. If the size change induces a variation in the composition distribution of the droplet, lower order resonances shift significantly more than higher order resonances, while maintaining nearly the same separation distance between the TE and TM mode resonances of the same order. The positions of resonances of various orders and modes have been used to discriminate

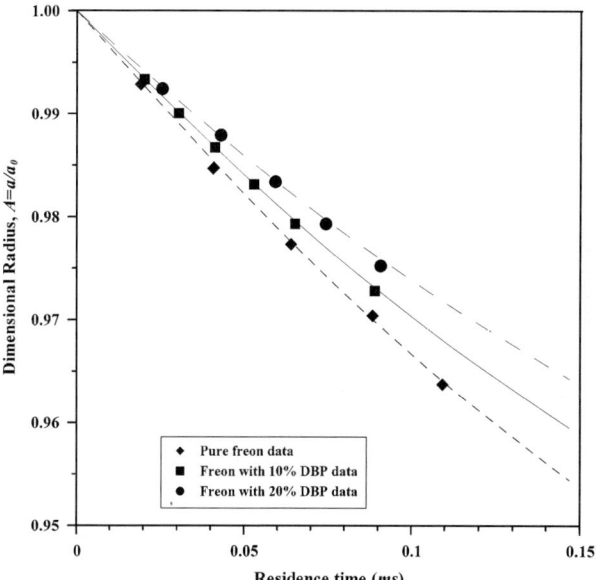

Fig. 8 Droplet radius *vs.* time results obtained from analysis of resonances from pure freon droplets, and from droplets initially containing 10 and 20 vol% DBP in freon.

microdroplets of various characteristics, and provide the basis for the determination of the size and the radial composition distribution inside a microdroplet.

Rapid evaporation of droplets containing a non-volatile solute in a volatile solvent creates non-uniformities in the concentration distributions inside the droplets due to the droplet phase diffusion. The composition variation in such a droplet depends on the evaporation rate and the droplet phase diffusion coefficient. The resonances observed in Raman scattering spectra from evaporating binary droplets of DBP and freon show that various resonances shift differently with the droplet residence time, as predicted by the light scattering as well as by diffusion theory. By aligning the observed resonances with the resonances computed from the theory for given size and composition distribution, we have determined the error in the alignments between the observed and calculated resonances. From the global minimum in the alignment error we have established the optimum droplet size and the associated radial composition distribution for a given set of resonances observed at a particular residence time. The size and composition profiles determined at various times from the resonances agree with the results calculated from an evaporation model.

Acknowledgements

The authors gratefully acknowledge the financial support of the National Science Foundation (grant #CTS-0130778), the Kentucky Science & Engineering Foundation (grant #KSEF-456-RDE-005) and the DOE National Institute for Climatic Change Research (grant #06-SC-NICCR-1066).

References

1 A. K. Ray, B. Devakottai, A. Souyri and J. L. Huckaby, *Langmuir*, 1991, **7**, 525.
2 A. K. Ray and J. L. Huckaby, *Langmuir*, 1993, **9**, 2225.
3 V. Dhariwal, P. G. Hall and A. K. Ray, *J. Aerosol Sci.*, 1993, **24**, 197.
4 J. L. Huckaby and A. K. Ray, *Langmuir*, 1995, **11**, 80.
5 A. K. Ray and S. Venkatraman, *AIChE J.*, 1995, **41**, 938.
6 A. K. Ray and R. Nandakumar, *Appl. Opt.*, 1995, **34**, 7759.

7   H. Tu and A. K. Ray, *Appl. Opt.*, 2001, **40**, 2522.
8   H. Tu and A. K. Ray, *Chem. Eng. Commun.*, 2005, **192**, 474.
9   K.-Y. Li, H. Tu and A. K. Ray, *Langmuir*, 2005, **21**, 3786.
10 H. Tu and A. K. Ray, *Appl. Opt.*, 2006, **45**, 7652.
11 H. Tu and A. K. Ray, *Appl. Phys. Lett.*, 2006, **89**, 131921.
12 V. Devarakonda, A. K. Ray, T. Kaiser and G. Schweiger, *Aerosol Sci. Technol.*, 1998, **28**, 531.
13 R. N. Berglund and B. Y. H. Liu, *Environ. Sci. Technol.*, 1973, **7**, 147.
14 H. B. Lin, J. D. Eversole and A. J. Campillo, *Rev. Sci. Instrum.*, 1990, **61**, 1018.
15 V. Devarakonda and A. K. Ray, *J. Colloid Interface Sci.*, 2000, **221**, 104.
16 V. Devarakonda and A. K. Ray, *J. Aerosol Sci.*, 2003, **34**, 837.
17 H. C. van de Hulst, *Light Scattering by Small Particles*, Dover, New York, 1981.
18 M. Kerker, *The Scattering of Light and Other Electromagnetic Radiation*, Academic Press, New York, 1983.
19 C. F. Bohren and D. R. Huffman, *Absorption and Scattering of Light by Small Particles*, Interscience, New York, 1983.
20 J. D. Eversole, H. B. Lin, A. J. Campillo, P. T. Leung, S. Y. Liu and K. Young, *J. Opt. Soc. Am. B*, 1993, **10**, 1955.
21 H. Chew, P. J. McNulty and M. Kerker, *Phys. Rev.*, 1976, **13**, 396.
22 N. Velesco and G. Schweiger, *Appl. Opt.*, 1999, **38**, 104.
23 J. L. Huckaby, *Elastic and inelastic light scattering by microdroplets*, PhD thesis, University of Kentucky, 1991.
24 S. C. Hill, C. K. Rushforth, R. E. Benner and P. R. Conwell, *Appl. Opt.*, 1985, **9**, 2380.
25 J. L. Huckaby, A. K. Ray and B. Das, *Appl. Opt.*, 1994, **33**, 7112.
26 A. K. Ray, A. Souyri, E. J. Davis and T. M. Allen, *Appl. Opt.*, 1991, **30**, 3974.
27 A. L. Aden and M. Kerker, *J. Appl. Phys.*, 1951, **22**, 1242.
28 H. Tu and A. K. Ray, *J. Aerosol Sci.*, 2000, **31**, S328.
29 P. J. Wyatt, *Phys. Rev.*, 1962, **127**, 1837.
30 J. R. Wait, *Appl. Sci. Res.*, 1963, **10**, 441.
31 A. K. Ray and D. Bhanti, *Appl. Opt.*, 1997, **36**, 2663.

General Discussion

Professor Jaenicke opened the discussion of Professor Chang's paper: Focal surfaces exist in cloud droplets. You can see that in supernumerous rainbows. Do those focal points change the chemical reactions in cloud droplets?

Professor Chang replied: Sorry, but I do not know much about supernumerous rainbows and the connection between that and morphology dependent resonances and whispering gallery modes. Therefore, I cannot answer your question properly.

Professor Ray asked: What is the width of the 3rd order resonance in Fig. 3 of your paper? Our experience shows that resonances of widths less than 10^{-5}, can not be detected in elastic scattering. What is the effect of collecting scattered light over a large angle such as 13° in Fig. 3?

Professor Chang responded: The scattered coefficient must not exceed 1 based on the fact that the scattered intensity can never exceed the input intensity. Thus the area under the peak of the individual WGM is proportional to $\Delta\omega$. When we use a white light source, the scattering of the 3rd order WGMs is not able to be detected. The 4th order peaks have a Q of about 10_5 for spheres with a radius on the order of 10 μm. However, with a New FOCUS tunable diode laser, we were able to see values of Q of 10^6 or 10^7 WGMs associated with the cladding of the communication fiber (125 μm cladding diameter). In our calculations we had included 13° acceptance angle with the observation angle set at 90° relative to the laser beam. The actual solid angle used in the calculation is equal to the angles set in the experiment.

Dr Reid asked: Firstly, as a comment on the possible importance of the distribution of the internal field on the properties of cloud droplets made by Prof. Jaenicke, Jacobson has examined the impact of the mixing state on the radiative forcing of black carbon in the atmosphere, highlighting the importance of considering black carbon as an internal mixture.[1] Further, Chylek *et al.* have considered the impact of the spatial distribution of components in the internal mixture on the absorption of light by mixed component aerosol, demonstrating that increased absorption can be expected when the black carbon inclusion is at the top or close to the bottom of the droplet due to the strong focusing effect.[2] Secondly, as a question, what potential is there for determining composition from your large scale scattering measurements through a determination of refractive index?

1 M. Z. Jacobson, *Nature*, 2001, **409**, 695–697.
2 P. Chylek, G. B. Lesins, G. Videen, J. G. D. Wong, R. G. Pinnick, D. Ngo and J. D. Klett, *J. Geophys. Res.*, 1996, **D18**, 23365–23371.

Professor Chang answered: Thank you for making me aware of the M. Z. Jacobson article. You and I both know how quickly complications can set in when the index of refraction is no longer homogeneous. We have been attempting to extract the composition of the spherical droplet by using the index of refraction as a variable. The index of refraction is known to have a rather profound effect on the scattering intensity *versus* wavelength measurements. In particular, the intensity ratio of the peaks to valleys is sensitive to the index of refraction. How we relate the deduced index of refraction to the chemical composition is not so specific. Simultaneous measurements of the elastic scattering over a large angular range (LATAOS) have been performed by us at Yale in cooperation with the Lincoln Lab Group by using two wavelengths, one absorbing and the other non-absorbing (both in the mid-IR range). The differences between the absorbing and the non-absorbing in the TAOS patterns were quite dramatic. This was reported in *Opt. Lett.* a few years ago.[1]

1 Kevin B. Aptowicz, Yong-Le Pan, Richard K. Chang, Ronald G. Pinnick, Steven C. Hill, Richard L. Tober, Anish Goyal, Thomas Jeys, and Burt Bronk, Two-Dimensional Angular Optical Scattering Patterns of Microdroplets in the Mid Infrared with Strong and Weak Absorption, *Opt. Lett.*, 2004, **29**, 1965.

Professor Rühl commented: Can you explain what factors affect the huge enrichment factors in the rapid aerodynamic sorting system you mentioned in your presentation? Is this explained by the scattering geometry or the conditions the puffer is operated in?

Professor Chang responded: The factors that affect the huge enrichment factor in our fluorescence spectrum system are as follows:

(1) the entrance nozzle must be carefully designed in order that the microparticles (in the 1–10 μm range) are tightly focused for a downstream distance of 1.5 cm or more;

(2) the number of particles per second that transit through the trigger volume must not exceed 90 000 which is set by the onboard digital processor of the fluorescence spectrum gathering apparatus; and

(3) the on-demand puffing rate cannot exceed 1000 particles s^{-1}. The high enrichment factor is higher in the lower particle concentration, for example, at a few 10^3 particles s^{-1}. In the particle stream, there will be ample distance between the preceding and the following particles for the air puff to push the "wanted particle" to the direction where the collection substrate is situated. After collecting 10^4 to 10^5 deflected particles on a glass slide, we then looked at the magnified image to count how many of the unwanted particles were unintentionally deflected and then collected. This is how we determined the enrichment factor.

Professor Jaenicke said: For atmospheric aerosols one most probably needs an enrichment factor of 10^6. Could this be achieved?

Professor Chang answered: The probability of attaining an enrichment factor of 10^6 or greater needs to be explored further. In my answer to Professor Rühl's question, there are three items in our experiments that affect the enrichment factor. Each of these three items could be optimized with the help of fluid mechanics computation. Thus, in my judgment, it is possible to attain an enrichment factor of 10^6 or greater. There are some developments on the way for electrostatic deflection which thereby achieve enrichment in a totally different manner. It is not obvious whether the aerodynamic method (puffer) or the electrostatic method is better for enrichment.

Dr Hudson opened the discussion of the Professor Wolf's paper: This paper has described a method of discriminating between the presence of biological molecules and polycyclic aromatic hydrocarbons in an aerosol using a novel fluorescence measurement that is sensitive to the differences in the excited-state dynamics. A related strategy is used in the biological sciences to identify fluorescent groups (such as tryptophan), and obtain some information on the environment, by measuring multi-exponential decay profiles for the fluorescence using time-correlated single-photon counting.[1] It is appreciated that this technique would not be appropriate for applications in remote sensing (the main objective in the paper). Nevertheless, could it be adapted for laboratory studies of droplet streams or local measurements in air?

1 J. R. Lakowicz, *Principles of Fluorescence Spectroscopy*, 3rd edn, Springer, New York, 2006.

Professor Wolf answered: Lifetime measurements provide information about the radiative part of relaxation, as, for example fluorescence between two electronic potential sheets. However, coherence between quantum states and information about vibronic dynamics on the femtosecond timescale are lost. Lifetime can therefore provide interesting discrimination information for *in-situ* studies, but will not have the unique discrimination potential of coherent control.

Professor Signorell asked: Is it possible to improve depletion by using a UV/VUV laser as λ_{probe} (Fig. 3 in your paper)? If so, how can the improvement of depletion be explained?

Professor Wolf answered: Yes, it can. A main relaxation process that decreases the excited state population is ionization by the second laser pulse. It is well known that the ionization cross-section of aromatic molecules in solution increases significantly with the photon energy.

Dr Davies asked: I have a general question about the relationship of the PPD technique to stimulated emission pumping. Presumably you have to avoid SEP to make the PPD technique work? In Fig. 9 of your paper, can you give an idea of the number and dimensions of the elements in the mirror array?

Professor Wolf replied: As the second pulse wavelength (800 nm) is significantly longer than the first (270 or 400 nm), SEP is not in competition with excited state absorption (ESA). The dimensions of each mirror element is 40×40 microns.

Professor Rühl said: You mentioned in the discussion that pulse shaping further enhances the selectivity to identify biological species by using short pulse laser pump–probe schemes. The field of pulse shaping has developed over the last few years from an empirical pulse design to a rational pulse shaping, as shown for small systems from a combination of experiments and theory. In this work pump–dump-sequences play a crucial role. Is a similarly rational pulse design applicable to enhance the selectivity to probe biological species that are bound in micro-particles?

Professor Wolf answered: Intrapulse excitation schemes involving pump–dump sequences is an attractive approach, which we intend to follow up. Open-loop methods, involving theoretical predictions of the dynamics within the potential hypersurfaces may find their limits in these large molecular systems.

Professor Kaye asked: One of the key 'desirables' on the wish list of defence agencies is a bio-aerosol detection method which is capable of differentiating viable from non-viable organisms. Does ultrafast depletion spectroscopy hold any promise for achieving this in the future?

Professor Wolf responded: This is a question that we are currently exploring, with the help of microbiologists.

Professor Rudich remarked: What is the typical coherence times in the environmental system in which the chromophore is embedded in a very similar environment (lipids, proteins *etc.*)?

Professor Wolf answered: Typical decoherence times in these systems range from 10 fs to 1 ps, depending on the excited modes (and their respective coupling to the environment). Intrapulse excitation and stimulated Raman schemes are therefore very attractive in order to increase specificity, when the laser pulse duration is an issue.

Professor Arnold stated: Identifying tryptophan is not identifying a protein.

Professor Wolf responded: Right. This is why we have developed coherent control schemes instead of pump–probe approaches.

Professor Jaenicke communicated: What was the particle concentration in your artificial cloud? The atmospheric aerosol consists of 25% of biological particles (by mass and by number). Do you think that you could detect specific microorganisms in that matrix?

Professor Wolf communicated in reply: In our recent paper,[1] we showed that the detection limit in the cloud was of the order of 10 particles cm^{-3} with a depth resolution of 10 m (typ. release). For larger volumes and depths, the detection limit decreases accordingly.

The selectivity in the widespread organic and biological aerosols is difficult to predict. We demonstrated that it was possible to discriminate organic particles emitted by traffic from bioaerosols. Our next step will be to discriminate pollens from bacteria.

1   G.Méjean J. Kasparian, J. Yu, S. Frey, E. Salmon and J. P. Wolf, *Appl. Phys. B*, 2004, **78**, 535–537.

Professor Arnold opened the discussion of Professor Signorell's paper by: To what extent are your measurements reliant on uniformity of particle shape and monodispersity in size? Do you need to account for these in your comparison of experimental measurements and computation?

Professor Signorell replied: In the size range between about 10 and 100 nm, the band structure in the infrared spectra is almost independent of the size of the particles (The absorbance simply scales with the total volume of the particles.) Therefore uniformity in size is not so important here. A certain degree of uniformity in shape, however, is important. If the shape distribution is too broad it is almost impossible to assign the shape unambiguously. In principle the same argument also holds for larger particles, although elastic scattering complicates the situation. For particles with sizes below 10 nm the size and shape uniformity are very important. Here different sizes also lead to different band structures.

Professor Davidovits remarked: What is the range of particle size that the technique can detect the particle morphology dependance? Specifically on the large end, how is one to separate the spectrum from scattering?

Professor Signorell answered: On the experimental side there is "no" limit. Particles with sizes from subnanometers up to microns and more can be detected. Direct simulations/predictions with the exciton model can be made for particles up to 50 nm. For larger particles scattering is important in the infrared region. This cannot be calculated directly with the exciton model. However, the exciton model can be used to determine refractive index data. This data can later be used as input data for calculations with classical scattering theory (Mie *etc.*) which accounts for the scattering.

Dr Reid asked: To what extent does polydispersity compromise the distinctiveness of the spectral fingerprint and your ability to interpret it?

Professor Signorell responded: As long as the size distributions are not too broad (geometric standard deviations less than 2) polydispersity is not important for the size range under investigation in this contribution. For our measurements the geometric standard deviation lies clearly below a value of 2 and the mean size of our particles lies in the region between about 10 and 100 nm where the spectral features (band shapes) are independent of the size. In this size range the spectral features are sensitive to the particles shape only. We can thus obtain information on the particle shape from the spectral features.

Professor Ray asked: What are the differences between particles generated by premixed SO_2 : CO_2 gas mixtures, and those from twin pulses of SO_2 and CO_2? Can you give some idea about growth mechanisms of particles by these two methods?

Professor Signorell responded: Our expectation was that premixed SO_2 : CO_2 gas mixtures would form statistically mixed (mixed on a molecular level) particles. But the experiment shows that a core of SO_2 is formed and CO_2 forms a layer on this core. Our explanation is that SO_2 condenses first because the melting point of SO_2 is 70 K higher than the sublimation point of CO_2. In addition the intermolecular interactions of SO_2 and CO_2 are expected to be less favorable than for the pure substances (thus no mixing on a molecular level). In the twin pulse experiments substance A is introduced first and forms particles. Substance B is introduced with a time delay after particles A have been formed. Substance B thus forms a layer on particles A.

Mr Hunt commented: The lower the temperature the more amorphous the ice crystals are. Does the crystal shape depend upon the temperature or upon the temperature gradient? If it is dependant upon the temperature gradient, can the gradient be changed for a given final temperature, and if so, how would this affect the morphologies produced?

Professor Signorell replied: In general the crystal structure can depend both on the temperature gradient and on the absolute temperature. Comparison with bulk phase experiments for SO_2 show, however, that in the case of SO_2 it is the absolute temperature that determines the crystal structure. It is in principle possible to change the gradient. But for SO_2 we do not expect a major change.

Professor Mason asked:
(1) In Fig. 5 of your paper you show a spectrum of mixed CO_2 : SO_2 aerosol from a 1 : 1 premixture: Can you determine the composition of the aerosol itself; is it also 1 : 1 or time dependant?

(2) In your experiments you are cooling rapidly such that you get separate 'condensation' *e.g.* SO_2 then CO_2. In 'real' astrochemistry, molecules are deposited on a precooled [10 K] surface. We see similar IR spectra to your particulate spectra but on the surface *i.e.* we are forming small particles on the surface (*e.g.* for NH_3 see Davies *et al.*[1]). It would be interesting to do your experiments under slower cooling conditions and at lower temperatures. *And* use your data to seek structural information on these surface structures.

1 A. Dawes, R. J. Mukerji, M. P. Davis, P. D. Holtom, S. M. Webb, B. Sivaraman, S. V. Hoffman, D. A. Shaw and N. J. Mason, *J. Phys. Chem.*, 2007, **126**(24), 244711.

Professor Signorell responded:
(1) With the help of exciton calculations the concentration can be determined from the spectrum directly. The concentration is not time-dependent.

(2) We can do our experiment at temperatures down to 4 K. This is planned and is certainly interesting. The cooling conditions can, however, not be changed very

significantly compared with conditions in space. Because SO_2 and CO_2 do not like to mix on a molecular level we do not expect to find major differences compared with our results. We expect that SO_2 and CO_2 would separate with time.

Professor Rudich asked: How sensitive are the line shapes to the dimension of the core/shell? What "handle" do you have on controlling the thickness of the core or shell?

Professor Signorell replied: The band shapes in the infrared spectra are very sensitive to the dimensions of the core and shell. Different core/shell ratios lead to different band shapes. A discussion (experiments and modeling) of this aspect is given in Signorell et al..[1] The thickness of the core or shell is controlled by the concentration of the gas mixtures used.

1 R. Signorell, M. Jetzki, M. Kunzmann and R. Ueberschaer, *J. Phys. Chem. A*, 2006, **110**, 2890.

Professor Kersten asked: What would happen if you insert a third partner into the cooling system—*e.g.* sulfur or carbon particles as in Io's volcanoes? Would there be condensation of CO_2 and SO_2 onto these particles—or would there be the same CO_2/SO_2 shell condensation?

Professor Signorell replied: We would expect that first SO_2 condenses on these carbon particles and then CO_2. We do not expect that they would co-condense and form molecularly mixed layers on the carbon particles.

Professor Ray commented: What happens if you first introduce CO_2 and cool down to the particle formation temperature, and then introduce SO_2 gas?

Professor Signorell replied: We have actually performed the experiment. If the temperature is cold enough (below sublimation point of CO_2) CO_2 particles form. As soon as SO_2 is introduced it forms a layer on the CO_2 particles.

Dr Grothe remarked: What is the phase composition of the particles? Did you consider lattice mode vibrations?

Professor Signorell replied: The mid-infrared spectra show that the CO_2 particles are crystalline (see for example characteristic band splitting of the bending vibration and exciton calculations) and the SO_2 particles form a partially amorphous structure at higher temperatures and an amorphous structure below 10 K (see spectra and exciton calculations). We did not consider lattice modes because our light source and detector cannot cover the far infrared region. Such measurements, however, would be very interesting.

Mr Tuckermann asked: As well known from other examples *e.g.* CO_2, H_2O or NH_3, the size, the shape and the state of the nanoparticles generated in a collisional cooling cell depends strongly on the bath gas temperature and pressure, the concentration of the sample gas and the time evolution of the particles. In which range have you changed these four parameters? And do you always get the same structure of the SO_2 band, so that you always have amorphous SO_2?

Professor Signorell replied: This question refers to our previous investigations on similar particulate systems, where we have found a pronounced dependence of the particle properties on various experimental parameters (see Firanescu et al.[1]). The ranges in which the parameters have been varied can be found in the experimental section of the paper (bath gas temperature 4–80 K, bath gas pressure: 100 mbar–

1bar, sample concentration: some ppm to several ten thousands of ppm). For different bath gas pressures and sample gas concentrations we got the same structure for the SO_2 bands. At temperatures above 10 K the SO_2 particles are only partially amorphous and below this temperature they are amorphous.

1 George Firanescu, Dana Hermsdorf, Roman Ueberschaer and Ruth Signorell, *Phys. Chem. Chem. Phys.*, 2006, **8**, 4149.

Dr Hudson opened the discussion of Professor Arnold's paper: The cross sensitivity of antibodies to different antigens can restrict the use of certain types of biosensor to the screening of samples in clinical chemistry (as opposed to confirmatory testing). Also, the sensitivity of antibodies to pH and temperature, as well as the necessary conditions for storage to prevent denaturing of the protein, can also have an impact. Could the authors comment on the breadth of potential applications for the proposed microfluidic device?

Professor Arnold replied: I have nothing to add.

Professor Chang asked: I am fully aware of why you must excite only the very high Q modes in the equatorial plane of the slightly deformed sphere at the end of the fiber. I am also aware of the walk-off problems with the incident beam not being perpendicular to the fiber axis. However, this is the same problem that one has in not exciting at the equatorial plane, which will cause inhomogeneous broadening.

Why not consider the possible use of the same communication fiber with illumination perpendicular to the fiber axis? By using a cylindrical lens geometry one can illuminate 2 cm along the length of the fiber with a tunable diode laser. Additionally, by using a linear array detector, one can measure simultaneously the scattering intensity with 512 pixels with a spatial resolution of 2/512 cm. The spectral resolution is presently limited to 0.0001 nm at 675 nm. The Q that we have observed is 10^6 which is 10 times worse than our spectral resolution. Higher Q may be reached by proper cleaning of the sidewalls and fire polishing the surface to achieve a smoother surface. Using a 2 cm length fiber that is coated with different concentric discrete bands of different antibodies one may achieve the objective of seeing specific attachment to one of the bands of an antigen. Thereby one can obtain a discrete shift of WGM peak along the length of the fiber axis.

Professor Arnold answered: I think this is a very interesting idea.

Professor Ray asked:
(a) How good is the assumption that the field at particular bio-particles is not influenced by the contribution from its neighbours?
(b) Why does eqn (25) not show any effect on the mode of order number of the resonance?
(c) Prior experiments and calculations show that the shift of a resonance is sensitive to its identity: How did you find Q in Fig. 8!

Professor Arnold replied:
(a) This question has been investigated in one of our recent theoretical papers.[1] We find that the measured shift should be perturbed by less than 6% as the surface coverage approaches the maximum allowed by Random Sequential Adsorption, due to nearest neighbor dipole–dipole interactions.

1 I. Teraoka and S. Arnold, *J. Appl. Phys.*, 2007, **101**, 023505.

(b) This interesting property is a consequence of the 1st order expansion of the integral in eqn (19), which is independent of mode order. Eqn (19) first appeared in ref. 14. It works well for a monolayer of Rayleigh particles. In our experiments with protein and DNA monolayers we find that resonance dips shift by the same fraction in wavelength independent of mode order.

(c) The Q of 9×10^5 in Fig. 8 is based on measuring the linewidth of the resonance and dividing it into the resonance wavelength.

Dr Davies asked:
(a) Could you explain the meaning of eqn (21)? It is not clear how the LOD relates to the right hand side of eqn (21).

(b) Can you explain what "Random Sequential Adsorption" involves? Is it a form of Langmuir–Blodgett deposition?

Professor Arnold answered:
(a) N_{LOD} is the least number of adsorbed bio-particles at the equator of the microsphere that are detectable. It is obtained by summing the individual shifts in eqn (20) and setting this sum equal to $|\Delta\omega|_{min}$ in eqn (1).

(b) Random Sequential Adsorption corresponds to a sequence of adsorption events caused by spheres raining down on a surface with spatial randomness. The Random Sequential Adsorption Limit corresponds to the largest density of spheres that can fill a monolayer in this fashion without geometrical hindrance. Nearest neighbor interactions are ignored and surface diffusion is not allowed. It is not a form of Langmuir–Blodgett deposition.

Dr Reid asked: Early in the paper you establish that non-specific binding can allow the identification of a specific virus due to the relationship between the wavelength shift and molecular weight/size. What prospects are there for this being a viable strategy avoiding the need to use specific binding?

Professor Arnold responded: I cannot find a place in the paper in which we "establish that non-specific binding can allow the identification of a specific virus . . . " Although the wavelength shift can be used to estimate the molecular weight of the adsorbing species, this is not specific sensing, since bio-particles having similar molecular weights often show considerably different biological activity.

Dr Krieger asked: In Fig. 6 and Fig. 7 of your paper you show the exceptional sensitivity of the method being less than a single HIV virus particle. However, what is the limitation of the device in terms of experimental noise induced by the fluctuations in the number of adsorbing and desorbing molecules while the laser scans the transmission spectrum?

Professor Arnold replied: The most reasonable way to see single events would be to lower the concentration in solution in order to insure that the time between event is larger than the sampling time. A complication, not mentioned in the paper, is that the extension of the evanescent field allows the detection of Brownian particles in solution, however the temporal behavior of the resonance wavelength fluctuations for Brownian particles is sufficiently different from that of an adsorption or desorption events for discrimination to be possible.

Professor Signorell asked:
(a) Is it *a priori* obvious that the mutual interaction between adsorbed species can be neglected?

(b) Concerning the quantitative analysis of the dip traces: You use the values in solution for the polarizabilities of the adsorbate to estimate coverage from absolute wavelength shifts. Is it questionable that the adsorbed species has the same properties (polarizability) as the free species in solution?

Professor Arnold replied:

(a) No. However, this question has been investigated in one of our recent theoretical papers.[1] We find that the measured shift should be perturbed by less than 6% as the surface coverage approaches the maximum allowed by Random Sequential Adsorption, due to nearest neighbor dipole–dipole interactions.

1   I. Teraoka and S. Arnold, *J. Appl. Phys.*, 2007, **101**, 023505.

(b) It is questionable. Some bio-particle distortion is expected to occur due to the surface bio-particle interaction. A more easily detected effect seems to be that nonspherical bioparticles take specific orientations at a surface. We have recently been able to identify specific orientations by using TE/TM polarization discrimination in the measured shifts (see ref. 18 of our paper).

Mr Taylor asked: The discussion in your paper assumes a uniform shell of virus particles on the sphere. Given the high Q of the resonance, will an inhomogeneous coating (and in the extreme case, the presence of a single virus particle) broaden the resonance or affect the magnitude of the shift seen?

For example, even if the virus can be considered a point particle, it could be positioned in a very different environment depending on whether it is at a resonance node or antinode [the resonant field distribution should, I think, be stable over long periods of time shouldn't it?]

Professor Arnold answered: The resonance excitation by a guided wave generates a traveling wave resonator. So in theory there are no nodes or antinodes. However, a particle adsorbed on the surface allows energy to be coupled into a counter propagating mode that will lead to variations in intensity along the equator, and mode splitting. For a typical virus sized particle and $Q_s \sim 10^6$ this effect is small with the broadening far less than a linewidth. However, for much larger $Q_s \sim 10^8$ I suspect that this effect can be important.

Professor Rühl asked: Spatial localization of the antibodies mean the equation seems to be a current limitation. It is briefly mentioned in the paper that this limitation can be overcome. Can you elucidate on this point?

Professor Arnold replied: I have actually mentioned how this limitation may possibly be overcome in the conclusion section. Beyond this I have nothing to add at this time.

Professor Ray asked: Do you see any effect of non-uniform distribution of bio-particles on the surface of the silica particle? Prior experiments on nanoparticle seeding on microdroplets show that resonances are distorted by the deposition on the surface.

Professor Arnold answered: Yes. For a $Q \sim 10^6$ *E. coli* bacterial cells adsorbed on the surface of a 150 µm radius silica sphere produce a considerable change in the shape of a resonant dip (principally broadening) due to scattering of the WGM energy by the cell. *E. coli* is rod shaped and more than an order of magnitude larger than a virus. For a virus resonance broadening of this sort would be expected to appear at considerably higher Qs, since the cross section for scattering of a virus is much smaller than that of a bacterial cell.

Professor Chang stated: I am sure you are aware of the fact that silver and gold colloids have been coated on spherical surfaces such as the ones you use. This coating will give the Raman and all of the fluorescence emission an additional boost in the enhancement. Have you heard of any additional field intensity enhancement when the laser is tuned to the plasmonic frequency of the coated spherical surfaces?

Professor Arnold replied: Yes. I. M. White and X. Fan have reported[1] just such an additional SERS enhancement.

1 I. M. White and X. Fan, Demonstration of composite microsphere cavity and surface enhanced raman spectroscopy for improved sensitivity, Chemical and Biological Sensors for Industrial and Environmental Security, *Proc. SPIE*, 2005, **5994**, 59940G.

Dr Davies opened the discussion of Professor Ray's paper: How can we be confident that the peaks in Fig. 5 of your paper can be tracked through the diagram? There seems to be multiple opportunities for mis-assignment.

Professor Ray replied: We are confident about the peaks in Fig. 5. We input the peak frequencies at each generator position, and the peaks are numerically tracked by the program. The appearance (*i.e.*, shapes) of the peaks may distort somewhat due to the small fluctuations of the droplets, but the various peaks appear sequentially in each of the spectra.

Professor Signorell remarked: Concerning Fig. 5: I wonder how it is possible to unambiguously assign specific resonances in these spectra? Some assignments (lines) do not even coincide with a peak (*i.e.* C in the top trace). How high is the noise level in these spectra? I suggest including a calculated spectrum in Fig. 5 (for at least one of the experimental spectra) for comparison.

Professor Ray answered: The lines in Fig. 5 are a visual aid and have nothing to do with the peak positions. Each line is used as a reference position to show how the same peak shifts by different amounts at different positions. Fig. 5 contains Raman scattering spectra, and we cannot theoretically compute Raman scattering spectra. Input and output resonances in Raman scattering can, however, be calculated from elastic scattering theory. In addition, an experimental elastic scattering spectrum can be compared with theory (see reference no. 12 of our paper). Please also see my earlier response.

Dr Krieger remarked: How do you actually do the initial mode assignment if you have a gradient in the refractive index?

Professor Ray responded: For a small variation in the refractive index, we can align the observed resonances with theoretical resonances assuming constant refractive index. The process will generate high alignment error, but will yield correct mode assignments for the observed resonances. We can then improve the alignment error by considering an appropriate variation in the refractive index.

Professor Bain remarked: It seems to me that there is only one experimental unknown—the evaporation rate. Provided that you know the heat capacities, thermal conductivities and diffusion coefficients as a function of temperature, then you can calculate the refractive index profile as a function of time. How well do these computed profiles compare with those deduced from the positions of the resonances?

Professor Ray answered: We can, in principle, calculate the composition profile from the gas and liquid phase properties, and vapor liquid equilibrium data (*i.e.*, thermodynamics). For the droplet phase, we need to know the properties as functions of composition and temperature. All these create a lot of uncertainties in the computed results. The experimental results on the refractive index profile agree well with the predictions from the droplet evaporation model.

Professor Signorell commented: Concerning the polynomial to describe the refractive index in eqn (4) of your paper: How many terms are necessary to achieve good results? Is there any physical meaning behind this polynomial?

Professor Ray responded:
There is no physical meaning associated with the polynomial. Any functional form can be assumed for the variation of the refractive index with radial position, but a polynomial form simplifies the numerical analysis. We have used a 4th order, but a 3rd order polynomial is sufficient for most cases.

Dr Reid asked: How sensitive are your measurements to the refractive index at the centre of the droplet given that even the high mode orders are spatially distributed near/towards the surface?

Professor Ray replied: The resonance-based method is sensitive to the refractive index at the center. Low order resonances provide information about the region near the surface, while higher order about the interior of a droplet.

Professor Chang asked: In your estimation of the evaporation rate of a given droplet, do you take into consideration the wake effect caused by the preceding evaporating droplets?

Professor Ray answered: In our experiments with a linear array of droplets, because of the low separation distance between the droplets, the evaporation rate of a droplet is retarded by the inter-particle interactions. These interactions can reduce the evaporation rate by 65% compared to an isolated droplet (see ref. 16 of our paper).

Dr Krieger commented: How do you deal with the problem of partial pressure (of Freon) saturation in the gas phase due to the continuous evaporation of Freon?

Professor Ray answered: We circulate dry air through the chamber at a flow rate that prevents partial saturation of the environment surrounding the droplet. We measure evaporation rates at varying flow rates, and establish the minimum flow rate from the rate above which evaporation rate does not change. We operate the system 50% above the minimum flow rate.

Professor Davidovits commented: Your experiments are performed at atmospheric pressure: Why not reduce the pressure and reduce problems associated with transport limitations?

Professor Ray replied: We can enhance the gas phase mass transfer rate by lowering the ambient pressure, but the present evaporation problem involves droplet phase transport limitations that create solute concentration distributions inside a droplet.

Dr Huthwelker said: For your experiments and the evaluation using Mie-Theory, it is most critical whether the droplets are perfectly round. How did you know whether your droplets were round? Can you give error estimates on the sphericities?

Professor Ray replied: Experiments of the present study involve 20 μm size droplets. Because of the surface tension forces it is almost impossible to distort the spherical shape of a droplet whose radius is less than 50 μm.

Professor Davidovits remarked: I notice that your work is sponsored by the DOE Climate Change Program. Do you see a direct connection between the work presented and atmospheric processes?

Professor Ray replied: The resonance-based light scattering techniques can be used to study fundamental processes associated with microdroplets, and some of these processes are relevant in the atmosphere. The experiments are, however, only suitable for laboratory studies.

Professor Shiratani opened the discussion of Professor Stoffels' paper: Are there any good ways to discriminate between the effects of negative ion formation and those of nanoparticles on the decrease in electron density just after turning on the plasma?

Professor Stoffels responded: Both negative ions and nanoparticles will decrease the electron density. Since nanoparticle formation is linked with negative ions, they will always both be present during nanoparticle formation. Nanoparticles might be slightly less electronegative than negative ions. Thus during the particle formation process, the electron density might decrease as negative ions are formed, then slightly increase due to the reduced electronegativity of nanoparticles, before decreasing again when particles become positively charged. A careful analysis of the temporal electron density behavior during particle formation might reveal this effect.

Professor Shiratani commented: You may discriminate nanoparticles from negative ions using photodetachment, because ionization potential depends on the size of the nanoparticle and the kind of negative ion.

Professor Stoffels responded: That is correct, however the differences become very small for larger molecules and nanoparticles. Alternatively mass spectrometry can be used to determine the presence of larger molecules and/or nanoparticles.

Professor Goree said: You identified four states of growth: radicals, nanospheres, agglomerates of nanospheres, and agglomerates that are overgrown with deposited material. Between these four states are three transitions, with different mechanisms. You also mentioned that higher gas temperatures result in slower growth to the final stage. Can you speculate which transition is slowed the most by a higher temperature?

Professor Stoffels responded: From the measurements of the electron density, presented in the manuscript, it is clear that the temperature has an impact on the plasma and presumably the particle formation process from a very early stage. Measurements of particle size and density,[1] suggest however that the main obstacle occurs in the coagulation phase. Coagulation is significantly delayed with increasing temperature. For temperatures above 200 C, coagulation doesn't occur at all.

1 L. Boufendi, J. Hermann, A. Bouchoule, B. Dubreuil, E. Stoffels, W. W. Stoffels and M. L. de Giorgi, *J. Appl. Phys.*, 1994, **76**, 148.

Dr Reid remarked: Can you clarify the factors that govern the rate of coalescence and formation of larger clusters in step (b) of your mechanism?

Professor Stoffels replied: Coalescence is not fully understood. The charge on nanoparticles on average is negative [see the paper by Shiratani[1]], but it is only 0.1 electron. Moreover, the charge fluctuates rapidly. Therefore one and the same particle can be negative, neutral and even positive within a short time period. During coagulation, particles can therefore interact by attractive Coulomb and dipole forces resulting in a fast coagulation. Once particles reach a certain size (about 20 nm) they become permanently negatively charged and do not coagulate further. These particles will however continue to absorb any smaller particles. This explains why the particles have a very small size distribution during the subsequent growth: newly formed nanoparticles are absorbed by the existing larger particles and do not grow themselves.

1   Masaharu Shiratani, Kazunori Koga, Shinya Iwashita and Syota Nunomura, *Faraday Discuss.*, 2008, **137**, DOI: 10.1039/b704910b

Professor Goree remarked: Following up on a previous question regarding coagulation of charged particles, do you think that your 1–10 nm size nanospheres have a charge that fluctuates sometimes between positive and negative values, so that there are sometimes both positive and negative particles of this size present?

Professor Stoffels answered: I believe there will be a charge distribution. At any time most nanoparticles are neutral some will be negative and a few will be positive. Charging and decharging are rapid processes, so the charging kinetics of a single particle will significantly alter its kinetic behavior. Therefore we cannot simply look at the average charge value when we want to understand the particle kinetics in general and nanoparticle coagulation in particular.

Dr Reid asked: In your paper, well-ordered arrays of microparticles are observed. In the size regime you are examining, such ordering through interparticle interactions is not apparent. What governs this transition from collective ordering to inhomogeneous dispersion? Is it governed by particle density/size?

Professor Stoffels answered: Ordering of particles in Coulomb structures can occur when there is a strong particle–particle interaction. In particular when the potential energy of the electrostatic coupling is larger than the kinetic energy of the particles. The Coulomb coupling parameter Γ is defined as the ratio between the potential electrostatic energy and the kinetic energy of the particles. When $\Gamma \geq 2$ a gas to liquid transition occurs. For $\Gamma \geq 170$ a solid crystal like structure is formed. The Coulomb interaction can be increased by a higher particle density as it reduces the interparticle distance and by using larger particles as the particle charge scales roughly with the particle size.

Professor Kersten remarked: Can the temperature-dependent behaviour (see Fig. 12 of your paper) be explained by adsorption–desorption equilibria or by chemical reactions (Arrhenius)?

Professor Stoffels replied: There are several models explaining the temperature delay in particle formation. Perrin[1] claims that the temperature dependent electron attachment rate to Si_xH_y is important. Using an extensive chemical model Bhandarkar *et al.*[2] show that also increased diffusion is important in reducing the particle formation.

1   Jerome Perrin, in *Dusty Plasmas: Physics, Chemistry and Technological Impacts in Plasma Processing*, ed. by A. Bouchoule, Wiley, New York, 1999
2   Upendra Bhandarkar, Uwe Kortshagen and Steven L Girshick, *J. Phys. D: Appl. Phys.*, 2003, **36**, 1399–1408.

Professor Kersten remarked: Is there any chance of influencing the charge of the small particles in the early growth stage by using additional UV irradiation?

Professor Stoffels replied: UV radiation will cause photoelectrons to detach from the particles. Thus the particles will become—on average—slightly more positive. The changed particle charge distribution will also affect the particle growth. Positive particles are more likely to escape the plasma, which might reduce particle formation. On the other hand agglomeration might be enhanced as Coulomb–Coulomb interaction increases and the formation of permanently negatively charged particles is delayed. Apart from changing the particle charge, also the plasma is affected since a reduced particle charge results in a higher electron density. In conclusion, it is a very interesting proposal and it is not *a priori* clear what the effects will be.

Professor Mason remarked: Can you alter polarity in your discharge to modify anion/nanoparticle abilities to gauge rates of (small) molecular anions to larger clusters?

Professor Stoffels responded: The particles are produced in a radiofrequency (13 MHz) plasma. Thus the polarity changes every 70 ns. The electrons can follow these changes. Ions and particles are too heavy, so they can only follow time averaged fields. Since a plasma is typically positive with respect to its environment, this provides the trapping force of the negatively charged particles. It is possible to add a DC voltage to change the plasma potential. Howling *et al.*[1] have done this to extract negatively charged particles and ions to a mass spectrometer to study particle formation channels. Kersten *et al*[2] have introduced a segmented electrode, where the plasma potential can be locally changed. So far it has only been used to move larger particles into desired plasma regions, however the system could also be used to influence particle formation processes on a local scale.

1 A. A. Howling, L. Sansonnens, J.-L. Dorier, Ch. Hollenstein, *J. Phys D. Appl. Phys.*, 1993, **26**, 1003–1006.
2 Gabriele Thieme, Ralf Basner, Ruben Wiese and Holger Kersten, *Faraday Discuss.*, 2008, **137**, DOI: 10.1039/b703733n

Professor Mason asked: What is the effect of flow rate on the nanoparticle formation?

Professor Stoffels replied: The particle formation itself is not changed by flow, since it is on a different time scale. The subsequent growth is affected since it is largely limited by the available source gas (SiH_4). Increased flow therefore results in an increased growth.[1] In addition particles can move relatively freely within the plasma glow, so additional flow will drag them towards the plasma boundary.

1 W. W. Stoffels, E. Stoffels, G. M. W. Kroesen, and F. J. de Hoog, *J. Appl. Phys.*, 1995, **78**, 4867.

Professor Mason said: In interstellar medium, spectroscopic studies (*e.g.* Diffuse interstellar bands) suggesting PAH's may be *negatively* charged (C_6H^-, C_8H^-, *etc.*) have been observed recently. This may be due to photoelectron or cosmic ray secondary electron formation.

Professor Goree stated: As a comment, I'll mention that in many space plasmas the density of free electrons and ions is low, so that they contribute little to the charging of particles, whereas the flux of UV radiation can be high when a star is near, so that photoemission is often the dominant mechanism to charge a solid particle.

Professor Arnold said: How difficult is it to measure Thomson scattering from electrons in the plasma? This could be important in connection to Astrophysics.

Professor Stoffels responded: Thomson scattering is used as a diagnostics to determine the electron density and temperature in laboratory plasmas. The technique and all relevant cross-sections are well known. However, measuring Thomson scattering in a plasma is extremely difficult, due to the limited plasma size. Typically only 1 in every 10^{15} photons is scattered and these photons have to be distinguished from regular plasma emission and Rayleigh scattering. To obtain any signal strong lasers are used, which in turn cause photoionization or even breakdown. Thus, unless extreme care is taken, the measurements are performed on a laser induced discharge rather than the intended laboratory plasma. In Astrophysics objects are larger and therefore more scattering occurs as light travels through.

Dr Davies communicated: With reference to Fig. 5 of your paper: In most molecular plasmas at low pressure, cations seem more dominant than anions, based on *in situ* IR spectroscopy. What is special about silane discharges that seems to favour anions? Is it connected with ion–molecule reactions, electron affinity of SiH_4 or what?

Professor Stoffels communicated in reply: Plasmas contain electrons and they are assumed to be quasi neutral. The negative charge of electrons anions and negatively charged particles has to be balanced by positive ions. Thus cation density always exceeds anion density. The anion density depends on the electronegativity of the source gas. Gases like SF_6 and CCl_4 are extremely electronegative since there is no energy threshold for electron attachment ($SF_6 + e \rightarrow SF_5 + F^-$). In this type of plasma the anion density can be a 1000 times higher than the electron density (but equal to the cation density). SiH_4 is moderately electronegative. In absence of particles the electron and anion densities are in the same order of magnitude (about 10^{15} m^{-3}). Since a plasma is positive with respect to its surroundings, anions are trapped inside the plasma. Thus their lifetime can be very long. Polymerization reactions using neutrals or positive ions are not fast enough to form particles as these particles are rapidly lost to the wall. The long lifetime and trapping of negative ions inside the plasma provides a polymerization channel from which nanoparticles can grow.

Dr Pavlů communicated: At the very early stage of the particle formation (before agglomeration occurs), is there any possibility to monitor the particle growth, *i.e.*, temporal evolution of size? What's the threshold size of particles (in your particular case) detected using Mie scattering?

Professor Stoffels communicated in reply: Mie scattering can be used for particles larger than about 10 nm. We[1] have used laser induced evaporation to measure nm sized particles. Other methods include collecting particles and measuring their size *ex situ* using electron microscopy; mass spectrometry and cavity ring down spectroscopy. All methods are troublesome and give only partial information. For process control typically indirect measurements are favored (bias voltage, harmonic voltage generation, electron density) as these are easy to implement and have been shown to adequately reflect nm particle formation in the discharge.[2]

1 L. Boufendi, J. Hermann, A. Bouchoule, B. Dubreuil, E. Stoffels, W. W. Stoffels and M. L. de Giorgi, *J. Appl. Phys.*, 1994, **76**, 148.
2 R. Ghidini, C. H. J. M. Groothuis, M. Sorokin, G. M. W. Kroesen, W. W. Stoffels, *Plasma Sources Sci. Technol.*, 2004, **13**, 143–149.

PAPER www.rsc.org/faraday_d | Faraday Discussions

Charge and charging of nanoparticles in a SiH_4 rf-plasma

W. W. Stoffels,* M. Sorokin and J. Remy

Received 5th February 2007, Accepted 14th February 2007
First published as an Advance Article on the web 16th July 2007
DOI: 10.1039/b701763d

Plasma-produced nanoparticles are of interest for many applications. They have very specific properties that can vary greatly from those in the atomic and bulk materials, including thermodynamic properties, such as a reduced melting temperature, and optical properties, such as blue-shifted blackbody radiation. Since a plasma is dominated by free electrons, charge related properties, like a reduced work function for electron attachment and increased work function for photo ionization, are of major importance.
In situ detection of nanoparticles inside a plasma is difficult. Hence, indirect methods including changes in optical emission, plasma voltage and currents are used to study particle growth. Here, electron density measurements in the plasma are presented using an ultrafast microwave resonance technique (time resolution below 1 μs). This technique allows studying the charge and charging kinetics of nanoparticles within the initial milliseconds of their growth.

1. Introduction

The discovery in the late 1980s that dust particles, formed inside a processing discharge, contaminate semiconductor surfaces has sparked a novel research area.[1,2] Together with the existing research on astrophysical dust clouds, this has developed into the complex plasma research we see presently. It is known that the gases that yield high deposition rates in Plasma Enhanced Chemical Vapor Deposition (PECVD) applications are also prone to particle formation. However, today the presence of these particles is not feared, but welcomed as their specific properties are used to improve material properties. The silane chemistry has been actively researched by various groups. It has been shown that the incorporation of nanocrystals in an amorphous silicon layer both improves the efficiency and enhances the long term stability of solar cells.[3] Also, a variety of other applications using different chemistries are currently under investigation.[4,5] In future, this field will only expand further as accurately tailored nanoparticles will be used as drug delivery systems in medicine or as quantum devices in next-generation semiconductor applications.

Particle growth in a plasma strongly depends on the plasma environment. Micrometre-sized particles can be studied directly. Their size can be determined by Mie scattering and their composition by Fourier transform infrared spectroscopy. Fig. 1 shows the time- and angle-dependent Mie scattering signal of a single particle

Eindhoven University of Technology, Department of Applied Physics, PO Box 513, 5600 MB, Eindhoven, The Netherlands. E-mail: w.w.stoffels@tue.nl; Fax: +31 40 245 6050; Tel: +31 40 247 5753

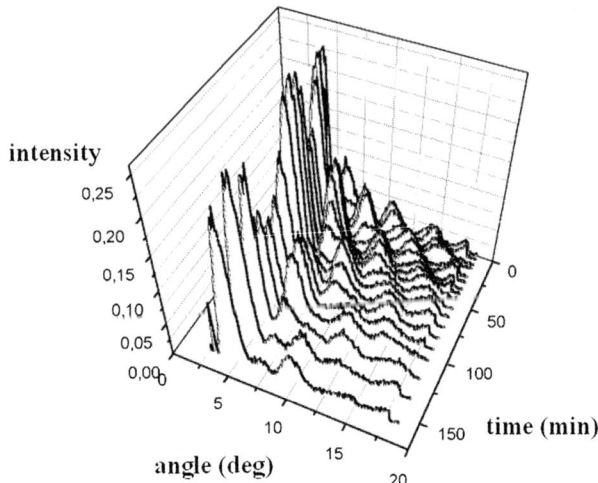

Fig. 1 Time-dependent and angle-resolved Mie scattering intensity for a single particle treated in 200 mTorr, 5 W oxygen plasma. The angular separation between maxima and minima increases in time, indicating that the particle size decreases continuously. The etch rate determined from these data is 1 Å s^{-1}.[6]

suspended in a plasma. From this, the absolute size and size variations can be determined with nm accuracy, as well as the particle's refractive index; in this example, a single melamine–formaldehyde particle was etched at 0.1 nm s^{-1}.[6] Nanoparticles are too small to be visualised by Mie scattering and other methods are required. Many present day techniques are indirect. Plasma parameters are monitored to understand the nature and growth processes of nanoparticles.[7]

In this contribution, we show that by combining various plasma diagnostics, a lot can be learned on the plasma–particle interaction of growing nanoparticles. We choose low pressure rf silane discharges as a model system, as much is known for these systems already. Particular attention is paid to the electron density in the plasma during the early stages of particle growth. Before going into the details of this experiment, we first review some nanoparticle properties, showing their unique properties.

2. Properties of nanoparticles

In order to understand nanoparticles and their behavior in a plasma, it is important to realize that nanoparticles are mesoscopic systems. They have properties that vary significantly from bulk properties, whereas they also cannot be accurately described using atomic and molecular theories.[8] Let us review some properties.

2.1 Melting temperature

For most materials, the melting starts at the gas–solid or liquid–solid interface, because at the surface the atoms are slightly less bound to the bulk material. Since small particles have a relatively large surface-to-volume ratio it can be expected that small particles have a reduced melting temperature. Theory predicts the temperature reduction to be inversely proportional to particle size (Ref. 2 and 9, page 95). The result is that a nm-sized gold particle melts already at room temperature, as shown in Fig. 2.[10,11] Even though this inverse relation of temperature and diameter is not valid for all materials, a lowered melting temperature has been reported for a large number of species.[12] It's remarkable to note that, for small cluster of only several tens of atoms, the melting temperature is often above the bulk melting temperature.

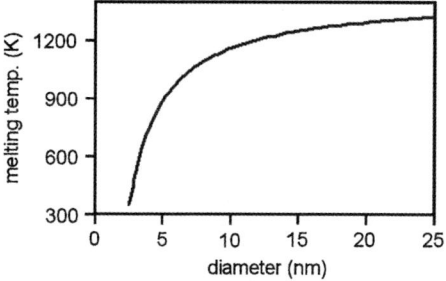

Fig. 2 Melting temperature of gold spheres as a function of the diameter.[2]

In this case, the bulk structure is lost and the cluster arranges itself in stable molecular configurations.[13]

In a plasma, particles are subject to a bombardment of plasma species, depositing energy on the particle. In the case of *in situ* growth, reaction energy is also absorbed by the particles. This can result in an increased particle temperature. In combination with the reduced melting temperature, it is likely that, in many cases, nanoparticles in the plasma are in a liquid phase. This explains why these particles very often turn up in the thermodynamically stable crystalline phase in deposited surfaces, even though deposited layers at the same temperature are completely amorphous.

2.2 Optical properties

As the particle size becomes comparable to the optical wavelength, also the optical properties of particles change. For light scattering this is described by the Mie theory. The strong dependence of scattered light on particles is well known and actively used to determine particle size and composition (see Fig. 1). Also, light emission and absorption properties are changed. In the plasma this has an effect on the energy balance of a particle. Bulk material emits black body radiation according to Planck's emission formula:

$$P(\lambda) = \frac{2hc^2}{\lambda^5} \frac{1}{\exp\left(\frac{hc}{\lambda kT}\right) - 1}$$

Here, h, c and k denote Planck's constant, the velocity of light and Boltzmann's constant, respectively. In practice, the emitted intensity $I(\lambda)$ is reduced by the emissivity of the material $\varepsilon(\lambda)$: $I(\lambda) = \varepsilon(\lambda) P(\lambda)$.

For bulk materials, the emissivity is close to, but smaller than, 1. For small particles the emissivity is described by the absorption efficiency Q_{abs}, derived from Mie theory:[14]

$$Q_{abs} = \frac{8\pi r}{\lambda} \operatorname{Im} \frac{n^2 - 1}{n^2 + 2} = \varepsilon(\lambda)$$

where n is the complex refractive index of the material. The dependence on r shows that emission—and consequently radiative cooling—is greatly suppressed with decreasing r. Secondly, the inverse dependence on wavelength shows that the black body spectrum of small particles is shifted to smaller wavelength.

Serious attempts have been made to utilize this effect in a novel lighting scheme. Most materials can only be heated to a few thousand degrees before decomposition. Even though the material is white hot, most radiative emission is still in the infrared. Tungsten is used in incandescent lamps because of its high melting temperature, but even there most energy is lost in infrared heat. The novel lighting concept utilizes small particles, suspended in a discharge, which, due to the blue shift, emit mainly visible rather than infrared light.

2.3 Charging of nanoparticles

One of the most important aspects of particles suspended in a discharge is their charge. In most complex laboratory plasmas, the charge is responsible for the levitation and trapping of the particle in the plasma. Furthermore, the charge on the particle plays an important role in the growth process of small particles in chemically active laboratory plasmas *e.g.* silane and methane used for thin film deposition, as well as in the atmosphere and in space (interstellar clouds, stellar nebula, planetary rings). Charged particles also play an important role in the earth's middle atmosphere, where they are assumed to be responsible for noctulescent clouds and mesospheric summer echoes. Presently, the Orbital Motion Limited theory (OML) is generally used to estimate the charge on a particle in a plasma environment.[2] It is based on the fact that, under steady state conditions, the electron and ion fluxes to a particle are equal. Since electrons are more mobile than ions, standard OML theory assumes that a Coulomb field must exist such that the electrons are sufficiently repelled and ions attracted, in order to arrive at equal fluxes. The theory can be expanded to allow for photo and secondary electrons. Since both electron and ion fluxes scale with particle surface area, the particle potential is independent of the particle size. For a spherical particle of radius r and Z elementary charges, the potential is given by:

$$V_{\text{particle}} = \frac{-Ze}{4\pi\varepsilon_0 r}$$

which, incidentally, shows that the charge on a particle is expected to scale with the particle radius. In addition to the Coulomb field, OML theory assumes that all charges arriving at the particle surface are actually absorbed. For nanoparticles, both assumptions may be incorrect.

In astrophysical clouds or in interstellar space, there are typically very few electrons. However, UV photons, capable of photo ionization, are abundant. As a result, particles in space environments are often positively charged. Combining the photon and electron fluxes results in an estimate of the particle charge. However, for nanoparticles, the photo ionization cross section also changes from its well-known bulk value.

In order to understand the work function of nanoparticles, we have to take into account the image charge, which is responsible for a lowering of the work function of neutral nanospheres. This reduces the amount of charge a particle can carry. Moreover, it increases the efficiency and lowers the threshold for photons and ions to release electrons from the surface. Finally, the results show that the potential that electrons and ions from the plasma encounter is not simply a Coulomb interaction. This implies that the particle charge, calculated using standard OML theory, is incorrect for small particles.

In this section, we shall briefly review the work function for a small neutral particle. We shall follow the simple approach of Wood,[15] who derived this work function based on a classical approach of a metal sphere. Errors and improvements to this argument have been pointed out by several authors (see *e.g.* ref. 2 and 8), but apart from pre-factors they yield essentially the same result, which agrees well with experimental data.

The work function of a material is defined as the energy needed to remove an electron from the bulk material to infinity. It is generally known that one of its major contributions is the interaction due to the image force between the released electrons and the induced surface charge. This is the principal interaction for an electron more than a few angstrom away from a surface. For a planar surface the image potential is[16]

$$\Phi^{\text{plane}}_{\text{image}} = \frac{-e^2}{4\pi\varepsilon_0\, 4x}$$

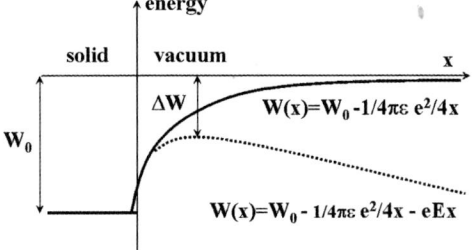

Fig. 3 Potential diagram of an electron near a solid with work function W_0. The electron feels an attractive potential that can be described by the image charge (solid line). If the surface is charged negatively, the electron will also feel a repulsive Coulomb force, resulting in a lowered work function (dotted line).

for $x > 0$, where x denotes the distance to the surface of a test charge e. The solid line in Fig. 3 graphically denotes the situation. A test charge arriving from infinity to an uncharged surface has a potential energy equal to the work function W_0, and feels an attractive force towards the surface. In case of a surface charge, there is an additional term indicated by the dotted line in Fig. 3. For a sphere with radius R, the image potential at $r > R$ is[16]

$$\Phi^{\text{sphere}}_{\text{image}} = \frac{-e^2 R^3}{4\pi\varepsilon_0 2r^2(r^2 - R^2)}$$

Close to the surface, the equations become comparable for $r = R + \delta$ and $x = \delta$. Assuming that there is no change in the material properties for small particles, the difference in work function for a sphere in comparison to bulk material is thus the difference in the work needed to move an electron from infinity to the surface

$$\lim_{\delta \to 0} \frac{-e^2 R^3}{4\pi\varepsilon_0 2(R+\delta)^2((R+\delta)^2 - R^2)} - \frac{-e^2}{4\pi\varepsilon_0 4\delta} = \frac{5}{8} \frac{e^2}{4\pi\varepsilon_0 R}$$

Thus, the work function for a neutral sphere W_0—the electron affinity—will be *reduced* by this amount relative to the work function for a plane W_{bulk}.

$$W_0 = W_{\text{bulk}} - \frac{5}{4\pi\varepsilon_0 8} \frac{e^2}{R} = W_{\text{bulk}} - \frac{0.9}{R(\text{nm})} \text{ eV}$$

Note, however, that in the case of (photo-)ionization of an initially neutral particle, for example by photo-ionization, there remains a positive charge on the particle if an electron is removed. This positive charge, which can be assumed to be at the center, results in an additional attractive interaction increasing the work function by $e^2/4\pi\varepsilon_0 R$. In this case, the work function becomes

$$W_0 = W_{\text{bulk}} + \frac{3}{4\pi\varepsilon_0 8} \frac{e^2}{R} = W_{\text{bulk}} + \frac{0.54}{R(\text{nm})} \text{ eV}$$

Various corrections have been proposed to these formulas, to incorporate quantum and material effects.[2,8] Nevertheless, these predictions are close to the measured values even for non-conducting spheres of hydrogenated silicon.[2]

The argument can be easily extended to a multiply charged particle. Let us consider a particle with Z elementary negative charges remaining at the particle. The potential around the particle, experienced by an incoming electron, will be a combination of the classic repulsive coulomb interaction and the attractive image charge.

$$\Phi_Z = \frac{Ze^2}{4\pi\varepsilon_0 r} - \frac{e^2 R^3}{4\pi\varepsilon_0 2r^2(r^2 - R^2)}$$

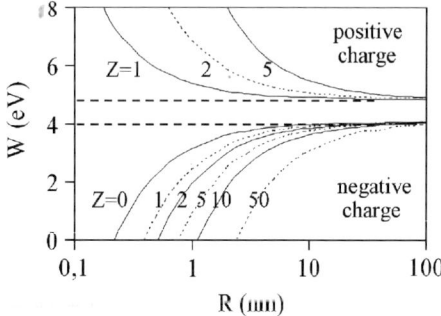

Fig. 4 The electronegativity and ionization potential as a function of particle radius, for various values of charge on the particle.

This will result in an effective lowering of the potential barrier around the particle, as shown in Fig. 3. Fig. 4 depicts the calculated attachment and ionization energies of nanoparticles as a function of radius for a variety of particle charges. The difference between the bulk work function for ionization and electron attachment is the band gap. An attaching electron will remain in the conduction band, while for ionization an electron from the valence band must be released. In addition to a change in work function and ionization potentials, the local field surrounding particles will deviate significantly from a simple Coulomb field as is generally assumed in OML theory. This affects both the maximum charge as well as the charging rate of nanoparticles. The above examples clearly show the large differences between bulk material and nanoparticles.

3. Nanoparticle growth

The charge of a dust particle in a plasma can vary from plus one elementary charge for nanometre and sub-nanometre particles, to minus many hundreds and thousands for micron and sub-micron particles (Fig. 4). While growing in the plasma, dust particles collect more and more electrons, strongly depending on the dust particle concentration, and the larger the particles, the greater the charge they obtain.[2] The currently accepted mechanism of dust formation in silane plasmas consists of four phases (ref. 2, page 77). In the first phase, primary clusters are formed as a result of molecular and ion polymerisation chemistry. Under the right conditions, the primary clusters grow into nanometre-sized particles. Once the nanoparticles have reached a critical density of 10^{15}–10^{18} m^{-3}, a rapid agglomeration takes place.[17–19] At the end of this phase, the plasma contains negatively charged clusters with a diameter of 20–50 nm at a density of 10^{14}–10^{15} m^{-3}.[17–19] After the agglomeration phase, the particles grow steadily by deposition of plasma species on their surface. This is shown graphically in Fig. 5.

After agglomeration, particles can be visualized by Mie scattering. An example is shown in Fig. 6. The figure shows the place- and time-resolved scattering signal for particles formed *in situ* in a 600 mTorr He/SiH$_4$—16 sccm gas flow plasma.

For the early growth phases the signal is too weak. Plasma emission can be monitored, which also correlates with particle growth. But, since charge is the driving force in particle growth, following the electron density evolution gives a more direct way to follow different particle growth stages. Numerous experimental studies of particle formation in silane discharges have been performed,[20–24] including studies of the electron density evolution using the microwave resonance photodetachment technique.[22–24] Measurements of the electron density in the first second after the discharge ignition have also been reported in ref. 22–24. However, no results have been published on fast early stage electron density measurements combined with measurements of electrical characteristics of the discharge.

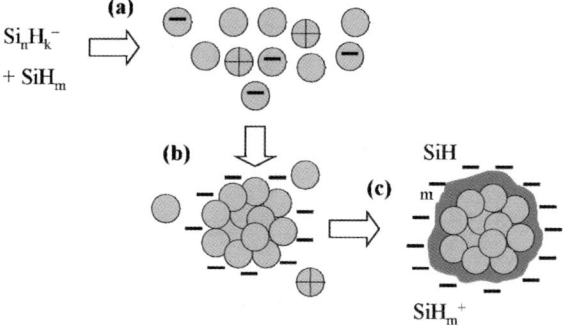

Fig. 5 Schematic view of the four phases in the formation of particles in silane plasmas: (1) plasma polymerization by negative ions; (2) nucleation of primary particles—nanometer-sized crystallites with time-varying charge; (3) coalescence phase and formation of larger clusters with permanent negative charge; (4) growth of clusters due to deposition of plasma species.

Moreover, no well-resolved data have been found for the electron density evolution in the first millisecond after the discharge ignition. By applying a fast and non-intrusive microwave resonance technique, we are able to measure the electron density with a time resolution around 700 ns. This provides us new interesting results on the electron density evolution during the initial dust growth in the plasma.

4. Experimental

The experiments are performed in an adapted low pressure parallel plate radio-frequency (rf) discharge reactor. The plasma chamber is mounted inside a vacuum vessel, which consists of a grounded cylindrical stainless steel box of 300 mm external diameter and 500 mm height. The plasma is confined in an internal aluminium chamber of 140 mm internal diameter, shown in Fig. 7. The distance between the electrodes is 40 mm. The upper electrode is powered capacitively by an

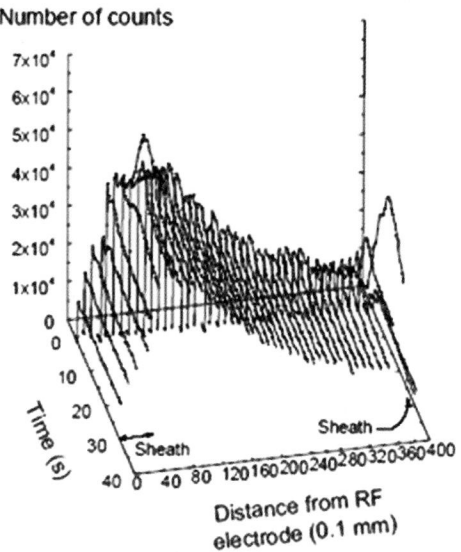

Fig. 6 Mie scattering signal on *in situ*-formed particles in a 600 mTorr He/SiH$_4$—16 sccm gas flow parallel plate rf plasma. The data are taken as a function of time after switching the discharge on and at various heights in the discharge. The maximum in the lower sheath, near the rf electrode, indicates that, under these conditions, particles are mainly produced locally.

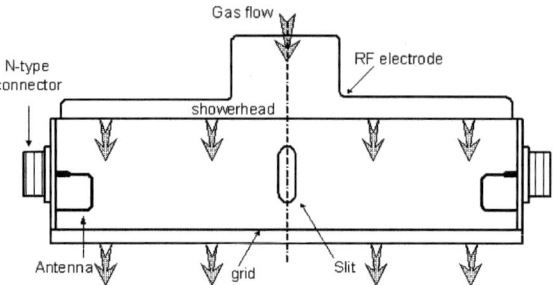

Fig. 7 A schematic view of the plasma chamber. It has a cylindrical geometry (140 mm diameter, 40 mm height), gas is introduced through a showerhead in the rf-powered top electrode and leaves through a grounded mesh in the bottom. The diagnostic microwave antennas are mounted in the side wall, which also has four viewing slits.

rf power supply at 13.56 MHz, and the sidewalls and the bottom are grounded. Two thermocouples (PT-100) and four heating elements located symmetrically around the upper electrode allow variations of gas and electrode temperature from room temperature to 180 °C. The gas flow is 4 sccm of 5% SiH_4/He mixture with 3.5 sccm of Ar added to it. With this gas composition, 23 W of 13.56 MHz rf power provides a stable plasma ignition for each measurement (in pure SiH_4/He, no ignition is possible below 60 W). The pressure used is 13.3 Pa (100 mTorr). A mass-flow controller monitors the gas flow. The pressure is measured outside the discharge chamber by a Baratron capacitive gauge. Slits in the inner cylindrical chamber and three windows on the vacuum chamber allow optical observations of the central region of the plasma. Observation slits and windows are aligned and placed at 90° from one another. In order to avoid spreading of the discharge outside the discharge chamber, a mesh grid covers each slit. The pumping system consists of a pre-vacuum pump and a turbo-molecular pump achieving a base pressure below 10^{-6} mbar.

To probe different electric characteristics we use two consecutively installed probes: the current/voltage sensor (IV sensor) from the Smart PIM (Plasma Impedance Monitor) of Scientific Systems and a plasma impedance probe designed and constructed in the GREMI laboratory of Professor Boufendi, Orléans University/CNRS, France.[25] The sensor of Scientific Systems provides signals proportional to the rf current and voltage. It allows sampling the undisturbed signal and analysing it later with a Fast Fourier Transformation (FFT) procedure to get information about any harmonic of voltage and current, being only limited by the speed of the Transient Recorders (TRC) used to sample the signal. In our case, the standard (slow) TRC is replaced by two 400 MHz TRCs to sample the signals from the PIM sensor, allowing, in principle and following the Nyquist theorem, to get information on the signals up to 200 MHz. That means that, in theory, with this system configuration we can get information on the behaviour of up to the 14th harmonic of the rf voltage and current, as well as on the phase shift between the signals. In practice, the amplitude of the harmonic signals can be too small to be resolved by the Transient Recorders' 12-bit analog–digital converters. The GREMI probe gives us directly real-time amplitudes of the rf current fundamental, third and fourth harmonic signals, so no extra calculations are needed for the acquired data.

The electron density is monitored using a microwave resonance technique.[26] Its working principle depends on the fact that the resonance frequency of an electromagnetic cavity varies with the electric permittivity of the medium inside. In the case of a plasma in the microwave range, this is mainly determined by the free electrons. In our case the plasma electrodes (see Fig. 7) are constructed in such a way as to simultaneously serve as a microwave resonator and antennas are used to inject and sample the microwaves inside the chamber. An increased electron density is reflected

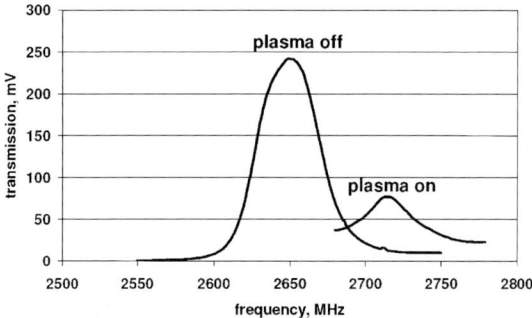

Fig. 8 Measured resonance frequencies inside the plasma reactor (Fig. 7) with the plasma off and on. The shift is proportional to the electron density inside the plasma.

by an increase of the resonance frequency of the plasma chamber, as shown in Fig. 8, by sampling the resonance curve of the chamber with and without plasma.

5. Electron density measurements

Since the same initial conditions before every experiment are crucial for the microwave resonance measurements, a fixed duty cycle for all the experiments with silane has been chosen. In each measurement the discharge was ignited for 10 s followed by a 60 s pause to restore the initial concentration of the gases in the discharge chamber before the next measurement. This procedure is necessary to avoid additional back-diffusion of silane from the vacuum reactor and pumps.[27] Fig. 9 shows the results for the electron density evolution in argon and silane in the first few ms after plasma ignition. As expected, in case of the pure argon discharge the electron density remains constant, as do the other plasma parameters. However, in case of silane discharge, a rapid concentration drop can easily be seen. Besides, the electron density in the case of silane reaches a relatively higher initial value than in the case of argon. This is due to the lower ionisation energy of silane as compared to argon and helium. The ionisation energies of the materials involved are: He −24.6 eV, Ar −15.8 eV and SiH_4 −12.0 eV. Hence, in the case of silane, the plasma is able to produce more free electrons rapidly, which results in the higher initial electron density. Presumably, after several hundreds of microseconds, when the electron density reaches the peak, formation of the negative ions of silane and its radicals starts to prevail, leading to the decrease in the value of the average electron density until it stabilises at a certain equilibrium value. This corresponds to the first two phases of particle growth, depicted in Fig. 5.

In each experiment, along with the electron density measurements the electrical data from the GREMI probe and from the Smart PIM are acquired. Fig. 10 shows

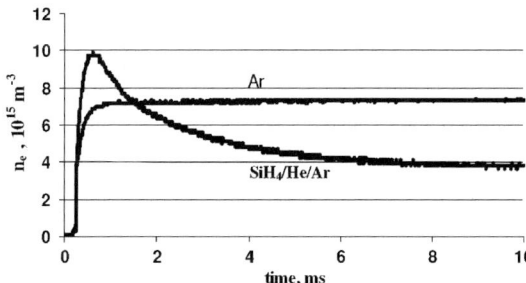

Fig. 9 The electron density in the plasma in the first milliseconds after plasma ignition in an argon plasma and a SiH_4/He/Ar mixture at 13.3 Pa and 23 W rf power.

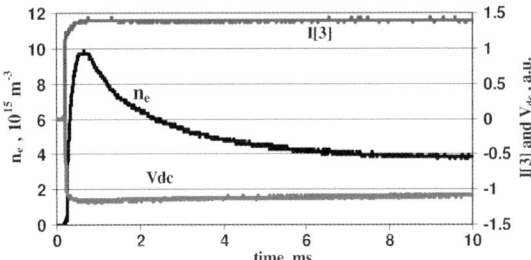

Fig. 10 The electron density, bias voltage and amplitude of the third harmonics in current in the plasma in the first milliseconds after plasma ignition in a SiH$_4$/He/Ar mixture at 13.3 Pa and 23 W rf power.

the evolution of the third harmonic and the DC bias voltage signals from the GREMI probe together with the evolution of the average electron density. Both electrical signals are characterised by a fast rise-time from the moment of the plasma ignition, but do not reflect the changes in electron density related with nanoparticle formation. Only the Vdc signal shows a slight decrease. Since the self-bias of the electrode and the current are collective effects of the electron temperature and the density, this suggests that an increase of the electron temperature must partially compensate the effect of the electron density drop. In the past, the authors showed that if a plasma changes from electropositive to electronegative, the electron density may go down by more than one order of magnitude, while the positive ion density goes up by a similar value.[26] Yet even then, the ion and electron current to the electrode, which determine Vdc, remains constant. Obviously, the plasma regulates itself to keep this flux constant, which is highly connected to the energy conservation, whereas the densities can vary greatly. It also shows that external electrical measurements are insufficient to study the internal plasma kinetics in detail.

After the electron density reaches its peak, it rapidly decreases with a characteristic time around 0.66 ms, which is attributed to the formation of negative ions in the silane plasma. Longer periods of measurements will also include agglomeration and homogeneous particle growth phases (Fig. 5) and, finally, particle expulsion (plasma burst) from the plasma, occurring, respectively, at about 0.6 s and 5.3 s after the discharge ignition. For these longer time scales shown in Fig. 11, the electrical measurements also show their previously reported behaviour.[28]

To see whether the fast initial phenomena of the electron density evolution during the plasma ignition are strongly influenced by the gas temperature, we have performed measurements at 120 °C. At this temperature, the agglomeration time moves from 0.6 s (at 20 °C) to 27 ± 4 s.[21] Fig. 12 shows the first 3 ms of the plasma

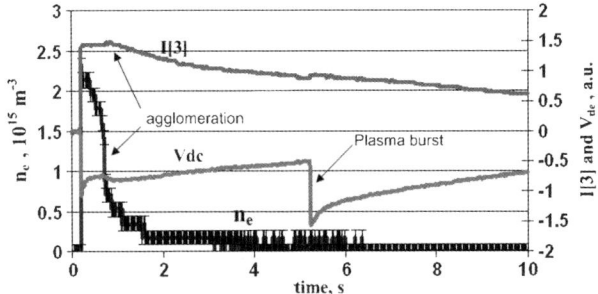

Fig. 11 The electron density (n_e), bias voltage (V_{dc}) and amplitude of the third harmonics in current (I^3) in the first seconds after plasma ignition in a SiH$_4$/He/Ar mixture at 13.3 Pa and 23 W rf power. The agglomeration phase and a plasma burst (expelling particles from plasma) are indicated by arrows.

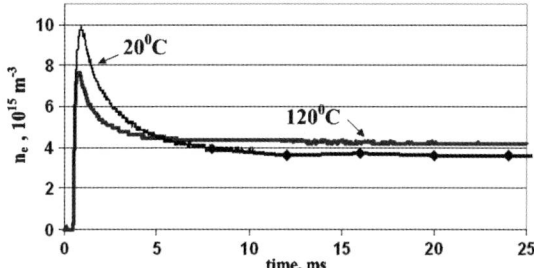

Fig. 12 The electron density in the first milliseconds after plasma ignition in a $SiH_4/He/Ar$ mixture at 13.3 Pa and 23 W rf power at 20 °C and 120 °C. At 20 °C, the electron density reaches a higher value and takes more time to reach the plateau value. This is attributed to a faster negative ion formation rate at elevated temperatures.

ignition. The characteristic rise-time of the electron density at 120 °C is around 60 μs (78 μs at 20 °C), and the characteristic decrease time is about 450 μs (660 μs at 20 °C). A closer look at the initial electron density increase shows that the two curves are identical in the first 100 μs. Thus, it seems the electron loss by negative ion formation is both faster and starts earlier at elevated temperatures. This also causes the lower peak in the electron density at higher temperatures. Note that this consequently results in a strong delay of the particle growth at elevated temperatures.

6. Conclusion

We have shown that nanoparticles have some peculiar properties that deviate widely from the material bulk properties, but also cannot be described using molecular theory. Low pressure plasmas are ideal tools to produce nanoparticles in a well controlled way. The total yield in grams is very limited, but the design freedom and reproducibility is very good, allowing to produce nanoparticles with carefully tailored properties.

In situ measurements of nanoparticles is extremely difficult. Therefore, it is easier to study the plasma parameters, from which the nanoparticle properties and growth dynamics can be derived.

Fast electron density measurements in $SiH_4/He/Ar$ mixture showed that the electron density experiences a peak at around 300 μs after the plasma ignition, with the characteristic rise time of 80 μs, followed by a rapid decrease with a characteristic time of 0.66 ms, which is attributed to the formation of the negative ions in the silane plasma. At a higher temperature (120 °C), the characteristic decrease times become smaller: 60 μs and 0.45 ms, respectively, implying a larger "loss term" of the electrons in the plasma at higher temperatures.

References

1. L. Boufendi and A. Bouchoule, *Plasma Sources Sci. Technol.*, 2002, **11**, 211.
2. *Dusty Plasmas: Physics, Chemistry and Technological Impacts in Plasma Processing*, ed. A. Bouchoule, Wiley, New York, 1999.
3. P. Roca, I. Cabarrocas, P. Stahel, S. Hamma and Y. Poissant, *Proceedings of the 2nd World Conference on Photovoltaics Solar Energy Conversion*, 1998, 355.
4. T. Nozaki, T. Goto, K. Okazaki, K. Ohnishi, L. Mangolini, J. Heberlein and U. Kortshagen, *J. Appl. Phys.*, 2006, **99**, 024310.
5. E. Kovačević, I. Stefanović, J. Berndt, Y. Pendleton and J. Winter, *Astrophys. J.*, 2005, **623**, 242.
6. G. H. P. M. Swinkels, *Optical studies of micron-sized particles immersed in a plasma*, PhD Thesis, Eindhoven University of Technology, 1999.
7. A. Mezeghrane, M. Jouanny, M. Cavarroc, M. Mikikian, O. Lamrous and L. Boufendi, *31st EPS Conference on Plasma Physics*, London, 28th June–2nd July 2004 ECA vol. 28G, O-1.09.

8 W. A. de Heer, *Rev. Mod. Phys.*, 1993, **65**, 611, and references therein.
9 N. Ichimose, Y. Ozaki and S. Kashu, *Superfine Particle Technology*, Springer-Verlag, London, 1992.
10 P. Buffat and J. P. Borel, *Phys. Rev. A*, 1976, **13**, 2287.
11 A. N. Goldstein, C. M. Echer and A. P. Alivisatos, *Science*, 1992, **256**, 1425.
12 K. M. Unruh, *Mater. Res. Soc. Symp. Proc.*, 1990, 195.
13 Zhong-Li Yu, Cai-Zhuang Wang and Kai-Ming Ho, *Phys. Rev. B*, 2000, **61**, 2329.
14 C. F. Bohren and D. R. Huffman, *Absorption and scattering of light by small particles*, John Wiley and Sons, New York, 1983.
15 D. M. Wood, *Phys. Rev. Lett.*, 1981, **46**, 749.
16 L. D. Landau and E. M. Lifschitz, *Electrodynamics of continous media*, Pergamon, New York, 1960, pp. 10–11.
17 L. Boufendi and A. Bouchoule, *Plasma Sources Sci. Technol.*, 1994, **3**, 262.
18 C. Courteille, C. Höllenstein, J. L. Dorier, P. Gay, W. Schwarzenbach, A. A. Howling, E. Bertran, G. Viera, R. Martins and A. Macarico, *J. Appl. Phys.*, 1996, **80**, 2069.
19 U. Kortshagen U and U. Bhandarkar U, *Phys. Rev. E*, 1999, **60**, 887.
20 A. Bouchoule, A. Plain, L. Boufendi, J. Ph. Blondeau and C. Laure, *J. Appl. Phys.*, 1991, **70**, 1991.
21 L. Boufendi, J. Hermann, A. Bouchoule, B. Dubreuil, E. Stoffels, W. W. Stoffels and M. L. de Giorgi, *J. Appl. Phys.*, 1994, **76**, 148.
22 W. W. Stoffels, E. Stoffels, G. M. W. Kroesen and F. J. de Hoog, *J. Appl. Phys.*, 1995, **78**, 4867.
23 L. Boufendi, A. Bouchoule and T. Hbid, *J. Vac. Sci. Technol., A*, 1996, **14**, 572.
24 E. Stoffels, W. W. Stoffels, G. M. W. Kroesen and F. J. de Hoog, *J. Vac. Sci. Technol., A*, 1996, **14**, 556.
25 L. Boufendi, G. Viera, J. Gaudin, S. Huet, M. C. Jouanny and M. Dudemaine, in *Proceedings of the XXV International Conference on Phenomena in Ionized Gases*, ed. T. Goto, 2001, **3**, p. 69.
26 E. Stoffels and W. W. Stoffels, *Electrons, Ions and Dust in a Radio-frequency Discharge*, PhD Thesis, Eindhoven University of Technology, The Netherlands, 1994.
27 M. Sorokin, G. M. W. Kroesen and W. W. Stoffels, *IEEE Trans. Plasma Sci.*, 2004, **32–2**, 731.
28 R. Ghidini, C. H. J. M. Groothuis, M. Sorokin, G. M. W. Kroesen and W. W. Stoffels, *Plasma Sources Sci. Technol.*, 2004, **13**, 143.

Rapid transport of nano-particles having a fractional elementary charge on average in capacitively-coupled rf discharges by amplitude-modulating discharge voltage

Masaharu Shiratani,*[a] Kazunori Koga,[a] Shinya Iwashita[a] and Syota Nunomura[b]

Received 2nd April 2007, Accepted 23rd April 2007
First published as an Advance Article on the web 30th July 2007
DOI: 10.1039/b704910b

We have observed transport of nano-particles having, on average, a fractional elementary charge in single pulse and double pulse capacitively-coupled rf discharges both without and with an Amplitude Modulation (AM) of the discharge voltage, using a two-dimensional laser-light scattering method. Rapid transport of nano-particles towards the grounded electrode is realized using rf discharges with AM. Two important parameters for the rapid transport of nano-particles are the discharge voltage and the period of AM. An important key of the rapid transport is fast redistribution of ion current over the whole discharge region; that is, fast change of spatial distribution of forces exerted on nano-particles. The longer period of the modulation is needed for rapid transport for the larger nano-particles. The higher discharge voltage of the modulation is needed for rapid transport of nano-particles having a smaller mean charge. Local perturbation of electric potential using a probe does not bring about global rapid transport of nano-particles, whereas it leads to their local transport near the probe.

I. Introduction

Many functionalities of materials on any length scale arise from the intricate interplay of its constituents.[1] The nanoscale is a particularly interesting regime and much attention has been given to the fabrication and properties of complex nanocomposite materials. Various kinds of nanocomposite films have a great potential for many applications such as catalysts,[2] sensors,[3] biointerfaces,[4] solar cells,[5,6] hydrogen storage,[7] Large Scale Integration (LSI),[8] and hard coatings.[9]

To fabricate such nanocomposite films at a low cost, we have proposed a novel method for synthesizing films composed of nano-particles, *i.e.*, nano-particle composite films.[6,8] The method uses nano-particles as nano-building blocks and radicals, produced in gas phase using reactive plasmas, as adhesives (nano-building block production phase); the blocks and adhesives are transported towards a substrate

[a] *Department of Electronics, Kyushu University, 744 Motooka, Fukuoka 819-0395, Japan. E-mail: siratani@ed.kyushu-u.ac.jp; Fax: +81-92-802-3734; Tel: +81-92-802-3733*
[b] *Research Center for Photovoltaics, National Institute of Advanced Industrial Science and Technology, Tukuba, Ibaraki 305-8568, Japan*

(transport phase); and then they are co-deposited on a substrate (nano-construction phase). Structure and properties of such films depend on size and structure of nano-particles incorporated into films as well as on their volume fraction in films. The size of nano-particles refers to their diameter in this article.

Up to now, systematic studies have been carried out on particle growth kinetics in a particle size range from sub-nm to μm in low-pressure, high-frequency discharges employed for depositing Si thin films.[10–13] Based on the results, we have developed a method for controlling the size and structure of nano-particles and their volume fraction in films.[6,8] The size of nano-particles is controlled by a gas residence time in the discharge region or the duration of the pulse discharge, and their structure is tailored by a gas mixture ratio. A Volume Fraction (VF) of nano-particles incorporated into films is controlled by the discharge power or the distance between the discharge region and the substrate. The method has advantages of precise control of the size of nano-particles and their VF value, and hence it offers easy control way of the structure of nano-particle composite films. Rapid transport of nano-particles to the substrate is required for realizing a high deposition rate of nano-particle composite films and for suppressing their coagulation during their transport, because the deposition rate is proportional to the density n_p of nano-particles and their speed v_p to a substrate, while the coagulation rate of nano-particles is proportional to n_p^2. To obtain information about transport of nano-particles, we have observed their transport in single pulse and double pulse capacitively-coupled rf discharges both without and with an Amplitude Modulation (AM) of the discharge voltage using a 2-Dimensional Laser-Light Scattering (2DLLS) method. In this article, we report these experimental results and discuss their transport mechanisms.

II. Experimental

Experiments were carried out using a capacitively-coupled rf discharge reactor. The reactor, together with a 2DLLS system, is shown in Fig. 1. A powered disc electrode (20 mm in diameter and 1 mm in thickness) was set in the middle of two grounded electrodes (60 mm in diameter with a separation of 40 mm) in the reactor (260 mm in inner diameter and 230 mm in height). $Si(CH_3)_2(OCH_3)_2$ gas diluted with Ar was supplied to the reactor. The flow rate of $Si(CH_3)_2(OCH_3)_2$ and that of Ar were 0.2 and 40 sccm, respectively. The total gas pressure was 133 Pa. The temperature of the reactor wall was kept at 358–373 K to avoid liquefaction of $Si(CH_3)_2(OCH_3)_2$ on the wall surface. To dissociate $Si(CH_3)_2(OCH_3)_2$ and generate nano-particles, we sustained a discharge by applying 816 peak-to-peak voltage of 13.56 MHz to the powered electrode for a discharging period T_{on} = 2–9 s. The self-bias voltage was −350 V. The corresponding discharge power was 75 W. For an AM discharge, the

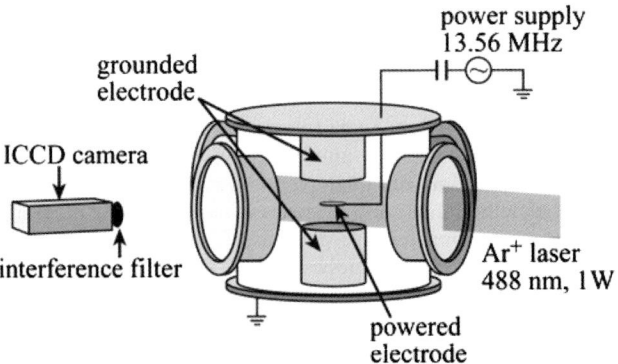

Fig. 1 Experimental setup.

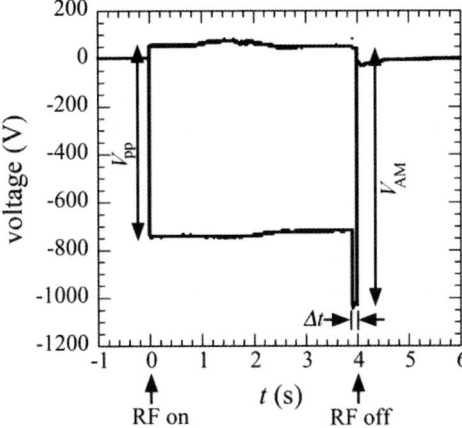

Fig. 2 Envelope of discharge voltage of single pulse discharge with AM.

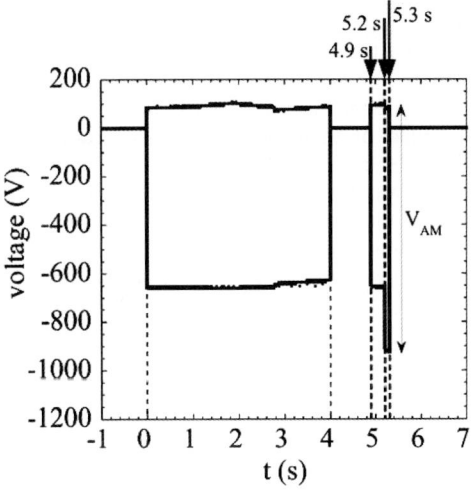

Fig. 3 Envelope of discharge voltage of double pulse discharge with AM.

discharge voltage was modulated as shown in Fig. 2. The modulation period Δt was 5–100 ms. The peak-to-peak voltage V_{AM} during the modulation was set in a range of 816 to 1193 V. A double pulse discharge with or without AM was employed to reduce the density of nano-particles, leading to an increase in the average charge on a nano-particle, in order to obtain information about effects of their charge on their transport. Fig. 3 shows an envelope of the discharge voltage of a double pulse discharge with AM.

Electron temperature, electron density, and ion density were measured with an rf-compensated stainless steel Langmuir probe (0.25 mm in diameter, 5 mm in length). Spatiotemporal evolution of optical emission intensity from the discharges was obtained with an ICCD camera having a gate width of 5 ms and an image frame rate of 30 Hz.

Spatiotemporal evolution of size and density of nano-particles was measured using a 2DLLS method[14] combined with a simple method for deducing their size and density.[13] A sheet beam of Ar^+ laser light of 1.0 W at 488 nm was passed parallel to the surface of the upper grounded electrode. The height and width of the sheet beam was 34 mm and 1 mm, respectively. The intensity of light scattered by nano-particles

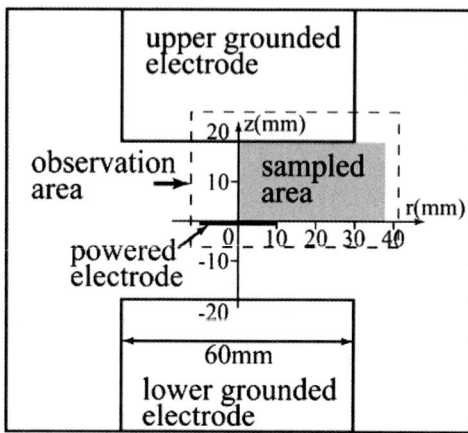

Fig. 4 Observation area of LLS measurements.

was detected at right angles with the ICCD camera equipped with an interference filter of a center wavelength of 488 nm, and FWHM of 1 nm. Spatiotemporal evolution of LLS intensity between the powered electrode and the lower grounded electrode was very similar to that between the powered electrode and the upper grounded electrode, because the mass of nano-particles was so light ($<5 \times 10^{-20}$ kg) that gravity had little effect on their transport. Therefore, LLS intensity in a region of $0 \leq r \leq 38$ mm and $0 \leq z \leq 20$ mm, shaded area in Fig. 4, was employed in this study, where $z = 0$ mm was the upper surface of the powered electrode. For high sensitivity measurements, the incident laser light was focused on a small observation volume, where the laser intensity had the FWHM of 0.15 mm. The scattered light from nano-particles was detected at right angles with a photomultiplier equipped with an interference filter of a center wavelength of 488 nm, and FWHM of 1 nm. The size and density of nano-particles were determined from their thermal coagulation, which took place after turning off the discharge.[13] The deduced size agreed fairly well with those obtained by SEM and TEM observations.

III. Results and discussion

A. Time evolution of plasma parameters and size, density, and charge of nano-particles

To obtain information about nano-particle growth kinetics in the discharges, we have measured the time evolution of plasma parameters with a Langmuir probe and size and density of nano-particles by high sensitivity LLS measurements using a photomultiplier, during a single pulse discharge without AM. We have deduced average charge Q_p of nano-particles from the electron density n_e, ion density n_i, and density n_p of nano-particles using a quasi-neutral condition, that is: $en_i - en_e - Q_p n_p = 0$. These results at $z = 5$ mm and $r = 0$ mm are shown in Fig. 5.

Electron and ion densities begin to increase at 1.4 s. Such effects of nano-particles on plasma parameters have been reported by many groups.[10] Appreciable effects take place when the electron loss rate on the nano-particle surfaces becomes comparable with that on the electrodes and on the reactor wall. To check this condition, time evolution of the total surface area of nano-particles suspended in gas in unit volume is deduced, as shown in Fig. 6. The total surface in unit volume increases with the discharging time. The total surface area of nano-particles in the whole discharge volume at 1.4 s becomes 20% of the surface area of electrodes, which probably meets the condition mentioned above.

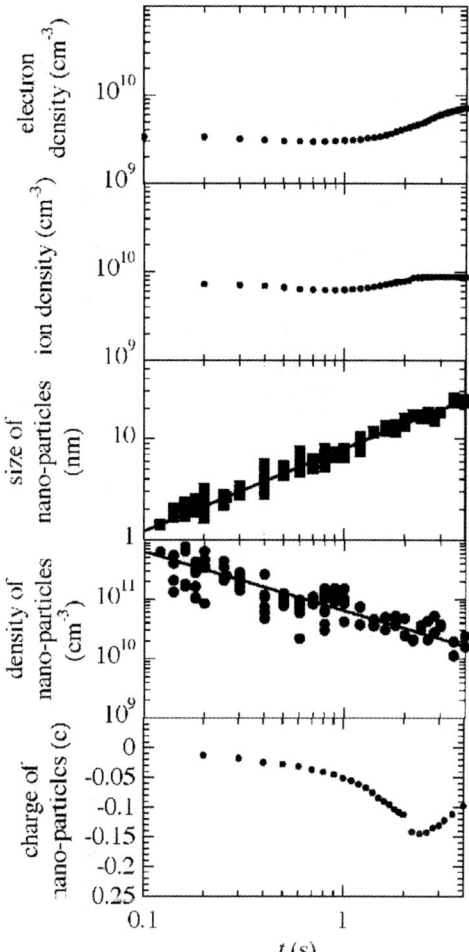

Fig. 5 Time evolution of electron density, ion density, size, density, and charge of nano-particles at $z = 5$ mm and $r = 0$ mm.

The nano-particle size increases from 1.4 nm to 25 nm and the corresponding density decreases from 6×10^{11} to 1.6×10^{10} cm^{-3} with increasing the discharge time from 0.12 s to 4 s. *Ex situ* TEM observations of nano-particles collected at the grounded electrode show that their size distribution is well expressed by a Gaussian form with a narrow size dispersion.[13] During the discharge, nano-particles are nucleated in the plasma-sheath boundary region around the powered electrode, and they tend to grow there.[15,16] After nucleation of nano-particles, there are two possible growth mechanisms; coagulation and Chemical Vapor Deposition (CVD). Coagulation is a process whereby nano-particles collide with one another due to a relative motion between them and adhere to form larger nano-particles. Coagulation-dominant growth keeps a volume fraction of nano-particles suspended in gas in unit volume constant. CVD is a process whereby nano-particles grow with time by collecting radicals on their surface. For CVD-dominant growth, the volume fraction increases due to accumulation of radicals on their surface. To identify the growth process of nano-particles in Fig. 5, time evolution of the volume fraction of nano-particles is deduced from their size and density. The results are shown in Fig. 6. The volume fraction increases monotonously with the discharging time. Such an increase indicates that nano-particles grow by collecting radicals on their surface, *i.e.*, via

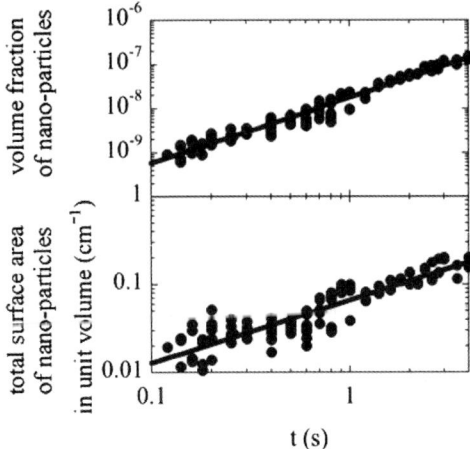

Fig. 6 Time evolution of volume fraction and total surface area of nano-particles suspended in gas in unit volume.

CVD. Because nano-particles are charged negatively during the discharge, as shown in Fig. 5, their coagulation is strongly suppressed by their mutual Coulomb repulsive force. The suppression of coagulation takes place even when a fraction of nano-particles are charged.[13,17] These results show that their size is controlled by the duration of the pulse discharge.

Since density of nano-particles surpasses positive ion density, 1–15% of nano-particles have a negative elementary charge, −e, and the rest are neutral. Hence, the average charge on a nano-particle is in the range −0.01e to −0.15e, namely a fractional elementary charge, as shown in Fig. 6. Although a fraction of nano-particles are charged negatively, almost all of them are electrostatically trapped in the discharge region. A quantitative criterion and detailed discussion of such trapping will be described elsewhere.[17] To deposit nano-particles on a substrate, they must escape from the discharge region by some means. One of the means is turning off the discharge to eliminate the trapping electrostatic potential profile in the discharge region. After turning off the discharge, however, nano-particles usually grow due to thermal coagulation, which is confirmed by the fact that their volume fraction is constant during the growth (not shown here).[13] In order to disperse nano-particles in an isolated way in films, and keeping their size identical to that just before turning off the discharge, they should be transported rapidly to a substrate before their coagulation taking place. Such transport is discussed in Section B.

B. Transport of nano-particles

In Section B, we focus on transport of nano-particles towards a substrate placed on the grounded electrode, namely, their transport in the z direction in Fig. 4. We describe their transport in single pulse discharges without and with AM in Section B 1 and that in double pulse discharges without and with AM in Section B 2.

1. Transport of nano-particles in single pulse discharges without and with amplitude modulation of discharge voltage. We have shown that size and transport of nano-particles are separately controlled using a pulse AM discharge having a discharge voltage (Fig. 2).[18] Their size is determined by the duration of the pulse discharge, whereas their transport is determined by the discharge voltage and the period of amplitude modulation. During the modulation period, nano-particles move in the z-direction, that is, towards the grounded electrodes, at a speed more than 60 cm s^{-1}, which is at least six times that after turning off the unmodulated discharges. One of the important features is the fact that the AM discharge can bring about the rapid transport of not only nano-particles charged negatively, but also of neutral nano-particles when nano-particles of these two kinds of charge states coexist. At the beginning of the modulation period, the absolute value of self-bias voltage increases within 20 μs. According to the corresponding increase in the voltage across the sheath near the powered electrode, the sheath width increases while most nano-particles tend to reside in the place where they resided just before the modulation because of their large inertia. Moreover, nano-particles residing in the sheath just after the modulation tend to have a negative charge as before the modulation, due to a delayed charging effect.[19] Therefore, a high electric field in the sheath drives such negatively charged nano-particles towards the grounded electrode. Although a part of the nano-particles must be neutral since n_p surpasses n_i, neutral nano-particles turn into negatively charged ones due to their charge fluctuation during the modulation period, and hence they also move towards the upper grounded electrode. Once nano-particles pass through the presheath region near the powered electrode, where ion drag force towards the powered electrode pushes them to the electrode, they can move to the plasma/sheath boundary region near the grounded electrode due to ion drag force. Because our reactor has a powered electrode much smaller than the grounded electrode, there is a large voltage drop across the sheath near the powered electrode. Due to the asymmetric potential profile, ionization mainly takes place around the sheath near the powered electrode. This consideration is supported by the fact that plasma emission intensity is the highest there. Since the ions flow from the ionization region towards the powered and grounded electrodes, there the direction of ion flow, and hence that of ion drag force, changes. Thus, the control of spatial distribution of ion flow is important to realize rapid transport.

Another important feature is the fact that the LLS intensity during the transport is nearly the same as that before AM for the discharge with AM, whereas the intensity increases appreciably during the transport for the discharge without AM, because nano-particles grow due to their coagulation. Therefore, nano-particles are transported rapidly towards the grounded electrode *without appreciable coagulation* using the pulse discharge with AM.

We have studied the dependence of nano-particle transport towards the grounded electrodes on their size in AM discharges of $V_{AM} = 1193$ V as a parameter of Δt. The results are shown in Fig. 7. The transport can be classified into three types: transport in region A in Fig. 7 is nearly the same as the slow transport in the discharge without AM; transport in region C is the rapid transport at a speed of

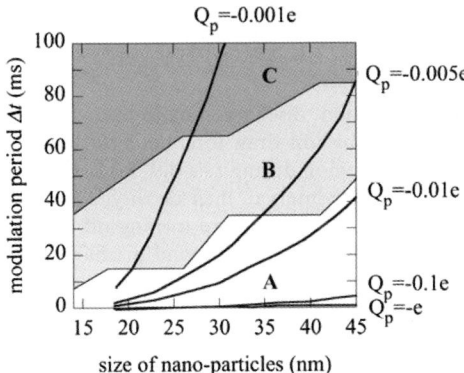

Fig. 7 Classification of transport of nano-particles in AM discharges. The transport can be classified into three types: transport in region A is nearly the same as the slow transport in the discharge without AM; transport in region C is the rapid transport at a velocity more than 60 cm s^{-1}; and transport in region B has behaviour between that in region A and region C. The solid line shows the inverse of dust plasma frequency.

more than 60 cm s^{-1}; and transport in region B has a behavior between that in region A and region C, namely, some of the nano-particles are driven rapidly, as in region C, and the rest of them are transported slowly, as in region A. The threshold Δt value between regions A and B as well as that between regions B and C increases with the size of nano-particles, probably because the lager nano-particles have the slower response time due to their inertia. The characteristic response time of nano-particles is evaluated as a parameter Q_p using the dust plasma frequency f_{pd}[20]

$$f_{pd} = \frac{1}{2\pi}\sqrt{\frac{Q_p^2 n_p}{\varepsilon_0 m_p}}, \quad (1)$$

where m_p is the mass of nano-particles. The inverse of the dust plasma frequency f_{pd} is shown as solid lines for several Q_p values in Fig. 7. The response time for $Q_p = -0.1e$, which is close to the experimental Q_p values and is much shorter than the experimental threshold Δt values, whereas both the threshold Δt values and the response time lines increase with the size of nano-particles. The threshold Δt value between regions A and B is close to the line for $Q_p = -0.005e$ in a size region of 25 nm to 35 nm, and that for $Q_p = -0.01e$ in a size region of 40 nm to 45 nm. We need a more sophisticated model to predict qualitatively the threshold Δt values.

2. Transport of nano-particles in double pulse discharges without and with amplitude modulation of discharge voltage. Fig. 8(a) and (b) shows two-dimensional spatial images of the LLS intensity, as a parameter of time t, in a double pulse discharge without and with AM. During the discharging period, nano-particles are mainly generated in the plasma/sheath boundary region near the powered electrode and they are distributed in a discharge space, as shown in the images at $t = 3$ s and 4 s.

After turning off the first discharge, they move from their generation region around the powered electrode towards the upper grounded electrode. Their motion in the interval $t = 4$–4.9 s is determined by the balance between thermophoretic force and gas viscous forces, and thermophoretic force drives nano-particles towards the grounded electrodes. Nano-particles in the region $r < 10$ mm move at a nearly constant speed of 7 cm s^{-1} in the z-direction. In the region $r > 10$ mm, their speed towards the grounded electrode decreases with increasing r, because the temperature gradient in the z direction for $r < 10$ mm decreases. As the sticking probability of nano-particles to the surface is unity,[21] nano-particles reaching the grounded electrode stick to the electrode. Since most nano-particles become neutral in the interval, they can coagulate with each other. Due to the coagulation, their size increases slightly from 25 nm at $t = 4.0$ s to 28 nm at $t = 4.9$ s, even in the region where their density is the highest.

After turning on the second discharge, nano-particles begin to be charged negatively and to be driven by ion drag force and electrostatic force, and hence they spread in the discharge region during $t = 4.9$–5.3 s. Due to the sticking loss of nano-particles to surfaces in the interval, their density, averaged over the discharge region, decreases from 10^{10} cm^{-3} just before turning off the first discharge at $t = 4.0$ s to 5×10^8 cm^{-3} at $t = 5.2$ s in the second discharge. Since the positive ion density surpasses density of nano-particles at $t = 5.2$ s, *most nano-particles are charged negatively in the second discharge.*

After turning off the second discharge without AM at $t = 5.3$ s, nano-particles move towards the upper grounded electrode with a speed of 7 cm s^{-1} because the thermophoretic force drives them. With the modulation, nano-particles rapidly accumulate around the plasma/sheath boundary region near the upper grounded electrode. Such rapid transport, being similar to that in region C in Fig. 7, is obtained for $V_{AM} \geq 925$ V, whereas transport similar to that in the region B in Fig. 7 takes place for $V_{AM} = 816$–925 V. In a single pulse discharge of 4.0 s duration with

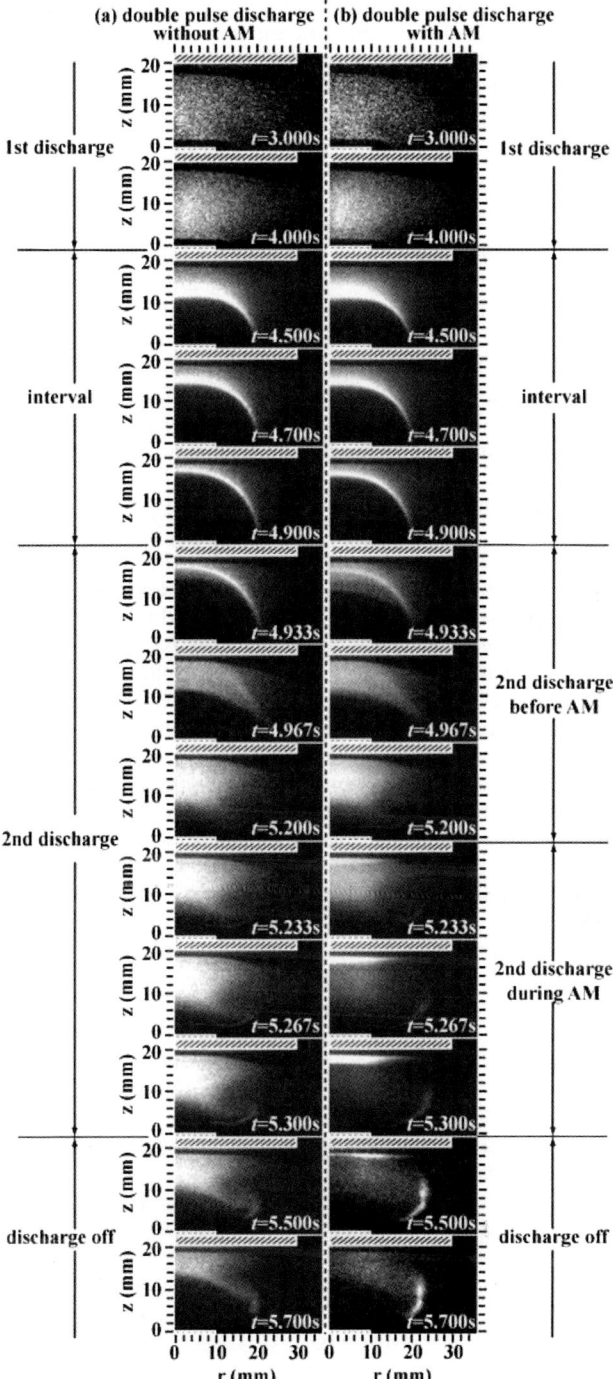

Fig. 8 Two-dimensional images of the LLS intensity in double pulse discharge without modulation (a) and with modulation of t = 100 ms and V_{AM} = 1193 V (b) as a parameter of time t.

AM of $\Delta t = 100$ ms, region B is obtained for $V_{AM} \geq 1112$ V, whereas region A is obtained for $V_{AM} = 1056$–1112 V. Because the average charge on a nano-particle in the second discharge with AM is higher than that in the single pulse discharge with AM, rapid transport in the second discharge is realized for a V_{AM} value that is low compared to that in the single pulse discharge.

C. Behavior of nano-particles around a biased probe

To obtain information about effects of an electric field on nano-particles in bulk plasma region, a stainless steel Langmuir probe (0.25 mm in diameter, 5 mm in length) was set at $z = 10$ mm and $r = 0$ mm and an amplitude-modulated 3.5 MHz voltage was applied to the probe through an rf matching network with a blocking capacitor of 51 pF. The modulation was carried out using a sine wave of 0.1–10 Hz. It is worth noting that DC voltage applied to the probe does not work well, since dielectric materials deposit on the probe. A two-dimensional image of LLS intensity for such experiments is shown in Fig. 9. A cylindrical sheath region is formed around the cylindrical probe, and, due to the electric potential profile, a hole in the nano-particle cloud is formed there. Time evolution of the sheath edge position and the hole edge position of the nano-particle cloud from the probe is shown in Fig. 10. There are five features in the results. First of all, although 85–90% of nano-particles are neutral, the perturbation of electric potential using the probe has an influence on all nano-particles, and hence the hole in the nano-particle cloud appears around the probe. The apparent behavior of nano-particles looks like a "collective" behavior despite the fact that coulomb coupling among nano-particles is quite weak. This "collective" behavior is obtained due to rather complicated interactions among nano-particles charged negatively, neutral nano-particles, electrons, ions, and electrostatic potential. Secondly, both the edges move in phase without an appreciable delay. Namely, the response time of nano-particles to the potential is less than the time resolution (33 ms) of the measurements. Thirdly, the amplitude of the hole edge position is smaller than that of sheath edge position. Fourthly, the center of oscillation of hole edge position decreases with time, whereas that of sheath edge position is at nearly the same position. Ion drag force pushing nano-particles towards the probe is proportional to the square of their size, whereas the electrostatic force pushing them away from the probe depends little on their size because n_p surpasses n_i. Therefore, the larger the size of the nano-particles, the larger ion drag force pushing them to the probe. Hence, the center of oscillation of hole edge position decreases with time. Last but not least, *no rapid "global" transport of nanoparticles towards the grounded electrode takes place*. This feature is different from the

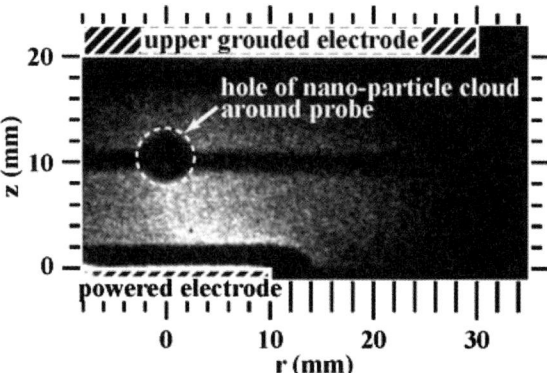

Fig. 9 Image of LLS intensity during discharge with a probe, to which amplitude modulated 3.5 MHz voltage was applied. A hole in the nano-particle cloud appears around the probe. Laser light is not irradiated in a region of $z = 9$–11 mm to avoid scattering on the probe.

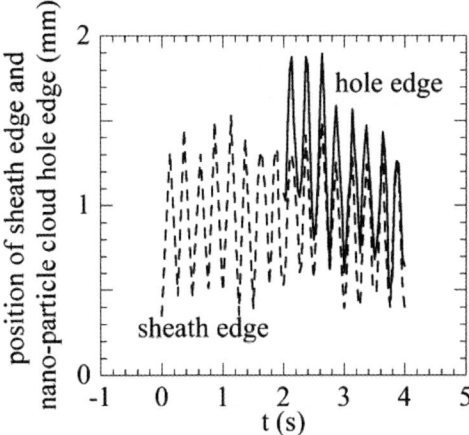

Fig. 10 Time evolution of sheath edge position and hole edge position of nano-particle cloud from the probe.

transport in rf discharges with an amplitude modulation of the discharge voltage described in Section B. Because the discharge is sustained by ionization in the plasma/sheath boundary region around the powered electrode, amplitude modulation of the discharge voltage changes the ion current distribution over the whole discharge region, which turns into redistribution of nano-particles (their rapid transport) due to ion drag force. Since the perturbation of the electric potential profile around the cylindrical probe has little effect on the ion current distribution over the whole discharge region, no rapid "global" transport of nano-particles towards the grounded electrode takes place, and effects of the perturbation on the density distribution of nano-particles is limited in the "local" region around the cylindrical probe. The experimental results in this section, together with Section B, show that an important key of the rapid transport of nano-particles towards the grounded electrode is rapid redistribution of ion current over the whole discharge region; that is, rapid change of spatial distribution of forces exerted on nano-particles. Here, rapid means that the change takes place within a time scale shorter than the response time of nano-particles to the forces.

IV. Conclusions

We have observed transport of nano-particles in single pulse and double pulse capacitively-coupled rf discharges without and with an amplitude modulation of the discharge voltage using the two-dimensional laser-light scattering method. The following conclusions are obtained in this study:

1. Two important parameters for the rapid transport of nano-particles are the discharge voltage and the period of AM.

2. An important key of the rapid transport is rapid redistribution of ion current over the whole discharge region; that is, rapid change of spatial distribution of forces exerted on nano-particles.

3. The longer period of the modulation is needed for rapid transport for the larger nano-particles.

4. The higher discharge voltage of the modulation is needed for rapid transport of nano-particles having the smaller mean charge.

5. Perturbation of electric potential using the probe does not bring about global rapid transport, whereas it leads to local transport near the probe.

From a scientific viewpoint, these results show that *electrostatic force and ion drag force are exerted on a nano-particle cloud having, on average, a fractional charge.*

From a technological viewpoint of nano-particle applications, the results indicate that size and transport of nano-particles can be separately controlled using a pulse rf discharge with AM. We expect that experimental as well as theoretical studies regarding such nano-particles having a fractional charge on average will open novel science and applications concerning nano-particles.

Acknowledgements

We are grateful to Dr Y. Watanabe, Emeritus Professor of Kyushu University, for helpful discussions. We also would like to acknowledge the assistance of Messers T. Kinoshita and H. Matsuzaki, who contributed greatly to the preparation of the experimental setup. This work was partly supported by New Energy and Industrial Technology Development Organization (NEDO) and the Japan Society of the Promotion of Science (JSPS).

References

1 G. Decher, *Polyelectrolyte Multilayers, an Overview in Multilayer Thin Films*, ed. G. Decher and J. B. Schlenoff, Wiley-VCH, Weinheim, 2003, pp. 1–2.
2 T. Giomelli, A. Lofberg and E. Bordes-Richard, *Thin Solid Films*, 2005, **479**, 64.
3 G. Barillaro, A. Diligenti, G. Marola and L. M. Strambini, *Sens. Actuators, B*, 2005, **105**, 278.
4 M. Tanaka, M. Takebayashi, M. Miyama, J. Nishida and M. Shimomura, *Bio-med. Meter. Eng.*, 2004, **14**, 439.
5 C. S. Solanki, R. R. Bilyalov, J. Poortmans, G. Beaucame, K. Van Nieuwenhuysen, J. Nijs and R. Mertens, *Thin Solid Films*, 2004, **451**, 649.
6 M. Shiratani, K. Koga, S. Ando, T. Inoue, Y. Watanabe, S. Nunomura and M. Kondo, *Surf. Coat. Technol.*, 2007, **201**, 5468.
7 K. Higuchi, K. Yamamoto, H. Kajioka, K. Toiyama, M. Honda, S. Orimo and H. Fujii, *J. Alloys Compd.*, 2002, **330–332**, 526.
8 S. Nunomura, K. Koga, M. Shiratani, Y. Watanabe, Y. Morisada, N. Matsuki and S. Ikeda, *Jpn. J. Appl. Phys.*, 2005, **44**, L1509.
9 J. Musil, *Surf. Coat. Technol.*, 2000, **125**, 322.
10 Y. Watanabe, *J. Phys. D*, 2006, **39**, R329, and references therein.
11 M. Shiratani, T. Fukuzawa and Y. Watanabe, *Jpn. J. Appl. Phys.*, 1999, **38**, 4525.
12 Y. Watanabe, M. Shiratani, T. Fukuzawa and K. Koga, *J. Tech. Phys.*, 2000, **41**, 505.
13 S. Nunomura, M. Kita, K. Koga, M. Shiratani and Y. Watanabe, *J. Appl. Phys.*, 2006, **99**, 083302.
14 Y. Matsuoka, M. Shiratani, T. Fukuzawa, Y. Watanabe and H. S. Kim, *Jpn. J. Appl. Phys.*, 1999, **38**, 4556.
15 T. Fukuzawa, M. Shiratani and Y. Watanabe, *Appl. Phys. Lett.*, 1994, **64**, 3098.
16 T. Fukuzawa, K. Obata, H. Kawasaki, M. Shiratani and Y. Watanabe, *J. Appl. Phys.*, 1996, **80**, 3202.
17 S. Nunomura, M. Kita, K. Koga, M. Shiratani and Y. Watanabe, submitted to *Phys. Rev. Lett.*
18 K. Koga, S. Iwashita and M. Shiratani, *J. Appl. Phys. D*, 2007, **40**, 2267.
19 S. Nunomura, T. Misawa, N. Ohono and S. Takamura, *Phys. Rev. Lett.*, 1999, **83**, 1970.
20 J. B. Pieper and J. Goree, *Phys. Rev. Lett.*, 1996, **77**, 3137.
21 S. Iwashita, K. Koga and M. Shiratani, *Surf. Coat. Technol.*, 2007, **201**, 5701.

Interaction between single dust grains and ions or electrons: laboratory measurements and their consequences for the dust dynamics

Jiří Pavlů,*[a] Ivana Richterová,[a] Zdeněk Němeček,[a] Jana Šafránková[a] and Ivo Čermák[b]

Received 23rd February 2007, Accepted 16th March 2007
First published as an Advance Article on the web 20th July 2007
DOI: 10.1039/b702843a

The present paper reviews our latest, and brings several new, results on charging of dust grains of various materials and sizes. Charging processes of dust in space and their influence on the dust dynamics are analyzed in laboratory simulations of secondary emission, field ion and electron emissions, and dust sputtering. Single micrometre-sized grains and grain clusters are stored in a hyperbolic quadrupole field under ultra-high vacuum conditions for long time periods. The charge state of the grain and its evolution are recorded while the grain is exposed to ion or electron beams of various energies and fluxes. The influence of the secondary electron emission on the charge state is measured and compared with a computer model. Limitations on the grain charge by the field electron and ion emission are considered next. The measurements allow analyzing field emission from conductive and dielectric grains. The existence of long-lived surface states on insulating materials, which are probably responsible for the anomalous behavior of field electron emission and the low threshold of the field ion emission, is indicated. The observation of sputtering by energetic ions showing a surprising anisotropic erosion of a conductive grain is analyzed. The sputtering and the field ion emission are discussed as possible sources of the so-called pick-up ions.

1. Introduction

The manifestations of dust in different parts of the solar system have been known and intensively studied for a long time, namely, cometary comae and their dusty tails, planetary rings, circumsolar dust rings, the interplanetary medium, interstellar molecular clouds, *etc.* In the context of interplanetary space, "dust" refers to particulate matter that does not manifest itself as isolated bodies but from some distance can be recognized as an ensemble of indistinguishable particles. The interplay between plasmas and charged dust grains has opened up a new research area: that of a dusty (or complex) plasma. Dusty plasmas are fully or partially ionized gases comprising neutral gas molecules, electrons, ions, and highly charged sub-micron- and micron-sized dust grains. There is a variety of mechanisms by which cosmic dust grains can be electrically charged: the capture of ambient

---

[a] *Charles University, Faculty of Mathematics and Physics, V Holešovičkách 2, 180 00, Prague, Czech Republic. E-mail: jiri.pavlu@mff.cuni.cz*
[b] *CGC Instruments, Hübschmannstr. 18, 09112, Chemnitz, Germany*

electrons and ions, Secondary Electron Emission (SEE) by energetic electron and ion impacts, photoemission due to short-wavelength electromagnetic radiation, field emission of electrons, triboelectric effects, and field evaporation of ions. Interplanetary and interstellar dust grains acquire a positive charge in the solar wind and can be strongly influenced by the Lorentz force as they pass through planetary magnetospheres. There, the charge on the grains changes rapidly when they pass through different plasma environments.[1]

In the case of interplanetary dust grains, the relevant charging processes are interactions with solar wind electrons and ions and photoemission by solar UV radiation. The flux of photoelectrons from a metal surface at 1 AU was estimated by Wyatt[2] to be equal to 2.5×10^{10} cm^{-2} s^{-1}. The photoelectron flux from silicate and graphite surfaces can be up to one order of magnitude lower; the same is expected for the flux from icy surfaces. On the other hand, mean ion and electron fluxes are of the order of 10^8 cm^{-2} s^{-1}. From these numbers it follows that charging of interplanetary dust grains is dominated by photoemission, which leads generally to positively charged particles with surface potentials of several volts. The value of the potential depends on the photoemission yield of the grain material, e.g., silicate grains attain potentials of 2.5–5 V, which are practically constant in the grain size interval 0.1–10 μm.[3] Potentials of grains from conducting materials might be somewhat higher. In the case of very small grains with dimensions comparable to the wavelength of light, the photoemission yield can be enhanced by a factor of 2–3, which results in higher surface potentials.[4]

In dense plasma regions where the electron flux is dominant, the sign and value of the dust grain surface potential will be determined by the energy of the impinging electrons. Electron attachment dominates in the eV range but, at electron energies above about 10 eV, SEE becomes important and results in a reduction of the negative potential. At energies above a few hundred eV, the SEE yield becomes larger than unity which causes a sign change of the potential from negative to positive. Generally, the SEE yield reaches a maximum value σ_{max} at energies E_{max} between 300 and 2000 eV. At higher energies, the SEE yield becomes again lower than unity, resulting in negative surface potential.

However, as the size of the grains becomes comparable with the mean free path of the primary electrons and with the diffusion length of excited electrons, the SEE yield may be substantially enhanced. Thus, the charge of micrometer-sized grains can be either positive or negative and the charge sign depends on their shape, dimensions, and material.[5]

Ion impacts may also induce the emission of secondary electrons, however, with a much lower yield and thus this effect is negligible in most environments.

The total electric charge of dust grains might be limited by field evaporation, or by field emission. These processes become dominant at field strengths $> 3 \times 10^{10}$ V m^{-1} (positive charge) and $> 10^9$ V m^{-1} (negative charge), respectively. However, even at much lower field strength, electrostatic repulsive forces can destroy the grains (electrostatic fragmentation) when they become higher than the tensile strength of the material. The latter process is particularly important in case of fluffy grains that may have tensile strengths as low as about 10^3 Pa. For such a fluffy grain with a radius of 1 μm, electrostatic fragmentation can occur already at surface potentials as low as 10 V. The other process leading to material losses from dust grains is sputtering due to impact of energetic ions.

These processes are very important in space plasma environments. However, their investigation in conditions resembling those of outer space is rather difficult. Usually several processes act in accord. Also, ambient conditions change rapidly. We can only study charging processes in a highly idealized laboratory setting. In our experiment, we store a single (charged) dust grain in an electrodynamic trap. The grain can be exposed to electron and/or ion beams of a variable energy. We can measure grain diameter, mass, and charge as well as the currents flowing from and to the grain. Each charging process can thus be studied separately.

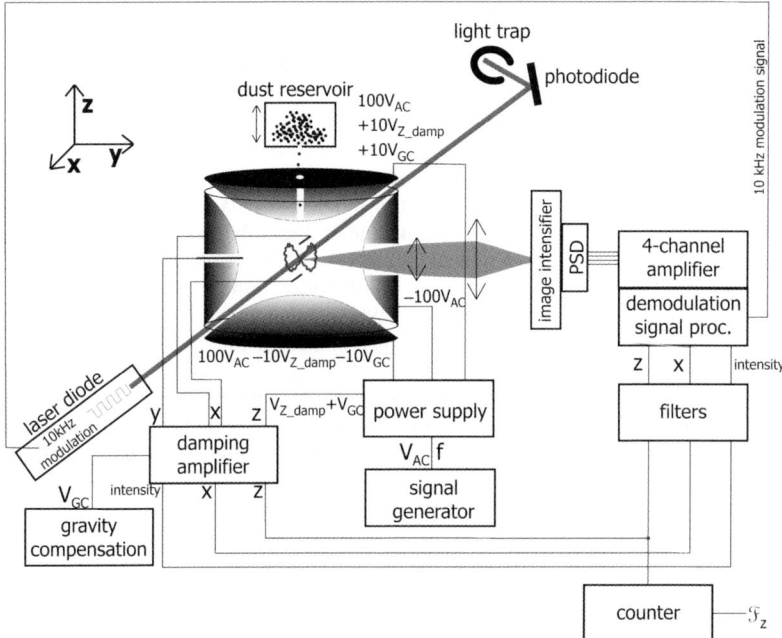

Fig. 1 Experimental setup. For details see text.

In this paper, we focus on electron-induced SEE, electron and ion field emissions, and sputtering of dust grains by energetic ions. By comparing these results with *in situ* observations, we provide a new view of the fundamental physical processes of dust in space.

2. Experimental

Details of the experimental setup are provided elsewhere,[6–9] thus, here we will only repeat the basic principles. The apparatus uses an electrodynamic quadrupole (a 3D quadrupole trap[10] with modified hyperbolic electrodes). As can be seen in Fig. 1, a grain trapped in the 3D quadrupole is irradiated by a 635 nm diode laser. The light scattered off the grain passes a small window in the ring electrode (electrically screened by a grid) and is magnified by a simple lens system. Since the intensity of the scattered light is rather low, an image intensifier is used to increase the detection sensitivity. The output of the image intensifier is directly coupled to a PIN diode serving as a 2D coordinate detector. The laser light is modulated at 10 kHz for a reduction of noise. The electrically amplified signal from the PIN diode is demodulated by the lock-in technique and the coordinates of the light spot are calculated. The grain oscillation frequency can be determined by a frequency counter or by Fourier analysis.

After several simplifications, theoretical considerations lead to the following relation between the grain oscillation frequency and its charge-to-mass ratio (specific charge, Q/m):

$$\frac{|Q|}{m} \cong \frac{2\pi^2 r_0^2 f F_u}{|\lambda_u| \, V_{\text{ef}}}, \quad (1)$$

where

$V_{\text{ef}} = V_{\text{AC}}/\sqrt{2}$ is the RMS value of the AC voltage on the quadrupole electrodes, V_{AC} is its amplitude, $f = \omega/2\pi$ the frequency of the applied AC voltage, $F_u = \Omega_u/2\pi$ the frequency of the particle motion in the u direction, r_0 denotes the inner radius of the quadrupole ring electrode, and λ_u is the weight factor of the electric field in the u direction.

This relation is based on the assumption of an adiabatic motion of the particle in the quadrupole field. This is valid for a sufficiently high-frequency ratio between the applied AC voltage and the particle oscillations. Further, the expression assumes an ideal quadrupole field. Any deviation from the ideal hyperbolic geometry manifests itself in a contribution of higher multipoles to the total field. The higher multipoles lead to an anharmonic potential, thus, the particle oscillation frequency becomes amplitude dependent. Since the deviation due to the anharmonicity increases with the oscillation amplitude, the amplitude must not exceed a certain value for a desired accuracy of the frequency determination. On the other hand, a reliable measurement of the oscillation frequency requires a sufficiently high oscillation amplitude. Therefore, to provide reproducible measurement conditions, the amplitude of the grain motion has to be stabilized.

The experiment was designed for operation under ultra-high vacuum conditions ($\approx 10^{-7}$ Pa or better). This is essential in order to reduce the interaction of the grain surface with molecules of the residual atmosphere. Interaction of the investigated grain with photons, electrons, or ions may lead to a substantial increase of its vibrational temperature, and thus, to an increase of the oscillation amplitude. Under ultra-high vacuum conditions, the collisional cooling of the grain by the residual gas is negligible and the grain can maintain this vibrational temperature for a very long time, or it can be even further heated by anharmonic effects. To control the vibrational temperature of a stored particle under any experimental conditions, we have developed an active control system. This system uses several auxiliary electrodes as well as the quadrupole electrodes themselves to produce an additional electric field along each coordinate. The electrodes are supplied by voltages derived from the coordinate signals. The coordinates of oscillating dust are monitored and the electric signal is used in an active feedback loop to control the amplitude of the dust motion. The damping signals are amplified and used to create auxiliary electric fields in the quadrupole that controls the grain oscillations.

The grain oscillation frequency is the only measurable quantity, and we have developed several techniques to determine the grain mass, charge, capacitance, and other parameters. The detailed description of these techniques can be found in Čermák et al.[8] and Pavlů et al.[9]

SEM images of the dust samples are shown in Fig. 2. Metallic grains are generally used as testing samples because their material properties are well known from previous studies, whereas SiO_2 and carbon are materials that can be frequently found in space. The ensemble of the used samples is complemented with grains from a Melamine Formaldehyde resin (hereafter MF) and with MF grains covered by a thin (≈ 20 nm) layer of nickel (hereafter MF/Ni). The dimensions of the SiO_2, MF, and MF/Ni samples are well defined, whereas the mass and the dimension of other grains have to be determined in the course of the experiment.

3. Secondary electron emission

Many papers on SEE from various materials have been published. Interest continues because of the important role that SEE plays in many fields of modern technology—insulator breakdown (damaging electronic devices), electron lithography, charging of spacecrafts and space dust grains exposed to cosmic radiation, and many others. As we noted above, secondary emission plays a prominent role when a portion of energetic (> 10 eV) electrons is present in the medium surrounding the grain.

One of the first theories published by Sternglass[11] describes experimental SEE from planar metal surfaces in the range of hundreds of eV to several keV. Primary electrons impacting the sample surface interact with the bulk material and lose their energy in many types of collisions. The energy losses often result in excitations of

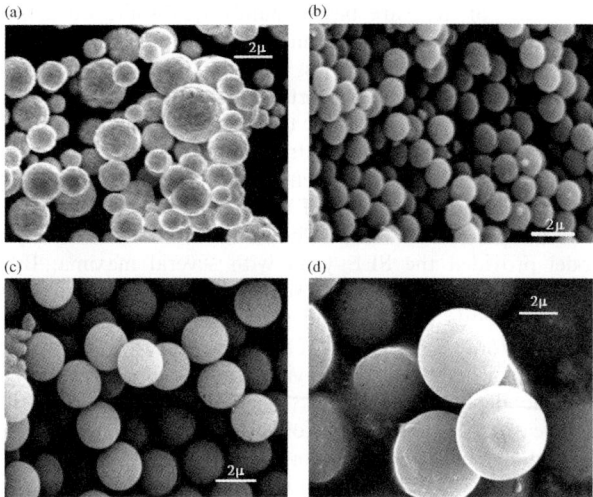

Fig. 2 Examples of grains—SEM photos: (a) gold grains, (b) monodispersed SiO$_2$ grains of 1.2 μm diameter, (c) precise-size MF grains of 2.35 μm diameter, (d) MF grains of 4.97 μm diameter.

material electrons and some of the excited electrons can leave the surface. These electrons, the so-called true Secondary Electrons (SE), have typical energies of a few eV. For large planar samples, the energy dependence of the SEE yield, $\delta(E)$, (defined as the mean number of SE per one primary electron) can be described by the Sternglass universal curve:[11]

$$\frac{\delta}{\delta_{max}} = \frac{E}{E_{max}} \exp\left(2 - 2\sqrt{\frac{E}{E_{max}}}\right), \qquad (2)$$

where δ_{max} is the maximum SEE yield, and E_{max} is the corresponding primary electron energy. For real materials, the curve exhibits a maximum at energies in the range from several hundred eV to a few keV and decreases to zero at very high and low beam energies. Its parameters, the maximum yield, δ_{max}, and the corresponding energy, E_{max}, depend only on the sample material at the incident angle. However, the validity of the universal curve is limited to $4E_{max}$. In a later work, Draine and Salpeter[12] found a new approximation by fitting to experimental data; this approximation is valid to higher energies.

The primary electrons undergo scattering and may be re-emitted from the solid without a significant loss of their initial energy. The yield of backscattered electrons, η, increases with the material density and the atomic number up to ≈ 0.5 for a normal incident angle. It grows only slowly with the beam energy above a few hundreds of eV. Thus, the total SEE yield, $\sigma = \delta + \eta$, and the yield of true SE δ varies in a similar way with the beam energy.[13]

Srama et al.[4] noted that the process of SEE is very sensitive to the grain size and to the physical properties of the dust grains. At electron energies of >1 keV, the mean free path of electrons in a compact dust grain is >0.1 μm, therefore, electrons can penetrate through small grains and can also cause SEE from the exit side. However, using laboratory simulations and the SEE model, Richterová et al.[14] has shown that this effect has only negligible consequences for the grain charge.

The SEE yield increases with increasing angle of incidence of primary electrons by up to one order of magnitude over that seen from plane surfaces.[12] Our preliminary estimate of this for spherical grains and a parallel electron beam suggests an increase of the SEE yield by a factor of 1.2–2.

Theoretical studies of SEE are based mainly on Monte Carlo simulations of electron trajectories.[15–19] However, the majority of models was applied to planar metal or insulating targets and much less attention was paid to spherical samples. The finite grain size plays an important role when primary electrons are energetic enough to penetrate through the grain. Theoretical considerations of SEE from spherical sub-micrometre oil drops were carried by Ziemann et al.[20] He achieved a good match with experiment up to a primary energy of 250 eV. Chow et al.[21] developed a model of SEE from spherical bodies. Since that model did not reproduce experimental data,[22] Chow et al.[23] published an improved model. The new model provided the SEE yield with several maxima. By varying the parameters of the model, the authors were able to fit the data but they had to use different sets of constants for low- and high-energy regions. Richterová et al.[24] developed a simple Monte Carlo model of SEE from spherical dust grains. Although the model does provide a typical SEE yield curve and can roughly describe the observed energetic dependencies of the dust grain equilibrium charge, it has numerous non-measurable parameters. For this reason, Richterová et al.[14] prepared a new model based on more realistic assumptions. The results of the improved model can be briefly summarized as follows: (1) the scattering of primary electrons inside the grain is critical to understanding the SEE, (2) the grain charge is determined not only by the SEE yield—the energy distribution of SEs is equally important, (3) the increase of the surface potential with decreasing grain size is predominantly caused by the increasing number of backscattered primary electrons, not by true SEs. All these statements were demonstrated experimentally on spherical gold and glass grains.

SEs charge the dust grain positively; negative potentials can be reached only if the total SEE yield, σ is lower than unity. It is generally true for low beam energies (tens of eV) and, as we will show later, in the keV range for grain materials with a specific combination of true and backscattered SEE yields. These considerations are based on the assumption of compact spherical dust grains. In space, however, irregularly shaped grains can significantly change their properties. In order to demonstrate the effects of the grain surface structure and shape, we have carried out measurements of surface potentials of grain clusters composed of different numbers of ≈ 1.2 μm SiO_2 spheres. The clusters were exposed to an 0.1–10 keV electron beam, thus SEE was the dominant process for grain charging. The mass of each investigated cluster was determined by the method of elementary-charge steps[9] and the number of individual grains in the cluster was estimated from the total mass. These numbers are shown as parameters of surface potential profiles in Fig. 3. The shape of this profile for a single grain was discussed by Richterová et al.,[25] thus, we will concentrate on differences caused by the clustering.

Fig. 3 shows that, whereas the initial parts of all profiles are nearly identical, the slopes of the high-energy tails decrease with the number of spheres in the cluster in a systematic manner. The equilibrium potential of the grain or cluster is determined by a balance of incoming and outgoing electrons. If the beam energy is sufficiently low, the electrons are emitted only from the exposed part of the cluster and the curves roughly follow the dependence found for planar surfaces. If the emission from the "opposite side" of a particular grain becomes important, the surface potential is a decreasing function of the number of grains in the cluster. We suggest that it is caused by the shielding of the SEs as well as backscattered electrons by other grains in the cluster, because the SEs leaving the material "inside" the cluster are captured by another surface. It is interesting to note that, even for a relatively high number of grains in the cluster, the charging characteristics differ from those measured for compact grains of a similar equivalent diameter. The curve was measured for one cluster consisting of about 110 grains, thus, its mass corresponded to a 5 μm sphere, on which no comparable effect of enhanced emission at high energies could be observed. The lowest curve in Fig. 3 shows the surface potential profile measured on a 3.6 μm compact glass grain. The curve resembles the features measured on clusters

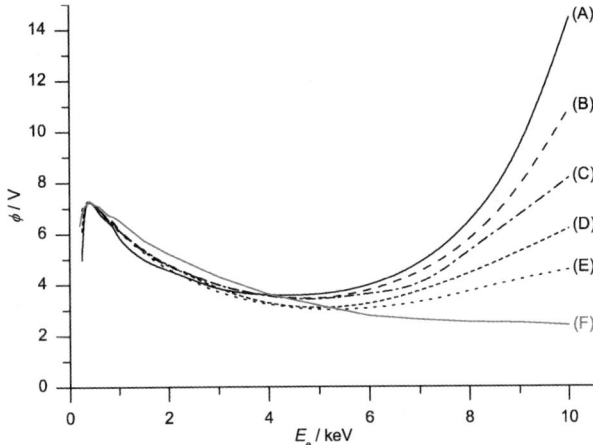

Fig. 3 Profiles of surface potential as a function of the primary electron beam energy measured for SiO$_2$ grains of 1.2 μm diameter and their clusters in comparison with a large glass grain. (A) Single grain, (B) cluster of three grains, (C) cluster of five grains, (D) cluster of about 34 grains, (E) cluster of more than 110 grains, (F) the glass grain of a diameter of about 3.6 μm.

in the low-energy range but the potentials follow the decreasing trend also at high energies. We assume that the grains on the edge of the cluster surface, which are not shielded, are responsible for the enhancement of the cluster potential at high beam energies.

This result can be applied to compact grains of an irregular shape. The surface potential of such grains will be determined by the part with the smallest characteristic dimension, thus, it would be significantly higher than that derived by a spherical approximation. As the cluster geometry in our experiment is unknown, we cannot reliably determine the tensile stress. However, the spherical approximation provides an upper limit of the order of 10 Nm^{-2}, *i.e.*, even below the value required for the destruction of fluffy aggregates. However, the tensile stress increases with decreasing diameter as the surface potential rises, and it could be sufficient to destroy clusters of nanoparticles.

It is generally expected that the surface potential of grains exposed to an electron beam would be a simple function of the SEE yield of a given material. In order to check this expectation, we have summarized the measurements of the equilibrium grain potential as a function of the electron beam energy carried out on grains of different diameters and materials in Fig. 4. The abscissa is divided into two regions: the high and the low energies. In the low-energy range, the profiles of the potential qualitatively follow the well known energy dependence of the SEE yield for planar samples because they rise with the beam energy until a maximum is reached. The beam energy corresponding to this maximum is several hundred eV. However, the SEE yield measured on planar samples would fall in the high-energy range since it is approximately equal to the yield of the backscattered electrons for highest energies. However, we could find even a rise of the grain potential in this energy range. These effects are connected to the finite size of the dust grain. They were broadly discussed in Richterová *et al.*[26] The computer simulation of Richterová *et al.*[14] showed without a doubt that for small dust grains, the size effects are predominantly caused by the increase of the yield of backscattered electrons. We would like to point out that the 3.7 μm carbon and the 2.35 μm MF grains, in the primary energy range from 3 to 7 keV, exhibit potentials lower than the detection limit of our setup. The use of a beam of energy between 4 and 6 keV would lead to negative grain potentials. Grains from all other investigated materials are charged positively in our energy range (0.1–10 keV).

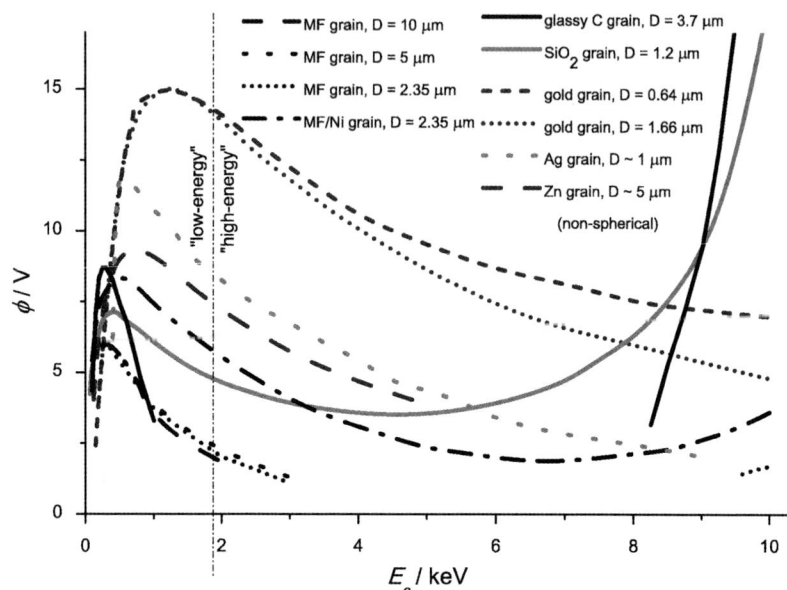

Fig. 4 Surface potential profiles as a function of the primary electron beam energy—comparison of various materials.

The quantitative interpretation of the low-energy part of the potential profiles is more difficult. One would expect that a larger SEE yield would result in a larger surface potential, but this is true only partially, as can be seen from Fig. 5. The surface potential at the SEE maximum, ϕ_{max} increases with σ_{max} (values from Bronstein[13] are used) for all depicted materials except SiO_2. SiO_2 has the largest SEE yield, but a rather small maximum potential ϕ_{max}. Taking into account that the equilibrium potential is reached when the current of primary electrons is balanced with the current of all true secondary and backscattered electrons with energy sufficient to overcome the grain potential, we can conclude that the principal difference between SiO_2 and other analyzed grains is in the energy spectrum of the

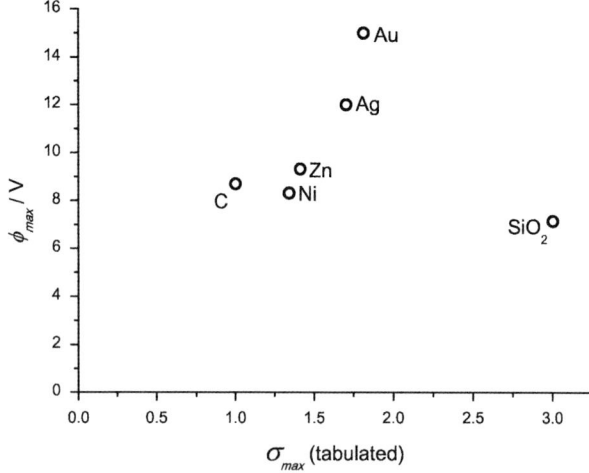

Fig. 5 Surface potential at the maximum of SEE as a function of the tabulated maximum total yield of SEE, σ.

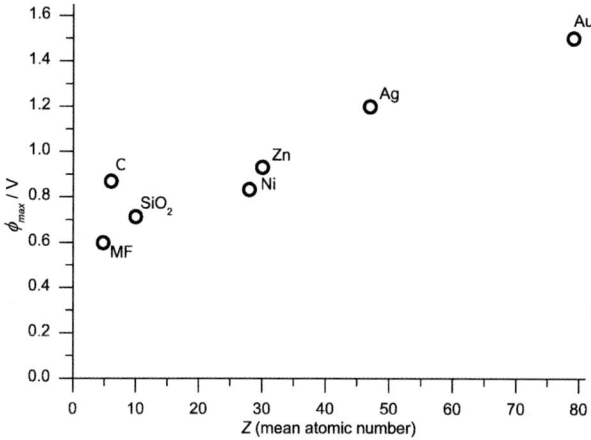

Fig. 6 Surface potential at the maximum of SEE as a function of atomic number, Z.

SEs. However, the reason for this difference is not clear. All materials in the first group are conductive elements, whereas SiO_2 is an insulator and a chemical compound. This problem can be elucidated with the data in Fig. 6, where the surface potentials are plotted as a function of the atomic number, A (the mean atomic number was used for compounds). This plot allows us to implement the measurements on MF that were not used in Fig. 5 because the SEE yield of MF is unknown. One can note that the atomic number sorts the data rather well. Since the atomic number influences predominantly the scattering of the electrons inside the grain, we can tentatively conclude that this scattering plays an important role in the formation of the energy spectrum of SEs. However, confirmation of this hypothesis requires precise measurements of the energy spectra of SEs in the eV range. This task, for single isolated microparticles, is a rather difficult experiment.

4. Field electron emission

Field emission of electrons from negatively charged dust grains occurs at high electric field strengths. This process was described in the 1950s for metals.[27] An explanation of this phenomenon is based on the inflection of the electron bands in the material, which enlarges the probability of electron tunneling from the surface to the vacuum level.

The field emission current is described by the Fowler–Nordheim equation:[27,28]

$$i \sim \frac{F^2}{\varphi t^2(F, \varphi)} \exp\left(\frac{\varphi^{\frac{3}{2}}}{F} v(F, \varphi)\right), \quad (3)$$

where F stands for the electric field strength, φ for the work function, and the variables t and v are weak functions of F and φ. If we plot the dependence of $\log i/F^2$ vs. $1/F$, the work function can be easily derived from the slope of this nearly linear plot:

$$\text{slope} = -2.96 \times 10^7 \varphi^{\frac{3}{2}} s(F, \varphi), \quad (4)$$

where the function s weakly depends on F and φ and varies in most cases within the range of 0.83–1, i.e., its change is usually smaller than the scatter of the experimental points. All three functions s, t, and v are described by Good and Müller,[27] their values are tabulated.

Since measuring the work function of insulators is difficult, the Fowler–Nordheim plot could be used for this task. Eqn (3) was derived for metals but one can expect

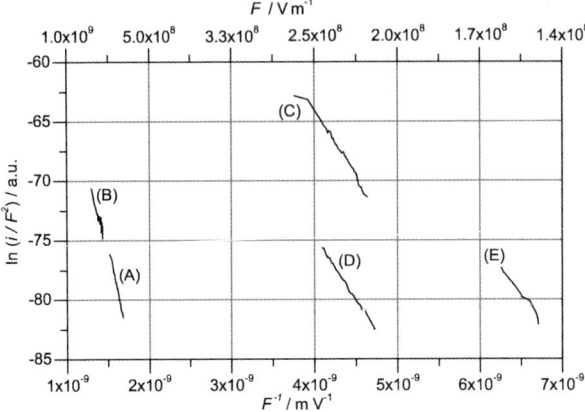

Fig. 7 Fowler–Nordheim plot of grain discharging due to a field electron emission for various materials. (A) Gold grain, $\varphi \approx 5.2$ eV, (B) MF grain (of a diameter of 4.97 μm), $\varphi \approx 4.6$ eV, (C) amorphous carbon grain, $\varphi \approx 2.5$ eV, (D) MF grain covered by a thin layer on nickel, $\varphi \approx 2.5$ eV, (E) SiO$_2$ grain, $\varphi \approx 2$ eV.

that a highly charged dust grain from any material has enough free electrons, thus, this method can be applied.

In order to observe the field emission current, the grain has to be charged to a sufficiently negative potential. As can be seen in Fig. 5, SEE tends to charge the grains positively. Negative surface potentials can be reached if the SEE yield is lower than unity. This is true at very low primary energies when electron attachment dominates (up to several tens of eV for insulators) and in the keV range for grains with a specific combination of the grain material and size (see below). Examples of such grains are carbon and MF particles in Fig. 5. We would like to point out that the application of high beam energies is further complicated by the enhancement of the SEE yield due to the large electric field at the grain surface.[29]

We have recorded field emission currents from several grains of different materials. The corresponding Fowler–Nordheim plots are shown in Fig. 7. These plots reveal several surprising facts: (1) The field emission current is able to fully compensate the primary beam current at field strengths as low as 1.5×10^8 V m^{-1} (the scale at the top edge of the figure), (2) the work functions that are given as parameters beside the curves vary in a broad range. Values for gold ($\varphi = 5.2$ eV) and MF ($\varphi = 4.6$ eV) roughly correspond to the expectations but those for MF/Ni, carbon, and SiO$_2$ are too low. However, values obtained on MF grains are questionable for several reasons described further below.

This effect was found by Pavlů et al.[30] for SiO$_2$ and attributed to the emission of electrons from surface states of SiO$_2$. However, such an explanation can be hardly applied to MF/Ni grains because the electrons are emitted from a metallic surface of the grain. If one would assume that the Ni layer on the grain surface is oxidized and insulating, the presence of surface states can be expected. The last sample exhibiting a small work function is amorphous carbon. Since the internal structure of carbon can vary from insulating diamond to conducting graphite and the exact structure of the samples is unknown, we can probably safely assume that the emission occurs from surface states with a low work function. Moreover, carbon is a highly reactive species that tends to saturate the dangling bonds at the material surface. In most cases, the bonds are saturated by hydrogen, however, other, more complicated chemical compounds can be chemisorbed, which can act as a source of the tentative surface states.

In contrast to the low work functions measured for the aforementioned samples, Fig. 7 shows measurements on an MF grain that exhibit a work function equal to 4.6 eV. This is in good agreement with the work functions of insulators found by

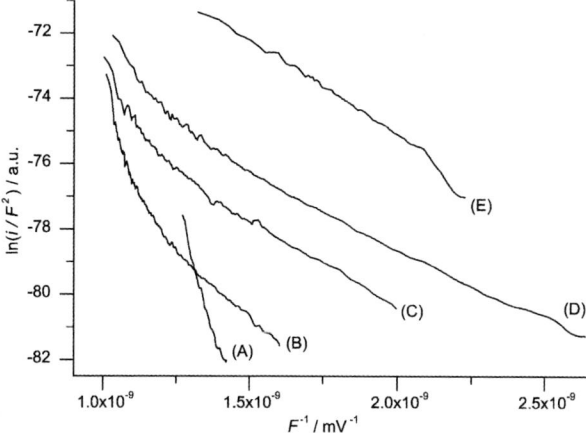

Fig. 8 Fowler–Nordheim plots of one MF grain. Characteristics were measured consecutively in the order from (A) to (E) using a 5 keV electron beam. Estimated work functions thus vary over 1.4–7.5 eV.

Sternovsky et al.[31] using triboelectric charging. The measurements on MF and SiO_2 were compared in Pavlů et al.[29] The authors suggest that the way of charging (i.e., use either low- or high-energy electrons) is the reason for the substantial differences in the work functions of these materials.

As noted above, a negative grain charge can be achieved by applying either low- or high-energy electron beams. Whereas the eV electrons cannot modify the grain structure and can be captured in the surface states, the high-energy electrons penetrate deeper and their energy can destroy the electronic structure of the original insulating material with a clear band gap. In the experiments of Pavlů et al.,[29] SiO_2 samples were charged by a low-energy beam, whereas a 5 keV beam was used for MF grains. However, our new measurements on carbon samples charged by the 5 keV beam seem to contradict the above hypothesis. For this reason, we have repeated the measurements on MF grains and the results are plotted in Fig. 8. Different curves in this figure were measured on the same sample and they differ in the duration of bombarding (charging) the grain with 5 keV electrons. One can note a gradual evolution of the sample structure leading to a decrease of the work function from $\phi = 7.5$ eV to $\phi = 1.4$ eV. This suggests that the electron bombardment strongly affects the internal structure and that long-lived surface states are populated by the exposure to energetic electrons. Moreover, the measurements in a broader range of discharging currents clearly show that the Fowler–Nordheim theory cannot be applied to the MF samples, since the plots are strongly non-linear. This is probably caused by several effects: first, the surface states are discrete, thus, the description by the Fowler–Nordheim theory based on a band structure cannot be correct, and second, the finite lifetime of the surface states leads to their depopulation, thus, to a redistribution of the density of states.

Since such materials are relevant for investigations of interplanetary space, an explanation of the observed effects is desirable. Moreover, MF spheres are frequently used as calibration objects for their well defined diameters in electron microscopy and other experimental techniques. However, some of our measurements suggest a significant change in their dimensions during electron bombardment might be occurring.[32]

5. Field ion emission

Current understanding of field emission of ions from dust grains is based on experiments and theoretical considerations connected to field ion microscopy (see, e.g., review of Forbes[33] and references therein). The studies are based on the assumption of a sharp conductive tip biased by an external voltage source. The mechanism of field ion emission from an isolated particle, however, may be different since the strong electric field at the surface is a consequence of charge accumulated in the grain due to ion bombardment.

The emission of positive ions may be caused by three main processes: field desorption, field evaporation, and field ionization.[27,34] All these processes are based on tunneling of electrons from atoms situated above the sample surface into or toward the sample. The tunneling probability has a sharp maximum at a critical distance that is a function of the electric field strength. Consequently, the tunneling can occur only in a very thin layer above the dust grain. The three processes differ in the source of the atoms entering this layer. Atoms and molecules of the surrounding gas represent the source for the field ionization, whereas an "internal" source—atoms of the adsorbed gas or atoms of the grain material—acts in the field desorption and in the field evaporation, respectively. Since the adsorbed atoms are weakly bound in comparison with the bulk atoms, the field desorption should be the dominant process that releases the charge from positively charged dust grains in space or under experimental UHV conditions.

Field desorption proceeds in two phases. Adsorbed gas atoms leave the surface and are ionized at (or slightly beyond) a critical distance and are then pushed away by the repulsive electric force. The positive charge from the adsorbed atom leaves in the form of the ion; the released electron is attracted by the grain, and thus does not change the net charge. Pavlů et al.[35] and Jeřáb et al.[36] performed laboratory experiments where impacts of energetic ions led to a deposition of a positive charge onto a spherical grain. In these experiments, the accumulated charge is spontaneously released and the discharge is measurable when a certain value of the electric field at the grain surface is reached. This value was experimentally found to be on the order of 10^9 V m^{-1}, which is rather low, however, the field strength was determined under the assumption of spherical samples, thus, a local deformation of the sample surface could enhance the field by a factor of 3–5. By analyzing the plots of the discharge currents vs. the electric field strength, they found the field desorption to be the main discharging process. The authors suggested that the beam ions recombine at the grain surface and create a layer of atoms that then desorb during the discharging process.

In order to check this suggestion, we have carried out a series of measurements on one gold grain. The results are plotted in Fig. 9. In this experiment, the grain was exposed to a 5 keV ion beam for a time that is called "time of treatment" in the figure. After the beam was switched off, the discharge current was recorded together with the grain surface potential for several hours. The value of the discharge current at 1 kV surface potential is plotted on the ordinate. Although the spread of the experimental points is rather large, the plot shows a clear increase of the discharge current with the time of treatment. We would like to note that the uncertainty of the discharge current is principally due to the charge quantization—the measured current is rather small (number of elementary charges per second is used as the current unit in the figure) and the measurement time is limited by the rapid discharging at the origin.

The increase of the discharge current with the time of treatment is surprising because one would expect that grain charging is stopped at the point when the number of incoming and outgoing atoms are in equilibrium. Our experiments suggest that there is a mechanism that stores and gradually releases a portion of incoming ions.

Our tentative explanation is based on the implantation of beam ions into the grain material. However, this hypothesis must be checked further. Since the diffusion of the implanted ions toward the surface is probably a very slow process, it opens the question whether it can considerably contribute to the discharge current.

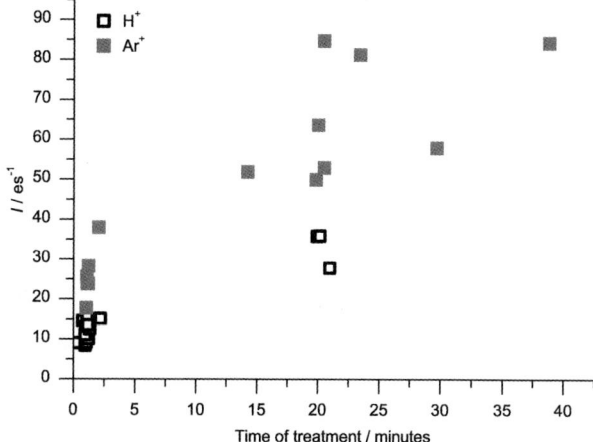

Fig. 9 Field ion emission currents from one gold grain after different treatment times, and for two ion species. Currents were measured at 1 kV of the grain surface potential. The full squares denote the Ar^+ bombardment, while the open circles show bombardment by H^+.

The implantation of solar wind ions and their subsequent release in the form of neutral atoms was suggested as a possible source of the so-called pick-up ions in the solar system (*e.g.*, Mann *et al.*[37]). Thus, the laboratory investigation of the rates of these processes is of great interest and will be followed up on in our future experiments.

6. Dust grain sputtering

Investigations of the sputtering rate require continuous monitoring of the grain mass evolution. By a precise measurement of the grain's oscillation frequency, we can determine the change in the specific charge Q/m induced by one elementary charge and, consequently, we can obtain the total charge and mass of the grain. Although the precision of the mass determination is better than 10^{-6}, the noise in the secular frequency restricts this measurement to weakly charged grains only, typically not larger than 10^5 elementary charges.

An alternative method is the determination of the grain's specific capacitance C/m that acts as a scaling constant between the specific charge and the surface potential:

$$\frac{Q}{m} = \frac{C}{m}\phi. \tag{5}$$

The specific capacitance C/m can be determined from the specific charge at a known surface potential, ϕ.[9] Assuming a spherical grain, $C = 4\pi\varepsilon_0 R$, the specific capacitance depends on the grain radius R and the mass density of its material only. From the known mass density, the grain radius, and consequently its mass, can be obtained.

In the experiment on grain sputtering, a spherical gold grain of an initial radius $R_0 = 0.57$ μm and a corresponding initial mass $m_0 = 1.48 \times 10^{-14}$ kg was used. The sputtering was initiated by an Ar^+ beam with an energy $E_0 = 5$ keV and a current density $i_p = 4 \times 10^{-4}$ Am^{-2}.

The primary experimental task is the investigation of the grain sputtering, but the beam ions charge the grain and the grain charge influences the energy and the flux of the beam ions impinging the grain. Without any other processes, the grain will charge up to the surface potential numerically equal to the energy of the beam ions (their effective energy, respectively) and the ion flux will drop to zero. However, as we have shown in the previous section, the grain charge is limited by the field ion

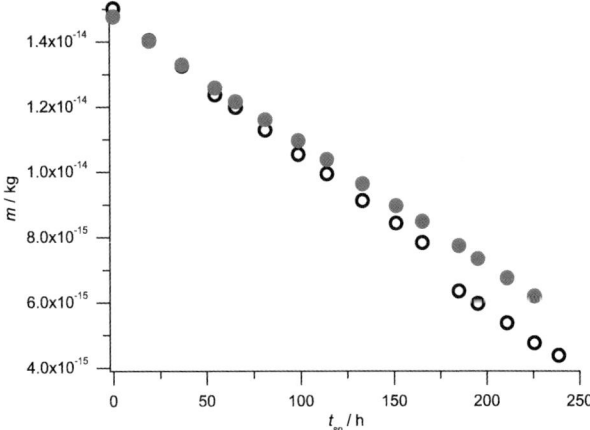

Fig. 10 Reduction of the grain mass in the course of ion bombardment. The full circles show the mass estimated by the elementary charge method; open circles stand for the mass estimated from the grain capacitance.

emission. Since the emission current increases rapidly with the electric field strength at the grain surface,[36] the grain's surface potential can be considered as constant for our experiment. We have obtained the value of the corresponding electric field: $F = 1.95 \times 10^9$ V m^{-1}. The constant surface field strength implies changes of the surface potential during the grain sputtering due to the changes of the grain diameter. Indeed, this potential was $\phi \approx 1.05$ kV at the beginning of the experiment and $\phi \approx 0.75$ kV at the end. The energy of the beam ions with respect to the grain surface thus changed from 3.95 to 4.25 keV. According to Behrisch,[38] the sputtering yield, Y, is roughly constant in this energy range with the following value: $Y \approx 5$. This value was determined for Ar$^+$ ions impacting perpendicular to a planar surface. The yield increases with decreasing incidence angle,[38] and we have estimated[39] that a factor of 1.2–2 should be considered for sputtering of spherical samples.

The temporal evolution of the mass of the treated Au grain with an initial radius of 0.568 μm is shown in Fig. 10. For direct mass measurements, we applied the two techniques mentioned above—the method of elementary charge steps and the calculation of the specific capacitance. Both methods lead to a sputtering yield of $Y \approx 20$, *i.e.*, four times larger than that for a planar uncharged surface. We suggest that this enhancement can be divided into two multiplicative parts. One part is attributed to the shape of the sample, the other part is probably caused by the presence of a strong electric field at the grain surface. This field is achieved by ion bombardment in the specific conditions of our experiment, however, very small grains in space (≈ 20 nm), when sufficiently charged by photoemission, can reach similar field strengths. Moreover, we used a gold grain to be able to make comparison with known data, but the dust materials in space are insulators and the yield enhancement can be much larger due to the penetration of the electric field into the grain. These processes will be the subject of our further studies.

As can be seen in Fig. 10, the two methods of grain mass determination provide results that systematically differ. Since the method of the elementary charge steps does not require any additional assumption, we assume that it provides reliable mass estimates. The determination of the grain capacitance, however, is based on the known grain shape. Comparing the results of the two methods, the changes in the grain shape can be followed. The increasing difference between the measurements lets us conclude that the originally spherical grain shape is changed during the sputtering. The resulting shape is unknown, but the difference can be explained if an increase of the grain capacitance by a factor of 1.2 is assumed.[39] Such an increase corresponds to a change of the grain shape from a sphere to a rotational ellipsoid

Table 1 Rough approximation of sputtering times computed for micron-sized gold grain in different space locations for laboratory and space conditions (Earth orbit)

Conditions	Ion	1/2 mass	1/2 radius
Laboratory	Ar^+	190 hr	450 hr
Space (1 AU)	H^+	8×10^2 yr	2×10^3 yr
	He^{2+}	2×10^3 yr	5×10^3 yr
	Heavy ions	8×10^2 yr	2×10^3 yr

with an axis ratio of $\approx 1 : 2$. This result is surprising because one would expect that the sputtering, together with random rotation of the grains (in our experiment as well as in space), would result in well spherical grains. We suggest that the space charge of the ion beam interacts with the quadrupole term of the electric field of the grain, which can be described as a charged conductive rotational ellipsoid. Since the minimum potential energy of such a charged ellipsoid is reached if its axis is identical with the beam direction, the interaction leads to cancellation of any rotational motion and to orientation of the ellipsoid along the ion beam. As the surfaces parallel to the beam undergo a more rapid erosion, the sputtering causes a positive feedback. This leads to enlarging the axis ratio of the grain, thus to increasing the quadrupole term and, consequently, to a stronger orientation of the grains with respect to the ion beam.

In order to check the possible relevance of the measurements to interplanetary space, we have calculated[40] the sputtering times (Table 1) of a gold grain in typical solar wind conditions. At 1 AU, we assume that the solar wind has a velocity of 400 km s^{-1} and a number density of 10 cm^{-3}. In our estimates, we consider all solar wind ions as singly ionized atoms with an energy of 5 keV. However, Insepov *et al.*[41] show that the sputtering yield of multiply ionized ions can be much larger (by an order of magnitude). Indeed, all species (except protons) in the solar wind are in high ionization states. A typical mass composition of the solar wind was considered, however, the abundance of heavy ions can be significantly enhanced during coronal mass ejections, and these species are very effective for dust sputtering.

We would like to point out that the values in Table 1 are based on the tabulated values of the sputtering yield for gold corrected to the assumed spherical grain shape. The real times of the dust grain sputtering can therefore easily vary within an order of magnitude. Taking into account the observed orientation of conductive grains with respect to the direction of the bombarding ions and the consequently enhanced sputtering yield, we expect that our estimates represent the upper limit of the expected sputtering times.

Since the sputtering rate decreases with the square distance from the Sun, the lifetime of dust particles is expected to be rather short in the inner solar system and the sputtering can be, beside the ion field emission, the next important source of the pick-up ions.

7. Concluding remarks

The paper analyzes new results of dust grain charging under laboratory conditions and discusses the contribution of these simulations to processes in interplanetary space. A single spherical grain, trapped in ultra-high vacuum conditions for long time intervals, is exposed to ion and/or electron beams, and their mutual interactions are investigated. Several charging processes—secondary emission, field ion and electron emissions, and dust sputtering—are described in detail. For these simulations, micrometer-sized grains of various materials and different energies and fluxes of ion and electron beams were used. As a result, we can conclude:

1. The charge and corresponding potential of dust grains affected by the electron beam are determined by secondary emission. We have shown that the energy spectrum of secondary electrons plays an important role in this process and that scattering of electrons inside the grain influences the resulting spectrum of secondary electrons. However, confirmation of this hypothesis requires precise measurements of the SE electron spectra in the eV range.

2. Our study of field electron emission brings a few surprising facts to light—the field emission current is able to fully compensate the primary beam current at field strengths as low as 1.5×10^8 V m^{-1}, and the work functions of some materials (MF/Ni, amorphous C, and SiO_2) are too low. A tentative explanation based on the emission from surface states needs further detailed investigations.

3. We assume that the field desorption is the dominant process that releases the charge from positively charged dust grains in our simulations, thus, we have prepared long-term experiments where impacts of energetic ions led to a deposition of a positive charge onto a grain. An increase of the discharge current with the time of the grain exposure to a 5 keV ion beam was noted. Our tentative explanation of this observation is based on the implantation of beam ions into the grain material. However, this hypothesis should be further checked and the experimental results should be compared with a theoretical simulation.

4. Dust grain sputtering is an important process in space. Our simulations use Au spheres. We found that the sputtering yield can be significantly enhanced by the presence of a high electric field at the grain surface. Moreover, the shape of grains exposed to a collimated ion beam evolves from sphere to ellipsoid, and this change further enhances the sputtering yield. The consideration based on the mean abundance of heavy ions in the solar wind leads to a conclusion that these minor species can considerably contribute to the sputtering rate. However, a typical interplanetary dust is composed of insulators. The sputtering of such materials should be the subject of further studies, because the field-induced sputtering enhancement can be significantly larger due to penetration of the electric field into the grain.

Acknowledgements

The present work was financially supported by research plan MSM 0021620860, which is financed by the Ministry of Education of the Czech Republic.

References

1. J. E. Colwell, M. Horányi and E. Grün, in *Physics of Dusty Plasmas, Seventh Workshop of AIP Conference Proceedings*, ed. M. Horanyi, S. Robertson and B. Walch, American Institute of Physics, Melville, New York, 1998, vol. 446, pp. 299–306.
2. S. P. Wyatt, *Planet. Space Sci.*, 1969, **17**(2), 155–171.
3. P. L. Lamy, J. Lefevre, J. Millet and J. P. Lafon, in *IAU Colloq. 85. Properties and Interactions of Interplanetary Dust of ASSL*, ed. R. H. Giese and P. Lamy, 1985, vol. 119, pp. 335–339.
4. R. Srama, T. Ahrens, N. Altobelli, S. Auer, J. Bradley, M. Burton, V. Dikarev, T. Economou, H. Fechtig, M. Görlich, M. Grande, A. Graps, E. Grün, O. Havnes, S. Helfert, M. Horanyi, E. Igenbergs, E. Jessberger, T. Johnson, S. Kempf, A. Krivov, H. Krüger, A. Mocker-Ahlreep, G. Moragas-Klostermeyer, P. Lamy, M. Landgraf, D. Linkert, G. Linkert, F. Lura, J. McDonnell, D. Möhlmann, G. Morfill, M. Müller, M. Roy, G. Schäfer, G. Schlotzhauer, G. Schwehm, F. Spahn, M. Stübig, J. Švestka, V. Tschernjawski, A. Tuzzolino, R. Wäsch and H. Zook, *Space Sci. Rev.*, 2004, **114**, 465–518.
5. I. Richterová, J. Pavlů, Z. Němeček, J. Šafránková and P. Žilavý, *Adv. Space Res.*, 2006, **38**(11), 2551–2557.
6. I. Čermák *Laboruntersuchung elektrischer Aufladung kleiner Staubteilchen*, PhD thesis, Naturwissenschaftlich-Mathematischen Gesamtfakultät, Ruprecht-Karls-Universität, Heidelberg, 1994.
7. P. Žilavý, Z. Sternovský, I. Čermák, Z. Němeček and J. Šafránková, *Vacuum*, 1998, **50**(1–2), 139–142.

8   I. Čermák, J. Pavlů, P. Žilavý, Z. Němeček, J. Šafránková and I. Richterová, in *WDS'04 Proceedings of Contributed Papers Part II—Physics of Plasmas and Ionized Media*, ed. J. Šafránková, Matfyzpress, Prague, 2004, pp. 279–286.
9   J. Pavlů, A. Velyhan, I. Richterová, Z. Němeček, J. Šafránková, I. Čermák and P. Žilavý, *IEEE Trans. Plasma Sci.*, 2004, **32**(2), 704–708.
10 W. Paul and H. Steinwedel, *Apparatus for separating charged particles of different specific charges*, German Patent 944 900, Jun. 28, 1956. U.S. Patent 2 939 952, Jun. 7, 1960.
11 E. Sternglass, *Theory of secondary electron emission under electron bombardment*, Scientific Paper 6-94410-2-P9, Westinghouse Research Laboratories, Pittsburgh 35, 1957..
12 B. T. Draine and E. E. Salpeter, *Astrophys. J.*, 1979, **231**, 77–94.
13 I. M. Bronstein and B. S. Fraiman, *Secondary Electron Emission*, Nauka, Moskva, 1969.
14 I. Richterová, J. Pavlů, Z. Němeček and J. Šafránková, *Phys. Rev. B*, 2006, **74**(23), 235430.
15 D. C. Joy, *J. Microsc.*, 1987, **147**(1), 51–64.
16 R. Shimizu and Z.-J. Ding, *Rep. Prog. Phys.*, 1992, **55**(4), 487–531.
17 A. Dubus, J.-C. Dehaes, J.-P. Ganachaud, A. Hafni and M. Cailler, *Phys. Rev. B*, 1993, **47**, 11056–11073.
18 R. Renoud, F. Mady, C. Attard, J. Bigarré and J.-P. Ganachaud, *Phys. Status Solidi A*, 2004, **201**, 2119–2133.
19 J. Cazaux, *Nucl. Instrum. Methods Phys. Res., Sect. B*, 2006, **244**, 307–322.
20 P. Ziemann, P. Liu, D. Kittelson and P. McMurry, *J. Phys. Chem.*, 1995, **99**, 5126–5138.
21 V. Chow, D. Mendis and M. Rosenberg, *J. Geophys. Res., [Atmos.]*, 1993, **98**(17), 19065–19076.
22 J. Švestka, I. Čermák and E. Grün, *Adv. Space Res.*, 1993, **13**(10), 199–202.
23 V. Chow, D. Mendis and M. Rosenberg, *IEEE Trans. Plasma Sci.*, 1994, **22**(2), 179–186.
24 I. Richterová, Z. Němeček, J. Šafránková and J. Pavlů, *IEEE Trans. Plasma Sci.*, 2004, **32**(2), 617–622.
25 I. Richterová, J. Pavlů, Z. Němeček and J. Šafránková, *Adv. Space Res.*, 2007, submitted.
26 I. Richterová, Z. Němeček, J. Šafránková, J. Pavlů and M. Beránek, *IEEE Trans. Plasma Sci.*, 2007, **35**(2).
27 R. H. Good and E. W. Müller, in *Electron-Emission Gas Discharges I of Encyclopedia of Physics*, Springer-Verlag, 1988, vol. XXI ch. 2, pp. 176–231.
28 R. Fowler and L. Nordheim, *Proc. R. Soc. London, Ser. A*, 1928, **119**(781), 173–181.
29 J. Pavlů, Z. Němeček, J. Šafránková and I. Čermák, *IEEE Trans. Plasma Sci.*, 2004, **32**(2), 607–612.
30 J. Pavlů, Z. Němeček, J. Šafránková and I. Čermák, *Czech. J. Phys.*, 2003, **53**(2), 151–162.
31 Z. Sternovsky, S. Robertson, A. Sickafoose, J. Colwell and M. Horanyi, *J. Geophys. Res., [Atmos.]*, 2002, **107**(E11), 5105.
32 I. Richterová, J. Pavlů, Z. Němeček, J. Šafránková and M. Jeřáb, in *New Vistas in Physics of Dusty Plasmas of AIP Conference Proceedings*, ed. L. Boufendi, M. Mikikian and P. K. Shukla, American Institute of Physics, Melville, New York, 2005, vol. 799, pp. 395–398.
33 R. G. Forbes, *Ultramicroscopy*, 2003, **95**, 1–18.
34 R. Gomer, in *Field Emission and Field Ionization*, Harvard University Press, Cambridge, Massachusetts, Harvard monographs in applied science, 1961, vol. 9.
35 J. Pavlů, A. Velyhan, I. Richterová, J. Šafránková, Z. Němeček, J. Wild and M. Jeřáb, *Vacuum*, 2006, **80**(6), 542–547.
36 M. Jeřáb, I. Richterová, J. Pavlů, J. Šafránková and Z. Němeček, *IEEE Trans. Plasma Sci.*, 2007, **35**(2).
37 I. Mann, H. Kimura, D. A. Biesecker, B. T. Tsurutani, E. Grün, R. B. McKibben, J.-C. Liou, R. M. MacQueen, T. Mukai, M. Guhathakurta and P. Lamy, *Space Sci. Rev.*, 2004, **110**, 269–305.
38 R. Behrisch, *Sputtering by Particle Bombardment I, II*, Springer-Verlag, Berlin-Heidelberg-New York, 1981.
39 J. Pavlů and J. Wild, in *WDS'00 Proceedings of Contributed Papers: Part II—Physics of Plasmas and Ionized Media*, ed. J. Šafránková, Matfyzpress, Prague, 2000, pp. 238–242.
40 J. Pavlů, I. Richterová, Z. Němeček, J. Šafránková and J. Wild, *IEEE Trans. Plasma Sci.*, 2007, **35**(2).
41 Z. Insepov, J. P. Allain, A. Hassanein and M. Terasawa, *Nucl. Instrum. Methods Phys. Res., Sect. B*, 2006, **242**, 498–502.

PAPER

Microparticles in plasmas as diagnostic tools and substrates†

Gabriele Thieme,[a] Ralf Basner,[a] Ruben Wiese[a] and Holger Kersten*[b]

Received 12th March 2007, Accepted 16th April 2007
First published as an Advance Article on the web 13th July 2007
DOI: 10.1039/b703733n

An interesting aspect in the research of complex (dusty) plasmas is the experimental study of the interaction of nano- and micro-particles with the surrounding plasma for diagnostic purpose. From the behaviour of the particles, local electric fields can be determined ("particles as electrostatic probes"), the energy fluxes towards the particles ("particles as thermal probes"), or reactive processes on surfaces ("particles as micro-substrates") can be studied. The behaviour of particles in front of an adaptive electrode, which allows for an efficient confinement and manipulation of the grains, has been experimentally studied in dependence on the discharge parameters and on different bias conditions of the electrode. The effect of the biased surface on the charged micro-particles has been investigated by novel particle falling experiments, which were observed by a fast camera. Furthermore, preliminary experiments on the excitation of whispering gallery modes of micro-particles trapped in a plasma sheath have been demonstrated for the first time.

1. Introduction

The idea that externally injected small particles can serve as micro-probes in complex (dusty) plasmas has been under recent investigation by several groups.[1-3] In this paper, the behaviour of test particles will be described under variation of plasma parameters and conditions of the surrounding surfaces. Since the micro-particles can easily be observed in the plasma sheath, they can serve as electrostatic probes for the characterisation of the potential planes and electric fields in this plasma region.[4,5] Usually, the plasma sheath—which is an important zone of energy consumption and, hence, often the essential part of a discharge for application—is difficult to monitor by common plasma diagnostics as Langmuir-probes or optical spectroscopy.

The *PULVA-INP* device, which is available at the INP Greifswald, is superbly suited for these investigations because the sheath structure can be specifically influenced by means of an Adaptive Electrode (AE). Furthermore, the influence of additional plasma sources (*e.g.*, external ion beam source or sputter magnetron) on the behaviour of micro-particles in the rf-plasma can be investigated.[6,7]

---

[a] INP Greifswald, F.-L.-Jahn-Str. 19, 17489, Greifswald, Germany
[b] Institute of Experimental and Applied Physics, University of Kiel, Leibnizstr.19, D-24098, Kiel, Germany. E-mail: kersten@physik.uni-kiel.de

† The HTML version of this article has been enhanced with colour images.

Another opportunity of using particles as less invasive micro-probes is the determination of temperature fields in the plasma. From the temperature-dependent fluorescence of special micro-particles, it is possible to determine their equilibrium temperature resulting from the balance of the energy influxes and losses of the particles in the plasma.[8,9] The analysis of such measurements can provide information about the temperature distribution in the plasma volume and the different energy fluxes to small floating solids in the plasma.

On the surface of micro-particles in a process plasma, not only charging and heating processes are of importance but also chemical reactions.[10–12] With suitable diagnostics (scattering spectroscopy, Raman spectroscopy) one can obtain information about the mechanisms at the surfaces of microscopic solids which are, for example, of great interest in catalyst research and particle coating, respectively.

Hence, micro-particles can be employed in process plasmas as "electrostatic probes", as "thermal probes" and as "chemical probes/micro-substrates". In the present paper, we will show the perspectives of these methods and we will provide (preliminary) results and applications, especially for the first and third method.

2. Particles as electrostatic probes

If dust particles are injected into a plasma, they become negatively charged up to the floating potential V_{fl} by the electron and ion currents (j_e, j_i) towards the particles, and can be confined in the discharge. The spatial distribution and movement of the dust particles in a low-temperature plasma is a consequence of several forces acting on the particles.[13–15] The charged particles interact with the electric field in front of the electrode or wall, respectively, whereas the electrostatic force has to be balanced by various other forces in order to confine the dust grains.

2.1 Forces on dust particles in plasma

According to the balance of gravitational force, electrostatic force, ion drag, neutral drag, and Coulomb interaction, micro-particles disperse in a relatively small region of the plasma sheath depending on their size and charge.[15,16] Usually, only some of these forces play a role in laboratory complex (dusty) plasmas under certain conditions. Commonly, the electrostatic field and the gravitational force are of most importance. Superposition of both forces results in a parabolic potential trap at equilibrium position.[17]

As an example of the effects of the neutral or ion drag, the influence of an external ion beam supplied by an ECR ion source has been demonstrated. The deformation of a levitated 2D dust particle cloud under broad beam ion operation has already been observed and described elsewhere.[7,18] If the beam is switched on, the shape of the trapped particle cloud changes in characteristic manner due to inhomogeneities and divergence in the ion beam.

The effect of the different forces on the particles is shown in Fig. 1. When the hole of the ion beam tube is closed by a shutter (Fig. 1a), the common situation of complex plasma is realized. The particles are levitated in this plane due to the force balance between gravity F_g and electrostatic force F_{el} by the electric field in front of the rf-electrode. If the shutter is removed and, thus, the hole is open, there exists a strong pressure gradient between the rf-plasma region (3 Pa) and the beam region inside the tube (0.1 Pa). The result is a neutral drag F_n by the gas flow that changes the shape of the originally flat dust cloud into a dome-like structure, see Fig. 1b. The additional force F_n acts in radially as well as in a vertical direction due to the pressure gradient, and the dust cloud structure can be explained by flow patterns. Finally, if the ion beam is switched on the dome is distorted again by the pushing ion drag force F_{ion} (Fig. 1c). Since the electric field in the rf-sheath varies strongly with

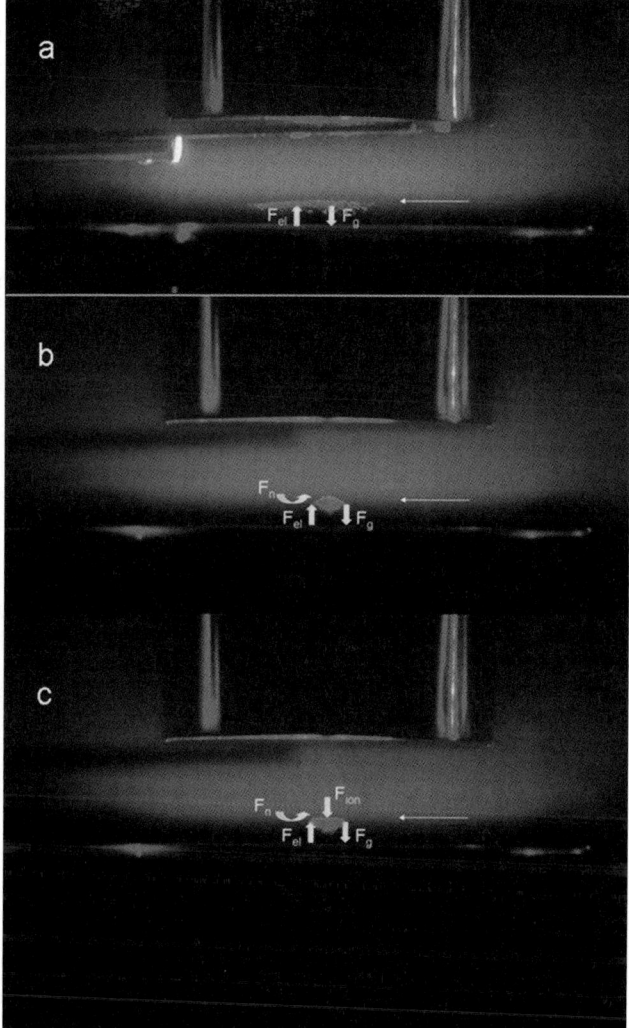

Fig. 1 Interaction of the gas/plasma and the external ion beam with injected test particles at $p = 3$ Pa and $V_{bias} = 150$ V. (a) Hole covered, (b) hole open, (c) ion beam ($V_{beam} = 1300$ V) on.

varying distance z from the electrode, the upper layers of the particle cloud are more likely influenced by the ion beam. Therefore, the displacement of the particle dome (e.g., top of the cloud), which looks like an indentation, has been taken as measured quality.

The knowledge and the manipulation of the spatial particle distribution is, for example, of great interest for sorting of particles and surface modification of powders.[18–21]

However, the interaction of the plasma with injected dust particles is of interest not only regarding their spatial distribution; vice versa, from the particle behaviour conclusions about the surrounding plasma properties can be obtained. With knowledge of the charge and size of the particles, they can be used as electrostatic "microprobes" for measuring the field strength and determining field structures in the plasma and the sheath.[22]

There is only a relatively small knowledge, but a lot of uncertainty on these issues in the plasma sheath. Hence, charged micro particles could provide additional information on the sheath structure. In a few papers, this approach has been successfully demonstrated in front of the powered electrode of a capacitively-coupled rf-discharge.[1,15] In contrast, we present measurements on the behaviour of charged dust grains in front of grounded or biased electrodes. These situations are of special interest for plasma processing of substrate surfaces[23,24] and in plasma chemistry.[25,26]

2.2 Particle behaviour in front of an adaptive electrode

In order to influence the particles and to simulate different electrostatic surface conditions, an AE[22,27] has been used as central part of the experimental setup *PULVA-INP*, see Fig. 2. A typical asymmetric, capacitively-coupled rf-plasma in argon (0.1–100 Pa) is employed to charge the particles, which are spherical Melamine-Formaldehyde (MF) particles of 0.5, 1, 5, and 10 μm in diameter. The rf-power (5–100 W) is supplied by the upper electrode at a frequency of 13.56 MHz and amplitude up to 1000 V. In dependence on the discharge conditions, we measured electron densities of 10^9–10^{11} cm^{-3}, electron temperatures of 0.8–2.8 eV, and plasma potentials in respect to ground of 20–30 V for the pristine plasma.[28] The plasma is bounded to the surrounding surfaces by the self-organizing structure of the sheath, which has a characteristic potential slope and charge carrier profile, and this region shall be probed by the micro particles.

The particles are illuminated by a laser fan (532 nm), their position and movement have been observed by CCD camera and video recording, which were employed to obtain information on the powder particles distributed in the plasma.

At low pressure (<1 Pa), the sheath is collision-free, whereas at higher pressure (>10 Pa) it is dominated by collisions. Most experiments have just been carried out in the transition regime between (1–10 Pa), which is of interest for plasma processing. The transition behaviour can be clearly observed in the energy distribution (IEDF) of the ions hitting a grounded surface. The measured distributions, which are plotted in Fig. 3, have been performed with a EQP Plasma Diagnostic System (HIDEN Analytical). The maximum ion energies vary between nearly thermal energy (∼1 eV) due to thermalization and charge exchange by collisions at high pressure and nearly plasma potential (∼25 eV) due to collision-free transfer from plasma to the surface.

If the boundary wall (electrode, substrate) is without any external potential (*e.g.* floating), the electron and ion fluxes towards the surface are equal and the

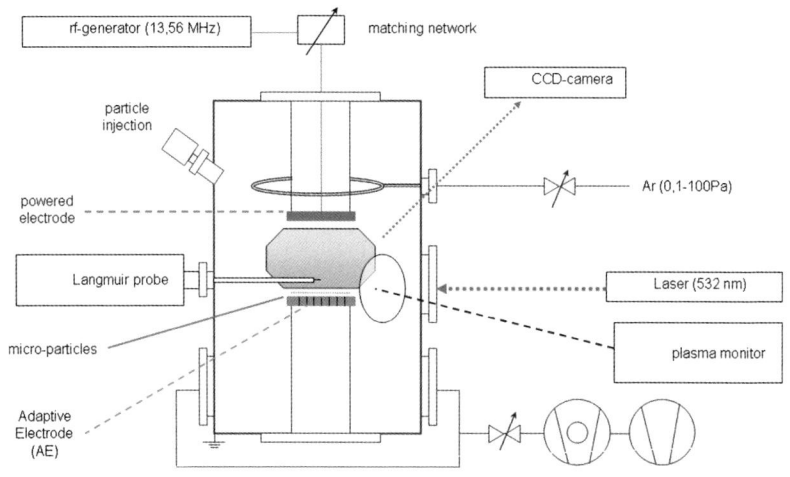

Fig. 2 Schematic of the set-up *PULVA-INP*.

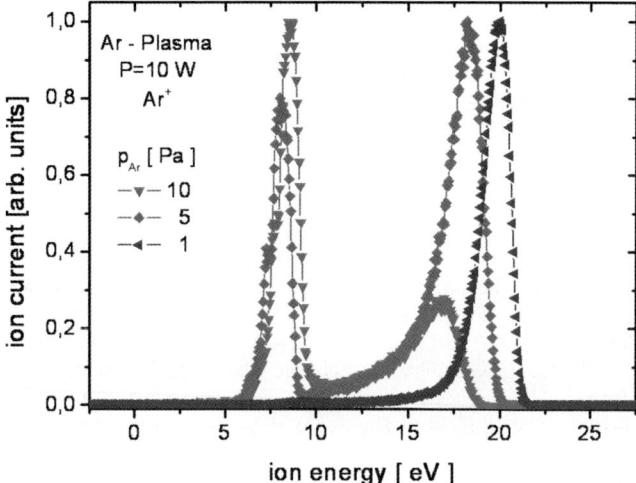

Fig. 3   IEDF of the Ar ions in the confining rf-plasma for different pressure.

floating potential V_{fl} reflects the internal plasma properties. In this case, the variation of discharge parameters results in a global influence of the plasma and induces changes of the whole plasma sheath. In contrast, an external biasing of selected surface elements results in a local change of the sheath and the plasma. This idea is just the basis for the concept of the AE,[22,27] by which a spatial and temporal change of the plasma sheath is possible. In our case, the AE consists of 101 identical square electrode segments (7×7 mm^2) surrounded by four larger segments in order to fit the circular geometry of the planar electrode. All 105 "electrode pixels" can be biased individually or in groups, respectively, by an external $\pm$ 100 V dc-voltage or ac-voltage (sinus, square, triangle shape) up to a frequency of 50 Hz and any phase. In addition, for three segments of the AE, an rf-power supply (13.56 MHz) up to 4 W is also possible. The whole ensemble of electrode segments is surrounded by a ring and a ground shield.

Due to the variable biasing, it is possible to tailor the potential and field distribution in front of the AE in order to confine and manipulate micro-particles and, *vice versa*, in order to study their interaction with the surrounding plasma. The negatively charged grains move horizontally and vertically due to their charge and the potential distribution along the AE. As a result, the arrangement of the trapped particles reflects the potential structure of the sheath above the AE. As an example, in Fig. 4 a photograph of a typical 2D plasma crystal with a tailored sheath and, hence, with a resulting particle arrangement is shown. In this figure the construction of the AE with its pixel structure can also be recognized.

In order to obtain information on the electric field structure in front of a wall (here the AE) by particle probes, the charge of the particles has to be known. There exist several methods for the determination of charge, see, for example ref. 29, where different methods are summarized. One of the most frequently used methods is the superposition of a low-frequency voltage to excite harmonic oscillations of the trapped particles.[1,30] The evaluation is based on the determination of the resonance frequency for the oscillation of particles, Fig. 5. The particles have been confined above the centre pixel of the AE and the amplitude has been measured for different excitation frequencies.

The resonance frequency has been measured for different discharge conditions (*e.g.*, variation of pressure) by using particles of different sizes. Hence, a relation between resonance frequency and vertical position of the particles could be obtained, Fig. 6.

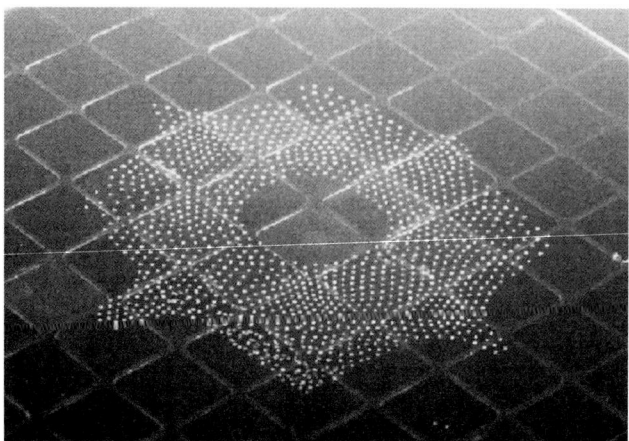

Fig. 4 The particle distribution across the AE reflects the sheath structure. The segments beneath the plasma crystal are on ground potential, whereas the center and the surrounding pixels are on negative potential of −30 V.

The harmonic oscillation of a particle can be described by

$$F_0 \cos(\omega t) = m \frac{d^2 z}{dt^2} - Qz \frac{dE(z)}{dz} \qquad (1)$$

where $E(z)$ is the electric field at vertical position z, and m and Q are the mass and the charge of the particle, respectively. The resonance frequency ω is attributed to the

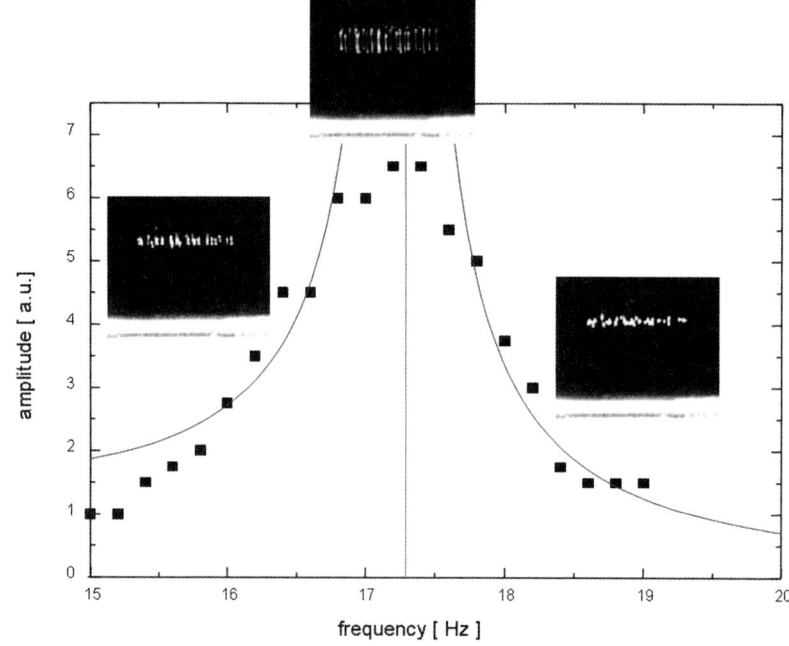

Fig. 5 Particle oscillation above the centre pixel of the AE, on which a sinusoidal bias voltage has been superposed.

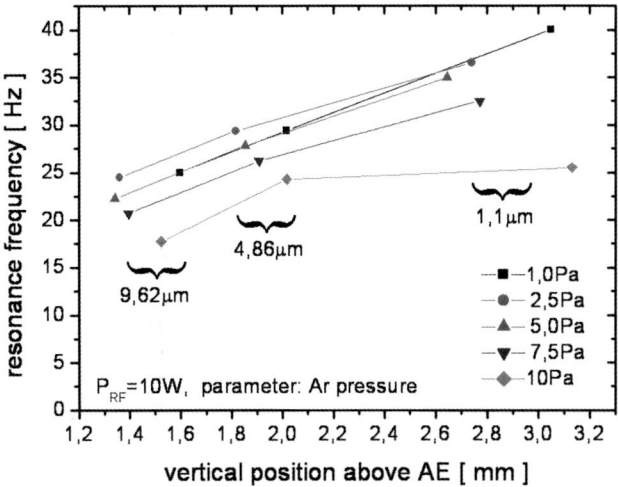

Fig. 6 Relation between resonance frequency and vertical position of the particles in the sheath at different pressure.

charge and the field gradient:

$$\omega_0^2 = -\frac{Q}{m}\frac{dE(z)}{dz} \quad (2)$$

By considering the equilibrium position where $QE = mg$, one obtains

$$\frac{\omega_0^2(z)}{g} = -\frac{-dE(z)/dz}{E(z)} \quad (3)$$

as the relation for the relative field structure.[15] By taking into account a linear ansatz for the frequency and integration

$$\omega = 2\pi(az + b) \quad (4)$$

the vertical field distribution $E(z)/E(0)$ can be written as

$$\frac{E(z)}{E(0)} = \exp\left(-\frac{2\pi}{g}\left(\frac{a^3}{3}z^3 + abz^3 + b^3z\right)\right) \quad (5)$$

The electric field at the surface of the electrode segment ($E(0)$) can now be determined from the integral over the electric field across the sheath, which is equivalent to the sheath voltage. Eqn (5) can be used under the assumptions of neglecting ion and neutral drag forces. Finally, the vertical electric field could be determined. The result is shown in Fig. 7.

The experiments show clearly that the electric field structure can be determined by means of charged micro-particle probes also in front of grounded or additionally biased surfaces. Under equilibrium conditions, the particle charge can be directly obtained from the field strength. Unfortunately, in the very near neighbourhood of the AE we have no experimental values. In future experiments heavier particles should also be used in order to probe this region. At higher pressure, the friction (neutral drag) force also needs to be considered.

2.3 Falling particles in an electric field of biased pixels of the AE

The construction of the AE allows for experiments on the sinking and oscillation behaviour of individual particles. For this purpose, a single particle has been trapped above the centre pixel (E5) of the AE, as depicted in Fig. 8.

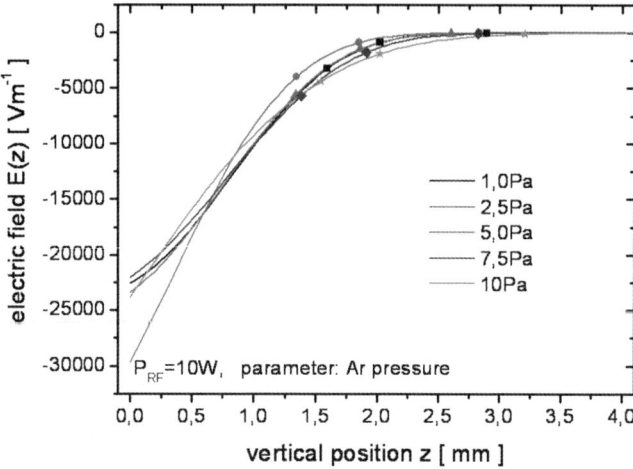

Fig. 7 Vertical field distribution in front of the AE.

If only the forces are considered that act in the vertical direction, the net force on the particle, as discussed in Section 2.1, is given by :

$$F(z) = F_{el}(z) + F_n(z) - F_g - F_{ion}(z) = 0 \Rightarrow F(z) \approx 0 - m_d g + F_n(z) = m_d a \quad (6)$$

If the confining plasma is switched off and the pixel E5 is on ground potential ($V_{bias} = 0$), the forces F_{el} and F_{ion} become zero. In the vacuum case, eqn (6) is much

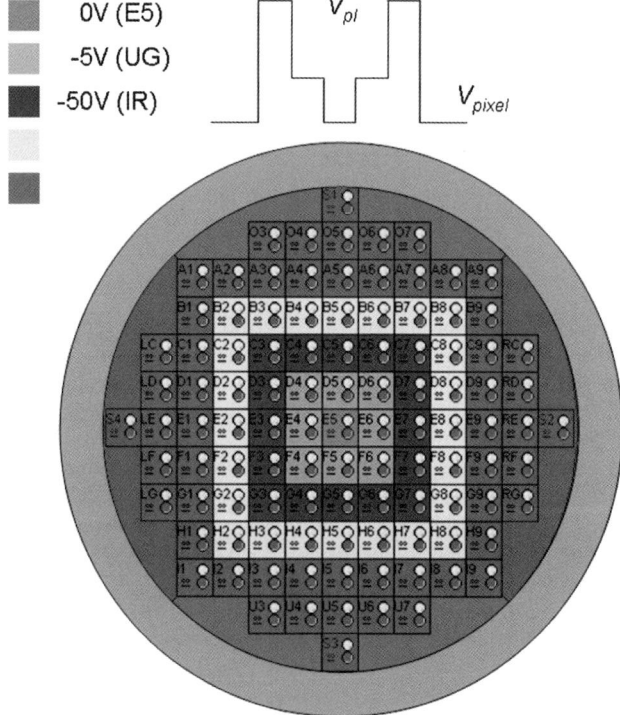

Fig. 8 Structure of the AE. The centre pixel (E5) is biased in order to influence the sheath for particle trapping. The surrounding pixels are biased due to the particle confinement potential trap.

simplified and the free fall of the particle is:

$$F(z) = 0 - m_d g + 0 = m_d a \Rightarrow z(t) = \frac{g}{2} t^2 \qquad (7)$$

However, under real experimental conditions with gas pressure (5 Pa), there is no free fall regime, but the gas friction (neutral drag) has to be considered:

$$F(z) \approx 0 - m_d g + F_n(z) = -m_d g + \beta v = m_d a \qquad (8)$$

Solution of this differential equation results in

$$z(t) = \frac{mg}{\beta}\left(t - \frac{m}{\beta}\left(1 - e^{-\frac{\beta}{m}t}\right)\right). \qquad (9)$$

The damping constant β is about 2×10^{-11} kg s^{-1}. The diagrams for the free fall and the real fall of the particles after switching-off the plasma are plotted in Fig. 9. The starting position (distance = 0) means the equilibrium position where the particle is trapped. The sheath in front of the AE (pixel E5) is about 1.6 mm.

A more interesting case which yields information on field distribution $E(z)$ in front of the AE or the particle charge Q_d, respectively, is given when the pixel is biased and the plasma is still in operation. Then, the charged particle expires a bias potential which is the sum of the plasma potential ($V_{pl} \sim 20$ V) and the pixel potential (e.g. -50 V).

By taking account the position of the trapped particle, the sheath width and the bias potential—and under the assumption of a linear field slope—we obtain a typical value of about 10^4 V/m for the electric field. Due to the "suddenly" added bias potential the density of the ions in the sheath changes remarkably and, obviously, a field reversal occurs which causes the motion of the dust particles towards the electrode. Now, the force balance results in eqn (10) and the position of the falling particle as a function of time depends on gravity, gas friction and electric field (see eqn (11)):

$$F(z) \approx Q_d E(z) - m_d g + F_n(z) = -Q_d E(z) - m_d g + \beta v = m_d a \qquad (10)$$

$$z(t) = \frac{mg}{\beta}\left(t - \frac{m}{\beta}\left(1 - e^{\frac{\beta}{m}t}\right)\right) - \frac{Q_d}{m_d} E(z) t^2 \qquad (11)$$

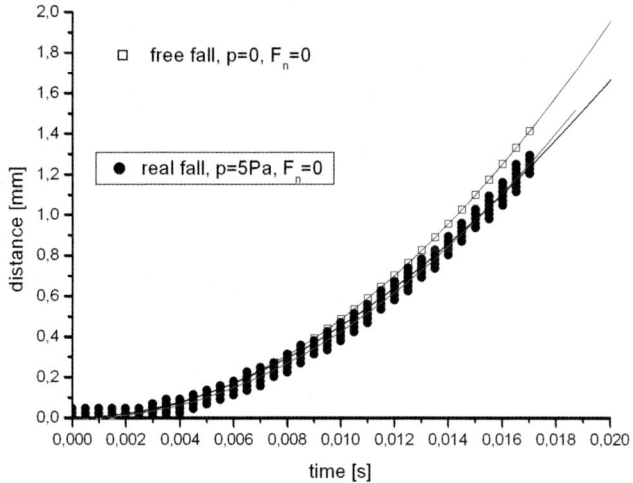

Fig. 9 Particle behaviour after switching-off the plasma (free fall regime ($p = 0$) and real fall regime ($p = 5$ Pa)).

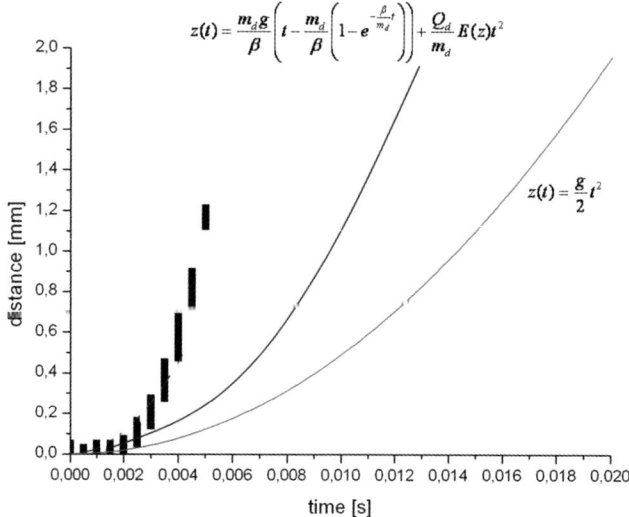

Fig. 10 Experimental data (squares) and model for the free fall and real fall + field effect, respectively.

A comparison of the measured values with the calculated curve from eqn (11) shows a relatively good accordance if we assume a grain charge of 16 900 e_0. The experiments (squares) and the calculated curves are plotted in Fig. 10. The experimental data can be interpreted in a first attempt by this simplified field assumption. Also, the charge value is in quite good agreement with the particle charge given by Allen et al.,[31] who did experiments under comparable conditions.

Observations of the particle behaviour after changing the pixel potential have also shown an oscillation of the particles around their equilibrium position. These experiments will be continued in order to obtain additional information on gas friction and particle charge.

3. Particles as micro-substrates

On the surface of micro-particles in a process plasma, not only charging processes are of importance but also chemical processes like molecule association and radical reactions. With suitable diagnostics like scattering spectroscopy and Raman spectroscopy, one can obtain information about the mechanisms at the surfaces of microscopic solids, which are of great interest e.g., for catalyst research and particle coating.

Cavity enhanced spectroscopy is a powerful diagnostic technique for the characterisation of micron-sized spheres. It has been used very successfully for aerosol droplets.[32] In the present investigation, the feasibility for applying this technique to solid micron-sized particles levitated in an rf-plasma is investigated.

A pulsed laser is used to excite cavity resonances (whispering gallery modes) in individual micro particles, leading to enhanced Raman scattering at characteristic wavelengths. This non-invasive method gives direct access to the size and also the chemical composition of the micro-spheres, and is potentially a very interesting tool for the characterisation of growing layers deposited on micro particles e.g., in molecular plasmas.

3.1 Cavity-enhanced spectroscopy

If a droplet of a liquid or a micro-particle is struck by a laser beam, under suitable incident conditions and at particular resonant wavelengths, the light can undergo total internal reflection and become trapped inside the droplet for long periods. The term "long" here means of the order of nanoseconds, with the result that the light travels a few metres inside the particle. The trapped laser light leads to stimulated Raman scattering at particular resonant wavelengths within the Raman spectrum and an enhanced Raman scattering signal can be detected. The resonant behaviour can be accurately described by Mie scattering theory.[33] Thus, micro-particles can act as optical cavities, greatly enhancing light whose wavelength is coincident with resonant modes of the cavity.[34] Fig. 11 shows a spectrum of a water droplet illuminated with a laser at a wavelength of 590 nm. Sharp, equally spaced peaks appear at a Raman shift of *ca.* 3400 cm^{-1}, corresponding to the O–H stretching band of water. Those resonant modes can be thought of as the light forming standing waves within the particle. They can be assigned a mode number, which corresponds to the number of wavelengths in the standing wave, and a mode order, which corresponds to the number of radial intensity maxima. Resonances of the same mode order and polarisation are approximately equally spaced. From the line spacing we readily arrive at the particle size. The following formula gives an approximation for the particle radius r:[32]

$$r \approx \frac{\tan^{-1}\sqrt{m^2-1}}{\sqrt{m^2-1}} \frac{1}{2\pi \Delta k} \qquad (12)$$

Here, m is the refractive index of the material and Δk the line spacing between resonances of consecutive mode numbers. Particle sizes can be calculated from eqn (12) with an accuracy of a few hundred nanometres.

3.2 Optical measurements on micro-particles

The investigations were carried out in the plasma reactor *PULVA-INP* described in Section 2.2. In order to excite cavity-enhanced scattering, a pulsed tuneable dye laser centred at 590 nm with pulse energies of *ca.* 1.5 mJ per pulse, pumped by a Nd:YAG-laser (532 nm). The latter was also used directly for some experiments. The spectra were resolved and captured by a spectrograph and an ICCD-camera. The signal was sent to the entry optics of the spectrograph *via* an optical fibre that could be positioned at a variable angle with respect to the laser beam.

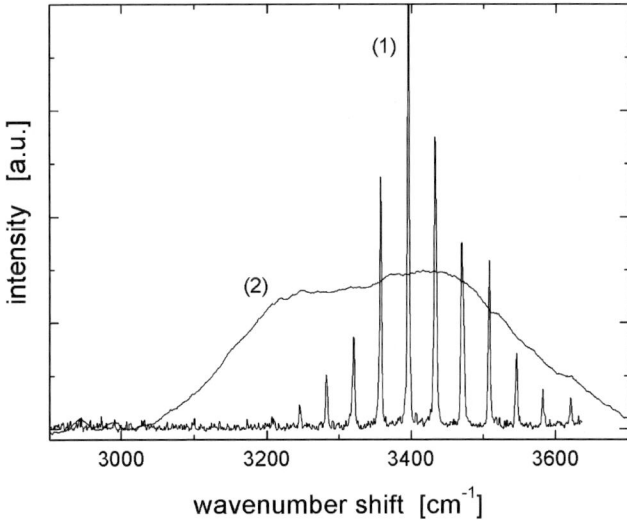

Fig. 11 Cavity enhanced Raman scattering of a water droplet (1) in comparison to the spontaneous Raman scattering of bulk water (2).

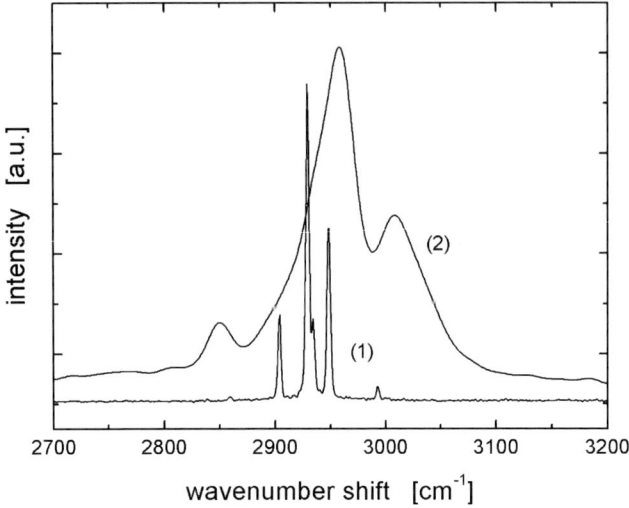

Fig. 12 Cavity-enhanced Raman scattering of PMMA particles (1) and their corresponding spontaneous Raman scattering (2).

Before investigating solid micro-particles in the plasma, experiments were performed on particles in air, where they could be manipulated more easily. Spherical particles of different sizes and chemical composition were investigated. Cavity-enhanced Raman scattering was successfully detected from polymer spheres. Fig. 12 shows a spectrum measured from a PMMA-particle (polymethylmethacrylate) with a diameter of 50 μm. Clear peaks appear at a Raman shift of 2900 cm^{-1}, which is in the C–H stretching region. Calculating the particle size from the spectrum gives a value of 51 μm, which is in good agreement with the value specified by the sphere manufacturer (Microparticles GmbH).

To test whether it is possible to detect a cavity-enhanced signal only from the particle surface, particles were investigated coated with a fluorescent dye (Rhodamine B). In an analogous way to cavity-enhanced Raman scattering, the fluorescence signal is enhanced at wavelengths corresponding to cavity resonances. The resulting

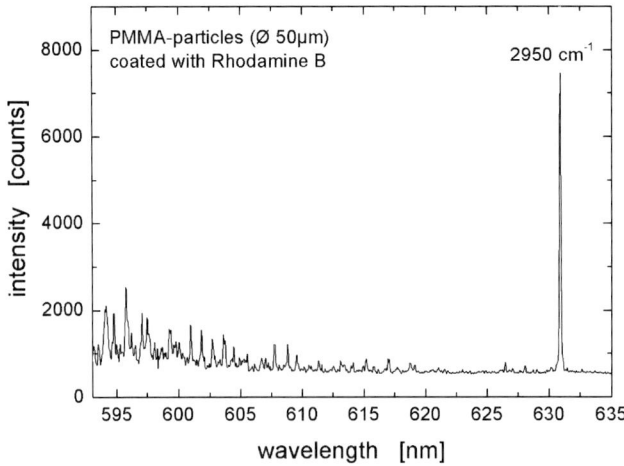

Fig. 13 Spectrum of a 50 μm-PMMA-sphere coated with Rhodamine B. Both cavity-enhanced Raman scattering and cavity-enhanced fluorescence can be seen.

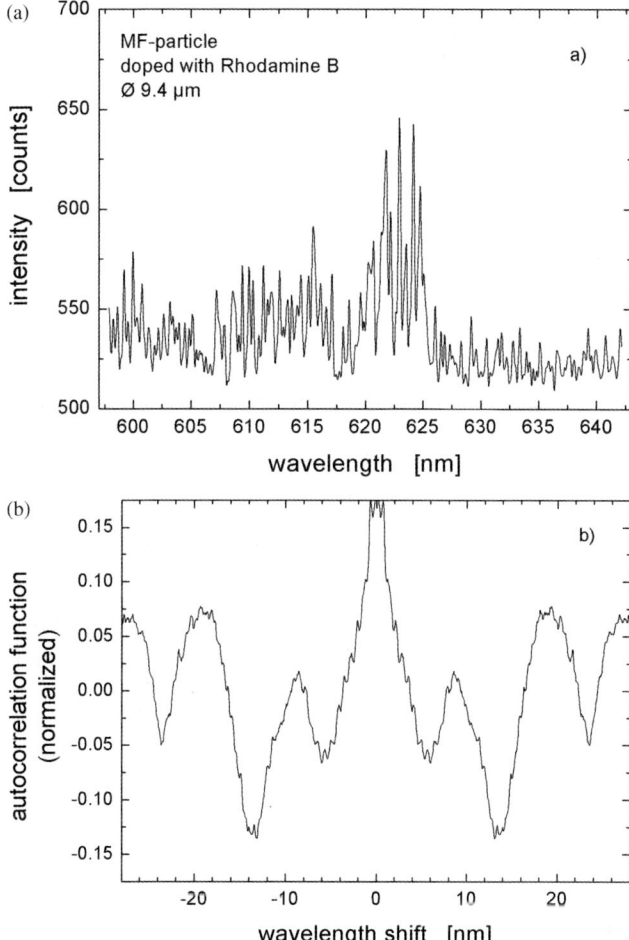

Fig. 14 (a) Cavity enhanced fluorescence spectrum of a Rhodamine B-doped MF particle levitated in an Argon plasma, (b) autocorrelation function of the signal.

spectra show peaks superimposed on the broad fluorescent band, where the signal is enhanced through coupling into cavity resonances.

Fig. 13 shows a spectrum from a Rhodamine B-coated PMMA-particle where both the cavity-enhanced Raman signal from the particle material and the fluorescence signal originating from the surface can be seen. These results demonstrate that cavity-enhanced spectroscopy is a surface-sensitive technique, making it suitable for the characterisation of growing layers in molecular plasmas.

The AE of the plasma reactor proved to be an excellent tool for controlling the position of levitated micro-spheres. For the measurements, micro-particles were trapped above the central segment of the adaptive electrode. To achieve this condition, the segments of the electrode surrounding the central segment were set to a negative bias voltage, thus forming a potential well that confined the particles. The particles are trapped at the position where all acting forces balance. They appear to the eye to be very stationary, but if viewed with a camera with high spatial and temporal resolution it becomes clear that they move around their equilibrium position by distances at least of the order of their diameter, which complicates the alignment of the laser beam to the particles.

The first cavity-enhanced spectra were obtained from MF-particles (Melamin Formaldehyde) that were doped with Rhodamine B. The particles had a diameter of 9.4 μm and a very smooth surface.

Fig. 14a shows the fluorescence spectrum obtained from such a particle levitated in an Argon rf-plasma. The signal level is not much above the noise level, but it shows a clear periodicity of ca. 10 nm. This becomes much more obvious if we look at the autocorrelation function of the signal (Fig. 14b). The periodicity that can be determined experimentally (9.3 nm) is in very satisfactory agreement with the theoretically-expected separation between cavity resonances of consecutive mode numbers for micro-particles of this size.

4. Conclusion

The interaction between plasma and injected micro-disperse test particles has been used for the study of plasma phenomena. By observing the position and movement of the particles in dependence on the discharge parameters, we obtained information on the electric field in front of the AE. The sheath, which is an important region in regard to plasma-enhanced surface processes, has been surveyed by the confined particles.

In addition, the use of micro-particles as small substrates on which surface processes occur has been successfully demonstrated. Especially, experiments on the excitation of whispering gallery modes of micro-particles trapped in a plasma sheath have been demonstrated for the first time. We have shown that it is possible to detect cavity resonances from the surface of a single particle, making it a suitable technique for the characterisation of growing layers in plasmas. After the encouraging results obtained from micro-particles in air, the investigations have been continued for particles levitated in an rf-plasma.

Acknowledgements

This work has been supported by the Deutsche Forschungsgemeinschaft under SFB-TR24/B4. Part of this work was sponsored by the Air Force Office of Scientific Research, Air Force Material Command, USAF, under grant number FA8655-07-1-3033. The US Government is authorized to reproduce and distribute reprints for Governmental purpose notwithstanding any copyright notation thereon.

The authors like to thank Paul Davies and Jonathan Reid as well as Jürgen Röpcke and Jörg Ehlbeck for their support and fruitful discussions.

References

1 S. V. Vldimirov, K. Ostrikov and A. A. Samarian, *Physics and Applications of Complex Plasmas*, Imperial College Press, London, 2005.
2 J. E. Allen, *Phys. Scr.*, 1992, **45**, 497.
3 H. Kersten, H. Deutsch, M. Otte, G. H. P. M. Swinkels and G. M. W. Kroesen, *Thin Solid Films*, 2000, **377–378**, 530.
4 G. E. Morfill, H. M. Thomas, U. Konopka, H. Rothermel, M. Zuzic, A. Ivlev and J. Goree, *Phys. Rev. Lett.*, 1999, **83**, 1598.
5 J. Goree, *Plasma Sources Sci. Technol.*, 1994, **3**, 400.
6 K. Matyash, M. Fröhlich, H. Kersten, G. Thieme, R. Schneider, M. Hannemann and R. Hippler, *J. Phys. D: Appl. Phys.*, 2004, **37**, 2703.
7 H. Kersten, R. Wiese, M. Hannemann, A. Kapitov, F. Scholze, H. Neumann and R. Hippler, *Surf. Coat. Technol.*, 2005, **200**, 809.
8 J. E. Daugherty and D. B. Graves, *J. Vac. Sci. Technol., A*, 1993, **11**, 1126.
9 G. H. P. M. Swinkels, H. Kersten, H. Deutsch and G. M. W. Kroesen, *J. Appl. Phys.*, 2000, **88**, 1747.
10 J. Winter, A. P. Fortov and A. P. Nefedov, *J. Nucl. Mater.*, 2001, **290–293**, 509.
11 I. Herrmann, V. Brüser, S. Fiechter, H. Kersten and P. Bogdanoff, *J. Electrochem. Soc.*, 2005, **152**, A2179.

12 E. Stoffels, H. Kersten, G. H. P. M. Swinkels and G. M. W. Kroesen, *Phys. Scr., T*, 2001, **89**, 168.
13 J. E. Daugherty, R. K. Porteous and D. B. Graves, *J. Appl. Phys.*, 1993, **74**, 1617.
14 *Dusty Plasmas*, ed. A. Bouchoule, J. Wiley & Sons, New York, 1999.
15 A. A. Samarian and B. W. James, *Plasma Phys. Controlled Fusion*, 2005, **47**, B629.
16 T. Nitter, *Plasma Sources Sci. Technol.*, 1996, **5**, 93.
17 A. Homann, A. Melzer and A. Piel, *Phys. Bl.*, 1996, **52**, 1227.
18 H. Kersten, R. Wiese, D. Gorbov, A. Kapitov, F. Scholze and H. Neumann, *Surf. Coat. Technol.*, 2003, **173–174**, 918.
19 L. Boufendi and A. Bouchoule, *Plasma Sources Sci. Technol.*, 2002, **11**, A211.
20 H. Kersten1, G. Thieme, M. Fröhlich, D. Bojic, D. H. Tung, M. Quaas, H. Wulff and R. Hippler, *Pure Appl. Chem.*, 2005, **77**, 415.
21 H. Kersten, R. Wiese, G. Thieme, M. Fröhlich, A. Kopitov, D. Bojic, F. Scholze, H. Neumann, M. Quaas, H. Wulff and R. Hippler, *New J. Phys.*, 2003, **5**, 93.1.
22 B. M. Annaratone, M. Glier, T. Stuffler, M. Raif, H. M. Thomas and G. E. Morfill, *New J. Phys.*, 2003, **5**, 92.
23 H. Kersten, H. Deutsch, H. Steffen, G. M. W. Kroesen and R. Hippler, *Vacuum*, 2001, **63**, 385.
24 G. S. Selwyn, *Plasma Sources Sci. Technol.*, 1994, **3**, 340.
25 D. Samsonov and J. Goree, *J. Vac. Sci. Technol., A*, 1999, **17**, 2835.
26 S. Stoykov, C. Eggs and U. Kortshagen, *J. Phys. D: Appl. Phys.*, 2001, **34**, 2160.
27 R. Basner, H. Fehske, H. Kersten, S. Kosse and G. Schubert, *Vak. Forsch. Prax.*, 2005, **17**, 259.
28 M. Tatanova, G. Thieme, R. Basner, M. Hannemann, Yu. B. Golubovskii and H. Kersten, *Plasma Sources Sci. Technol.*, 2006, **15**, 507.
29 H. Kersten, H. Deutsch and G. M. W. Kroesen, *Int. J. Mass Spectrom.*, 2004, **233**, 51.
30 E. B. Tomme, B. M. Annaratone and J. E. Allen, *Plasma Sources Sci. Technol.*, 2000, **9**, 87.
31 J. E. Allen, B. M. Annaratone and U. deAngelis, *J. Plasma Phys.*, 2000, **63**, 299.
32 R. Symes, R. M. Sayer and J. Reid, *Phys. Chem. Chem. Phys.*, 2004, **6**, 474.
33 G. Mie, *Ann. Phys.*, 1908, **25**, 337.
34 R. J. Hopkins, R. Symes, R. M. Sayer and J. P. Reid, *Chem. Phys. Lett.*, 2003, **380**, 665.

PAPER

Oxidation of biogenic and water-soluble compounds in aqueous and organic aerosol droplets by ozone: a kinetic and product analysis approach using laser Raman tweezers

Martin D. King,*[a] Katherine C. Thompson,[b] Andrew D. Ward,[c] Christian Pfrang†[a] and Brian R. Hughes[b]

Received 12th February 2007, Accepted 20th March 2007
First published as an Advance Article on the web 30th August 2007
DOI: 10.1039/b702199b

The results of an experimental study into the oxidative degradation of proxies for atmospheric aerosol are presented. We demonstrate that the laser Raman tweezers method can be used successfully to obtain uptake coefficients for gaseous oxidants on individual aqueous and organic droplets, whilst the size and composition of the droplets is simultaneously followed. A laser tweezers system was used to trap individual droplets containing an unsaturated organic compound in either an aqueous or organic (alkane) solvent. The droplet was exposed to gas-phase ozone and the reaction kinetics and products followed using Raman spectroscopy. The reactions of three different organic compounds with ozone were studied: fumarate anions, benzoate anions and α pinene. The fumarate and benzoate anions in aqueous solution were used to represent components of humic-like substances, HULIS; α-pinene in an alkane solvent was studied as a proxy for biogenic aerosol. The kinetic analysis shows that for these systems the diffusive transport and mass accommodation of ozone is relatively fast, and that liquid-phase diffusion and reaction are the rate determining steps. Uptake coefficients, γ, were found to be $(1.1 \pm 0.7) \times 10^{-5}$, $(1.5 \pm 0.7) \times 10^{-5}$ and $(3.0–7.5) \times 10^{-3}$ for the reactions of ozone with the fumarate, benzoate and α-pinene containing droplets, respectively. Liquid-phase bimolecular rate coefficients for reactions of dissolved ozone molecules with fumarate, benzoate and α-pinene were also obtained: $k_{fumarate} = (2.7 \pm 2) \times 10^5$, $k_{benzoate} = (3.5 \pm 3) \times 10^5$ and $k_{\alpha\text{-pinene}} = (1–3) \times 10^7$ dm^3 mol^{-1} s^{-1}. The droplet size was found to remain stable over the course of the oxidation process for the HULIS-proxies and for the oxidation of α-pinene in pentadecane. The study of the α-pinene/ozone system is

[a] *Department of Geology, Royal Holloway University of London, Egham, Surrey, UK TW20 0EX. E-mail: m.king@gl.rhul.ac.uk; Fax: +44 (0)1784 414038*
[b] *School of Biological and Chemical Sciences, Birkbeck University of London, Malet Street, London, UK WC1E 7HX*
[c] *Central Laser Facility, CCLRC Rutherford Appleton Laboratory, Chilton, Oxon, UK OX11 0QX*

† Present address: Department of Chemistry, University of Reading, P.O. Box 224, Whiteknights, Reading, UK RG6 6AD.

the first using organic seed particles to show that the hygroscopicity of the particle does not increase dramatically over the course of the oxidation. No products were detected by Raman spectroscopy for the reaction of benzoate ions with ozone. One product peak, consistent with aqueous carbonate anions, was observed when following the oxidation of fumarate ions by ozone. Product peaks observed in the reaction of ozone with α-pinene suggest the formation of new species containing carbonyl groups.

Introduction

Particulate matter is abundant in the troposphere and significantly influences both its chemical and physical characteristics.[1] The effect of aerosol on radiative properties and cloud formation was identified by the intergovernmental panel on climate change, IPCC, to be the largest uncertainty in assessing the impact of particulate matter on climate change.[2] The impacts of inorganic aerosol on cloud formation has been examined and that of organic aerosol has started to be addressed (indirect aerosol effects were recently reviewed[3]), but are yet to be well explored. Atmospheric aerosol is not chemically inert; particles and droplets may provide sites for chemical reaction. Reactions will alter the chemical and physical properties of cloud droplets (and hence potentially their size, in line with adapted Köhler theory[4]) consequently altering how the droplets interact with solar radiation and thus influence the climate.

We explore the reactions between a ubiquitous pollutant, ozone, and aerosol droplets containing unsaturated organic species in aqueous and non-aqueous solutions. The oxidation of fumarate (1) and benzoate (2) ions in aqueous solution, as proxies for HUmic-LIke Substances, HULIS, and the terpenoid compound α-pinene (3) in organic aerosol, as a proxy for biogenic non-aqueous aerosol, were studied:

$$\text{fumarate} + O_3 \rightarrow \text{products} \quad (1)$$

$$\text{benzoate} + O_3 \rightarrow \text{products} \quad (2)$$

$$\text{α-pinene} + O_3 \rightarrow \text{products} \quad (3)$$

This research builds on our previous work studying droplet size and composition during the oxidation of a film of oleic acid on aqueous aerosol.[5]

The term HULIS has been used to describe the organic material found in rain, fog and aerosol that resembles the organic material in river/sea water and soils. HULIS are probably formed in aerosol by chemical reactions as opposed to humic material which is produced in river water and soils by the breakdown of biological substances. HULIS may be present in as much as 20–50% of aerosol particles.[6] The reports of HULIS in clouds, fog and aerosol have been reviewed by Graber and Rudich.[6] Humic and HULIS material can be difficult to chemically characterise, because it defies speciation and molecular definition owing to its complexity, size and non-uniformity. However, consensus is now forming on a model HULIS structure consisting of an aromatic core bearing substituted aliphatic chains with –COOH,

–CH$_2$OH, –COCH$_3$ groups.[7] Indeed one study found that HULIS was composed of n-alkanoic acids, ω-alkenoic, benzoic mono- di- and tri-carboxylic acids, methoxy and acetic and methoxy benzoic acid and a few nitrogen containing glycerine derivatives.[8] Kiss et al.[9] attempted to determine the "size" of a HULIS "molecule" and suggested 40–520 Da (where 1 Da is equal to 1/12 of the mass of an unbound atom of ^{12}C at rest and in its ground state). HULIS is produced from marine, biomass burning and small terrestrial sources, and in secondary aerosol formation (condensation, reaction and oligomerization). An oligomer is a molecule that consists of a finite number identical monomer units. The latter route to HULIS formation is intriguing and HULIS has been shown to be formed by reaction of OH radical with 3,5-dihydroxybenzoic acid on the time scale of hours to days[10] by oligomerization. The presence and/or oxidation of HULIS in clouds, fog and aerosol may lead to droplet activation (cloud formation):[11] droplets containing HULIS were found to activate at lower diameters and thus make cloud formation more facile. Dinar et al.[11] also found that HULIS extracted from daytime filter samples has a lower critical supersaturation diameter than night-time samples; strongly suggesting the reaction with daytime atmospheric oxidants to be important in oxidising HULIS and to increase the hygroscopic properties of cloud droplets. Thus the reactions between HULIS and atmospheric oxidants such as ozone need to be studied to determine if they (1) lower the critical supersaturation required for cloud formation and (2) lead to the formation of oligomers. The rate for these reactions must be quantified to establish if they are atmospherically important.

The biosphere produces the large majority of volatile organic compounds, VOCs, emitted into the atmosphere.[12] Oxidation of these biogenic VOCs tends to produce low vapour pressure compounds that partition to particulate matter to form secondary organic aerosol, SOA,[13–15] although some large biogenic VOCs (e.g. sesquiterpenes) may partition directly to aerosol particulate matter. Secondary organic aerosol may affect the critical saturation ratios of clouds, lead to global dimming and provide a surface for further heterogeneous chemistry to occur.[16] The surface chemistry on SOA leads to the formation of large compounds and oligomers with very low vapour pressures.[17–22] The precursor biogenic VOC typically contains a carbon–carbon double bond or an aromatic moiety. The composition and aqueous oxidation chemistry of a SOA particle influences subsequent oligomer formation.[23] Two of the largest components of the biogenic VOC budget are isoprene and α-pinene.[12] In this study we use α-pinene in a long chain alkane to represent a biogenic VOC in an "organic" particle. We selected α-pinene as its oxidation in the gas-phase has been previously studied.[23–27]

Some of the first studies were performed by Thurn and Kiefer, Biswas et al. and Omori et al.[28–30] Zellner and co-workers have used optical levitation to study the phase behaviour of inorganic acids relevant of the atmosphere[31–34] and recently employed optical levitation to measure hygroscopic growth curves for ammonium sulfate and glutaric acid solutions in an aqueous droplet.[35] The optical leviatation technique traps a particle by balancing the gravitational and the scattering/radiation pressure forces on the particle. The optical tweezers traps a particle using the gradient force and provide a "3-D" trap unlike optical levitation. The history and recent developments of optical tweezers techniques in relation to aerosol trapping has been reviewed by McGloin[36] and will not be repeated in the framework of this paper. Reid, Ward and co-workers have used laser tweezers to study hygroscopic growth,[37,38] and coagulation of two droplets.[39,40] Reid has also studied morphological Raman resonances (cavity enhanced Raman scattering) to size particles[41,42] and together with McGloin to develop new optical traps.[43] Ward and co-workers have also used laser tweezers to study aerosol chemistry relevant to atmospheric chemistry.[5]

In the work presented here we used the laser Raman optical-tweezers technique to trap aqueous or organic droplets containing benzoate ions, fumarate ions or α-pinene in a gaseous flow of humidified oxygen. The particles were subjected to a flow of ozone (ppm range) whilst the particle size was monitored using optical microscopy and changes in particle composition were detected with Raman spectroscopy. Particles of 2–10 microns were trapped in the focus of an Ar-ion laser and could be held for up to 6 h during oxidation, scattered laser light was collected and analysed to produce a Raman spectrum of the organic reactant and products. The aim of this work was to demonstrate the performance of laser-Raman optical-tweezers for studying atmospheric reactions relevant to cloud droplet size and dynamics.

2. Experimental

2.1 Laser Raman tweezers

The set-up of the laser Raman tweezers consisted of an Ar-ion laser source (Coherent Innova 90-5-UV), with an emission wavelength of 514.5 nm. The beam is passed through two sets of beam expansion optics[44] which are anti-reflection coated for the laser wavelength. The optics expand the laser beam to slightly overfill the aperture of the water-immersion objective lens (×63, NA 1.2, Leica Microsystems) and also creates a conjugate focal plane at a steering mirror that is used to manipulate the optical trapping position. Before entering the Leica DM-IRB microscope the beam is passed through a custom-made dichroic (Ingcrys Ltd) that transmits 514.5 nm but reflects from 520 nm to 630 nm (*i.e.* the Stokes Raman scattering). The laser beam is directed upwards into the objective lens, using a second dichroic mirror that reflects 500 nm to 630 nm (*i.e.* laser and Raman lines) and transmits the remaining visible light for optical imaging. The laser beam is tightly focused, forming the optical trap, and backscattered Raman shifted light from the trapped aerosol droplet is collimated by the objective lens and passes back along the same optical pathway. The signal is reflected from both dichroic mirrors and through a 514.5 nm notch filter (Kaiser Optical Systems, HNF-5145) to remove any traces of the Rayleigh scattering. It is then focussed into a spectrograph (Acton Research Corporation SP500i, 1200 groove blazed at 500 nm) and imaged onto a deep depletion CCD camera (Princeton Instruments Spec10:400 BR/LN). In a typical experiment, the power of the source was attenuated to 12 mW at laser focus and Raman spectra were collected continuously with a 10 s scan time for each trapped droplet. In all cases, background spectra were obtained when the droplet was released from the trap, and these signals were subtracted from the droplet spectra. To allow observation of the trapped particles, visible light from the microscope lamp is used to obtain brightfield images of the sample, recorded using a CCD camera.

2.2 Aerosol generation and reaction

The droplets were trapped inside a small aluminium cell that had two cover-slip glass windows to allow the passage of the laser beam from below and to image the particle from above. The cell had a simple stainless steel tube exhaust and entrance. Three gas streams were combined and entered the cell: a flow of dry oxygen that passed through an ozone generator (where oxygen was exposed to the emission from a mercury pen-ray lamp), a flow of dry nitrogen which was humidified by bubbling through pure water, and a flow from an ultrasonic nebulizer which provided the source of the initial droplet. All flows were at atmospheric pressure and controlled by needle valves. The oxygen and nitrogen flows, typically in the range 0.1–1 $cm^3 s^{-1}$, were altered to change the relative humidity and ozone concentration in the cell, but were kept constant during any one experiment. The relative humidity was measured at the entrance and exhaust of the reaction chamber. The ozone concentration (∼0.5–2 ppm in the cell) was calibrated by bubbling through potassium iodide solutions and titrating the molecular iodine formed with thiosulfate.

Aerosol was generated by nebulizing aqueous solutions of sodium benzoate, sodium fumarate, or solutions of α-pinene in dodecane or in pentadecane, using a commercial ultrasonic nebulizer. The nebulizer was switched on until a droplet was trapped in the optical tweezers and then the flow from the nebulizer was switched off. The droplet was allowed to equilibrate with the relative humidity in the cell. Collisions with the cell walls removed other droplets. The sodium benzoate solutions (0.086 mol dm^{-3}) were prepared by dissolving sodium benzoate in pure water. Sodium fumarate solutions (0.086 mol dm^{-3}) were prepared by dissolving fumaric acid in a dilute solution of sodium hydroxide. The pH of the solutions of sodium benzoate and sodium fumarate were 8.6 and 10.

A typical experiment would start with trapping a particle within the reaction cell and the scattered laser light would be collected over 10 or 30 s to obtain the Raman spectrum. A digital image of the particles was also collected every 10 or 30 s. The laser power and optics would be adjusted to bring the particle into focus, so that it was optimally trapped. The humidity in the reaction chamber was allowed to equilibrate with the entering gas-flows for 5–30 minutes to allow the particle size to stabilise. A constant humidified gas stream ensured the particle quickly attained its equilibrium size. The particle was normally held about 50 μm above the surface of the cell. The relative humidity 50 μm above the cell surface was assumed to be the same as that of the humidified gas stream. The ozone was then allowed to enter the cell. Some experiments were conducted with the ozone present before the equilibration. At the end of the experiment the particle was released from the optical trap by blocking the laser beam momentarily and Raman spectra were continued to be collected in case of any spurious measurement—this was never the case. The digital photography was size-calibrated by photographing a microscope stage graticule with 10 μm spacing. Seven pixels on the CCD camera approximately correspond to 1 μm, thus the precision of the particle sizing may be as good as 0.13 μm, however we quote an error in the size of the particle of ~0.25 μm. The particle images were sized manually using in-house image analysis software. The Raman spectra were used to follow the loss of benzoate, fumarate and α-pinene and the growth of any products. The amount of benzoate, fumarate and α-pinene was measured by subtraction of a proportion of a reference spectrum for these compounds. Subtraction of a complete spectrum overcomes the interfering effect of morphological Raman resonances (or cavity-enhanced Raman scattering) that sometimes appear in the spectra and affect the intensity of the peaks (NB Fig. 4 demonstrates such a peak). Morphological Raman resonances (or cavity-enhanced Raman scattering)[30,45–47] were not used to size the particles as they were observed in some spectra but not all.

3. Results

From the laser Raman tweezers experiments we obtained information on reaction kinetics, particle size and products, and shall report each set of results separately.

3.1 Reactive uptake model and kinetic analysis

The reaction of gas-phase ozone with an organic molecule in an aqueous droplet proceeds *via* several consecutive and simultaneous processes. The ozone must first diffuse to the droplet, accommodate at the surface of the particle, and then incorporate itself within the bulk phase of the droplet, or alternatively ozone may react with the organic molecule at the surface of the droplet. The system can be described by a set of coupled differential equations which can be solved for limiting cases.[48–51] A resistance model is commonly used to analyse gas-particle reactions in laboratory studies of atmospheric chemistry.[52–55] Each process is treated as a "conductor" and can be added in series or parallel like resistors in an electronic circuit to give the overall uptake coefficient, γ, of a gas-phase species (*i.e.* ozone) on a particle or droplet. This approach assumes that the processes can all be treated independently. Each conductance, Γ, is normalised to the rate of gas-particle collisions. The conductances are gas-phase transport to the surface, Γ_g, accommodation at the surface of the particle, α, solubility/incorporation into the bulk of the particle, Γ_{sol}, and reaction in the bulk aqueous phase of the particle, Γ_{rxn}. The conductances yield the following equation:

$$\frac{1}{\gamma} = \frac{1}{\Gamma_g} + \frac{1}{\alpha} + \frac{1}{\Gamma_{rxn} + \Gamma_{sol}} \quad (I)$$

Table 1 demonstrates that in our experimental conditions, the diffusive transport of ozone to the surface of the particle and mass accommodation of ozone on the aqueous droplet are typically fast, not rate-limiting processes, and can be neglected in the determination of the uptake coefficient, γ. The aqueous-phase diffusion and reaction are the rate-determining steps. Table 1 lists characteristic times for the separate process described above. The characteristic time of gas-phase transport of ozone to the particle can be calculated for a particle of radius, $r = 4$ µm, and a gas-phase diffusion constant of ozone in air, $D_g(O_3) = 1.78 \times 10^{-5}$ m^2 s^{-1} at $T = 293$ K.[56] The mass accommodation characteristic time is given by the mass accommodation coefficient for ozone on water, $\alpha = 1 \times 10^{-2}$,[57] Henry's law constant, H, for O_3 in the solution, the gas constant, $R = 8.205 \times 10^{-5}$ m^3 atm K^{-1} mol^{-1}, the temperature, $T = 293$ K, the liquid-phase diffusion constant, $D_l(O_3)$ and the average molecular speed in the gas-phase, $\bar{v}$. For ozone in aqueous solution $H = 12.17$ mol m^{-3} atm^{-1} was used taking into account ionic strength and temperature corrections[57] to replicate experimental conditions. Two values of Henry's law constant for ozone in organic liquid were found in the literature: 80 and 480 mol m^{-3} atm^{-1}. A value of $D_l(O_3) = 1.19 \times 10^{-9}$ m^2 s^{-1} was used for the liquid-phase diffusion constant of ozone in an aqueous droplet and a value of $D_l(O_3) = 1.0 \times 10^{-9}$ m^2 s^{-1} was employed for the organic liquid media.[58] The mass accommodation coefficient for ozone on an organic droplet has not been reported, we thus had to use the value of $\alpha = 10^{-2}$ reported for aqueous droplets.[57] It is likely that the mass accommodation coefficient has a larger value for organic liquids as ozone is typically a factor of seven more soluble in organic compounds than in water.[59]

Table 1 considers the characteristic time for reaction of ozone in the bulk liquid phase of the droplet. Rate coefficients considered for the reaction of ozone with fumarate ions in aqueous solutions are $k_1 > 1 \times 10^5$ dm^3 mol^{-1} s^{-1} [60] and for reaction with benzoate ions $k_2 = 1.2$ dm^3 mol^{-1} s^{-1},[61] and 3.5×10^5 dm^3 mol^{-1} s^{-1} (this work). For the examples listed here [benzoate] $= 0.086$ mol dm^{-3}, and [fumarate] $= 0.086$ mol dm^{-3}. Table 1 lists another important quantity: the diffuso-reactive length, l. This parameter indicates how far an ozone molecule may diffuse before it reacts in the liquid phase. Large diffuso-reactive lengths suggest fast diffusion and/or slow kinetics, whereas small diffuso-reactive lengths indicate that reaction dominates over diffusion.[51] For the systems studied here, eqn (I) reduces to

$$\frac{1}{\gamma} = \frac{1}{\Gamma_{rxn} + \Gamma_{sol}} \quad (II)$$

i.e. the uptake of ozone on aqueous particles containing the organic compounds considered here will depend predominantly on the rate of change of the ozone concentration in the particle by reaction and diffusion. Smith *et al.*[58] have described this situation using the equation

$$\frac{\partial[O_3]}{\partial t} = D\nabla^2[O_3] - k[O_3], \quad (III)$$

Table 1 The rate-determining processes in the resistor model of gas uptake on aerosol droplets can be identified by considering the characteristic times associated with each process.[52] We consider the diffusion of ozone, with the diffusion constant $D_g(O_3) = 1.78 \times 10^{-5}$ m^2 s^{-1},[56] through gaseous air at a temperature of $T = 293$ K with an average molecular speed, $\bar{v} = 470$ m s^{-1},[81] to a liquid droplet of radius, $r = 4$ μm. The equilibrium at the gas–water surface depends on the liquid-phase diffusion coefficient for ozone, $D_l(O_3) = 1.19 \times 10^{-9}$ m^2 s^{-1},[81] the Henry's law constant for ozone above an aqueous solution, $H = 12.17$ mol m^{-3} atm^{-1},[57] and a mass accommodation coefficient, $\alpha = 1 \times 10^{-2}$.[57] The equilibrium at the gas–organic surface depends on the liquid-phase diffusion coefficient for ozone, $D_l(O_3) = 1.0 \times 10^{-9}$ m^2 s^{-1},[58] the Henry's law constant for ozone above an organic solution, H, which can be found in the literature: 480 mol m^{-3} atm^{-1},[59] and a mass accommodation coefficient, $\alpha = 1 \times 10^{-2}$.[57] For the examples listed here [benzoate] = 0.086 mol dm^{-3}, [fumarate] = 0.086 mol dm^{-3}, [α-pinene] = 3 mol dm^{-3}. It is clear for the examples listed here the liquid-phase diffusion and reaction are important and that gas-phase transport and mass accommodation are not important in the calculation of the uptake coefficient

Process	Calculation	Characteristic time/s			
		Benzoate $k = 1.2$ dm^{3}mol^{-1}s^{-1}[61]	Benzoate $k = 3.5 \times 10^5$ dm^3 mol^{-1}s^{-1} [this work]	Fumarate $k > 1 \times 10^5$ dm^{3}mol^{-1}s^{-1}[60]	α-pinene $H = 480$ mol m^{-3} atm^{-1}[58] $k = 6 \times 10^5$ dm^3 mol^{-1} s^{-1}
Gas-phase diffusion	$\dfrac{r^2}{\pi^2 D_g(O_3)}$	$\sim 9 \times 10^{-8}$	$\sim 9 \times 10^{-8}$	$\sim 9 \times 10^{-8}$	$\sim 9 \times 10^{-8}$
Mass accommodation	$D_l(O_3)\left(\dfrac{4HRT}{\alpha \bar{v}}\right)^2$	$\sim 7 \times 10^{-11}$	$\sim 7 \times 10^{-11}$	8×10^{-11}	1×10^{-7}
Aqueous phase diffusion	$\dfrac{r^2}{\pi^2 D_l(O_3)}$	$\sim 1 \times 10^{-3}$	$\sim 1 \times 10^{-3}$	$\sim 1 \times 10^{-3}$	2×10^{-3}
Aqueous phase reaction	$\dfrac{1}{k[\text{organic}]}$	~ 10	3×10^{-5}	$< 2 \times 10^{-4}$	6×10^{-7}
Diffuso-reactive length	$\sqrt{\dfrac{D_l(O_3)}{k[\text{organic}]}}$	110 μm	0.2 μm	< 0.4 μm	0.02 μm

where ∇^2 represents the Laplace operator. Smith et al.[58] demonstrated that two analytical solutions can be found to eqn (III), based on work by Worsnop et al.[51] One solution (eqns (IV) and (V)) is obtained for fast diffusion of ozone in the particle (i.e. the uptake is controlled by a 'slow' reaction):

$$[\text{organic}]_t = [\text{organic}]_{t=0} e^{-P(O_3)Hkt} \quad \text{(IV)}$$

and

$$\gamma = \frac{4HRTrk}{3\bar{v}} [\text{organic}], \quad \text{(V)}$$

where $P(O_3)$ is the partial pressure of ozone in the gas-phase and [organic] is the concentration of the organic compound in the liquid-phase droplet, all other variable are as before. This solution is valid when the diffuso-reactive length is greater than the particle radius ($l > r$).

A second solution has been suggested[51,58] for a situation when the diffusion of ozone in the droplet is the rate-limiting step and the uptake is controlled by 'fast' reaction, i.e. the diffuso-reactive length is small compared to the droplet radius ($l \ll r$):

$$\sqrt{[\text{organic}]_t} = \sqrt{[\text{organic}]_{t=0}} - \frac{3P(O_3)H\sqrt{kD_l(O_3)}}{2r} t \quad \text{(VI)}$$

and

$$\gamma = \frac{4RT}{\bar{v}} H\sqrt{kD_l(O_3)}\sqrt{[\text{organic}]} \quad \text{(VII)}$$

As noted by Smith et al.[58] a plot $\sqrt{\frac{[\text{organic}]_t}{[\text{organic}]_{t=0}}}$ as a function of time, t, has an intercept of unity and a gradient of $\frac{3P(O_3)H\sqrt{kD_l(O_3)}}{2r\sqrt{[\text{organic}]_{t=0}}}$. The quantity $H\sqrt{kD_l(O_3)}$ may thus be obtained and then used in eqn (VII) to calculate the experimental uptake coefficient for the reaction with ozone without having to determine the Henry's law or diffusion constant for ozone in a concentrated solution, or determine the rate coefficient for reaction of ozone with organic. If H and $D_l(O_3)$ are known, it is possible to calculate the bimolecular rate coefficient, k, for reaction of ozone with the organic species in solution, although this was not the primary aim of the work presented here.

Table 1 lists the diffuso-reactive lengths for the chemical systems studied (fumarate, benzoate and α-pinene). A comparison of the diffuso-reactive lengths with the particle radius suggests that the concentration-time data for fumarate and α-pinene should be fitted to $\sqrt{\frac{[\text{organic}]_t}{[\text{organic}]_{t=0}}}$ versus t and the concentration-time data for benzoate to $\ln\left(\frac{[\text{Organic}]_t}{[\text{Organic}]_{t=0}}\right)$ versus t for $k_2 = 1.2$ dm^3 mol^{-1} s^{-1} and to $\sqrt{\frac{[\text{organic}]_t}{[\text{organic}]_{t=0}}}$ versus t for $k_2 = 3.5 \times 10^5$ dm^3 mol^{-1} s^{-1}. In practice, the concentration-time data were fitted both to $\ln\left(\frac{[\text{Organic}]_t}{[\text{Organic}]_{t=0}}\right)$ versus t and to $\sqrt{\frac{[\text{organic}]_t}{[\text{organic}]_{t=0}}}$ versus t and the best fit was then analyzed.

3.1.1 Fumarate-ozone system.
Fig. 1 shows a plot of particle radius and $\sqrt{\frac{[\text{Fumarate}]_t}{[\text{Fumarate}]_{t=0}}}$ as a function of time, t, obtained when a droplet containing aqueous fumarate ions is exposed to gas-phase ozone. From the gradient of the plot of $\sqrt{\frac{[\text{Fumarate}]_t}{[\text{Fumarate}]_{t=0}}}$ versus t, the uptake coefficient, $\gamma = (1.1 \pm 0.7) \times 10^{-5}$, and the rate coefficient for reaction (1), $k_1 = (2.7 \pm 2) \times 10^5$ dm^3 mol^{-1} s^{-1}, were determined using eqn (VI). Table 2 gives the experimental conditions for the different systems investigated. The literature rate coefficient is 1×10^5 dm^3 mol^{-1} s^{-1} for a

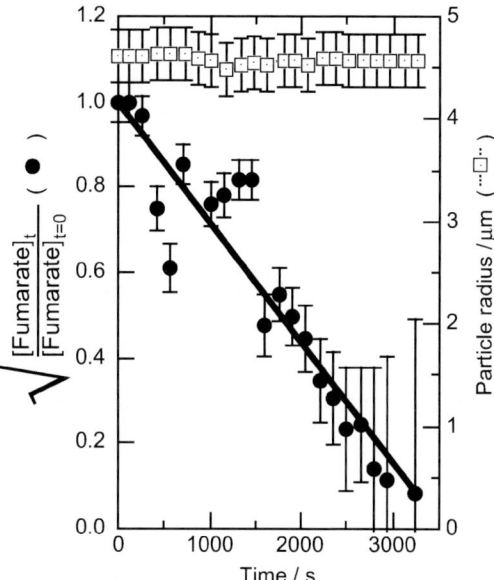

Fig. 1 A typical plot of $\sqrt{\frac{[\text{Fumarate}]_t}{[\text{Fumarate}]_{t=0}}}$ and the radius of levitated droplet as a function of time. The 4.5 μm radius particle was subjected to a relative humidity of 80% and an ozone mixing ratio of 1ppm. The scatter in the points representing $\sqrt{\frac{[\text{Fumarate}]_t}{[\text{Fumarate}]_{t=0}}}$ at t = 700–1200 s is due interference in the Raman spectrum from large intensity morphological Raman resonances. The error bars on the radius data represent 0.25 μm error in sizing the particle. The error on the values of $\sqrt{\frac{[\text{Fumarate}]_t}{[\text{Fumarate}]_{t=0}}}$ originates from the precision in the determination of [fumarate] by spectral stripping.

bulk aqueous solution at pH = 5 and the lower limit for pH = 8 was reported to be $>1 \times 10^5$ dm^3 mol^{-1} s^{-1}.[60] We consider these values to be in good agreement with our result and this gives us confidence in our application of laser Raman tweezers for studying aerosol–gas-phase reactions. This finding also confirms

Table 2 Uptake coefficients, γ, for the heterogeneous reaction between gaseous O_3 and aqueous droplets of sodium fumarate, sodium benzoate and organic droplets of α-pinene/dodecane and α-pinene/pentadecane mixtures. The second-order rate coefficients, k, for the solution-phase reaction between ozone and aqueous sodium fumarate, sodium benzoate and organic-phase α-pinene are also reported

System	r/μm	$P(O_3)$ (ppm)	Relative humidity (%)	Initial mole fraction	γ	k_2/dm^3 mol^{-1} s^{-1}	
Fumarate	4–5	1.0	80	>0.0015	$(1.1 \pm 0.7) \times 10^{-5}$	$(2.7 \pm 2) \times 10^5$	
Benzoate	3–4	1.0	60	>0.0015	$(1.5 \pm 0.7) \times 10^{-5}$	$(3.5 \pm 3) \times 10^5$	
						$H = 80$ mol m^{-3} s^{-1}	$H = 480$ mol m^{-3} s^{-1}
α-pinene/ Dodecane	4–8	0.7	30	~0.59	$(4.0 \pm 0.7) \times 10^{-3}$	$(1.6 \pm 0.6) \times 10^7$	$(5 \pm 2) \times 10^5$
α-pinene/ Pentadecane	4.7–4.9	0.4	60	~0.66	$(3.0 \pm 0.9) \times 10^{-3}$	$(1 \pm 0.5) \times 10^7$	$(3 \pm 1) \times 10^5$
α-pinene/ Pentadecane	5–7.5	0.9	10	~0.93	$(7.5 \pm 1.3) \times 10^{-3}$	$(3 \pm 1) \times 10^7$	$(9 \pm 3) \times 10^5$

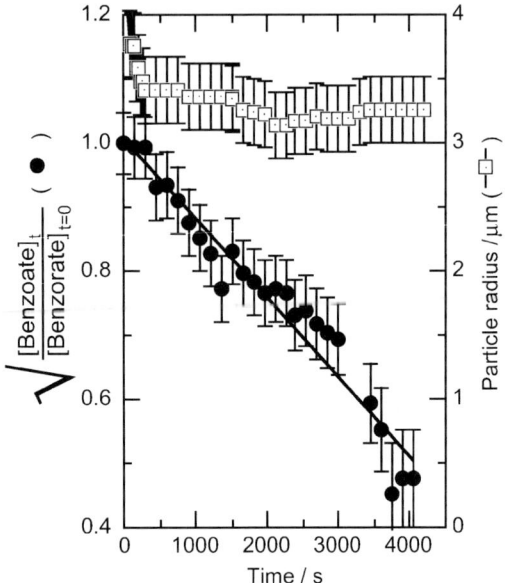

Fig. 2 A typical plot of $\sqrt{\frac{[\text{Benzoate}]_t}{[\text{Benzoate}]_{t=0}}}$ and the radius of levitated droplet as a function of time. The 3.8 μm radius particle was subjected to a relative humidity of 60% and [O_3] of 1ppm. The error bars on the radius data represent 0.25 μm error in sizing the particle. The error on the values of $\sqrt{\frac{[\text{Benzoate}]_t}{[\text{Benzoate}]_{t=0}}}$ originates from the precision in the determination of [benzoate] by spectral stripping.

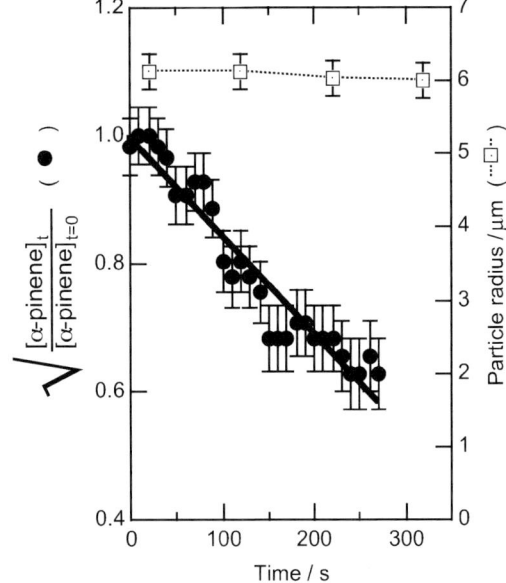

Fig. 3 A typical plot of $\sqrt{\frac{[\alpha\text{-pinene}]_t}{[\alpha\text{-pinene}]_{t=0}}}$ and the radius of levitated droplet as a function of time. The 6.1 μm radius particle was subjected to a relative humidity of 10% and [O_3] of 0.9 ppm. The error bars on the radius data represent 0.25 μm error in sizing the particle. The error on the values of $\sqrt{\frac{[\alpha\text{-pinene}]_t}{[\alpha\text{-pinene}]_{t=0}}}$ originates from the precision in the determination of [α-pinene] by spectral stripping.

that the reaction between fumarate and ozone occurs in the bulk of the droplet and can be described by eqn (VI) and (VII). Our results are based on the analysis of four experiments in detail; altogether we studied 20 droplets, with each showing similar behaviour and giving similar values. The errors in the uptake coefficient and the value of k_1 are equal to a standard deviation of all the measurements of these quantities.

3.1.2 Benzoate-ozone system. The uptake of ozone on aqueous droplets containing benzoate ions was measured as $\gamma = (1.5 \pm 0.7) \times 10^{-5}$. The uptake coefficient was obtained from analysing six droplets in detail and using eqn (VI) and (VII) in conjunction with a plot of $\sqrt{\frac{[\text{Benzoate}]_t}{[\text{Benzoate}]_{t=0}}}$ as a function of time, exemplified in Fig. 2. In all 25 droplets were studied. Eqn (VI) also allows the aqueous bimolecular rate coefficient for reaction (2) to be calculated as $k_2 = (3.5 \pm 3) \times 10^5$ dm^3 mol^{-1} s^{-1}. There is one reported literature value for the rate coefficient for reaction (2): $k_2 = (1.2 \pm 0.2)$ dm^3 mol^{-1} s^{-1}.[61] The difference between our value for k_2 and the published value is considered in the Discussion section. The errors in the uptake coefficient and the value of k_2 are equal to a standard deviation of all the measurements of these quantities.

3.1.3 α-pinene-ozone system. The uptake coefficient of ozone on a droplet which is a mixture of long-chain alkane and α-pinene was measured to be 3.0–7.5×10^{-3}. Table 2 lists the values obtained when the organic solvent was changed from dodecane to pentadecane and the mole fraction of α-pinene was altered. Fig. 3 gives a typical plot of $\sqrt{\frac{[\alpha\text{-Pinene}]_t}{[\alpha\text{-Pinene}]_{t=0}}}$ as a function of time for the reaction (3). Eqn (VI) was used to extract a rate coefficient for reaction (3), k_3, from plots such as Fig. 3. Table 2 lists the rate coefficients measured in a series of experiments studying 30 droplets, 10 droplets in detail. Values of k_3 were determined to be 3–9×10^5 dm^3 mol^{-1} s^{-1}, using a value for the Henry's law coefficient of $H = 80$ mol m^{-3} atm^{-1}.[59] For $H = 480$ mol m^{-3} atm^{-1},[58] k_3 was found to be 1–3×10^7 dm^3 mol^{-1} s^{-1}. The values of the rate coefficient k_3 and the uptake coefficient are reported as ranges to reflect the values measured over particles with different sizes and compositions, *i.e.* it may not be meaningful to simply average these values.

3.2 Particle size changes

During the oxidation of aqueous fumarate ions by ozone the droplet radius was not found to change significantly. The maximum size change was 12%. Size changes are not correlated with the amount of fumarate in the droplet or the extent of the reaction. This result implies the hygroscopicity of the droplet, as described by Köhler theory,[4] has changed very little over the course of the oxidation of the aqueous organic anion.

Aqueous benzoate droplets trapped in the laser tweezers also showed no dramatic change in size upon oxidation by ozone. Over the course of the reaction about half of the experiments demonstrated a very slow monotonic decrease in droplet radius (typically <5%). This decrease was not considered to be significant. It was concluded that, similar to the fumarate system, the oxidation of benzoate in solution did not cause a large change in the hygroscopicity of the droplet.

The kinetics of the reaction of pure α-pinene droplets with gas-phase ozone could not be studied as pure α-pinene droplets (3–10 μm radius) evaporated rapidly. Reactions of droplets of α-pinene/dodecane mixtures with gas-phase ozone showed a gentle decrease in radius (about 1–2 μm) with time but monitoring the droplet size of α-pinene/dodecane droplets and pure dodecane droplets in the absence of ozone also gives a decrease in radius with time. Monitoring the Raman spectra with time for the α-pinene/dodecane mixtures demonstrates that the volatile α-pinene is being lost by evaporation. The droplets of α-pinene and the larger alkane pentadecane were subject to significantly lower evaporation rates. For all the experimental runs analysed, the loss of α-pinene due to reaction with ozone was at least a factor of four larger than the loss due to evaporation. Reactions of ozone with α-pinene/pentadecane droplets show no discernable size change ($\ll 0.5$ µm) during the oxidation process.

3.3 Products

In all cases exposure to gas-phase ozone caused a loss in the amount of organic reactant present in the trapped droplet, as followed by Raman spectroscopy. The products of the reactants that remained in the droplet were observed by Raman spectroscopy.

3.3.1 Fumarate-ozone system.
Only one product peak was observed for oxidation of fumarate ions by ozone in aqueous droplets. The peak (at 1065 cm^{-1}) was small and consistent with aqueous carbonate anions; other carbonate peaks (expected at $\sim$1436 and 684 cm^{-1}) are less intense and were not observed.[62] No other product peaks were observed suggesting that either volatilisation or consumption by secondary reactions may have occurred, as discussed later.

3.3.2 Benzoate-ozone system.
No products were observed for the reaction between ozone and aqueous droplets containing the benzoate anion. The final products of the reaction must have volatilised, be Raman inactive, or present in concentrations beneath the detection limit for the compound by Raman spectroscopy with the experimental set-up employed. To preserve electrical neutrality in the system one of the products must be an anion and hence non-volatile.

3.3.3 α-pinene-ozone system.
The Raman spectrum recorded for a α-pinene/pentadecane droplet before and after oxidation with ozone is given in Fig. 4. The figure shows product peaks for new functional groups between 1600 and 1800 cm^{-1}. In this region there are several new bands associated with the C=O bond deformation. Comparison of spectra (c) and (a) also demonstrates that the C=C bond stretch of α-pinene has been removed in oxidation by ozone. The bands associated with CH$_2$ and CH$_3$ deformation ($<$1500 cm^{-1}) have similar wavenumbers similar to those observed for α-pinene suggesting that the product structure may not be too different from the starting material.

4. Discussion

4.1.1 Uptake coefficients

To the Authors' knowledge the uptake coefficients for ozone on for aqueous droplets of fumarate and benzoate solutions have not been previously measured. It should be noted that the uptake coefficients reported here do not take into account any uptake of ozone owing to hydroxide anions in solution and should thus be interpreted as upper limits. Uptake coefficients for α-pinene have been determined previously.[63] Measurements of the uptake of ozone on α-pinene films at temperatures lower than -30 °C gave values of $\gamma = 2$–2.5×10^{-3}.[63] The values measured here (reported in Table 2) are broadly in agreement with the earlier determination giving us confidence in our method.

4.1.2 Liquid-phase bimolecular rate coefficients

The kinetic scheme used to measure the uptake coefficients allows the liquid-phase bimolecular rate coefficient for the reaction between ozone and fumarate, benzoate

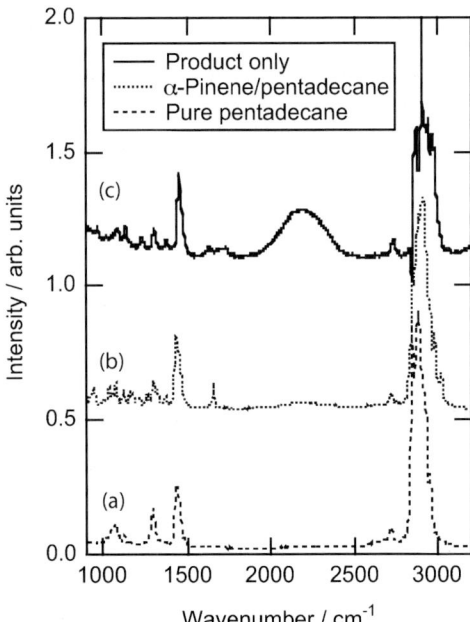

Fig. 4 Raman spectra (intensity as a function of the relative wavenumber) of (a) a pure pentadecane droplet, (b) a droplet of α-pinene and pentadecane before reaction with ozone, mole fraction of α-pinene is ~0.66, and (c) the Raman spectrum obtained after reaction with ozone, from which the reactants (α-pinene and pentadecane) have been subtracted. Spectrum (c) has several product peaks in the region 1600–1800 cm^{-1} characteristic of C=O stretches. The peaks characteristic of the carbon–carbon double bond stretches in α-pinene can be found in spectrum (b) at 1640 cm^{-1}, and the two peaks between 3000 and 3100 cm^{-1} are gone in spectrum (c). The broad feature centered on 2200 cm^{-1} in spectrum (c) is stray light used to image the droplet/particle and can be removed by subtraction of background signal. The sharp feature at 2850 cm^{-1} in spectrum (c) is a morphological Raman resonance, and is included to demonstrate its occurrence in some spectra. The particle studied had 4.7 μm radius, the relative humidity was 10% and [O$_3$] was at 0.4 ppm.

or α-pinene to be measured. This analysis is a secondary aim of the work presented here. It should be noted that the precision (and accuracy) with which such a rate coefficient can be determined relies on precise (and accurate) measurement or knowledge of the particle size, ozone partial pressure, Henry's law constant and diffusion constant for ozone in the droplet, *i.e.* the parameters in eqn (VI). From the slope of a graph such as Fig. 2, *i.e.* $\sqrt{\frac{[\text{organic}]_t}{[\text{organic}]_{t=0}}}$ versus t, the bimolecular rate coefficient can be determined as,

$$k = \frac{\sqrt{[\text{organic}]_{t=0}}}{D_l(O_3)} \left(\frac{2r \times \text{slope}}{3P(O_3)H}\right)^2. \quad \text{(VIII)}$$

The Henry's law coefficient for ozone and the partial pressure of ozone near the particle may not be well known and the determination of k depends on the square of these values. With these caveats in mind, it is worth recording that the value of the rate coefficient for the aqueous-phase reaction of fumarate with ozone determine in this work (k_1 = (2.7 ± 2) × 10^5 dm^3 mol^{-1} s^{-1}) is consistent with the previous measurement k_1 > 1 × 10^5 dm^3 mol^{-1} s^{-1}.[60] It should be noted that in order to obtain uptake coefficients from our data, it is not required to calculate H, D_l(O$_3$) or k.

The bimolecular rate coefficient measured for the aqueous-phase reaction between ozone and benzoate, k_2 = 3.5 × 10^5 dm^3 mol^{-1} s^{-1}, does not agree with the literature value of 1.2 dm^3 mol^{-1} s^{-1}.[61] However, it is not clear if the literature value is for the reaction of ozone with benzoic acid rather than the benzoate anion.[61] The rate coefficient for the reaction of an oxidant with the carboxylate anion in solution is often several orders of magnitude larger than that for the corresponding carboxylic acid, *i.e.* the reaction kinetics of organic acids are extremely pH dependent.[60,61] Inspection of Table 1 shows that for the literature value, the diffuso-reactive length would be much larger than the particle radius (*i.e.* fast diffusion of O$_3$ would occur throughout the particle). The work of Smith *et al.*[58] would then suggest that the loss of benzoate should be fitted to the following equation

$$[\text{benzoate}]_t = [\text{benzoate}]_{t=0}\, e^{(P(O_3)Hkt)} \qquad (\text{VIII})$$

and not to eqn (VI). However, eqn (VI) provides a better fit to the data than eqn (VIII), although both equations can be fitted. Using eqn (VIII) to obtain the rate coefficient, k_2, for the reaction between ozone and benzoate gives values of $\sim 1 \times 10^5$ dm^3 mol^{-1} s^{-1} and diffuso-reactive lengths that are smaller than the particle radius, so that application of eqn (VIII) becomes invalid. We thus conclude that the large value for the rate coefficient we derived for the reaction between ozone and benzoate is not an error due to incorrect application of the equations suggested by Smith *et al.*[58]

The bimolecular rate coefficient, k_3, for α-pinene reacting with ozone determined here depends on the value of Henry's law coefficient taken for ozone in the organic solvent. There is no previously reported measured value of k_3 for comparison. However, using the methodology of structure-activity relationships for predicting gas-phase kinetic rate coefficients[64–66] the value of k_3 should be similar or greater than that for ozone reacting with 2-methyl-2-butene. Williamson and Cvetanoic[67] report a value of $k = 2.6 \times 10^5$ dm^3 mol^{-1} s^{-1} for this reaction in a solvent of carbon tetrachloride, suggesting that $H = 480$ mol m^{-3} atm^{-1} is the most sensible choice of value of H for our system and the rate coefficient $k_3 = 3$–9×10^5 dm^3 mol^{-1} s^{-1}.

For the analysis of the decay of fumarate, benzoate and α-pinene with time, we have assumed a bulk reaction of ozone with these compounds, since ozone is reasonably soluble in water and alkanes. However, for completeness it is valuable to illuminate why the reaction may not be happening at the surface of the droplet. Provided that ozone diffusion and mass accommodation of ozone are not rate-limiting, the appropriate equation to describe the loss of organic species in the particle with time (due to reaction with ozone) is[58]

$$[\text{Organic}]_t = [\text{Organic}]_{t=0}\, e^{\left(\frac{-3\delta^2}{r} P(O_3) H k^{\text{surf}} t\right)}, \qquad (\text{IX})$$

where δ is the depth of the surface layer and k^{surf} is the second-order surface rate coefficient for reaction between organic species and ozone. In contrast to eqn (VI), the temporal decay of the reactant in the particle is exponential. For the reactions of benzoate and fumarate, eqn (VI) provides a convincingly better fit to the experimental data than eqn (IX). For α-pinene eqn (VI) only provides a slightly better fit than eqn (IX). However, although the droplet sizes studied form a reasonably narrow size distribution (see Table 2), there is no dependence of the uptake coefficients on droplet size for the any of the reactions studied here, strong evidence that bulk reactions as opposed to surface reactions are occurring. The diffusion of the co-reactant (*i.e.* benzoate, fumarate, and α-pinene) in the droplet has not been considered in this work and may have a slight effect on the value obtained for the uptake coefficients, as discussed by Smith *et al.*[58,68]

4.2.1 Size changes

The particles studied here did not undergo a large size change during reaction. In our previous work the same laser Raman tweezers technique recorded a striking size change following the oxidation of a thin layer of oleic acid on a saltwater droplet.[5] Reactions in aerosol particles can alter the chemistry and thermodynamics of the particles and they might lose or gain water vapour from the atmosphere in line with Köhler theory.[4] For the reactions of ozone with aqueous droplets of benzoate or fumarate, a large size change is not expected from Köhler theory as the number of anionic solutes is not changing and only the mass of these solutes is decreasing. A large change in size of the droplet during the oxidation of α-pinene might have been expected, as the production of carboxylic acids would have likely caused a hydrophilic layer to form at the organic–air interface of the particle. However, a size change was not recorded and this finding gives some evidence that there may be some oligomerisation occurring within the particle. Oligomerisation reactions would prevent the formation of free hydrophilic carboxylic acids.

Very recently submitted work by Reid *et al.* [in press[69]] suggests the occurrence of possible errors in measuring particle size by bright-field microscopy owing to uncertainties in the position of the trapped aerosol in the optical trap. These potential errors were minimised in this work by ensuring that a proportion of the measurements of the particle size were recorded on sharply focussed images of the particles. By making small adjustments in the collimation optics, used in conditioning the laser beam prior to optical trapping, the particle was brought into the focal plane (calibrated with a graticule) ensuring that sizing had no optical artefact.

It should be noted that a series of "blank" experiments were undertaken to characterise chemical and size changes of the particles. The nascent aqueous particles of fumarate and benzoate in the absence of ozone would normally shrink (or occasionally grow) in size as water evaporated (or condensed) from (onto) the particles to equilibrate the relative humidity within the cell in line with Köhler theory.[4] This equilibration normally took a few minutes, and the particle size would remain stable for 90 minutes *i.e.* for time scales greater those required for a kinetic experiment. The intensity of the Raman signal would stay constant whilst the size was stable (in the absence of morphologic Raman resonances). The particle in the cell may experience a local relative humidity that is higher than the relative humidity of the gas entering the cell as some particles held for extended periods of time (>3 hours) tended to shrink further. For these particles it is assumed that the particle size initially responds to a local relative humidity in the cell that is wet from aerosol impacting on the cell walls. On long time scales the cell equilibrates with the lower relative humidity of the gas flow through the cell and thus the particle experiences a lower relative humidity and evaporates accordingly. On the time scales of the kinetics experiment this was not a detrimental effect. However, no work is presented here using Köhler theory as the relative humidity is not accurately known. Problems originating from the measurement of the relative humidity in laser tweezers cells have been reported before.[37]

4.3.1 Products and mechanism of fumarate oxidation by ozone

The liquid-phase reaction between the fumarate ions and ozone has been previously studied[70] and the direct products of the ozonolysis have been shown to be the glyoxylate anion and the 2-hydroperoxy-2-hydroxyacetate anion:

(4)

The 2-hydroperoxy-2-hydroxyacetate decays rapidly to formic acid (and hence formate in basic solutions) and carbon dioxide:

$$\text{HO-C(OOH)(H)-C(O)O}^- \rightarrow CO_2 + HC(O)OH + OH^- \quad (5)$$

or overall

$$CO_2CH=CHCO_2^{2-} + O_3 \rightarrow CHOCO_2^- + CO_2 + HCO_2H + OH^- \quad (6)$$

The Raman spectra of glyoxylate and formate are not observed in our experiments, but there is a small sharp feature in the product spectrum that has been tentatively assigned to the carbonate anion, CO_3^{2-}. The 2-hydroperoxy-2-hydroxyacetate may also decompose to hydrogen peroxide and glyoxylate, but these two products can react rapidly to form formic acid, and carbon dioxide. Test experiments with nonanoic acid on a seawater particle with a 4 μm radius demonstrated that the nonanoic acid could evaporate to below the detection limit on the time scale of minutes. However, in the pH conditions of these experiments any formic acid formed will exist as formate and not be volatile. Thus it appears that the formic acid or formate ion is consumed in a secondary reaction. Glyoxylate and formate are unlikely to react with ozone.

$$O_3 + HCO_2^- \rightarrow \text{products}, \ k_7 = 100 \ dm^3 \ mol^{-1} \ s^{-1} \ ^{61} \quad (7)$$

$$O_3 + HC(O)CO_2^- \rightarrow \text{products}, \ k_8 = 1.9 \ dm^3 \ mol^{-1} \ s^{-1}. \ ^{61} \quad (8)$$

However, the reaction of ozone with hydroxyl anions can produce OH radicals:[71]

$$O_3 + OH^- \rightarrow HO_2^- + O_2 \quad (9)$$

$$HO_2^- + O_3 \rightarrow HO_2 + O_3^{-\cdot} \quad (10)$$

$$O_3^{-\cdot} \rightarrow O^{-\cdot} + O_2 \quad (11)$$

$$O^{-\cdot} + H_2O \rightarrow OH^\cdot + OH^- \quad (12)$$

Hydroxyl radicals react rapidly with formate and may react rapidly with glyoxylate ions,

$$OH + HCO_2^- \rightarrow \text{products}, \ k_{13} = 2\text{–}5 \times 10^9 \ dm^3 \ mol^{-1} \ s^{-1} \ ^{72} \quad (13)$$

$$OH + HC(O)CO_2^- \rightarrow \text{products} \quad (14)$$

and thus the presence of hydroxyl radical would reduce the concentration of formate and glyoxylate in the droplet to below the detection limit of our system. It should be noted that the first-order loss of ozone owing to reaction with the hydroxyl anion in the basic solution is slow compared to first-order loss with the fumarate anion. This secondary chemistry does thus not affect the determination of the kinetics of the reaction between ozone and formate, as shown below:

$$O_3 + CO_2CH=CHCO_2^{2-} \rightarrow \text{products},$$
$$k_1 = 1 \times 10^5 \ dm^3 \ mol^{-1} \ s^{-1},^{60} \ [CO_2CH=CHCO_2^{2-}] \sim 0.09 \ mol \ dm^{-3},$$
$$k_1 \times [CO_2CH=CHCO_2^{2-}] \sim 9000 \ s^{-1} \quad (1)$$

$$O_3 + OH^- \rightarrow HO_2^- + O_2, \ k_9 = 48 \ dm^3 \ mol^{-1} \ s^{-1},^{70}$$
$$[OH^-] \sim 1 \times 10^{-4} \ mol \ dm^{-3}, \ pH = 10,$$
$$k_9 \times [OH^-] \sim 0.005 \ s^{-1} \quad (9)$$

4.3.2 Products and mechanism of benzoate oxidation by ozone

No products were observed for the reaction between ozone and benzoate. The oxidation of phenolic and benzoic species in aqueous solution by ozone produces intermediate compounds of fumaric and maleic acids and the resulting compounds are glyoxylic and formic acids.[73–78] These species were not detected. Rate coefficients for reaction of fumarate and maleate with ozone are five orders of magnitude larger than that for reaction of benzoate with ozone, so that fumarate and maleate are unlikely to be observed. Reasons for the lack of product peaks in the Raman spectra owing to formate or glyoxylate production have been discussed earlier for the reaction of fumarate with ozone.

4.3.3 Products and mechanism of α-pinene oxidation by ozone

The liquid-phase reaction between α-pinene and ozone has been studied previously[79,80] and the major product detected was a hydroperoxide followed by verbenol, verbonone and pinoic acid. The Raman spectrum presented in Fig. 4 is not pinoic acid. Our investigation of the reaction of α-pinene in pentadecane with ozone is analogous to work performed by Mochida et al.[81] who studied the oxidation of methyl oleate in a mixture of dioctyl adipate and myristic acid. Mochida et al.[81] found that the products of these reactions were hydroperoxides, secondary ozonides, peroxides and oligomers generating high molecular weight species. The product peaks in Fig. 4(c) around 1600–1800 cm^{-1} are indicative of several C=O carbonyl stretches. Not shown in Fig. 4 is the region of the spectrum between 800 and 1000 cm^{-1}. This region was investigated in separate experiments to test for the presence of the peroxide stretch around 900 cm^{-1}—a strong signal was not observed, but in a few experiments there was a weak peak that may be tentatively assigned to peroxide. Our results are consistent with other studies of alkene oxidation by ozone suggesting the formation of several different carbonyl species and possibly peroxides. A study of the gas-phase oxidation of α-pinene by ozone in the presence of aqueous inorganic particles has been undertaken by Gao et al.[27] The study by Gao et al.[27] has parallels with our work with respect to the production distribution but differs from our study in two aspects: (a) Gao et al. study examined the gas-phase oxidation followed by partitioning of the product to the aerosol phase and then by further reaction; and (b) the particle phase was inorganic whereas ours is organic. Gao et al.[27] found by mass spectral analysis that four types of organic compounds were formed: organic acids, di-acid alkyl esters and hydroxy di-acids. They also obtained evidence for oligomer formation. Formation of oligomers was enhanced by an increase in particle acidity. In our study, we observe several different bands associated with C=O stretches that may be indicative of groups similar to those seen by Gao et al. The occurrence of the tentative O–O stretch may suggest secondary ozonide or peroxide formation. Czoschke et al.[20] present FTIR spectra of the aerosol products of the oxidation of α-pinene by ozone. They report that CH$_2$ and CH$_3$ stretches of the product species are similar to those found for α-pinene and the existence of C=O stretches and OH carboxylic stretches was also reported.

Conclusions

In this work we have shown that laser Raman tweezers can be used to measure uptake coefficients of gaseous oxidants on aqueous and organic droplets, whilst monitoring the size and composition of the droplet. Laser Raman tweezers are thus a powerful technique for the investigation of oxidation processes of atmospheric importance. The ability to study single droplets and not the average kinetics and chemistry of a distribution of droplets is a unique advantage of this technique.

The uptake of ozone on aqueous droplets containing benzoate or fumarate ions was found to be small. The droplets did not appear to change size following oxidation, suggesting cloud droplet sizes will not be influenced by these reactions, although the Authors fully acknowledge that only a limited range of relative humidities were probed. No evidence was found for oligomerization leading to HULIS formation at these high pHs. We have demonstrated that the laser Raman tweezers can be used successfully to study gas-aerosol reactions and we intend to study more complicated systems, such as those involving substituted phenols in the future.

The ozone-initiated oxidation of α-pinene in an alkane was the first study using organic seed particles. The products did not differ from those found in earlier gas-phase studies of this reaction. Interestingly, no water uptake was observed, so that the hygroscopicity of the particle does not seem to have dramatically increased over the course of the reaction. Further studies are required to determine if the products formed are water-soluble or if oligomerisation is the dominant process.

Acknowledgements

The authors wish to thank the CCLRC under direct access grants CM11E2/05 and CM15C2/04 which have enabled use the laser tweezers apparatus at the Central Laser Facility, and Jonathan Reid for discussions on morphological resonances and particle sizing strategies. MDK wishes to thank NERC for support under NER/B/S/2003/00289. CP wishes to thank the Leverhulme Trust for support under the visiting scholar programme (grant F/07537/W).

References

1 U. Pöschl, *Angew. Chem., Int. Ed.*, 2005, **44**, 7520.
2 J. T. Houghton, Y. Ding, D. J. Griggs, M. Noguer, P. J. van der Linden and D. Xiaosu, *Climate Change: The Scientific Basis*, Cambridge University Press, Cambridge, 2001.
3 U. Lohmann and J. Feichter, *Atmos. Chem. Phys.*, 2005, **5**, 715.
4 M. L. Shulman, M. C. Jacobson, R. J. Carlson, R. E. Synovec and T. E. Young, *Geophys. Res. Lett.*, 1996, **23**, 277.
5 M. D. King, K. C. Thompson and A. D. Ward, *J. Am. Chem. Soc.*, 2004, **126**, 16710.
6 E. R. Graber and Y. Rudich, *Atmos. Chem. Phys.*, 2006, **6**, 729.
7 S. Decesari, M. C. Facchini, S. Fuzzi and E. Tagliavini, *J. Geophys. Res., [Atmos.]*, 2000, **105**, 1481.
8 A. Gelencser, T. Meszaros, M. Blazso, G. Kiss, Z. Krivacsy, A. Molnar and E. Meszaros, *J. Atmos. Chem.*, 2000, **37**, 173.
9 G. Kiss, E. Tombacz, B. Varga, T. Alsberg and L. Persson, *Atmos. Environ.*, 2003, **37**, 3783.
10 A. Gelencser, A. Hoffer, G. Kiss, E. Tombacz, R. Kurdi and L. Bencze, *J. Atmos. Chem.*, 2003, **45**, 25.
11 E. Dinar, T. F. Mentel and Y. Rudich, *Atmos. Chem. Phys.*, 2006, **6**, 5213.
12 A. Guenther, C. N. Hewitt, D. Erickson, R. Fall, C. Geron, T. Graedel, P. Harley, L. Klinger, M. Lerdau, W. A. McKay, T. Pierce, B. Scholes, R. Steinbrecher, R. Tallamraju, J. Taylor and P. Zimmerman, *J. Geophys. Res., [Atmos.]*, 1995, **100**, 8873.
13 J. H. Seinfeld and J. F. Pankow, *Annu. Rev. Phys. Chem.*, 2003, **54**, 121.
14 S. Hatakeyama, K. Izumi, T. Fukuyama and H. Akimoto, *J. Geophys. Res., [Atmos.]*, 1989, **94**, 13013.
15 D. Grosjean, *Atmos. Environ.*, 1992, **26**, 953.
16 T. Hoffmann, J. R. Odum, F. Bowman, D. Collins, D. Klockow, R. C. Flagan and J. H. Seinfeld, *J. Atmos. Chem.*, 1997, **26**, 189.
17 C. N. Cruz and S. N. Pandis, *Atmos. Environ.*, 1997, **31**, 2205.
18 H. J. Tobias and P. J. Ziemman, *Anal. Chem.*, 1999, **71**, 3428.
19 M. S. Jang, N. M. Czoschke, S. Lee and R. M. Kamens, *Science*, 2002, **298**, 814.
20 N. M. Czoschke, M. Jang and R. M. Kamens, *Atmos. Environ.*, 2003, **37**, 4287.
21 A. Limbeck, M. Kulmala and H. Puxbaum, *Geophys. Res. Lett.*, 2003, **30**, 4.
22 Y. Iinuma, O. Boge, T. Gnauk and H. Herrmann, *Atmos. Environ.*, 2004, **38**, 761.
23 S. Koch, R. Winterhalter, E. Uherek, A. Kolloff, P. Neeb and G. K. Moortgat, *Atmos. Environ.*, 2000, **34**, 4031.

24 M. P. Tolocka, M. Jang, J. M. Ginter, F. J. Cox, R. M. Kamens and M. V. Johnston, *Environ. Sci. Technol.*, 2004, **38**, 1428.
25 M. Kalberer, J. Yu, D. R. Cocker, R. C. Flagan and J. H. Seinfeld, *Environ. Sci. Technol.*, 2000, **34**, 4894.
26 J. Z. Yu, D. R. Cocker, R. J. Griffin, R. C. Flagan and J. H. Seinfeld, *J. Atmos. Chem.*, 1999, **34**, 207.
27 S. Gao, M. Keywood, N. L. Ng, J. Surratt, V. Varutbangkul, R. Bahreini, R. C. Flagan and J. H. Seinfeld, *J. Phys. Chem. A*, 2004, **108**, 10147.
28 R. Omori, T. Kobayashi and A. Suzuki, *Opt. Lett.*, 1997, **22**, 816.
29 A. Biswas, H. Latifi, R. L. Armstrong and R. G. Pinnick, *Opt. Lett.*, 1989, **14**, 214.
30 R. Thurn and W. Kiefer, *Appl. Spectrosc.*, 1984, **38**, 78.
31 C. Mund and R. Zellner, *ChemPhysChem*, 2003, **4**, 638.
32 C. Mund and R. Zellner, *ChemPhysChem*, 2003, **4**, 630.
33 J. F. Lubben, C. Mund, B. Schrader and R. Zellner, *J. Mol. Struct.*, 1999, **481**, 311.
34 C. Mund and R. Zellner, *J. Mol. Struct.*, 2003, **661**, 491.
35 N. Jordanov and R. Zellner, *Phys. Chem. Chem. Phys.*, 2006, **8**, 2759.
36 D. McGloin, *Philos. Trans. R. Soc. London, Ser. A*, 2006, **364**, 3521.
37 L. Mitchem, J. Buajarern, R. J. Hopkins, A. D. Ward, R. J. J. Gilham, R. L. Johnston and J. P. Reid, *J. Phys. Chem. A*, 2006, **110**, 8116.
38 L. Mitchem, R. J. Hopkins, J. Buajarern, A. D. Ward and J. P. Reid, *Chem. Phys. Lett.*, 2006, **432**, 362.
39 J. Buajarern, L. Mitchem, A. D. Ward, N. H. Nahler, D. McGloin and J. P. Reid, *J. Chem. Phys.*, 2006, **125**.
40 L. Mitchem, J. Buajarern, R. J. Hopkins, A. D. Ward, R. J. J. Gilham, R. L. Johnston and J. P. Reid, *J. Phys. Chem. A*, 2006, **110**, 8116.
41 J. P. Reid and R. M. Sayer, *Chem. Soc. Rev.*, 2003, **32**, 70.
42 R. J. Hopkins and J. P. Reid, *J. Phys. Chem. B*, 2006, **110**, 3239.
43 M. D. Summers, J. P. Reid and D. McGloin, *Opt. Express*, 2006, **14**, 6373.
44 E. Fallman and O. Axner, *Appl. Opt.*, 1997, **36**, 2107.
45 R. D. B. Gatherer, R. M. Sayer and J. P. Reid, *Chem. Phys. Lett.*, 2002, **366**, 34.
46 R. M. Sayer, R. D. B. Gatherer and J. P. Reid, *Phys. Chem. Chem. Phys.*, 2003, **5**, 3740.
47 R. Symes, R. M. Sayer and J. P. Reid, *Phys. Chem. Chem. Phys.*, 2004, **6**, 474.
48 P. V. Danckwerts, *Trans. Faraday Soc.*, 1950, **46**, 300.
49 P. V. Danckwerts, *Trans. Faraday Soc.*, 1951, **47**, 1014.
50 P. V. Danckwerts, *Gas-liquid reactions*, McGraw-Hill, New York, 1970.
51 D. R. Worsnop, J. W. Morris, Q. Shi, P. Davidovits and C. E. Kolb, *Geophys. Res. Lett.*, 2002, **29**, 4.
52 B. J. Finlayson-Pitts and J. N. Pitts, *Chemistry of the Upper and Lower Atmosphere*, Academic press, San Deigo, 2000.
53 C. E. Kolb, D. R. Worsnop, M. S. Zahniser, P. Davidovits, T. R. Keyser, M. T. Leu, M. M. J. D. R. Hanson and A. R. Ravishankara, *Progress and Problems in Atmospheric Chemistry*, world Scientific Publishing, Singapore, 1995, p. 771.
54 D. R. Hanson, *J. Phys. Chem. B*, 1997, **101**, 4998.
55 S. E. Schwartz and J. E. Freiberg, *Atmos. Environ.*, 1981, **15**, 1129.
56 *Chemical and Physical Handbook*, C. R. C. Press.
57 S. P. Sander, R. R. Friedl, D. M. Golden, M. J. Kurylo, G. K. Moorgat, H. Keller-Rudek, P. H. Wine, A. R. Ravishankara, C. E. Kolb, M. M. J. B. J. Finlayson-Pitts, R. E. Huie and V. L. Orkin, *Chemical Kinetics and photochemical data for use in Atmospheric studies: Evaluation Number 15*, Jet Propulsion Laboratory, Pasedena, 2006.
58 G. D. Smith, E. Woods, C. L. DeForest, T. Baer and R. E. Miller, *J. Phys. Chem. A*, 2002, **106**, 8085.
59 T. Moise and Y. Rudich, *J. Geophys. Res., [Atmos.]*, 2000, **105**, 14667.
60 J. Hoigne and H. Bader, *Water Res.*, 1983, **17**, 185.
61 J. Hoigne and H. Bader, *Water Res.*, 1983, **17**, 173.
62 J. D. Frantz, *Chem. Geol.*, 1998, **152**, 211.
63 J. A. de Gouw and E. R. Lovejoy, *Geophys. Res. Lett.*, 1998, **25**, 931.
64 M. D. King, C. E. Canosa-Mas and R. P. Wayne, *Phys. Chem. Chem. Phys.*, 1999, **1**, 2231.
65 M. D. King, C. E. Canosa-Mas and R. P. Wayne, *Phys. Chem. Chem. Phys.*, 1999, **1**, 2239.
66 C. Pfrang, M. D. King, C. E. Canosa-Mas and R. P. Wayne, *Atmos. Environ.*, 2006, **40**, 1180.
67 D. G. Williamson and R. J. Cvetanovic, *J. Am. Chem. Soc.*, 1968, **90**, 3668.

68 G. D. Smith, E. Woods, T. Baer and R. E. Miller, *J. Phys. Chem. A*, 2003, **107**, 9582.
69 K. J. Knox, J. P. Reid, K. L. Hanford, A. J. Hudson and L. Mitchem, *J. Opt. A: Pure Appl. Opt.*, 2007, **9**, S180.
70 A. Leitzke, E. Reisz, R. Flyunt and C. von Sonntag, *J. Chem. Soc., Perkin Trans. 2*, 2001, 793.
71 L. Forni, D. Bahnemann and E. J. Hart, *J. Phys. Chem.*, 1982, **86**, 255.
72 G. V. Buxton, C. L. Greenstock, W. P. Helman and A. B. Ross, *J. Phys. Chem. Ref. Data*, 1988, **17**, 513.
73 T. Poznyak and J. Vivero, *Ozone-Sci. Eng.*, 2005, **27**, 447.
74 T. Poznyak and B. Araiza, *Ozone-Sci. Eng.*, 2005, **27**, 351.
75 F. J. Beltran, F. J. Rivas and O. Gimeno, *J. Chem. Technol. Biotechnol.*, 2005, **80**, 973.
76 T. N. Murakami, M. Takahashi and N. Kawashima, *Chem. Lett.*, 2000, 1312.
77 A. Mokrini, D. Ousse and E. Esplugas, *Water Sci. Technol.*, 1997, **35**, 95.
78 E. Gilbert, *Z. Naturforsch., B: Chem. Sci.*, 1977, **32**, 1308.
79 J. M. Encinar, F. J. Beltran and J. M. Frades, *Chem. Eng. Technol.*, 1993, **16**, 68.
80 J. M. Encinar, F. J. Beltran and J. M. Frades, *Chem. Eng. Technol.*, 1994, **17**, 187.
81 M. Mochida, Y. Katrib, J. T. Jayne, D. R. Worsnop and S. T. Martin, *Atmos. Chem. Phys.*, 2006, **6**, 4851.

General Discussion

Dr Reid opened the discussion of Professor Shiratani's paper: You stated that there is a temperature gradient of 25K cm^{-1}. How do you measure this or is it based on simulation?

Professor Shiratani responded: The electrode temperature was measured with a thermocouple. The gas temperature between the electrodes was deduced using tiny thermo-labels hanging between the electrodes.

Professor Goree said: You mentioned that particles rise towards your upper horizontal electrode due to a thermophoresis force. This force arises from a temperature gradient in the gas. Did you purposefully apply a temperature difference to the two electrodes to produce this temperature gradient, or does it arise for another reason?

Professor Shiratani replied: We did not give intentionally produce the temperature gradient. The powered electrode was heated by bombarding ions flowing out of the plasma, whereas the temperature of the grounded electrodes was kept at 358 K. The temperature rise of the powered electrode led to the temperature gradient between the powered electrode and the grounded electrodes.

Professor Rühl asked: How did you determine the particle size and the width of the size distribution? How are the charges distributed in the particles? Are there multiply charged particles?

Professor Shiratani responded: The average size of particles was measured by an *in-situ* laser light scattering method. The particle size distribution was obtained by *ex-situ* TEM observation of particles collected on TEM mesh.

We have not measured the charge distribution of nanoparticles. However, we can deduce the average charge on a nanoparticle from quasi-neutral conditions of plasmas. The average charge is in a range of -0.01 e to -0.15 e as shown in Fig. 5 of our paper, and therefore most nanoparticles are neutral and the rest have a charge of e$^-$. We expect few multiply charged particles in our case.

Professor Goree remarked: How did you determine the size of particles in your time-series graphs of particle diameter? Did you do this by collecting them after turning the plasma off and sizing them using electron microscopy?

Professor Shiratani answered: We determined the size of particles using an *in-situ* laser light scattering method. The experimental setup and the method are described in ref. 13 of our paper.[1]

1 S. Nunomura, M. Kita, K. Koga, M. Shiratani and Y. Watanabe, *J. Appl. Phys.* 2006, **99**, 083302.

Dr Reid asked: How do you determine the size of the nanoparticles? Can you give us more detail?

Professor Shiratani answered: The size and density of the nanoparticles are determined from their thermal coagulation that takes place after turning off the discharge. Thermal coagulation usually takes place after turning off the discharge; because during the discharge, particles are charged negatively and their coagulation is strongly suppressed by their mutual Coulomb repulsive force. The details were described in ref. 13 of our paper.[1]

1 S. Nunomura M. Kita, K. Koga, M. Shiratani and Y. Watanabe, *J. Appl. Phys.* 2006, **99**, 083302.

Professor Kersten commented: Do you have any information on the time evolution of the electron temperature during particle growth? (see Fig. 5 of your paper).

Professor Shiratani answered: At the time of the discussion, we had not carried out optical emission spectroscopic measurements which give information on the time evolution of the electron temperature. After the Faraday Discussion, we carried out such measurements. The results show that intensity of Ar emissions from the plasma tend to increase with time. Moreover, a ratio of two Ar emission lines varies with time, which indicates the electron temperature tends to increase during particle growth.

Professor Stoffels stated: A remark on the question by Professor Kersten: Considering the electron density decreases, it is expected that the electron temperature increases. In Ar/SiH_4 mixtures this is observed as a significant increase of emission of Ar lines. Determination of the excitation temperature, based on the Ar emission lines also shows an electron temperature increase.

Professor Shiratani answered: We have carried out some optical spectroscopic measurements and based on the results the time evolution of the excitation temperature during particle growth in our study is similar to that in Ar/SiH_4 mixture discharges.

Professor Stoffels commented: Is the average charge enough and/or fluctuating fast enough that collective effects are important? Does a cloud of 1–2 nm particles behave as a Coulomb liquid?

Professor Shiratani answered: Fast fluctuation of charge on particles is important to obtain collective effects of particles. For instance, the suppression of thermal coagulation occurs even when a fraction of particles are charged. It is known that the charge on particles always fluctuates due to the random attachment of electrons and ions, therefore, particles undergo Coulomb repulsion in a certain period of time. The suppression conditions for the thermal coagulation are considered to be that the charging frequency f_{ch} of particles is higher than the collision frequency f_{p-p} among them.

Professor Goree opened the discussion of Dr Pavlů's paper: You reported that your spherical gold particles become elongated as they are reduced in size by sputtering when exposed to a directional ion beam. I would have expected that rotation of particles suspended in your Paul trap would average out the directional effects of the beam. Can you speculate why this occurs?

Dr Pavlů replied: The question is very complex starting with the uncertain shape of the grain—the only parameter we observe is the temporal change of the grain capacitance with respect to the sphere of the same mass (Fig. 1), thus the ellipsoidal shape is a speculation so far.[1] We can consider two possibilities:

(i) A very simple intuitive mechanical approach is that the irregular (for any reason) grain is turning towards the ion beam with the smallest area (like a weathervane in a wind).

(ii) The second approach is taking into account the interaction of the space charge of the ion beam with the quadrupole term of the electric field of the grain. However,

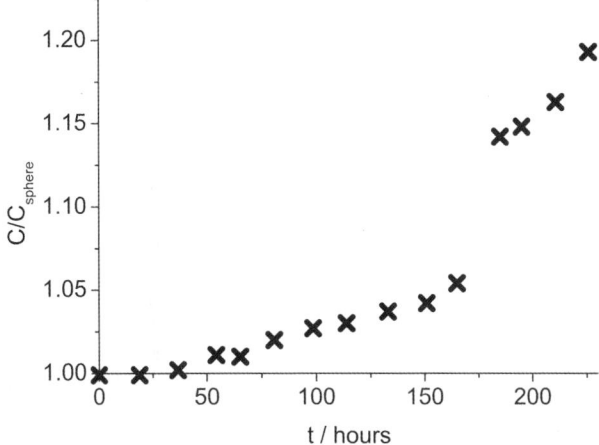

Fig. 1 Temporal evolution of the grain relative capacitance–the indication of shape change.[1] Modified with permission © 2007 IEEE.

there is a key question of the conductivity of the grain—the described idea is valid for metal grains but for dielectrics it would probably lead to an increased rotation of the grain because of the dipole moment. This approach is described in our paper, however, without any quantitative measurements in our case. The question which of these alternatives dominates remains open so far. Nevertheless, our preliminary calculations show that the second interaction (ii) could be reasonably strong to stop grain rotation.

Note here that, no matter whether (i) or (ii) is right, a similar effect would be observed wherever the grain is exposed to the directional stream of ions/atoms, *e.g.*, the dust grains in the solar wind.

1   J. Pavlů, I. Richterova, Z. Nemecek, J. Safrankova, and J. Wild, *IEEE Trans. Plasma Sci.*, 2007, **35**(2), 297–302.

Professor Signorell asked: Would a comparison with sputtering of deposited particles help to elucidate the origin of the shape change?

Dr Pavlů replied: The origin of the grain shape change is known—the sputtering process is more efficient for grazing impacts. The question is how can the grain be stopped from rotating. Though we suppose we can carry out such an experiment, we don't think it will bring any new light to the changing shape feature we've observed.

The major role of the grain–beam interaction would be overprinted by a rigid connection of the grain to the substrate.

Professor Davidovits asked: What is the astrophysical evidence for the sputtering results you observed?

Dr Pavlů answered: There exists some indirect observation of sputtering. Or, perhaps it is better to say, there exist phenomena that are likely to be explained by the sputtering of dust grains. One of these is the presence of so called pick-up ions. They are ions of heavy species which do not fit to the solar wind neither by their abundances nor by their energies. One of the possible sources of these ions could be sputtering of dust followed by ionization.[1] Mukai and Schwehm[2] calculated the critical distance from the Sun where sputtering of grains dominates their sublimation. They have shown that life-times of interplanetary dust grains are controlled by sputtering by solar wind.

Another indirect evidence comes from modeling of Saturn's E-ring where sputtering is supposed to play a significant role. Our calculations, based on sputtering of 1 μm ice grain by O^+ ions (number density 5 cm^{-3}, mean energy 100 eV), predict sputtering to half mass in the range of tens of years, sputtering to half diameter in less then hundred years, and total grain erosion in about one or two hundred years. The model of Jurac et al.[3] clearly shows that the speed of grain erosion due to sputtering would be similar—1 μm grain in about 50 years.

1 I. Mann, H. Kimura, D. A. Biesecker, B. T. Tsurutani, E. Grün, R. B. McKibben, J.-C. Liou, R. M. MacQueen, T. Mukai, M. Guhathakurta and P. Lamy, *Space Sci. Rev.*, 2004, **110**, 269–305.
2 T. Mukai and G. Schwehm, *Astron. Astrophys.*, 1981, **95**, 373–382.
3 S. Jurac, R. E. Johnson and J. D. Richardson, *Icarus*, 2001, **149**(2), 384–396.

Professor Rühl asked: Is the emission of secondary electrons the only and the most important charging mechanism? Would you expect to obtain comparable results from a photon impact experiment?

Dr Pavlů replied: Definitely not, however, in several regions of space or in particular laboratory experiments hot electrons are present and secondary electron emission becomes important.

Recent experiments by Grimm et al.[1] have shown that secondary emission plays an indispensable role when dust grains are charged by UV or soft X-ray irradiation because most of the electrons leaving the irradiated surface are secondary electrons excited inside the target by primary photoelectrons or Auger electrons.

The photoemission and secondary emission become similar if the grain size is comparable with the penetration depth of the beam electrons into the grain. In such a case, a great majority of beam electrons deposits a part of their energy inside the grain and leaves it without changing its charge (like the photons do). However, there is an important difference in maximum potential of the grain that can be achieved by charging with a photon or electron beam of the same energy. Whereas an upper limit of the grain potential charged by photons is slightly less than the photon energy,[1] it is only one half of the energy of the electron beam. The upper limit of the grain potential is given by the highest energy of the electrons leaving the grain. This energy is roughly equal to the photon energy in case of photoemission. On the other hand, the primary electron from the beam brings its charge into the grain and, in an ideal case, it can spend half of its energy for excitation of the secondary electrons and both of them can then escape from the grain charged below a half of the beam energy.

Our estimation of the equilibrium potential of ≈ 1 μm glass grain as a function of the energy of primary particles (electrons or photons) is sketched in Fig. 2. The thin lines show theoretical limits of the grain potential for photons (dashed) and electrons (solid). The solid part of the thick gray curve shows the charging by photons estimated from Grimm et al.[1] The dotted continuation of this line is our estimation of further increase of the grain potential taking account of potential limitations by the ion field emission.[2] The solid part of the black curve is consistent with the measurement on 1.2 μm SiO_2 spheres shown in our paper and the dotted part is our estimation of a further potential evolution. We expect the same limit of the surface potential due to the ion field emission as for photon charging. From the sketch it follows that the above described "half-energy limit" cannot be reached for grains in the micrometer range of size.

1 M. Grimm, B. Langer, S. Schlemmer, T. Lischke, U. Becker, W. Widdra, D. Gerlich, R. Flesch and E. Rühl, *Phys. Rev. Lett.*, 2006, **96**(6), 066801.
2 M. Jerab, I. Richterova, J. Pavlů, J. Safrankova and Z. Nemecek, *IEEE Trans. Plasma Sci.*, 2007, **35**(2), 292–296.

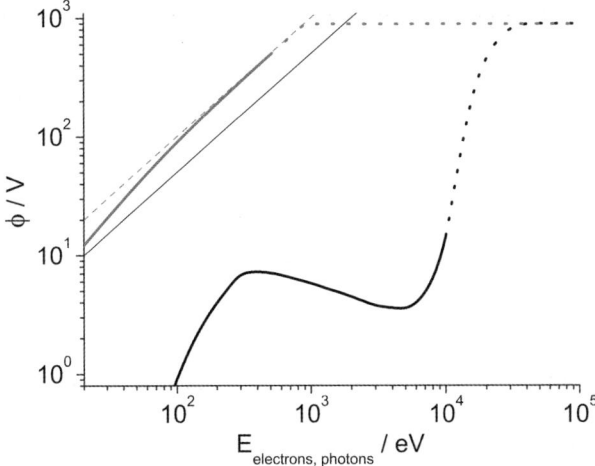

Fig. 2 A sketch of possible equilibrium surface potential of about 1 μm glass dust grain as a function of the primary electron (black line)/photon (gray line) energy.

Professor Rudich asked: How would surface corrugation affect the process and conclusion with respect to the real space-borne grains?

Dr Pavlů answered: All of the aforementioned processes have their size dependences. Space-born grains are often highly irregular exhibiting a lot of tips and other features.

Most of these can be described by a characteristic dimension (curvature of the tip, thickness of flakes, *etc.*) which determines the charging process.

Experiments with the spherical grains of similar diameter to this characteristic dimension could shed light on these processes. Sharp edges will play a major role when dealing with emissions due to an intense electric field.

Also the secondary electron emission from thinner parts of dust grains can be significantly enhanced. Although we tend to use spherical grains, some of the results can be easily extended to irregular grains. We've reported several features of grain clusters that can simulate space-born dust.

Professor Mason said: Role of secondary electrons in Interstellar dust chemistry; production of secondary electrons from *ice* covered grains which subsequently leads to complex chemistry in the ice—such that UV + ion irradiation produce similar products.

Dr Pavlů replied: We fully agree with Professor Mason. The surfaces of grains serve as small chemical laboratories in space.

Professor Kersten asked: Do you observe the effect of ion drag (particle movement) if you switch on the ion beam? Do you have to tune the particle trap in order to stabilize the particles?

Dr Pavlů replied: The trap is tuned so that the secular frequency of the grain is well within the stability region.[1,2] Moreover, the grain motion is stabilized by a damping feedback loop which uses the position signal to reduce the amplitude of the grain oscillation in particular directions using the position signal.[3] When the ion beam is switched on we often observe grain "shaking" or being shifted a little bit outside the trap center. But it doesn't get far away from the center for a long time being firmly trapped in the potential well and actively stabilized thus the drag would be compensated by both of these.

1. W. Paul and H. Steinwedel, *Apparatus for separating charged particles of different specific charges*, Ger. Pat. 944 900, Jun. 28, 1956. US Pat. 2 939 952, Jun. 7, 1960.
2. R. F. Wuerker, H. Shelton and R. V. Langmuir, *J. Appl. Phys.*, 1959, **30**(3), 342–349.
3. I. Cermak, J. Pavlů, P. Zilavy, Z. Nemecek, J. Safrankova and I. Richterova, in *WDS'04 Proceedings of Contributed Papers: Part II–Physics of Plasmas and Ionized Media*, ed. J. Safrankova, Matfyzpress, Prague, 2004, pp. 279–286.

Professor Ray commented: How do you keep the particle stable? By adjusting frequency or AC voltage or both?

Dr Pavlů answered: In principle, both of these parameters can (and sometimes must) be changed.

However, the frequency is usually adjusted automatically in order to keep an optimum ratio of the quadrupole AC frequency and secular frequency of the grain motion that follows from the mathematical solution of the trap. In the present setup, we can tune the trap frequency over about two orders of magnitude and the voltage about one order of magnitude. Consequently, the range of Q/m we can trap spans over five orders of magnitude (approximately from 4×10^{-3} to 4×10^{1} C kg^{-1}). For details on the trap principle and setup see Wuerker *et al.*[1] and Cermak *et al.*[2]

1. R. F. Wuerker, H. Shelton and R. V. Langmuir, *J. Appl. Phys.*, 1959, **30**(3), 342–349.
2. I. Cermak, J. Pavlů, P. Zilavy, Z. Nemecek, J. Safrankova and I. Richterova, in *WDS'04 Proceedings of Contributed Papers: Part II –Physics of Plasmas and Ionized Media*, ed. J. Safrankova, Matfyzpress, Prague, 2004, pp. 279–286.

Professor Goree opened the discussion of Professor Kersten's paper: When you turn off the plasma and the particles then fall down, but more slowly than they would in free fall: how long does it take for the plasma to disappear in this afterglow, do you think that the particles retain a residual charge, and are they exposed to an electric field?

Professor Kersten replied: I am sure that the particles retain a relatively large residual charge. We can levitate the particles after the plasma is switched-off in front of our adaptive electrode due to a bias voltage at the pixels of this electrode. Hence, the still charged particles are trapped in the electric field of the biased surface even without plasma.

Professor Stoffels asked: How does the charge on a particle vary as it moves and is it a valid approximation to keep the charge constant in the derivation of eqn (10) and (11) of your paper?

Professor Kersten responded: This is a legitimate question. The charge might vary with the height above the electrode, *e.g.* with the particle position in the sheath. The assumption of a constant charge is only very rough. Therefore, the measurements of the falling particles differs remarkably from the calculation, see Fig. 10 of our paper. Only at the beginning—where the change in the field is rather small—the assumption of a constant particle charge is valid. In principle, for each position z in the sheath, a new charge has to be calculated. We hope that we can do this by further evaluation of the fall experiments in the near future. On the other hand, for the falling experiments, where the plasma is switched-off, the particles retain almost their original charge.

Dr Reid asked: In this paper and those preceding we have seen that spatial variations in parameters such as temperature, electron density, *etc.* can be pronounced. There are also different timescales associated with different physical and chemical processes. Can you describe the degree of spatial and temporal resolution required to probe the range of processes important in plasma chemistry?

Professor Kersten answered: There are different spatial and time scales in dusty plasmas. The rf-cycles of the plasma are in the order of nanoseconds and the particle charging is in the order of µs. The motion of the charged particles to follow changes in the field structure of our adaptive electrode is in the order of a few tens of Hz, which we can observe. The chemical processes resulting in etching or deposition of thin films at the particle surface is even longer. The spatial resolution of the particle position is about 10 to 100 µm compared to typical plasma structures such as the sheath in front of the electrodes or substrates which is commonly in the order of mm to cm.

Professor Goree commented: Is the slowest time scale of all the inertial time scale for particle motion?

Professor Kersten replied: The lowest time scale is given by the response of the charged test particles to changes in the electric field which the particles can follow, *e.g.* in the order of some Hz.

Dr Reid said: Temperature measurements were made in Professor Shiratani's paper[1] within the plasma with a thermocouple. You propose to use microparticles which will give a higher degree of spatial resolution in the sheath. How do these two length scales for measurements compare?

1 Masaharu Shiratani, Kazunori Koga, Shinya Iwashita and Syota Nunomura, *Faraday Discuss.*, 2008, **137**, DOI: 10.1039/b704910b

Professor Kersten replied: A thermocouple in the plasma acts like a macroscopic probe which is shielded by a sheath. By this method, we would certainly disturb the plasma. For some applications this is not a problem, since plasmas in contact with solids (*e.g.* substrate processing) always interact with the sheath.
However, if we are interested in the energy fluxes in the plasma or in the sheath, respectively, we want to know the temperature field at the probe position without any disturbance. Therefore, tiny test particles are more appropriate for this purpose.

Professor Davidovits commented: Although the particle is levitated it is still in (microscopic) motion about a mean position. Does this give us kinetic (temperature) information about the particle and plasma?

Professor Kersten responded: The kinetic temperature of a plasma crystal might be in the order of tens of eV and can be determined by the assessment of the microscopic particle motion about its equilibrium position. This motion can be excited by field oscillations, by laser radiation, gas flows, *etc*.
On the other hand, the real particle temperature (surface temperature) differs remarkably from the kinetic temperature and is much lower (*e.g.* ambient temperature to about 300 °C). We try to measure this temperature by fluorescence.
The energy of the plasma (*e.g.* mainly of the electrons), which is typically in the order of some eV, has to be determined by other methods like Langmuir-probe measurements.

Professor Goree said: As a comment, I'll mention that the experimenter can track the motion of particles and compute their velocities. This allows computing their mean-square velocity, which is interpreted as a kinetic temperature for the particles. This kinetic temperature can be as high as tens of thousands of Kelvin, even when the surface of the particles remain at room temperature. The particle's surface temperature can be vastly different from its kinetic temperature.

Professor Kersten replied: This is completely right.

Professor Shiratani commented: When we want to measure local electric field at a certain position, we must put a microparticle in the position Q not only in the z-direction, but also the y-direction. How accurate do you determine the position in 3D space?

Professor Kersten replied: If we assume a homogeneous field and potential distribution along the electrode, in the first approximation we have only to take into consideration the z-direction. In the 2D x–y-plane the change above the electrode is small compared to the z-direction. That means we can use a 1D-approximation. However, at the edge of the electrode or if we have a distinguished sheath structure of our adaptive electrode, the other directions also have to be considered. We can determine the particle position with an accuracy of about 10 to 100 μm. The pixels of the adaptive electrode have a diameter of 7 mm. Mostly, we apply a symmetric pattern in order to get a symmetric sheath structure for particle confinement and handling.

Professor Mason asked: A key question for chemistry of microparticles is at which size the particle may exhibit bulk like properties—do your experiments give any clue to this? 100 nm? 100 μm?

Professor Kersten answered: This really is always a key question. The transition between the different stages are not sharp. The synthesis of particles in a plasma is by radicals/molecules *via* clusters and nanoparticles to microparticles. In the stage of cluster/nanoparticle the charge can vary. They might be positively charged by secondary electron emission, negatively charged by electron attachment, or even neutral. These fluctuations are important for the coagulation of clusters/nanoparticles to larger ones. In this period, the particles do not exhibit real bulk properties. If they are grown to microparticles they are permanently negatively charged and now they are more likely to exhibit bulk properties.

Professor Chang asked: Are you aware of any work in the area of droplet spectroscopy that results in the changing of colors as a function of temperature? I remember there was some literature about an inorganic liquid that changes color as a function of temperature and this could be used as a temperature sensor in the droplet form.

Professor Kersten answered: Thank you for this interesting advice. We are looking for particles which exhibit a temperature-dependent fluorescence in order to use them as tiny temperature sensors in the plasma or plasma sheath, respectively.

Ms Kahan opened the discussion of Dr King's paper: How does the hygroscopicity of the unoxidized aqueous solutions compare to that of pure water?

Dr King replied: The best way to answer this question is to draw the Köhler curves for a pure water droplet and a 10 μm diameter droplet of an aqueous solution of sodium benzoate (0.086 mol dm^{-3}). As can be seen from Fig. 3 the Köhler curve for the sodium benzoate solution is much more hygroscopic than the pure water droplet and will form a stable micron sized droplet at a high RH, whereas the pure water droplet will evaporate without a supersaturation of water vapour. Please note the form of the Köhler equation[1] is very simple and any effect for surface tension,

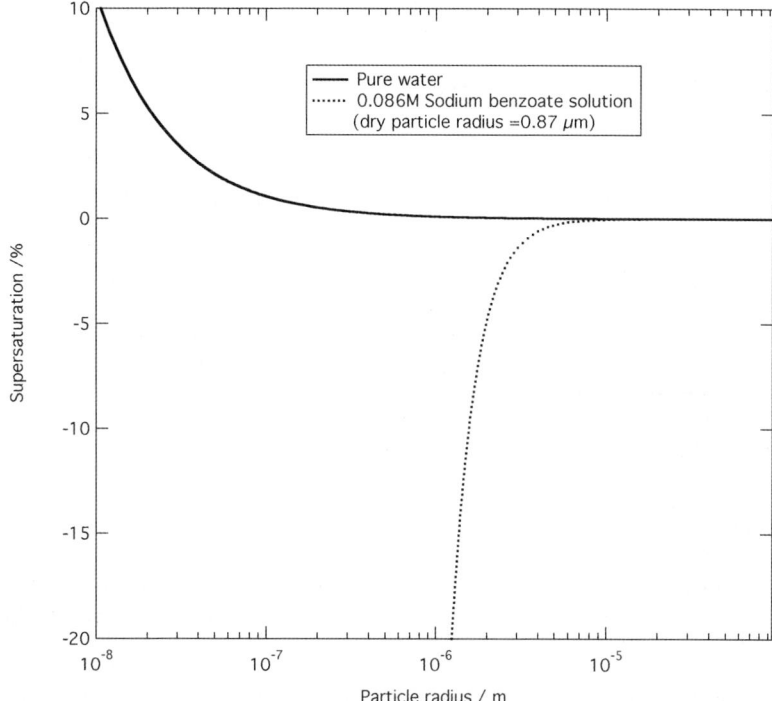

Fig. 3 Köhler curves for a pure water droplet and a 10 μm diameter droplet of an aqueous solution of sodium benzoate (0.086 mol dm^{-3}).

density, solution non-ideality, or insoluble mass at small diameters has not been considered.

1 John H. Seinfeld and Spyros N. Pandis, Atmospheric Chemistry and Physics: From Air Pollution to Climate Change, Wiley Interscience, New York, 1st edn, 1986.

Ms Kahan said: With respect to the lack of droplet growth observed during ozonation of α-pinene, is it possible that volatilization of oxidation products might be occurring?

Dr King answered: Yes, this is indeed possible and we at present do not investigate gas-phase products. As an example we monitored the size and Raman spectra of a mixed 8 μm particle of nonanoic acid and aqueous sodium chloride trapped in laser tweezers. The Raman spectra clearly demonstrated that within a few minutes the nonanoic acid had evaporated from the surface of the aqueous droplet; the particle size did not change after the nonanoic acid had evaporated.

Dr Reid asked: Knowledge of the gas phase composition is crucial in interpreting changes in involatile solute composition within the trapped droplet. Changes in relative humidity, can, for example, lead to significant changes in wet particle size and solute concentration. To what extent have you included in your analysis any potential errors in reactant concentration resulting from uncertainty in relative humidity?

Dr King responded: Kinetics of the reaction could in effect be carried out in two regimes of Relative Humidity (RH). In the first regime the generation of aerosol using an ultrasonic nebuliser wets all surfaces inside the cell. The local RH inside the cell is larger than the RH of the gas stream entering the cell. A trapped particle quickly equilibrates with the local RH in the cell (1–2 min). The particle size is then stable in the cell for approximately an hour as the local RH in the cell is maintained by evaporation from the wet cell walls. However, once the water on the cell walls is exhausted the particle size is controlled by the RH of the incoming gas-stream. Thus, as the cell dries out the particle often shrinks over a period of 5 min and then equilibrates with the new RH. In this second RH regime the particle is again stable. Kinetic runs are often committed in one of these regimes of RH, and any kinetic run that has a changing RH (observed visually by changes in the background on the microscope camera) is discarded. The concentration of reactant is "measured" by using the peak area of the Raman signals for the reactant and knowledge of the initial reactant concentration from the nebulising solution and particle size, both of which are monitored during the initial trapping process. Raman spectra are collected over a maximum of 10 s.

Dr Reid commented: As the cell dries out, particularly during the reaction, can this lead to errors in your assessment of changing reactant and product concentrations?

Dr King replied: Kinetics of the reaction could in effect be carried out in two regimes of Relative Humidity (RH). In first regime the generation of aerosol using an ultrasonic nebuliser wets all surfaces inside the cell. The local RH inside the cell is larger than the RH of the gas stream entering the cell. A trapped particle quickly equilibrates with the local RH in the cell (1–2 min). The particle size is then stable in the cell for approximately an hour as the local RH in the cell is maintained by evaporation from the wet cell walls. However, once the water on the cell walls is exhausted the particle size is controlled by the RH of the incoming gas-stream. Thus, as the cell dries out the particle often shrinks over a period of 5 min and then equilibrates with the new RH. In this second RH regime the particle is again stable. Kinetic runs are often committed in one of these regimes of RH, and any kinetic run that has a changing RH (observed visually by change in background on the microscope camera) is discarded. The concentration of reactant is "measured" by using the peak area of the Raman signals for the reactant and knowledge of the initial reactant concentration from the nebulising solution and particle size, both of which are monitored during the initial trapping process. Raman spectra are collected over a maximum of 10 s.

Professor Rühl asked: There are probably other oxidizing agents which might react with the organic compounds. This includes OH. What is the mixing ratio of OH radicals near the particles? Are the rate constants for the reactions with OH radicals known, and can these processes be of importance?

Dr King replied: In the experiments described in the paper there are no gas-phase hydroxyl radicals present. Hydroxyl radical production in the experiment was avoided by passing dry oxygen through the mercury discharge lamp. In the paper we raise the possibility that aqueous-phase hydroxyl radicals may be produced by secondary chemistry to explain the lack of observed products, but we also demonstrate in the paper that any hydroxyl radicals that may be formed in solution will not compete with the ozone reaction under study.

Dr Krieger commented: Suggestion: To use a test particle as a sensor for relative humidity and then repeat the measurements under identical conditions to ensure the relative humidity at the particle position.

Dr King responded: To use a reference particle trapped next to the reactant particle as a measurement and a probe of fluctuations in relative humidity is a good idea (*e.g.* Mitchem *et al.*[1]). However, this can limit the chemical systems that can be studied as it is extremely difficult to obtain two chemically different particles in two different traps without contaminating either particle during the nebulisation process. Such contamination may change the measurement of the uptake coefficient or Köhler behavior of the particle.

1   Laura Mitchem, Rebecca J. Hopkins, Jariya Buajarern, Andrew D. Ward and Jonathan P. Reid, *Chem. Phys. Lett.*, 2006, **432**(1–3), 362–366.

Ms Kahan asked: We have observed that aromatic hydrocarbons in aqueous solutions and in organic films undergo ozonation at the surface rather than in the bulk. Could a surface reaction help explain the fast reaction observed for benzoate in water?

Dr King replied: We find no evidence for a surface reaction in our studies, but further experiments could be undertaken, as highlighted by Dr Ammann (see below), for further confirmation. A difference between your experiments[1] and ours is the solubility of the aromatic species studied. Benzoate is very soluble, especially relative to PAHs. Benzoate also has negligible surface activity, and diffuses easily through water.

1   T. F. Kahan, N.-O. A. Kwamena, D. J. Donaldson, *Atmos. Environ.*, 2006, **40**(19), 3448–3459.

Dr Ammann said:
(1) With regard to the experiments with sodium benzoate, it is assumed that the particles are liquid at the relative humidities of the experiment. How was this assured? Have the hygroscopic properties of sodium benzoate been explored? What is the deliquescence humidity?
(2) Attempts are presented to model ozone uptake to the particles based on different assumptions to allow conclusions about the limiting process. Especially to differentiate bulk *vs.* surface kinetics, measuring uptake kinetics as a function of ozone partial pressure would allow a better resolution of this issue than the approach as presented.

Dr King responded:
(1) Previous experiments where we purposely dehydrated concentrated aqueous salt solutions held in an optical trap clearly show phase transitions. Using the microscope we observe the previous spherical liquid particles develop angular features and can image trapped particles which are a crystalline solid covered unevenly with some liquid. The resulting solid particles often leave the trap quickly and result in a crystalline solid on the cell window. The experimental Köhler curve for sodium benzoate was not measured by this method as it was not necessary and is time-consuming (~ 24 h). A calculated curve was used to predict concentrations of the nebulising solution that would result in stable particles of the right size to trap in the laser tweezers.
(2) The authors agree that this is one strategy to elucidate a surface *versus* bulk mechanism. Another strategy would be to use the suggestion of Smith *et al*[1] and to measure the uptake coefficient as a function of the particle size and thus particle area. In the experiments we undertook we did not have enough experimental runs at different particle sizes or ozone partial pressures to fully elucidate the mechanism. It should be noted that the kinetic decay of the concentration of the organic reactants studied all followed the mathematical formulation for a bulk reaction. As unpublished data for other compounds reacting with ozone show kinetic decays of reactant that clearly indicate surface only reactions, it is possible to distinguish between these two mechanisms with this type of experiment.

1 Geoffrey D. Smith, Ephraim Woods, III, Cindy L. DeForest, Tomas Baer and Roger E. Miller, *J. Phys. Chem. A*, 2002, **106**, 8085–8095.

Dr McGloin remarked: Is fluorescence from the glass in the sample cell a problem when collecting Raman spectra?

Dr King answered: We have had no problems with fluorescence from the glass windows on the cell. These are thin microscope glass cover slips from VWR International. They are pre-treated in conc. nitric acid and glued to the aluminium cell. During experimental runs the windows are regularly (every 1–2 h) cleaned in Decon-90 solutions and rinsed with high purity solvents: water, acetone and methanol. The cell and windows are regularly conditioned with high levels of ozone gas (>1 ppm) in oxygen.

Dr Reid asked: With the extension of optical trapping to droplets containing aromatic chromophores, can you estimate the degree of heating of your droplets, or do you see the results of heating?

Dr King answered: Heating of the samples is an important aspect to consider when selecting chemical systems to study. Any compound (or reaction product) which absorbs at the laser wavelength is unsuitable for study. Experiments with compounds that do absorb light at the laser wavelength (substituted phenols in basic solutions can be problematic) are not stably trapped and will evaporate (often quickly) under conditions that extended Köhler theory would suggest are stable. Non-aqueous pure compounds trapped in an optical trap which absorb at the laser wavelength rapidly and obviously degrade. In another comment I suggested that the trapping of leuco dyes (which exhibit thermochromism) may demonstrate laser heating as a problem or not. Detailed laser heating calculations have not been attempted as the absorption cross-section at the laser wavelength would need to be known for the compounds we have studied. These absorption cross-sections are lower that any practical means we have of measuring them.

Professor Jaenicke commented: To what extent would your results apply to the atmosphere? You have claimed influence on cloud properties!

Dr King answered: The systems studied were meant to be proxies for ozone reactions with HULIS (benzoate in water) and biogenic aerosol (α-pinene in alkane). To the authors' knowledge this is the first paper demonstrating the use of Laser Raman tweezers to measure kinetics and uptake coefficients and the systems are kept reasonably simple so they may be understood. Future studies will include more atmospherically complex systems. It should be noted that in a previous study we demonstrated that the oxidation of a droplet of oleic acid and sea water by ozone lead to a dramatic increase in hygroscopicity and thus a change in cloud microphysical properties.[1]

1 M. D. King, K. C. Thompson and A. D. M. Ward, *J. Am. Chem. Soc.*, 2004, **126**(51), 16710–16711.

PAPER

The influence of small aerosol particles on the properties of water and ice clouds

T. W. Choularton,*[a] K. N. Bower,[a] E. Weingartner,[b] I. Crawford,[a] H. Coe,[a] M. W. Gallagher,[a] M. Flynn,[a] J. Crosier,[a] P. Connolly,[a] A. Targino,[a] M. R. Alfarra,[b] U. Baltensperger,[b] S. Sjogren,[b] B. Verheggen,[b] J. Cozic[b] and M. Gysel[b]

Received 21st February 2007, Accepted 16th April 2007
First published as an Advance Article on the web 9th August 2007
DOI: 10.1039/b702722m

In this paper, results are presented of the influence of small organic- and soot-containing particles on the formation of water and ice clouds. There is strong evidence that these particles have grown from nano particle seeds produced by the combustion of oil products. Two series of field experiments are selected to represent the observations made. The first is the CLoud-Aerosol Characterisation Experiment (CLACE) series of experiments performed at a high Alpine site (Jungfraujoch), where cloud was in contact with the ground and the measuring station. Both water and ice clouds were examined at different times of the year. The second series of experiments is the CLOud Processing of regional Air Pollution advecting over land and sea (CLOPAP) series, where ageing pollution aerosol from UK cities was observed, from an airborne platform, to interact with warm stratocumulus cloud in a cloud-capped atmospheric boundary layer. Combining the results it is shown that aged pollution aerosol consists of an internal mixture of organics, sulfate, nitrate and ammonium, the organic component is dominated by highly oxidized secondary material. The relative contributions and absolute loadings of the components vary with location and season. However, these aerosols act as Cloud Condensation Nuclei (CCN) and much of the organic material, along with the other species, is incorporated into cloud droplets. In ice and mixed phase cloud, it is observed that very sharp transitions (extending over just a few metres) are present between highly glaciated regions and regions consisting of supercooled water. This is a unique finding; however, aircraft observations in cumulus suggest that this kind of structure may be found in these cloud types too. It is suggested that this sharp transition is caused by ice nucleation initiated by oxidised organic aerosol coated with sulfate in more polluted regions of cloud, sometimes enhanced by secondary ice particle production in these regions.

Introduction

The last Intergovernmental Panel on Climate Change (IPCC) scientific assessment (IPCC, 2001) showed that the largest source of uncertainty in radiative forcing

[a] Centre for Atmospheric Science, School of Earth, Atmospheric and Environmental Sciences, University of Manchester, Manchester, UK
[b] Laboratory of Atmospheric Chemistry, Paul Scherrer Institut, CH-5232, Villigen PSI, Switzerland

estimates arises from clouds *via* the first aerosol indirect effects. It is generally assumed that this tends to offset global warming but the magnitude is very uncertain. Mixed phase and ice clouds are very important in the atmosphere but little progress has been made in understanding the aerosol indirect effects in these clouds. To improve these models, a much clearer understanding of the role of ice nuclei in ice formation is required. It is clear from recent projects, such as INCA studying cirrus clouds in the Northern and Southern Hemisphere, that anthropogenic effects on Chamber studies such as those carried out at AIDA are providing a better understanding of how and under what conditions various materials, *e.g.*, these clouds, are substantial. Desert dust and soot are able to act as ice nuclei. Recent airborne studies have suggested that in many mixed phase clouds, concentrations of Ice Nuclei (IN) are orders of magnitude too small to explain the numbers of ice particles observed, implying ice multiplication processes are important. The Hallett–Mossop process is the only multiplication mechanism that has been well quantified; it operates at temperatures between -3 and -9 °C. There is also the possibility of splinter production associated with droplet freezing and, at colder temperatures, ice crystal fragmentation during evaporation has been identified as a possible mechanism of secondary ice particle production. The role of ice nuclei, which operate in water subsaturated but ice supersaturated conditions, needs to be investigated. Many of the studies designed to investigate these processes have been performed from aircraft, although ground-based mountain cloud studies provide the opportunity for more detailed study over longer periods.

In this paper, we investigate the role of small organic particles, generally found to be internally mixed with soluble inorganic material as both cloud droplet and ice particle nuclei. These particles are thought to originate from the burning of oil products in industry and motor vehicles, where they are emitted as organic particles a few tens of nanometre in size or smaller. The primary material has a large fraction of black elemental carbon and has associated organic carbon. The majority of the organic carbon observed in the experiments reported here is secondary in nature. The role and fate of these mixed organic particles in the atmosphere is largely unknown.

The CLACE experiments

Several CLoud-Aerosol Characterisation Experiments (CLACE) have been performed in both summer and winter within international collaborations at the high alpine research station Jungfraujoch (JFJ, 3580 m asl; 46.55 °N, 7.98 °E), Switzerland. The experiments are designed to investigate the chemical composition of aerosol particles, their hygroscopic properties and their interaction with clouds. The presented results are based on findings from CLACE 2, CLACE 3 and CLACE 4 and were carried out in July 2002, February and March 2004 and February to March 2005 respectively.

Description of Jungfraujoch (JFJ)

Due to its exposed location on a mountain col, the JFJ experiences a high frequency of clouds (annual mean cloud frequency of 37%, [Baltensperger *et al.*, 1998[1]]) and has been the subject of many cloud scavenging experiments. The high alpine site is influenced by remote continental/marine air masses from the free troposphere, and on occasion from the convective boundary layer due to local convection. The observation of cloud and aerosol parameters at the JFJ allows measurements to be conducted under a wide range of anthropogenically-influenced conditions yielding very different variable aerosol concentrations and compositions. Such measurements have been carried out since 1988 and have established the site to be suitable for the long-term monitoring of Free Tropospheric (FT) background aerosol [Baltensperger *et al.*, 1997[2]], (Henning *et al.* 2002[3] and 2003[4]). The station is part

Table 1 Current Jungfraujoch aerosol/cloud measurement program

Instrument	Measured parameter	Time resolution	Inlet system/sampling technique		
			Total aerosol	Interstitial aerosol	No inlet i.e., in situ
Condensation particle Counter (TSI 3010)	Aerosol number conc $d \sim 0.010$–1 µm	1 min	x		
Aethalometer (AE-31)	Aerosol BC mass conc.	10 min	x		
	Aerosol absorption coefficient at 7 different wavelengths 350–950 nm				
Nephelometer (TSI 3563)	Aerosol scat. coeff. 450, 550, 700 nm	30 min	x		
Filters packs	Aerosol major ionic composition PM1 and TSP	1 d every 6th d	x		

of the Swiss National Monitoring Network for Air Pollution (NABEL) and the Swiss Meteorological Institute (SMI). Many of the cloud and aerosol parameters recommended by the WMO for measurement at GAW baseline stations are already being measured at the JFJ and are indicated in Table 1.

Instrumentation and inlets

The large suite of instruments deployed during these experiments sampled from the free stream or utilised a combination of three well characterised inlets that will sample the total residual particulate, the interstitial (unactivated) particulate within the cloud, and the ice crystal residuals within the cloud.

The Institut fur Tropospharische Research (IfT) operated a Counterflow Virtual Impactor (CVI), which has been successfully employed in artificially seeded mixed phase clouds (e.g. Mertes et al., 2001[5] and 2006[6]). The CVI was part of a new prototype sampling system (Ice-CVI) that allows for the separation of small ice particles from large ice crystals, cloud droplets and interstitial aerosol particles. The extracted ice particles are dried airborne in the system and the remaining residual particles that correspond to the former ice nuclei are analysed with a variety of different instruments (Table 2). Initial separation of cloud droplets, ice crystals and non-activated aerosol particles is performed under ambient conditions. In addition, the total aerosol is sampled with a heated inlet designed to evaporate all activated cloud droplets and ice crystals smaller than 30 µm at an early stage of the sampling process [Weingartner et al., 1999[7]]. An interstitial inlet sampled the non-activated or interstitial aerosol particles by the removal of larger cloud particles during cloudy events. Ice particle residues may be complex if the crystal is heavily rimed as the residue would be a mixture of IN and CCN from the frozen droplets. The degree of riming was quantified using the Manchester Cloud Particle Imager (CPI) and formvar replicas.

The sampled air from the three different inlet systems was then brought into the laboratory, where aerosol measurements were performed at low Relative Humidities (RH < 20%). By duplicating and/or triplicating identical measurements for the total aerosol, the interstitial aerosol and the residual aerosol particles, cloud activation and heterogeneous ice nucleation properties are derived by differencing. Finally, other instruments were used to characterize cloud droplets in situ (Table 2).

The main cloud physics instruments are the Cloud Particle Imager (CPI), the Aerosol Droplet Analyser (ADA), a Knollenberg 2D probe, and the Forward Scattering Spectrometer Probe (FSSP). The CPI, manufactured by Stratton Park

Table 2 Additional instrumentation deployed during the campaigns. Note that duplication of measurements using the different inlets will allow cloud droplet properties to be derived by differencing

Instrument	Measured parameter	Availability 2004	Availability 2005	Time resol	Flow rate lpm	Responsible person/institute	Ice CVI	Total aerosol	Interstitial aerosol	No inlet i.e., in situ
Aerosol mass spectrometer AMS	Chemical aerosol composition	Y	Y	1 min	0.12	MPI Mainz	x			
SPLAT (Single Particle Laser Ablation Time-of-flight mass spectrometer)	Single Particle Analysis	Y					x			
Aerosol mass spectrometer AMS	Chemical aerosol composition	Y	Y	1 min	0.12	Manchester		x	x	
Forward Scattering Spectrometer Probe PMS FSSP-100	Cloud and aerosol number Spectra d = 2–44 μm	Y	Y	1 min		Manchester				x
Phase doppler anemometry system (ADA)	Liquid droplet number size distribution	Y	Y	1 min		Manchester				
Cloud particle imager (CPI) and 2-D probe	Ice Crystal imaging (>10 μm)	Y	Y	40 s^{-1}		Manchester				x
Formvar replicas		Y	Y			ETH Zürich				x
Gerber PVM-100	Cloud LWC, PSA Effective droplet diameter	Y	Y	1 min		PSI/Uni Copenhagen				x
Cloud condensation nuclei counter (parallel plate type)	CCN concentration	Y				Uni Copenhagen		x		
Cloud condensation nuclei counter (expansion type)	CCN concentration	Y	Y	5 min		PSI		x		
HTDMA	Hygroscopicity at ambient temperature ($T \sim -10\ °C$)		Y	10 min		PSI			x	
SMPS	Aerosol number spectra d = 15–800 nm	Y	Y	5 min	0.3	PSI		x	x	
SMPS	Aerosol number spectra d = 15–800 nm	Y	Y	5 min	0.3	PSI	x			

Table 2 (continued)

Instrument	Measured parameter	Availability 2004	Availability 2005	Time resol	Flow rate lpm	Inlet system/sampling technique Responsible person/institute	Ice CVI	Total aerosol	Interstitial aerosol	No inlet i.e., in situ
OPC (Grimm)	Aerosol number spectra $d = \sim 0.3$–$7.5 \mu m$	Y	Y	1 min	1	Manchester		x	x	
OPC (Grimm)	Aerosol number spectra $d = \sim 0.3$–$7.5 \mu m$	Y	Y	1 min	1	PSI	x		x	
Aethalometer (AE-31)	Aerosol BC mass conc. Aerosol absorption coefficient at 7 different wavelengths $\lambda = 350$–950 nm	Y	Y	10 min	1	PSI				
Filter samples	Major Ions	Y	Y	h	2.3	IFT	x	x	x	
PSAP	IN BC concentration	Y	Y	3 min	2	IFT	x			
CPC 3010	Residual number concentration	Y	Y	1 min	1	IFT	x			
Ly-alpha hygrometer	condensed water content	Y	Y	1 min	0.1	IFT	x			
Impactor for environmental scanning electron microscopy (ESEM) and electron microscopy	Ice nucleation on residual particles, Particle morphology, Elemental composition	Y	Y		4	Manchester TUD	x	x	x	

Engineering, images ice particles larger than 10 µm onto a CCD chip at a rate of 40 images per second. The instrument is designed for aircraft use and the Manchester group have successfully deployed the instrument on the ARA Egrett aircraft for studying the crystal habits of cirrus clouds during the NERC Clouds, Water Vapour and Climate programme funded EMERALD experiments (Choularton et al., 2002[8]). The CPI was adapted to ground-based operation by fitting it with a high volume aspirator, a method previously adopted by Lawson et al., 2001.[9] The images produced by the CPI are far more detailed than has previously been possible and lead to a more reliable and better resolved classification of ice crystals at higher temporal resolution. This enabled us to obtain information on the origin of the crystals and the growth regimes that the crystals have been subject to, as the growth habit is a strong function of temperature and supersaturation with respect to ice. In addition, any riming of the crystals can be resolved using this instrument. This was operated alongside a Knollenberg 2D probe owned by Manchester, which provided absolute number concentrations of crystals at around 200 µm to correct the CPI sample volume. Precipitation may significantly affect the results, a present weather sensor was used to identify such periods.

The Airborne Droplet Analyser (ADA), manufactured by TSI Inc., is based on phase Doppler anemometry. It utilises an Argon Ion laser to generate two cross-directed laser beams with a small sample volume in the interfering region. The interference fringes are modified by the presence of cloud droplets and the size and velocity of the droplet can be determined. The technique can discriminate between spherical droplets and aspherical particles and so is capable of rejecting the small ice crystals in the cloud. In addition, changes in refractive index can be measured and used to reject non-liquid water particles. The ADA has a much larger size range than the FSSP and were operated with approximately 200 channels ranging from 1–200 µm.

The FSSP, modified by DMT to provide faster electronics, improved size binning and better rejection of multiple counting, measures the number size distribution of both liquid droplets and ice crystals and cannot discriminate between them. In principle, the three instruments allow number closure between the liquid and ice phases to be achieved.

The Paul Scherrer Institute provided measurements of aerosol size distributions alternately through the total and interstitial inlets using a combination of Scanning Mobility Particle Sizers (SMPSs) and Optical Particle Counter (OPC). This allows a comparison of the number of activated droplets and particles in a cloud with the activated droplet and ice spectra derived from the Manchester instruments (Verheggen et al., 2007[10]). The particle number concentrations and size distributions provide sound measures of the anthropogenic pollution observed during the cloud event. These measurements, coupled with those of black carbon and CO, NO, NO_x and ozone, serve to identify the extent to which the air mass is influenced by anthropogenic effects.

Manchester deployed an Aerodyne Aerosol Mass Spectrometer (AMS) (Allan et al., 2003[11]). The AMS measures the mass of volatile and semi-volatile components of submicron aerosol particles in real time as a function of their size. This information makes it possible to link the chemistry of aerosol particles with the gas phase, and particle microphysical measurements at a time resolution that is higher than changes in air mass occur. This has not been possible in the past using filter or impactor technology and helps to resolve current uncertainties in long range transport, phase partitioning of organic material, the direct aerosol radiative effect and aerosol–cloud interactions at the heart of the indirect effect. The Manchester and Mainz AMS instruments were used to sample from all three inlets throughout the experiment, allowing discrimination of the mass spectra, mass loadings and mass size distributions of organics, sulfate, nitrate and ammonium from ice crystal residuals, total cloud particle residuals and interstitial particles at a time resolution of around 6 min. The MPI Mainz single particle laser ablation system was in other

CLACE experiments used to sample the residuals of ice crystals selected by the CVI. The instrument delivers mass spectra of single particles but is unable to deliver mass loadings due to the single stage ablation/ionisation process being non-reproducible. It is able, however, to deliver a mass spectral fingerprint of the residuals of the ice crystals measured. The instrument is sensitive to metals, crustal material and inorganic salts but tends to significantly fragment the organic fraction of the particles.

Aircraft observations in an urban plume

The interaction of pollution from urban areas with stratocumulus cloud is important both for understanding the fate of gases and particulate emitted in the urban environment and for understanding the aerosol indirect effect of ageing mixed organic and inorganic particulate. In this paper, results will be presented from an airborne study to investigate these processes, known as CLOud Processing of regional Air Pollution advecting over land and sea (CLOPAP).

The interaction of ageing aerosol emitted from urban areas of the UK was investigated as the plumes advected away from the area over the sea in a stratocumulus-capped boundary layer. Detailed measurements of the size distribution and chemical composition of the aerosol were made on the Facility for Airborne Atmospheric Measurements (FAAM), the UK community's new BAe 146 research aircraft. These measurements were complemented by detailed measurements of the atmospheric liquid water content and cloud microphysics. Detailed measurements of precursor trace gases were also made by colleagues from the University of York and the University of East Anglia.

Methodology of CLOPAP

During a typical sortie, the aircraft would fly to a region immediately downwind of the urban source in conditions with a well marked boundary layer capped by stratocumulus cloud. The aircraft made a series of horizontal passes perpendicular to the line of the plume below cloud, within the cloud deck and above the cloud top. This series of passes was repeated at 50 km intervals moving downwind from the source (Fig. 1). Within cloud, an airborne CVI was used to separately measure the droplet residual aerosol components to investigate the nucleation scavenging of the particulate.

A key measurement of the aerosol composition was made using an Aerodyne Aerosol Mass Spectrometer (Jayne et al., 2000[12], Allan et al., 2004[13]). This

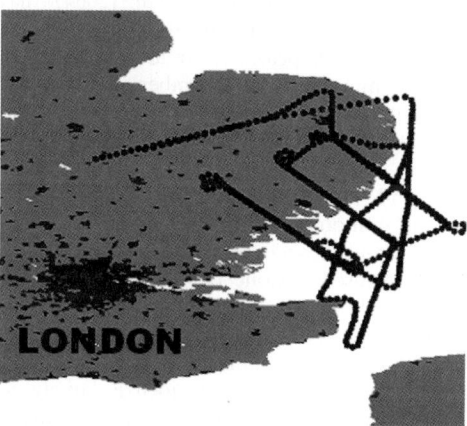

Fig. 1 An example of a typical sortie, in this case studying the London plume as it advects east.

instrument is able to provide size resolved information of the semi-volatile components of the aerosol, including major ions and organic material. Some information is available on the main functional groups in the organic material and the state of oxidation. Cloud and aerosol microphysical measurements were made using a series of outboard optical scattering probes to measure the size range 0.05 to 50 µm. These included particles that were nucleation scavenged and particles that remained interstitial to the cloud droplets.

Results from the CLACE experiments

It was found, from the AMS, that the aerosol consisted of a mixture of sulfate, organic and nitrate distributed as shown in Fig. 2. The similarity of the size distribution for all the species suggests that the particles are internally mixed. For the winter project, Fig. 2b (CLACE-3), the aerosol was mostly long range, transported from source regions over Europe and did not show a clear diurnal variation. Nevertheless, measurements of CO concentration on a 10-min timescale did show fluctuations and day-to-day variations as air from polluted sources and plumes arrived at the site. The influence of these variations in the amount of pollution aerosol on the cloud microphysics is discussed below. A similar result was found for warm clouds during CLACE 2, performed during the summer months, except that the composition of the aerosol was dominated by organic material that was internally mixed with sulfate and nitrate. There were marked variations in the loadings of these species on a diurnal basis and this could be attributed to boundary layer air being transported to the summit by convection during the day.

Composition of the organic fraction

The AMS delivers a mass spectrum of the aerosol ensemble as a function of time based on standard 70 eV ionisation after vaporization at 500 °C. The fragmentation pattern of the inorganic ions can be identified easily and signals due to air can be accounted for by referencing to the nitrogen ion signal at m/z 28 or the oxygen ion at m/z 32. Allan *et al.*, (2004)[14] have developed a method for accounting for such signatures and so are able to obtain a mass spectrum composed only of the organic mass. Major ion peaks in this spectrum occur at 41, 43, 55, 57, 69, 71 and at higher mass, separated by 14 mass units. These peaks represent alkyl chain fragments and are dominant when the particles are composed of aliphatic-rich compounds typically found in motor vehicle exhaust aerosol. Alfarra *et al.*, (2004)[15] have shown that after some distance from the source, mass fragments due to oxygenated ions become dominant. The main oxygenated peaks occur at m/z 43 ($C_2H_3O^+$) and m/z 55 ($C_3H_3O^+$), which occur commonly from a number of carbonyl functionalities, and m/z 44 (CO_2^+), which indicates mainly di- and multifunctional acids, as these undergo decarboxylation on the vaporizer forming neutral CO_2, which is ionised directly.

Mass fragments 44 and 43 dominate the mass spectrum, indicating that the organic fraction is mainly composed of highly oxygenated compounds of secondary nature, such as carbonyls and dicarboxylic and poly acids. The mass fragment 44, the AMS signature of oxidised organic compounds and humic-like substances, accounts, on average, for about 14% of the total organic mass at Jungfraujoch. The dominance of mass fragments 44 and 43 in the accumulation mode and the lack of significant contributions of mass fragments 57 and 43 in the small mode indicate that the aerosol organic fraction at Jungfraujoch is largely composed of highly oxidized compounds and that secondary organic aerosols are more significant than primary compounds during the sampling period. These results are supported by a previous report of oxygen-to-carbon atomic ratio of 0.55 at Jungfraujoch, which

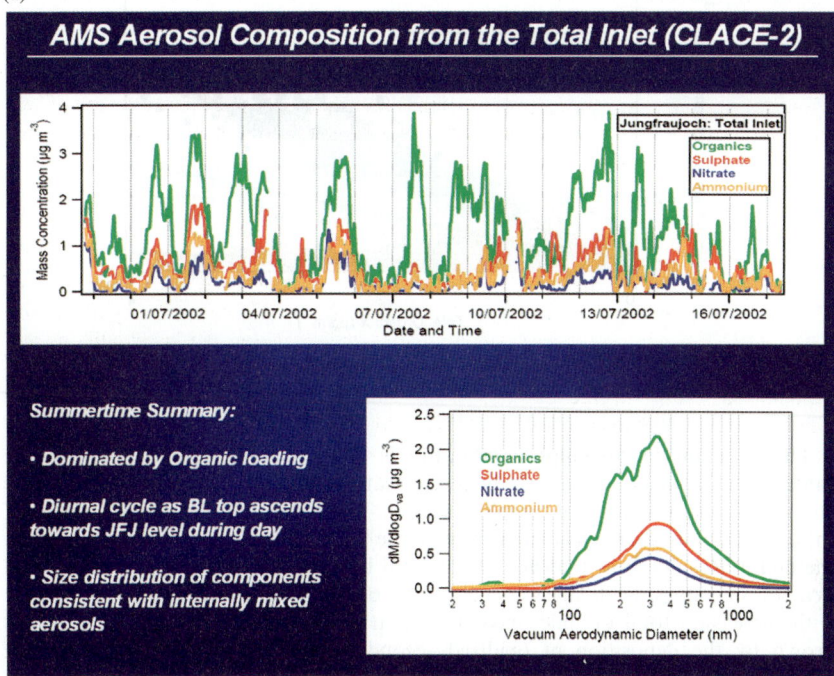
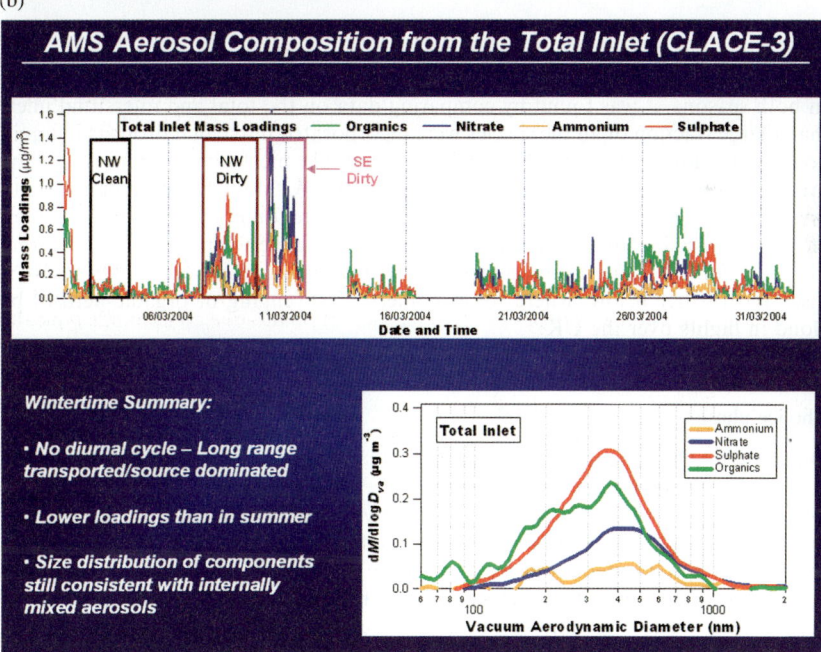

Fig. 2 A comparison between the Aerosol Mass spectrometer data in the (a) summer (CLACE-2) and (b) winter (CLACE-3).

implies that the organic compounds are highly oxygenated at this location (Krivacsy et al., 2001[16]).

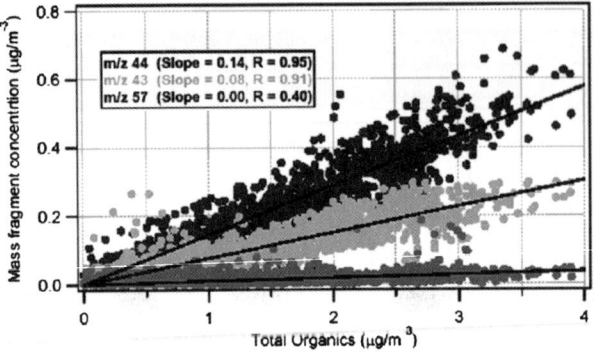

Fig. 3 A scatter plot of the mass concentration of the organic mass fragments 43, 44 and 57 *versus* the total organic mass concentration for the entire CLACE2 set.

These conclusions are emphasised in Fig. 3, where the mass concentrations of the three mass fragments discussed above are scatter-plotted against the total organic concentration for the entire sampling period. The slopes and Pearson's r values of the correlation of each mass fragment with the total organic mass are summarised in the figure. Results show that m/z 44 followed by m/z 43 have the highest slope and Pearson's r values, while mass concentration of m/z 57 does not appear to increase with increasing total organic mass. Hence, it would appear that the aerosol has grown by the deposition of oxidized secondary organic material on a seed of aliphatic material.

Aerosol scavenging in liquid clouds in CLACE

In both seasons, it was found by comparing data on the total and interstitial inlets that a large fraction of all the species were incorporated into cloud droplets in liquid dominated clouds. Results show that all four chemical components, including organics, appear to be scavenged by cloud droplets. Within the same in-cloud event, the scavenging ratios of the different components were found to be of similar values. However, variable mass scavenging ratios ranging between 0.3 and ~1 were observed for the same chemical species in different cloud events. This result is discussed further below, where we examine the scavenging of organic aerosol by cloud in flights over the UK.

The ice phase

During CLACE-3, temperatures were below freezing throughout and the cloud consisted of a mixture of supercooled water and ice crystals. It was observed, however, that rapid fluctuations occurred between cloud dominated by liquid water and ice crystals, often on timescales of tens of seconds. On some occasions it was found that ice and water co-existed in the same cloud at the same time, and at others there were rapid fluctuations between cloud regions consisting almost entirely of supercooled water and almost entirely of ice. Fig. 4 shows an example of this behaviour in a cloud at a temperature of -11 °C. In this figure, the CPI data shows large numbers of small spherical particles in the high liquid water content regions and smaller numbers of irregular particles in the regions dominated by ice crystals, note that the ice crystals are both larger (second panel) and less numerous than the water droplets. The third panel shows the number concentration of small droplets measured by the FSSP; again, these are much higher in the liquid water-dominated regions than the ice-dominated regions. Fig. 5 shows an example of the sharp

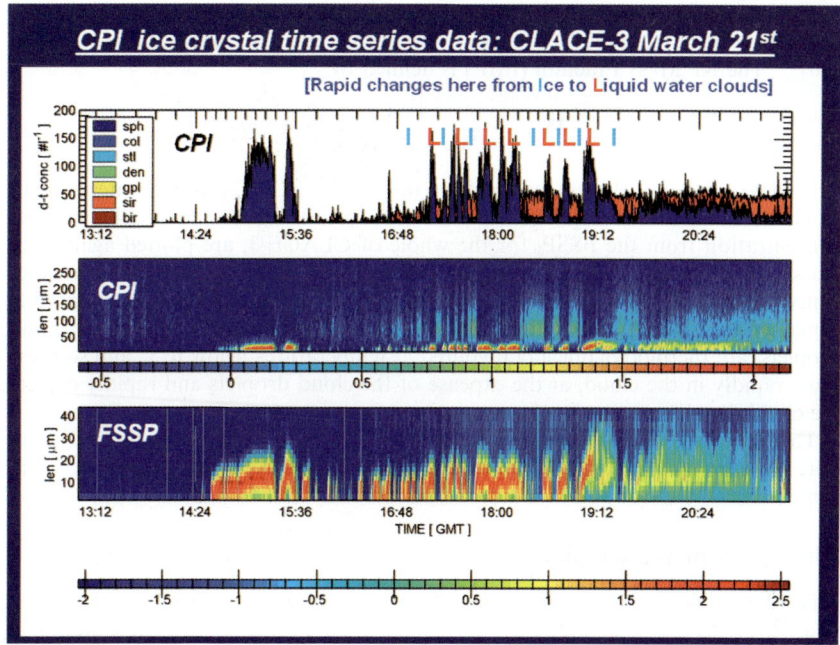

Fig. 4 Time histories of particle number concentration divided according to habit from the CPI (top panel), particle length from the CPI (middle panel), particle size and number concentration from the FSSP (lower panel) for 1 d during the CLACE-3 winter campaign.

transition between the ice- and water-dominated regions with examples of the images close to the interface.

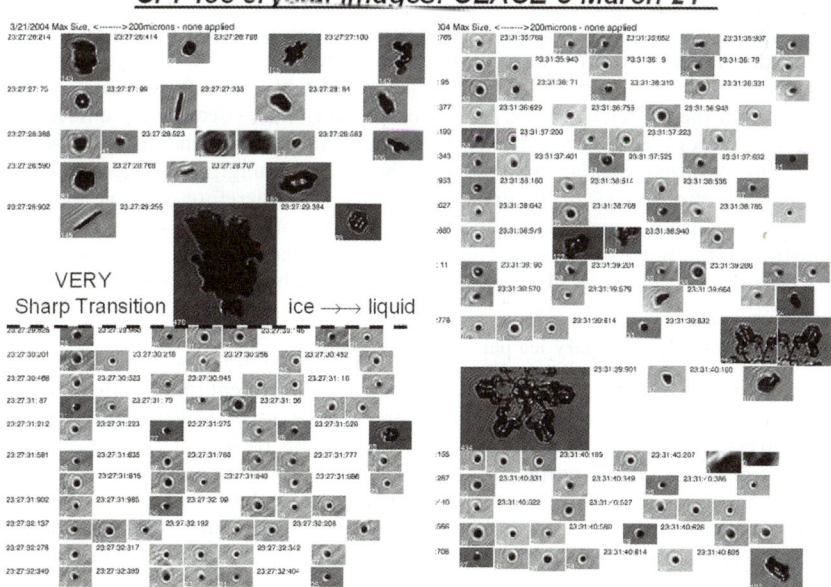

Fig. 5 Particle images recorded by the CPI showing the very sharp transition between water- and ice-dominated cloud.

The liquid water content in the cloud can be deduced from the PVM and the ice water content from the CPI images using the method described by Connolly et al. 2007.[17] The Ice Mass Fraction (IMF) is defined by

$$\mathrm{IMF} = \frac{\mathrm{IWC}_{\mathrm{CPI}}}{\mathrm{IWC}_{\mathrm{CPI}} + \mathrm{LWC}_{\mathrm{PVM}}}$$

where IWC is the Ice Water Content from the CPI, and LWC is liquid water content from the FSSP. When liquid water content from the PVM and droplet number concentration from the FSSP, for the whole of CLACE-3, are plotted against ice mass fraction it can be seen both the liquid water content and droplet number concentration fall with increasing ice water fraction Fig. 6. This can be readily explained by the fact that ice particles, which are fewer in number than the cloud droplets, are thermodynamically favoured at temperatures below 0 °C and so these grow rapidly in the cloud, at the expense of the cloud droplets and rapidly deplete the cloud droplets.

This process is likely to explain the rapid glaciation of parts of the cloud. It does not, however, account for the observation that ice particles are only observed in some parts of the cloud at the same temperature.[18]

The origin of the ice phase

It is not possible to use the AMS data to look at the short-term fluctuations in the ice content of the cloud as the resolution is not good enough. However, conditional sampling was carried out based on the presence of ice in the cloud. The result is shown in Fig. 7. It can be seen that regions containing ice have higher loadings, particularly of sulfate. The likely origin of this organic material, particularly in winter, is from combustion confirmed by its association with higher levels of sulfate and nitrate (from the AMS) and black carbon. At Jungfraujoch, it was found that ice crystal residues from the ice CVI consisted of black carbon and mineral particles (Walter et al., 2007,[19] Cozic et al., 2007[10]). The question that arises is whether these carbon-rich aerosols are likely to act as good ice nuclei. Work has been conducted in the AIDA chamber to investigate the role of black carbon and organic material as an ice nucleus (Möhler et al., 2005[20]). It was found that black carbon could act as a deposition nucleus at low temperatures but organic coatings inhibited this. However, it was shown that sulfate coatings considerably increased the effectiveness of the ice nuclei at water saturation when they first acted as CCN, then causing the droplets to freeze. In these circumstances, the carbon particles were effective as ice nuclei at −20 °C. Further recent work carried out at the chamber has shown that the presence of highly oxidized organic material (as observed at Jungfraujoch) can further enhance the effectiveness of the immersion nuclei. It is, therefore, possible that the source of the ice is emersion freezing by organic/carbon particles being carried to Jungfraujoch in plumes of polluted air. This would explain the very inhomogeneous glaciations of the cloud as these plumes arrive at the observation site.

A possible role for secondary ice particle production

A feature of the data presented above is that when ice is present there is sometimes a truly mixed phase cloud and at others it is purely an ice cloud. This does not seem to be clearly related to temperature. A likely explanation for this is the presence or absence of secondary ice particle production referred to in the introduction to this paper. As discussed above, the presence of ice will deplete the liquid water for thermodynamic reasons and this will tend to leave behind a relatively small number of larger ice particles, each ice particle being nucleated by an ice-forming nucleus. It is known, however, that at temperatures between −3 °C and −9 °C, a powerful process of secondary ice particle production by riming and splintering occurs called the Hallett–Mossop process. In a mixed phase cloud, ice particles will sweep up

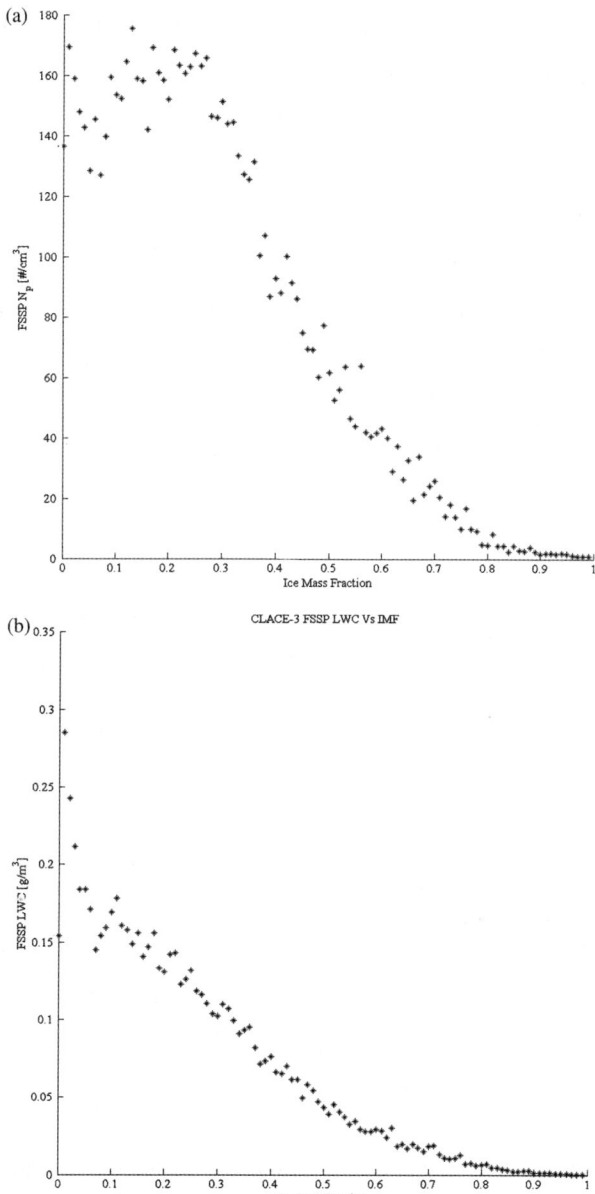

Fig. 6 (a) FSSP droplet number concentration plotted against ice mass fraction. (b) A comparison of the cloud droplet number concentration and cloud liquid water content measured by the FSSP with ice mass fraction.

supercooled cloud droplets, which will freeze on contact. As they freeze they throw off ice splinters which are then able to grow rapidly by vapour diffusion. This process could contribute to the relatively high concentrations of ice particles observed at $-11\ °C$ on 21 March (Fig. 4) as the cloud undoubtedly filled this temperature range between the summit site and the lower, warmer cloud base. Data from the ice CVI may help us to quantify this, as each ice particle should leave behind one ice-forming nucleus, whereas particles that were formed on secondary splinters may leave no residue. This work is not completed. As mentioned in the introduction, other, less

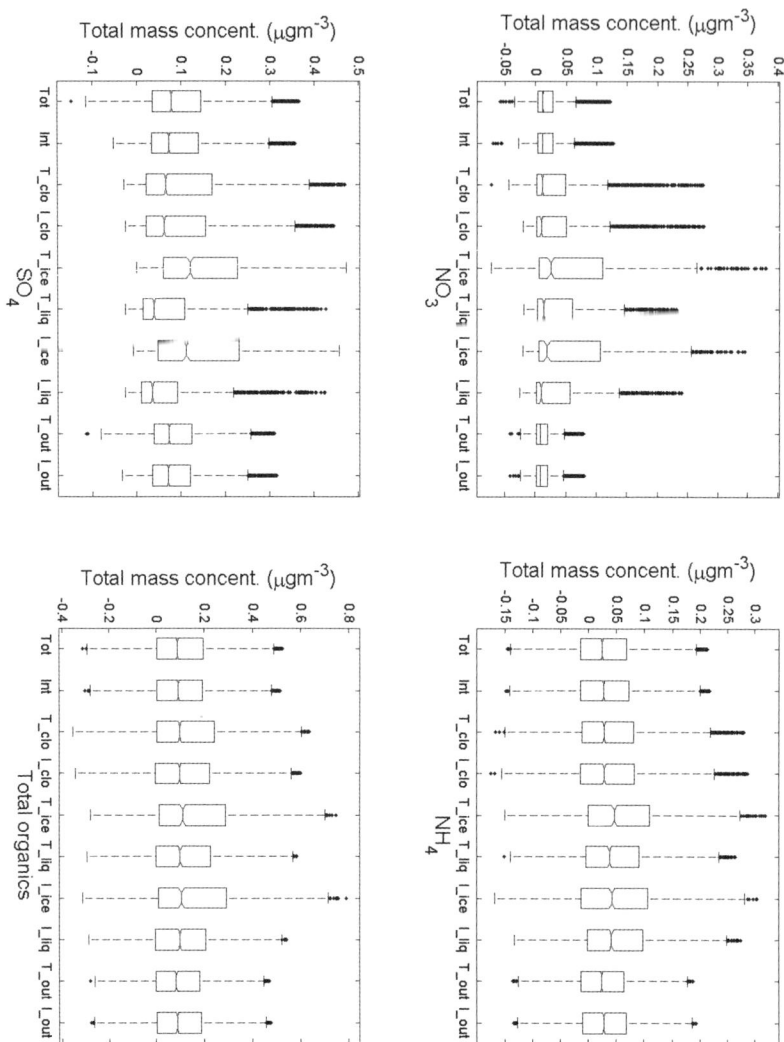

Fig. 7 Conditionally sampled AMS loading of major species showing higher values of sulfate, nitrate and organics in ice cloud than water cloud. Tot: All data (in and out of cloud) from total inlet. Int: All data (in and out of cloud) from Interstitial inlet. T_clo: Data from all cloud events, regardless of cloud phase (Total inlet). I_clo: Data from all cloud events, regardless of cloud phase. (Interstitial inlet). T_ice: Data from ICE cloud events (Total inlet). T_liq: Data from LIQUID cloud events (Total inlet). I_ice: Data from ICE cloud events (Interstitial inlet). I_liq: Data from LIQUID cloud events (Interstitial inlet). T_out: Data out of cloud (Total inlet). I_out: Data out of cloud (Interstitial inlet).

powerful secondary ice processes may operate at other temperatures, however, this remains to be quantified.

Results from the CLOPAP experiments in stratocumulus cloud

Several sorties took place during the summer of 2005, studying the plumes of several eastern UK cities as they evolved, as they were transported east. The changes in the chemical composition, size distribution and nucleation properties of the aerosol will be presented as a function of plume age for the different case studies.

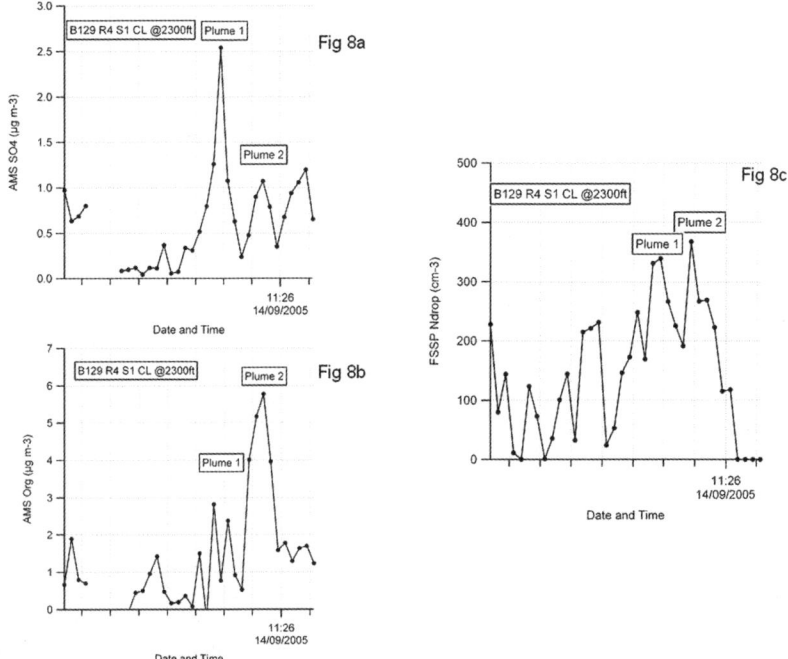

Fig. 8 CLOPAP case study from flight B129. Plume 1 contains predominantly sulfate aerosol internally mixed with some organic material, whereas plume 2 contains predominantly organic aerosol internally mixed with some sulfate (parts a and b). Part c shows that both aerosol plumes are effective as CCN increasing the droplet number in the cloud.

The compositions of cloud drop residuals were successfully measured with the AMS and shown to be composed mainly of a mixture of ammonium, sulfate, organics and nitrate. Changes as a function of age were noted, including the relative mass concentrations of inorganics to organics and the amount of oxidation in the organic fraction. The overall level of organic oxidation was higher than is typically seen in urban environments, and similar to that found in the winter CLACE experiments described above. A total of 12 case studies were flown around the UK. In this paper, we will concentrate on the results of one case study flown on 14 September 2005 over Kent and the Thames Estuary. The results in Fig. 8 show a transect through stratocumulus cloud about 100 km from London. The results presented are droplet residuals sampled through a CVI mounted on the aircraft and fed to an aerodyne AMS. It can be seen that one plume contained predominantly sulfate aerosol internally mixed with a small amount of organic material, whilst a neighbouring plume consisted of predominantly organic material internally mixed with sulfate. The organic material sampled shows evidence of being highly oxidised, with 10% m/z 44 fragments, similar to CLACE-2. Both plumes resulted in increased droplet number in the cloud to roughly the same degree. This confirms the picture from the CLACE experiments of aged organic aerosol acting as an excellent CCN, but much closer to the source. The likely scenario for the two plumes is that they had rather different sources. The following process probably formed plume 1: organic seeds of hydrophilic hydrocarbons produced from engine exhausts in the urban environment with a size of about 30 nm. Such particles are widely observed on urban areas. These would grow as additional oxidised secondary organic material was added. This would cause the particles to be weakly hydrophilic. At this point, nitric acid and then sulfuric acid would be deposited to the particles, the sulfuric acid tending to displace the nitric acid. Plume 2 probably formed in a similar way but evolved in a much more organic-rich environment, hence deposition of secondary organic material dominated. In this case, the atmospheric boundary layer was stratocumulus-capped throughout and hence it is not clear how much of a role cloud processing played in the evolution of the plume. Other flights in cloud free conditions did show plumes of similar composition at a similar distance for urban sources.

Discussion

In this paper, results have been presented of the influence of atmospheric aerosol particles in the size range 50 nm and upwards on the formation of water and ice clouds. Two series of field experiments were selected to represent the observations made. The first is the CLACE series of experiments performed at a high Alpine site (Jungfraujoch), where cloud was in contact with the ground and the measuring station. Both water, ice and mixed phase clouds were examined at different times of the year. The second series of experiments is the CLOPAP series where ageing pollution aerosol from UK cities was observed, from an airborne platform—the UK FAAM BAE146 aircraft—to interact with warm stratocumulus cloud in a cloud-capped atmospheric boundary layer.

It has been shown that aged pollution aerosol consists of an internal mixture of organics, sulfate, nitrate ammonium; the organic component is dominated by highly oxidized secondary material. The relative contributions and absolute loadings of the components vary with location and season. For example, during CLACE-2 in the summer at Jungfraujoch, it was observed that organic material dominated the aerosol, whereas during the CLOPAP experiments the observations were made much closer to sources, and hence the plumes had marked structure with very different aerosol composition in different parts of the plume. It was found, however, that most of the aerosols act as cloud condensation nuclei, irrespective of their composition and the ratio of the species found in the cloud droplets were very similar to those found in the total aerosol, confirming that the particles were internally mixed. Hence, much of the organic material along with the other species is incorporated into cloud droplets. It was observed, in CLOPAP, that by 100 km down wind of London the urban aerosol produced from that area has this characteristic internally mixed structure and was a good CCN. This is an important result, as the lifetime of organic aerosol will be limited, as by being incorporated into cloud droplets it will be readily removed by rainout and wet deposition. Further, however, it means that urban-produced aerosol will modify cloud structure only a short distance downwind of the urban environment.

In ice and mixed phase cloud, it is observed that very sharp transitions (just a few meters in horizontal extent) are present between highly glaciated regions and regions consisting of supercooled water. This is a unique finding, although unpublished results from aircraft observations in cumulus clouds suggest that a similar structure may exist. This observation is very important in understanding the structure of mixed phase clouds. Precipitation particles grow rapidly when ice and supercooled water are intimately mixed in cloud both by growth of the ice particles from the vapour and by the riming process. However, if the phases are separated on small scales then these processes will be inhibited.

It is suggested that this transition is caused by ice nucleation initiated by oxidised organic aerosol coated with sulfate in more polluted regions of cloud. This suggestion is supported by conditional sampling of the aerosol data, which shows that glaciated regions of cloud contain higher loadings of the major ions and organic material than regions dominated by supercooled water. The possible explanation of the glaciated regions often being short and intermittent is that they form as more polluted air is intermittently transported upwards from the atmospheric boundary layer below. Studies performed in the AIDA chamber have suggested that sulfate-coated organic material acts as an efficient condensation, freezing ice nuclei at temperatures encountered in these clouds in CLACE-3. Very recent results have suggested that the process is further enhanced if the organic material is highly oxidized. This interpretation is tentative at the present time, and further discussion and investigation is needed. This process does not necessarily explain the presence of completely glaciated regions of cloud at temperatures as high as −11 °C. Once ice is formed, it will grow rapidly by vapour diffusion at the expense of the supercooled water droplets that will tend to evaporate. However, a powerful mechanism of secondary ice particle production (the Hallett–Mossop process), which occurs at temperatures between −3 °C and −9 °C, may be playing an important role in completing the glaciation of the cloud at relatively high temperatures.

These results are potentially very important for understanding the role of man-made aerosol on the microphysics of mixed phase cloud and hence their interaction with solar radiation. These particles can also affect the precipitation production and lifetime of such clouds.

Acknowledgements

This work was funded by the Natural Environment Research Council.

References

1　U. Baltensperger, M. Schwikowski, D. T. Jost, S. Nyeki, H. W. Gäggeler and O. Poulida, *Atmos. Environ.*, 1998, **32**(23), 3975–3983.
2　U. Baltensperger, H. W. Gaggeler, D. T. Jost, M. Lugauer, M. Schwikowski, E. Weingartner and P. Seibert, *J. Geophys. Res., [Atmos.]*, 1997, **102**(D16), 19707–19715.
3　S. Henning, E. Weingartner, S. Schmidt, M. Wendisch, H. W. Gaggeler and U. Baltensperger, *Tellus Ser. B*, 2002, **54**(1), 82–95.
4　S. Henning, E. Weingartner, M. Schwikowski, H. W. Gäggeler, R. Gehrig, K.-P. Hinz, A. Trimborn, B. Spengler and U. Baltensperger, *J. Geophys. Res., [Atmos.]*, 2003, **107**, 4030.
5　S. Mertes, A. Schwarzenbock, P. Laj, W. Wobrock, J. M. Pichon, G. Orsi and J. Heintzenberg, *Atmos. Res.*, 2001, **58**, 267–294.
6　S. Mertes, B. Verheggen, S. Walter, M. Ebert, P. Connolly, E. Weingartner, J. Schneider, K. N. Bower, M. Inerle-Hof, J. Cozic, U. Baltensperger and J. Heinzenberg, *Environ. Sci. Technol.*, 2006, submitted.
7　E. Weingartner, S. Nyeki and U. Baltensperger, *J. Geophys. Res., [Atmos.]*, 1999, **104**(D21), 26809–26820.
8　T. W. Choularton, M. W. Gallagher, M. J. Flynn, D. Figueros-Nieto, K. N. Bower, J. Whiteway, C. Cook and J. Hacker. Observations of the cloud microphysics and dynamics in mid-latitude cirrus over South Australia, . Proceedings of the American Meteorological Society, 11th Conference on Cloud Physics & Precipitation, June 2002, Ogden, Utah, USA, pp. 249–250.
9　P. Lawson and B. Baker, *Second Annual Report, Evaluation of the Cloud Particle Imager and Investigations of Ice Multiplication*, NSF Grant No. ATM-9904710, For the Period:, 1 June 2000–31 May 2001, 2001.
10　J. Cozic, S. Mertes, B. Verheggen, U. Baltensperger and E. Weingartner, *Geophys. Res. Lett.*, 2007, to be submitted.
11　J. D. Allan, J. L. Jimenez, P. I. Williams, M. R. Alfarra, K. N. Bower, J. T. Jayne, H. Coe and D. R. Worsnop, *J. Geophys. Res., [Atmos.]*, 2003, **108**, 4090.
12　J. T. Jayne, D. C. Leard, X. F. Zhang, P. Davidovits, K. A. Smith, C. E. Kolb and D. R. Worsnop, *Aerosol Sci. Technol.*, 2000, **33**, 49–70.
13　J. D. Allan, K. N. Bower, H. Coe, H. Boudries, J. T. Jayne, M. R. Canagaratna, D. B. Millet, A. H. Goldstein, P. K. Quinn, R. J. Weber and D. R. Worsnop, *J. Geophys. Res., [Atmos.]*, 2004, **109**, D23S24.
14　J. D. Allan, H. Coe, K. N. Bower, M. R. Alfarra, A. E. Delia, J. L. Jimenez, A. M. Middlebrook, F. Drewnick, T. B. Onasch, M. R. Canagaratna, J. T. Jayne and D. R. Worsnop, *J. Aerosol Sci.*, 2004, **35**(7), 909–922.
15　M. R. Alfarra, H. Coe, J. D. Allan, K. N. Bower, H. Boudries, M. R. Canagaratna, J. L. Jimenez, J. T. Jayne, A. Garforth, S. Li and D. R. Worsnop, *Atmos. Environ.*, 2004, **38**(34), 5745–5758.
16　Z. Krivacsy, A. Hoffer, Zs. Sarvari, D. Temesi, U. Baltensberger, S. Nyeki, E. Weingartner, S. Kleefeld and S. G. Jennings, *Atmos. Environ.*, 2001, **35**, 6231–6244.

17 P. Connolly, M. W. Gallagher, M. J. Flynn, T. W. Choularton and Z. Ulanowski, *J. Tech. A*, 2007, in press.
18 B. Verheggen, J. Cozic, E. Weingartner, B. K. N. Bower, S. Mertes, P. Connolly, M. W. Gallagher, M. Flynn, T. Choularton and U. Baltensperger, *J. Geophys. Res., [Atmos.]*, 2006, in press.
19 S. Walter, J. Schneider, J. Curtius, S. Borrmann, S. Mertes, E. Weingartner, B. Verheggen, J. Cozic and U. Baltensperger, in preparation.
20 O. Möhler, S. Büttner, C. Linke, M. Schnaiter, H. Saathoff, O. Stetzer, R. Wagner, M. Krämer, A. Mangold, V. Ebert and U. Schurath, *J. Geophys. Res., [Atmos.]*, 2005, **110**(D11), 11210.

PAPER

Metastable nitric acid hydrates—possible constituents of polar stratospheric clouds?

Hinrich Grothe,[*a] Heinz Tizek[†a] and Ismael K. Ortega[b]

Received 14th February 2007, Accepted 23rd April 2007
First published as an Advance Article on the web 23rd July 2007
DOI: 10.1039/b702343j

Crystallization kinetics of the metastable modifications of Nitric Acid Dihydrate (NAD) was investigated by time-dependent X-Ray Diffraction (XRD) measurements. Kinetic conversion curves were evaluated adopting the Avrami model. The growth and morphology of the respective crystallites and particles were monitored *in situ* on the cryo-stage of an Environmental Scanning Electron Microscope (ESEM) under a partial pressure of nitrogen gas (0.5 Torr, 67 Pa). The morphologies were used to adapt the InfraRed (IR) extinction spectra by T-matrix calculation using respective optical indices of NAD. The results show a significant dependence of the band shapes on different morphologies.

Introduction

The knowledge of formation pathways and exact compositions of solid Polar Stratospheric Clouds (PSC), so-called *type Ia* and *II*, is still fragmentary. On one hand, several field studies provided evidence that Nitric Acid Trihydrate (NAT) exists in PSCs.[1–5] On the other hand, there are also field studies that are less conclusive regarding the composition of the dominating solid phases.[6–8] An explanation of this matter might be the nucleation of metastable Nitric Acid Dihydrate (NAD) from Supercooled Ternary Solutions (STS) and binary HNO_3/H_2O droplets at fluctuating stratospheric temperatures.[9–16] Finally, these metastable compositions transform into thermodynamic stable NAT, which can be observed in aged PSCs. This explanation is supported by the fact that the direct crystallisation of NAT in some laboratory studies was only observed at temperatures below 180 K and at a degree of supersaturation of gas-phase nitric acid with respect to NAT, which was larger than 45.[14,17] Such low temperatures and high supersaturation degrees are not reached in real synoptic-scale PSCs. Therefore, the contribution of homogeneous nucleation to the direct formation of NAT particles in PSCs is difficult to describe on the basis of the available laboratory models. Up to now, it is not clear whether NAT or NAD particles can directly be formed by homogeneous nucleation of STS droplets in the polar stratosphere.[18] However, NAD as a metastable phase should

---

[a] *Vienna University of Technology, Institute of Materials Chemistry, Veterinärplatz 1/GA, A-1210, Vienna, Austria. E-mail: grothe@tuwien.ac.at; Fax: +431 25077 3890; Tel: +431 25077 3809*
[b] *Department of Molecular Physics, Instituto de Estructura de la Materia, CSIC, C/ Serrano 121, E-28006, Madrid, Spain. E-mail: ikortega@iem.cfmac.csic.es; Fax: +34 915855184; Tel: +34 915901609*

† Present address: Vienna City Administration, Municipal Department for Environmental Protection—MA 22.

have the advantage of a rather low nucleation barrier. In order to understand the nucleation in more detail, Tizek et al.[19] have investigated respective model samples at non-equilibrium conditions (crystallized from quenched samples) and at different concentrations.

In the literature, several attempts have been made to study the nucleation of NAD by FTIR spectroscopy.[11,20–26] Differences between these spectra, obtained under more-or-less similar experimental conditions, could not be explained conclusively. Unlike vibrational spectroscopy, a diffraction technique is the more appropriate method for the required phase analysis. During our first experimental work on NAD,[19] the single crystal structures of two NAD modifications—denoted α-NAD and β-NAD—became available.[27,28] Thus, we could calculate the powder diffractograms of both phases, which helped us for the phase assignments. Both structures (monoclinic, $P2_1/n$) are very similar, but differ in the interconnection of layers by hydrogen bonds. The α-NAD modification is likely to be less ordered than β-NAD and presumably resembles the short range order of the liquid better than the high-temperature modification. Its nucleation barrier should thus be slightly lower than that for the β-NAD. However, the phase analysis carried out by X-Ray Diffraction (XRD) reveals a much more complicated phase distribution. On the basis of the classical nucleation theory, we have developed a stochastic model, which rather well explains the observations made.[19] The particular phase distribution is highly reproducible and is intimately related to the quenching procedure applied. In fact, the existence of the stable phase distribution initiated further studies,[29–32] where the vibrational spectra, microscopic pictures and crystallization mechanisms were assigned to the well defined hydrate phases.

The InfraRed (IR) spectra of NAD of Niedziela et al.[10] and Tisdale et al.[26] exhibit serious differences. They concluded that one possible explanation might be the existence of two NAD modifications; another one could be the distortion of the crystal lattice. However, the crystal structures of α- and β-NAD were not yet available and, therefore, we were the first to correlate the spectra with the two structures and thus present the first direct spectroscopic differentiation between the two modifications of solid NAD.[29] The respective samples were prepared by the above-mentioned quenching technique and were investigated by FTIR and Raman spectroscopy. Upon variation of temperature and concentration, several spectral changes were observed. Due to the phase compositions and transition temperatures corroborated by XRD,[19] the spectral changes were interpreted in terms of phase transitions between amorphous and crystalline, on one hand, and between α- and β-NAD, on the other.

With vibrational spectroscopy, both NAD modifications can be discriminated on the basis of seven characteristic features in the mid-IR region.[29] Additionally, below 200 cm^{-1}, lattice vibrations (phonon bands) are observed and Raman spectra of this region were found to be particularly well suited to investigate crystallisation and solid–solid phase transitions. This kind of low-frequency spectroscopy has been extended only recently.[31,33] The new NAD mid-IR spectra were also compared to those in the literature. It became obvious that the phases studied in the past were mostly that of the low-temperature phase α-NAD. Up to now, the only spectrum of pure β-NAD has been presented by us.[29] However, the spectra collection of Tisdale et al.[26] also exhibits some characteristics of β-NAD, which are probably due to an onset of phase transition. The dominance of α-NAD is very surprising, since α-NAD is less stable than β-NAD. Also, in a recent aerosol chamber study[34] the α-NAD modification instead of β-NAD has been observed. Wagner et al.[34] have also tested for the influence of the particle morphology on the NAD extinction IR spectra. The shape of the bands changed when the aerosol preparation procedure was altered from fast- to slow-cooling experiments. The different experimental conditions are reflected by changes in the band shape and band position. These spectra were fitted with calculations finding oblate morphologies for the respective particles with aspect ratios greater than five. The bands most sensible to the particle morphology were those at 1160 cm^{-1}, 1270 cm^{-1} and 1445 cm^{-1}. A phase transition—another possible explanation—can be excluded, since a comparison with the known NAD spectra reveals clearly that in all experiments exclusively α-NAD was present.[29]

Very recently, Maté et al.[35] have published transmission and grazing angle (polarized and non-polarized) Reflection Absorption IR spectra (RAIR) of NAD. The assignment of the spectra was based on the spectra of Grothe et al.[29] and α-NAD was unambiguously identified. The authors revealed strong differences between transmission and RAIR spectra, between s- and p-polarization and between samples of different thickness (400–700 nm). They compared the spectra with Fresnel model simulations and found important changes with the p-polarized spectra concerning the three bands mentioned above. The conclusion was that α-NAD films are anisotropic and oriented. A preferential growing of the α-NAD films with the (1 0 −1) crystallographic plane parallel to the metal substrate appeared to be the most appropriate explanation. Since Wagner et al.[34] and Maté et al.[35] found differences for the same three bands, it was concluded that anisotropy could also be present in the aspherical NAD particles.

In order to gather more detailed information on the crystallization mechanisms and the morphologies, as well as on their impact on the spectroscopic data, NAD was investigated in the form of a laboratory model. NAD particles were grown from amorphous samples situated in a cryostat. The exact phase composition, the kinetics of the phase transitions and the particle morphologies were analysed by XRD and Environmental Scanning Electron Microscopy (ESEM).

Experimental

The sample preparation for nitric acid hydrates is described elsewhere.[19,29,30] In short, it includes the following procedure: Liquid samples of well defined HNO$_3$/H$_2$O molar ratios were nebulized into an aerosol chamber. On opening the valve to the high vacuum cryostat chamber briefly, a part of the aerosol reaches the cold surface of a sample carrier held at 80 K and located at the end of the cold cryostat, where it freezes amorphously. The sample was then crystallized in the course of an annealing program which includes the following steps: (i) heating up the amorphous sample (5 K min^{-1}) to a temperature above the glass transition point; (ii) repeated isothermal measurements until the diffractogram shows no further changes; (iii) cooling back to 35 K and measurement over the whole angle range from 10 to 50° (2θ) controlling for the phase composition; (iv) subsequently, the sample can be used to study the α-/β-NAD phase transition at higher temperature. The preparation techniques for XRD and ESEM differ from each other. While in the case of XRD an aerosol ensemble was deposited, for ESEM only one droplet of 4–20 μl was frozen directly in a liquid nitrogen bath. The full procedure is described elsewhere.[36] A recent Raman study showed that different preparation techniques during the formation process of the amorphous solid have a minor impact on the later phase composition and the phase transition temperatures.[31] The XRD system (Seifert XRD 3000 TT, Germany) hosted the cryostat (Leybold RGD 210, Germany) in its centre. The ESEM system (FEI-Philips XL30 ESEM-FEG) was equipped with a cryo-transfer chamber (Gatan ALTO 2500). Therein, the frozen amorphous droplet was freeze-fractured horizontally, removing the upper section. The lower part, frozen onto the sample holder cup, was used for microscopy. The freeze-fracture creates a clean, virgin surface without any frost and minimises artefacts. Subsequently, the lock to the microscope chamber was opened and the sample transferred under vacuum, without annealing, onto the cooled cryo-stage (<133 K). For imaging, an electron beam accelerating voltage of 5 kV was used with 0.3–0.5 Torr (40–67 Pa) of nitrogen gas present in the chamber. The sample annealing was the same as mentioned above.

The phase transitions of amorphous/α-NAD, amorphous/β-NAD and α-/β-NAD were observed under isothermal conditions. The alterations in the diffractograms

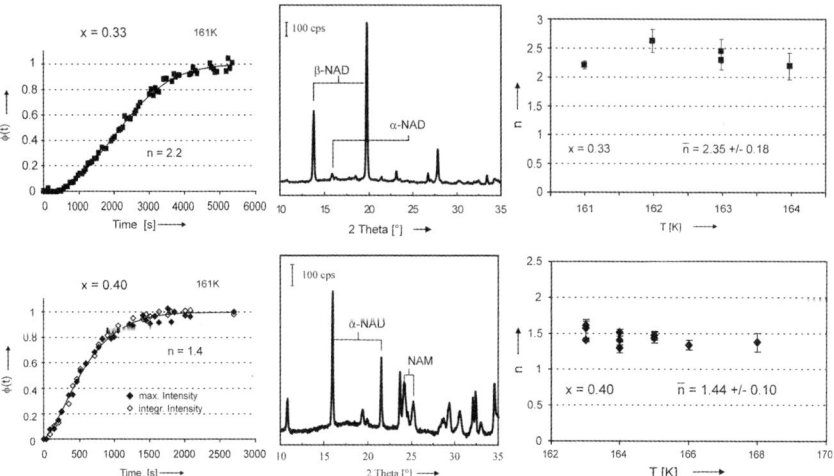

Fig. 1 The isothermal crystallization of α-NAD and β-NAD at 161 K, $x = 0.33$ and $x = 0.40$, respectively. Left: the S-shaped transformation–time curve. Middle: the diffractogram of the respective crystalline sample. Right: the Avrami exponent at different temperatures, $\bar{n}$ is the average Avrami parameter.

with time were used to calculate the conversion curves. XRD is a relatively slow technique; therefore, an alternative measurement procedure was needed. This involved scanning a particular interval of the diffractogram where the strongest changes are expected. In this range, only low resolution measurements ($\Delta(2\theta) = 0.05°$, time per step (1–3 s)) were carried out. The whole measurement of a data point on the conversion curve took on average 1 min. Measurements where the complete conversion took over 15 min were the most reliable. In faster cases, there were too few data points for a reasonable curve fit. Lower temperatures also had to be avoided, since the conversion time increases exponentially with decreasing temperature, causing an unacceptable time scale. For realization of isothermal measurements, (i) a suitable reflex was selected in a preliminary test, (ii) the intensity of this reflex $I(t)$ was measured as a function of time and (iii) the fraction of sample already converted into the new phase $\phi(t)$ was calculated (I_{end} is the final intensity at complete conversion):

$$\phi(t) = \frac{I(t)}{I_{end}} \quad (1)$$

$\phi(t)$ was plotted *versus* time exhibiting an S-shaped transformation–time curve. The chosen reflex has to meet with the following criteria: (i) enough intensity and thus a suitable signal-to-noise ratio, (ii) high specificity of the signal for the respective phase, this means no overlapping reflexes of other phases. These criteria were found for the α-NAD reflex at 16.0° (2θ) and for the β-NAD at 13.9° and 19.9° (2θ)—for more details see Table 1 in ref. 19. There was no difference in the result when the absolute intensity instead of the integral intensity was used for the calculation of the conversion curve (for comparison, see Fig. 1, bottom left).

We have also simulated extinction IR spectra of NAD aerosols of different particle shapes. For this purpose, we have performed T-Matrix calculations using an adaptation of the extended precision FORTRAN code of Mishchenko and Travis.[37] This code allows the use of different bodies of revolution to define the shape of the particles. The calculations have been carried out for α-NAD using the refractive indices from Nieziela et al.,[38] which were specifically derived from aerosol samples. The calculations assume a monodisperse random distribution of the particles. Unfortunately, a similar treatment could not be applied to β-NAD, since optical indices for this solid phase are not yet available.

Results

In the concentration range $0.25 < x < 0.5$, nitric acid dihydrate crystallizes from amorphous samples as two different modifications. Low- (α-NAD) and high-temperature modifications (β-NAD) have been differentiated by XRD. The phase distribution of the polymorphs exhibits a bimodal distribution for α-NAD and a monomodal distribution for β-NAD.[19] From these data, we have chosen one maximum of α-NAD crystallization (40 mol%) and one maximum of β-NAD (33.3 mol%) as well as two intermediate concentrations, where both phases are observed. In order to shed light on the differences of the crystallization mechanisms, a kinetic study of the transitions (amorphous/α-NAD (161 K) and amorphous/β-NAD (161 K)) was performed. A graph of $\phi(t)$ versus the reaction time t (see eqn (1)) was plotted, supplying an S-shaped curve. The graph was fitted by an exponential function,

$$\phi(t) = 1 - \exp[-(kt)^n] \quad (2)$$

from which the k and n values were determined. For $n = 1$, eqn (2) would describe a simple first order reaction being responsible for the transformation from the initial to the final phase with a temperature-dependent k as rate constant. Of course, the crystallization process is much more complex, involving a system-specific space filling tendency as well as mass transport in terms of diffusion. Therefore, Avrami[39] introduced the empirical exponent n and interpreted it as a parameter that characterizes roughly the type of mechanism involved in the crystallization and transformation process, respectively. The easiest way to estimate n was a curve fit of the plot, $\phi(t)$ versus t, carried out by the standard software Origin™ 4.10 Microcal Inc.

At the 1 : 2 mixing ratio of HNO_3/H_2O ($x = 0.33$) there is predominant formation of β-NAD, while the low-temperature modification α-NAD shows up only as a very minor content (Fig. 1, top middle). The reason for the preference of β-NAD is the formation of large, stable NAD *germ nuclei* during the quenching of the liquid with the 1 : 2 molar ratio and a subsequent dominance of these nuclei in the isothermal crystallization—for further details see ref. 19. During observation of the diffractograms in the course of time, the β-NAD formation comes up only very slowly. The maximum crystallization velocity is reached just below the half-value time (Fig. 1, top left). As in every other kinetic measurement, the rate decreases until complete conversion is achieved. Comparable to NAT, the progression of the NAD crystallization does not depend on the temperature (Fig. 1, top right). The arithmetic mean n-value at the 1 : 2 mixing ratio is 2.3 ± 0.2, which is significantly higher than the respective value of α-NAT at the 1 : 3 ratio ($n_{\alpha\text{-NAT}} = 1.7$), see ref. 32. At 40 mol% we find the reverse situation, NAD and NAM crystallize from the amorphous solid,[19] but α-NAD is the dominating phase with traces of β-NAD (Fig. 1, bottom middle). The crystallization behaviour of α-NAD differs considerably from the respective process of β-NAD, which can be recognized by comparing the transformation–time curves of both phases (Fig. 1, top and bottom left). The course of α-NAD crystallization starts rapidly and soon reaches the highest rates. Then, at $\phi \geq 0.3$, the conversion curve starts to flatten slowly. Altogether, this is an exemplification for the Avrami parameter, $n = 1.4$. Again, no change of n with the crystallization temperature was observed. The Avrami parameters only have small systematic and statistical errors (Fig. 1, bottom right). The arithmetic mean value of $n_{\alpha\text{-NAD}} = 1.4$ at 40 mol% is considerably lower than the n-values of β-NAD at 33.3 mol% ($n_{\beta\text{-NAD}} = 2.3$) and of α-NAT at 25 mol% ($n_{\alpha\text{-NAT}} = 1.7$).

In addition to the maxima in the phase distribution, there are two intermediate key concentrations where both NAD modifications show up with nearly the same

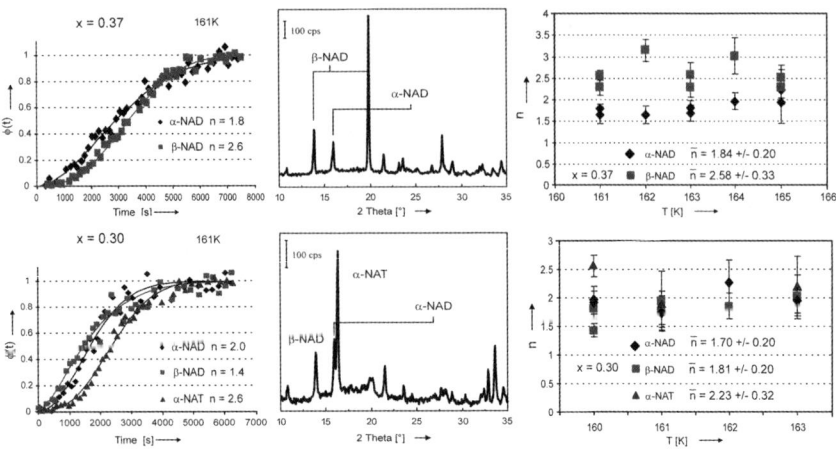

Fig. 2 The isothermal crystallization of α-NAD, β-NAD and α-NAT at 161 K, $x = 0.37$ and $x = 0.30$, respectively. Left side: the S-shaped transformation–time curve. Middle: the diffractogram of the respective crystalline sample. Right hand side: the Avrami exponent at different temperatures, $\bar{n}$ is the average Avrami parameter.

fraction. At 37 mol%, α- and β-NAD crystallize from the same amorphous sample, but with strongly different transformation–time curves (Fig. 2, top left). On one hand, α-NAD reflexes start to increase rapidly in intensity and the maximum crystallization velocity is soon reached. On the other hand, the intensity of β-NAD reflexes only slowly gain intensity and then the conversion curve increases rapidly, obtaining high conversion values and only at $\phi \geq 0.8$ does the curve flatten. Thus, at $x = 0.37$, α-NAD and β-NAD have very different Avrami parameters, which are $n_{\alpha\text{-NAD}} = 1.8$ and $n_{\beta\text{-NAD}} = 2.6$, respectively. The same results are gathered independently from the crystallization temperature (161–165 K), at which the isothermal measurements were carried out. At 30 mol% HNO_3, a similar coexistence of both NAD modifications is observed. However, the phase composition comprises α-NAD, β-NAD and α-NAT (Fig. 2, bottom middle). The same as for 37 mol% HNO_3, the NAD modifications do not have exactly the same transformation–time curves, but the curves exhibit rather similar shape. Thus, it is nearly impossible to estimate from the sum of all experiments a significantly different mechanism between α- and β-NAD crystallization (Fig. 2, bottom right). The Avrami parameter of α-NAT is slightly higher than that of α- and β-NAD, but the strong scatter of the values prevents a clear differentiation in the figure.

In Fig. 3, the average Avrami parameters of all investigated concentrations in the range 25 to 50 mol% have been presented. While the Avrami parameters of α-NAD are in the limited range 1.4–1.8, the parameters of β-NAD show a significant trend—they are increasing with concentration from $n = 1.8$ at $x = 0.30$ to $n = 2.6$ at $x = 0.37$ (Fig. 3, left side). Following the trend, at 40 mol% this difference should increase further. Unfortunately, the Avrami parameter of β-NAD at 161 K has not been quantified due to the very small amount of the β-phase and its nearly invisible diffraction pattern. Nevertheless, this sample concentration seems very interesting for a microscopic investigation since α-NAD crystallizes in its essentially pure form, and finally transforms at a temperature of about 200 K into β-NAD. The α/β transition starts rapidly in all measurements and slowly flattens until conversion is completed—the respective transformation–time curves are not shown here. The Avrami parameters are all situated between 0.5 and 1 (Fig. 3, right side). From the ESEM pictures, one can reveal that α-NAD particles are small needles with a length of about 5–10 μm and a diameter of about 1–2 μm, which means an aspect ratio of about 1 : 5 (Fig. 4, left side). In some cases, the needles have grown to sizes that are one order of magnitude larger. This observation is not surprising, since a bulk

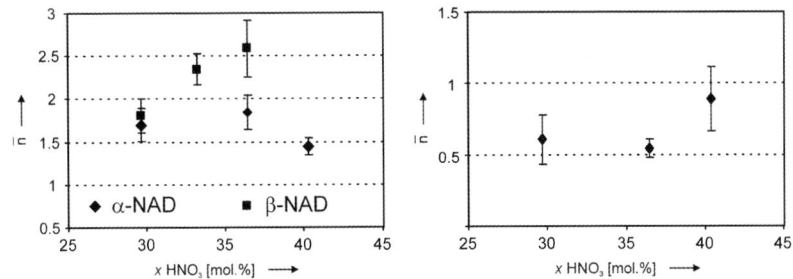

Fig. 3 Left side: mole fraction dependency of average *n*-values for α-NAD and β-NAD crystallization from the amorphous sample. Right side: mole fraction dependency of average *n*-values for the phase transition from α-NAD to β-NAD.

sample, and not a single aerosol particle, has been crystallized and the sample was annealed only slowly to 200 K, which favours the growth of the larger α-NAD particles. Nevertheless, the aspect ratio of all needles is rather similar. When the sample is kept at 200 K for a very long time, a remarkable change of the morphologies is observed, which we associate with the α-/β-phase transition.[19] Already, at the beginning of the process (Fig. 4, left side) small β-NAD grains can be recognized—these have slightly higher intensity in the ESEM pictures and appear with a brighter grey tone than the needles. During the time course, these grains increase in volume and exhibit an irregular, entangled shape (Fig. 4, right side, magnification). After 80 min, the overall transformation process is completed and cuboids have substituted the needles (Fig. 4, right side). Obviously, the cuboids are particularly formed during the phase transformation from α-NAD to β-NAD, while entangled particles are recognized for pre-existing β-NAD grains crystallizing directly from the amorphous substance. The smallest β-NAD particles have diameters of about 10–20 μm. From the shape of both NAD morphologies (needles and cuboids) one can deduce that these are not single crystals, but are polycrystalline. Therefore, we have estimated the full width at half maximum of the strongest reflexes in the respective X-ray powder diffractograms. We were able to calculate the crystallite sizes (Fig. 5), which exhibit rather large errors represented by the error bars, *via* the application of the Debye–Scherrer formula. Nevertheless, one can recognize significant differences in the crystallite sizes of β-NAD, α-NAD and NAM. Obviously, β-NAD crystals have a diameter of about 55 nm, which is larger than

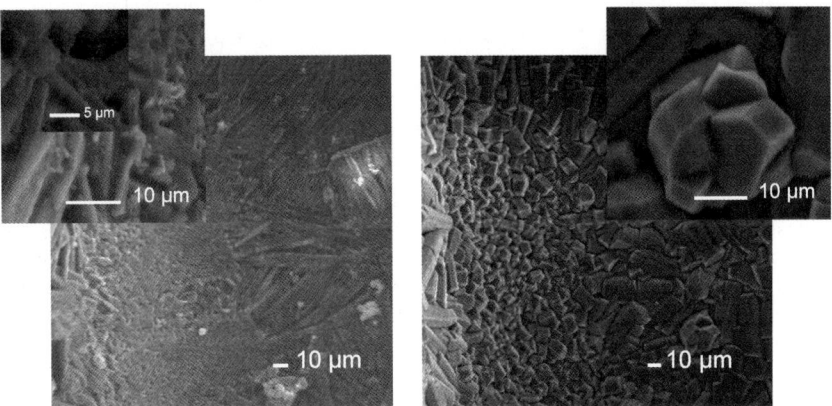

Fig. 4 Left side: ESEM images reveal prolate α-NAD particles in a sample of $x = 0.40$ at 200 K at $t = 0$ min. Right side: β-NAD particles in the same sample at the same temperature but at $t = 80$ min.

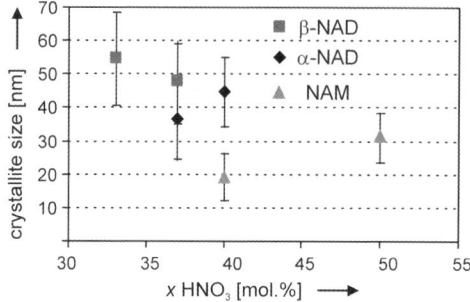

Fig. 5 The crystallite sizes of α-NAD, β-NAD and NAM as a function of mole fraction at 163 K.

α-NAD with about 45 nm, and NAM crystallites are the smallest with 30 nm. It is also observed that in samples with coexisting phases, the crystallites are about 10 nm smaller than in the essentially pure samples—see α- and β-NAD at 37 mol% and NAM at 40 mol%.

Finally, the shape of the α-NAD particles derived from the ESEM experiments have been used for the T-matrix simulations of extinction IR spectra of aerosols. Small particles, formed at the beginning of the crystallization process, were taken for the simulation since they are expected to reproduce better the size of the likely aerosol particles. The larger crystallites observed later gradually approach "bulk" samples and are not adequate for the simulation aiming at sub-micrometre particles. The calculations were performed for four different shapes: spheres (with a diameter of 0.78 μm); two types of spheroids with the same equivalent diameter; prolate spheroids (needles) with aspect ratio 1 : 5 and oblate spheroids (discs) with aspect ratio 5 : 1; and a Chebyshev particle of order 8 (see Fig. 3 of ref. 37), with the same equivalent diameter and a deformation factor of 0.1. The observed particles of α-NAD resemble closely the prolate spheroids and those of β-NAD are roughly similar to the Chebyshev particles. The results of these simulations are shown in Fig.

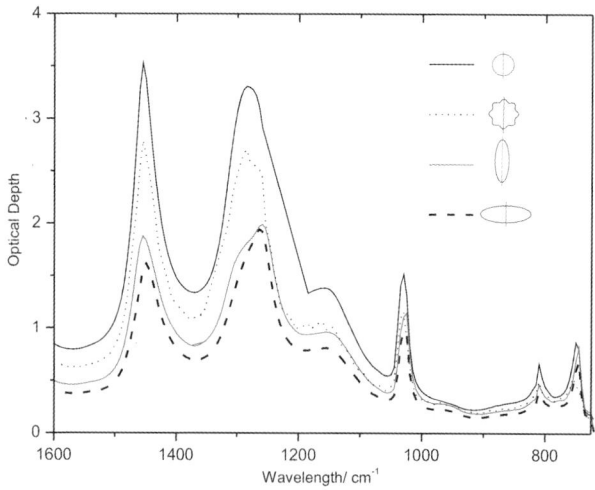

Fig. 6 Simulated extinction spectra for monodisperse aerosols of randomly oriented α-NAD particles of different shape. All of them correspond to an equivalent diameter of 0.78 μm. Solid black line: symmetrical spheres. Dotted line: Chebyshev particles ($T = 8$, deformation parameter = 0.1). Grey line: prolate spheroids with aspect ratio 1 : 5. Dashed line: oblate spheroids with aspect ratio 5 : 1.

6 over the relevant 700–1600 cm^{-1} range. Note that appreciably different extinction IR spectra are obtained for the various particle shapes.

Discussion

The Avrami model for the decay of a metastable phase is an empirical method used for understanding nucleation and crystal growth processes. In a former paper, we could show that for both NAT modifications the morphology of particles was correctly predicted by this method, which we have finally controlled by ESEM. In this paper, we have tried to apply the same procedure also to the NAD modifications, which are both metastable phases. Avrami suggested that for this type of system (a particular substance with a specific crystal habit), an isokinetic range of temperature and/or mole fraction exists, in which the fundamental mechanism of the phase change does not depend on temperature and mole fraction.[39] According to eqn (2) this necessarily implies that the Avrami exponent n is a constant for these conditions. Then the reaction rate can be separated into a nucleation dependent term and a crystallization term, where n is the sum of the nucleation exponent ν and crystallization exponent κ:

$$n = \nu + \kappa \qquad (3)$$

For spontaneous (athermal) nucleation the number of nuclei is time independent ($\nu = 0$). This is the case when nuclei of the new phase already exist at the beginning of the phase transition, e.g. in a quenched liquid. For nuclei formation during the transition, the process can be either linear ($\nu = 1$) or more complicated ($\nu \neq 1$). The growing volume of the new phase depends on the dimensionality d of the respective growth mechanism. Three-dimensional growth (spherical) is proportional to the third power of the reaction time ($\kappa = 3$). Two-dimensional growth (discoidal) is proportional to the second power ($\kappa = 2$) and one-dimensional growth (spicular) is directly proportional with time ($\kappa = 1$)—this means κ equals d. If, however, the transport of the reactants to the growing interface is the time-limiting step, then the volume growth depends on the square root of time and κ equals $0.5d$ (for further details see ref. 32, Table 1). Obviously, the same n value can follow from more than one set of parameters. Therefore, the interpretation of the Avrami exponent is ambiguous and in many cases is of more qualitative use. Very often, several reaction mechanisms seem to be possible and a decision can only be reached by taking into consideration experimentally observable parameters.

For the interpretation of NAD formation mechanisms, two additional facts have to be considered that are different from the crystallization of NAT: (i) both modifications of NAD can crystallize from the same amorphous sample, (ii) the phase distributions of both NAD phases vary with the molar ratio of the amorphous sample from which they crystallize. β-NAD shows a maximum at 1 : 2 mixing ratio HNO_3/H_2O ($x = 0.33$), while α-NAD has two maxima beside at $x = 0.40$ and $x = 0.30$.

During the quenching of 33.3 mol% nitric acid, i.e. the 1 : 2 mixing ratio, post-critical β-NAD nuclei are embedded in the amorphous solid and finally crystallize the sample into β-NAD. Thus, for the growth process the formation of the nuclei is negligible and no impact on the Avrami parameter n is expected ($\nu = 0$). On one hand, diffusion might have an impact due to the reduced temperature. On the other hand, the crystallization temperature of NAD (161 K) is about 6 K higher than that of α-NAT (155 K). With diffusion neglected, κ will equal the dimensionality of the growth and the average Avrami parameter $n \approx 2.35$ stands for the growth of pre-existing nuclei in a dimension between 2 and 3 (fractal dimension). This assumption is supported by the structural analysis of β-NAD that is build up by plane layers interconnected by hydrogen bonds, forming a three-dimensional structure. Therefore, the β-NAD crystals should grow perpendicularly to the layers with another velocity than in parallel. However, a control by the respective ESEM pictures is not yet available.

From 40 mol% nitric acid, NAD and NAM crystallize. The Avrami value is $n = 1.40$. Due to the concentration, which is aside of the 1 : 2 mixing ratio, the number of post-critical α-NAD nuclei, stabilized in the solid, should be rather low and might be insufficient to further support the crystallization. Thus, the impact of the nucleation on the n-value must not be neglected—ν is 1 for a constant nucleation rate or smaller than 1 if, during the growth, the nucleation rate decreases. Moreover, the diffusion of molecules to the interface of the growing phase plays a decisive role ($\kappa = 0.5d$). Thus, the crystallites should expand into one or two dimensions (eqn (3)). In fact, the ESEM pictures reveal needles, which support the assumption of one-dimensional growth. In addition to α-NAD, a few β-NAD particles exist at 40 mol%. ESEM pictures show these to be irregular and entangled, which would agree with the considerations made above for 33.3 mol% favouring a growth in fractal dimensions. When the sample is annealed to 200 K, α-NAD transforms into β-NAD. At this temperature the diffusion should be negligible ($\kappa = d$), and also consecutive nucleation should be without relevance ($\nu = 0$). Therefore, the deduced Avrami parameter $n \leq 1$ would mean a needle shape. However, ESEM has revealed cuboids, but the morphology is obviously different from β-NAD particles directly crystallized from the amorphous sample. The third phase present at 40 mol% is NAM. An Avrami parameter for NAM has not been determined. However, it has little impact on the growth of NAD, since the monohydrate crystallizes very slowly and has much smaller crystallites. NAM crystallites have a diameter of 20 nm, while α-NAD and β-NAD have diameters of 45 nm and 55 nm, respectively. Also the particles of NAM are smaller and have the shape of symmetrical spheres (4 μm), which can easily be distinguished from the non-spherical NAD particles (Fig. 4, upper magnification on the upper left side).

At 37 mol%, both NAD modifications crystallize from the same amorphous sample. The statements made above for 40 mol% can be transferred. The Avrami parameters are $n_{\beta\text{-NAD}} = 2.6$ and $n_{\alpha\text{-NAD}} = 1.8$. Due to the fact that the sample concentration is close to the 1 : 2 mixing ratio, an insufficient number of post-critical β-NAD nuclei is formed during quenching. Thus, the contribution of nucleation ν to the Avrami parameter n exhibits a value between 0 and 1—instead of 0 directly at the 1 : 2 mixing ratio. Therefore, the n-value of β-NAD slightly increases in comparison to 33.3 mol%. Due to the Avrami theory, this would mean that the β-NAD particles should exhibit fractal geometries (see above). Also, for α-NAD the Avrami parameter increases in comparison to the n-value at 40 mol%. However, in this case it is the smaller contribution of diffusion that causes the value to increase, since the impact of diffusion decreases in the direction of the 1 : 2 mixing ratio. From the Avrami theory, one can again deduce needles as the prevailing morphology of α-NAD particles—an ESEM picture at 37 mol% is not yet available.

At all concentrations a needle shape has been deduced for α-NAD, which has been corroborated by ESEM. Therefore, we have used prolate geometry (aspect ratio 1 : 5) for the adaptation of the respective extinction IR spectra. For comparison, we have also used a Chebyshev shape, an oblate spheroid (aspect ratio 5 : 1) and a symmetrical sphere. Our T-matrix calculations show distinctive differences between these spectra simulated with the various shapes (Fig. 6). Both prolate and oblate spheroids give similar spectra, with lower band intensities at ≈1450 cm^{-1} and ≈1250 cm^{-1} than the more isotropic spherical or order 8 Chebyshev-shaped particles. It is interesting to observe that the calculated spectra for both prolate and oblate spheroids are relatively similar and resemble more closely the spectra of α-NAD aerosols reported by Wagner et al.[34] from measurements in the AIDA facility. These authors concluded that the oblate shape provided a better fit to their data. However, the ESEM results and the spectral simulations of the present work indicate that α-NAD aerosol particles might have a prolate shape. The particles of β-NAD are more isotropic and correspond approximately to the order 8 Chebyshev shape. As mentioned above, refractive indices are not yet available for this crystalline phase. However, if they are not too different from those of α-NAD, the simulations of Fig. 6 suggest that extinction IR spectra could be helpful to distinguish between the two types of aerosols.

Conclusions

The mechanisms of α- and β-NAD crystallization are significantly different. This is not only true for molar ratios where the modifications are formed in their essentially pure form (33.3 and 40 mol%), but also at an intermediate concentration where both NAD modifications are formed (*e.g.*, 37 mol%). The crystallites of the different phases exhibit different sizes. From the Avrami theory, the morphologies of the crystallites have been deduced and transferred to the particle morphologies. Finally, these morphologies have been controlled by ESEM—prolate spheres for α-NAD and irregular cuboids for β-NAD have been corroborated. The extinction IR spectra of α-NAD have been calculated. At 200 K the low-temperature modification α-NAD slowly transforms into β-NAD. Finally, β-NAD decomposes into the stable phases NAM and β-NAT, which has not been investigated in the course of this paper.

This investigation poses several questions intimately related to nucleation and crystal growth. Probably the most important question is, how reliable is the Avrami theory? Moreover, the Avrami theory was set-up for crystals, but used also for particles (see Tisdale *et al.*[40]). Our work, this paper and a preceding article,[32] state accordance between theory and experiment for several cryo-particles. Was this just by chance or can we propose reliable predictions for particle morphologies in general? Finally, we have calculated the extinction IR spectra of α-NAD particles. However, the morphologies used for this procedure were taken from bulk samples and were not deduced from real aerosol particles. Can our results, including all the restrictions, be used, and can predicted morphologies help for a better phase analysis in field measurements? From our point of view, it is very important to corroborate the phase composition of any model sample, from which optical indices are determined. This also comprises the knowledge of the particle morphologies, for which the indices have to be adapted before they can be used, *e.g.*, in the evaluation of satellite measurements.

Acknowledgements

We thank Debbie Stokes and Debbie Waller for their help with the ESEM measurements during a stay of HG in Cambridge. We gratefully acknowledge financial support of the Austria–Spain Integrated Action *"Acciones Integradas"* and the Universities Jubilee Foundation of the City of Vienna, project H-1119.

References

1. C. Voigt, J. Schreiner, A. Kohlmann, P. Zink, K. Mauersberger, N. Larsen, T. Deshler, C. Kroger, J. Rosen, A. Adriani, F. Cairo, G. Di Donfrancesco, M. Viterbini, J. Ovarlez, H. Ovarlez, C. David and A. Dornbrack, *Science*, 2000, **290**, 1756.
2. C. Voigt, H. Schlager, B. P. Luo, A. D. Dörnbrack, A. Roiger, P. Stock, J. Curtius, H. Vössing, S. Borrmann, S. Davies, P. Konopka, C. Schiller, G. Shur and T. Peter, *Atmos. Chem. Phys.*, 2005, **5**, 1371.
3. D. W. Fahey, R. S. Gao, K. S. Carslaw, J. Kettleborough, P. J. Popp, M. J. Northway, J. C. Holecek, S. C. Ciciora, R. J. McLaughlin, T. L. Thompson, R. H. Winkler, D. G. Baumgardner, B. Gandrud, P. O. Wennberg, S. Dhaniyala, K. McKinney, T. Peter, R. J. Salawitch, T. P. Bui, J. W. Elkins, C. R. Webster, E. L. Atlas, H. Jost, J. C. Wilson, R. L. Herman, A. Kleinbohl and M. von Konig, *Science*, 2001, **291**, 1026.
4. T. Deshler, N. Larsen, C. Weissner, J. Schreiner, K. Mauersberger, F. Cairo, A. Adriani, G. Di Donfrancesco, J. Ovarlez, H. Ovarlez, U. Blum, K. H. Fricke and A. Dörnbrack, *J. Geophys. Res., [Atmos.]*, 2003, **108**, 4517.

5   M. Höpfner, B. P. Luo, P. Massoli, F. Cairo, R. Spang, M. Snels, G. Di Donfrancesco, G. Stiller, T. v. Clarmann, H. Fischer and U. Biermann, *Atmos. Chem. Phys. Discuss.*, 2005, **5**, 10685.
6   Y. Kim, W. Choi, K.-M. Lee, J. H. Park, S. T. Massie, Y. Sasano, H. Nakajima and T. Yokota, *J. Geophys. Res., [Atmos.]*, 2006, **111**, D13S90.
7   S. D. Brooks, D. Baumgardner, B. Gandrud, J. E. Dye, M. J. Northway, D. W. Fahey, T. P. Bui, O. B. Toon and M. A. Tolbert, *J. Geophys. Res., [Atmos.]*, 2003, **108**, AAC6/1–AAC6/11.
8   M. J. Northway, R. S. Gao, P. J. Popp, J. C. Holecek, D. W. Fahey, K. S. Carslaw, M. A. Tolbert, L. R. Lait, S. Dhaniyala, R. C. Flagan, P. O. Wennberg, M. J. Mahoney, R. L. Herman, G. C. Toon and T. P. Bui, *J. Geophys. Res., [Atmos.]*, 2002, **107**, SOL41/1–SOL41/22.
9   R. S. Disselkamp, S. E. Anthony, A. J. Prenni, T. B. Onasch and M. A. Tolbert, *J. Phys. Chem.*, 1996, **100**, 9127.
10 A. J. Prenni, T. B. Onasch, R. T. Tisdale, R. L. Siefert and M. A. Tolbert, *J. Geophys. Res., [Atmos.]*, 1998, **103**, 28439.
11 A. K. Bertram and J. J. Sloan, *J. Geophys. Res., [Atmos.]*, 1998, **103**, 3553.
12 A. K. Bertram, D. B. Dickens and J. J. Sloan, *J. Geophys. Res., [Atmos.]*, 2000, **105**, 9283.
13 D. Salcedo, L. T. Molina and M. J. Molina, *Geophys. Res. Lett.*, 2000, **27**, 193.
14 D. Salcedo, L. T. Molina and M. J. Molina, *J. Phys. Chem. A*, 2001, **105**, 1433.
15 D. A. Knopf, T. Koop, B. P. Luo, U. G. Weers and T. Peter, *Atmos. Chem. Phys.*, 2002, **2**, 207.
16 O. Stetzer, O. Möhler, R. Wagner, S. Benz, H. Saathoff, H. Bunz and O. Indris, *Atmos. Chem. Phys.*, 2006, **6**, 3023.
17 A. K. Bertram and J. J. Sloan, *J. Geophys. Res., [Atmos.]*, 1998, **103**, 13261.
18 O. Möhler, H. Bunz and O. Stetzer, *Atmos. Chem. Phys.*, 2006, **6**, 3035.
19 H. Tizek, E. Knözinger and H. Grothe, *Phys. Chem. Chem. Phys.*, 2002, **4**, 5128.
20 G. Ritzhaupt and J. P. Devlin, *J. Phys. Chem.*, 1991, **95**, 90.
21 B. G. Koehler, A. M. Middlebrook and M. A. Tolbert, *J. Geophys. Res., [Atmos.]*, 1992, **97**, 8065.
22 N. Barton, B. Rowland and J. P. Devlin, *J. Phys. Chem.*, 1993, **97**, 5848.
23 O. B. Toon, M. A. Tolbert, B. G. Koehler, A. M. Middlebrook and J. Jordan, *J. Geophys. Res., [Atmos.]*, 1994, **99**, 25631.
24 T. G. Koch, N. S. Holmes, T. B. Roddis and J. R. Sodeau, *J. Chem. Soc., Faraday Trans.*, 1996, **92**(23), 4787.
25 R. F. Niedziela, M. L. Norman, R. E. Miller and D. R. Worsnop, *Geophys. Res. Lett.*, 1998, **25**, 4477.
26 R. T. Tisdale, A. J. Prenni, L. T. Iraci, M. A. Tolbert and O. B. Toon, *Geophys. Res. Lett.*, 1999, **26**, 707.
27 N. Lebrun, F. Mahe, J. Lamiot, M. Foulon and J. C. Petit, *Acta Crystallogr., Sect. C*, 2001, **57**, 1129.
28 N. Lebrun, F. Mahe, J. Lamiot, M. Foulon, J. C. Petit and D. Prevost, *Acta Crystallogr., Sect. B*, 2001, **57**, 27.
29 H. Grothe, C. E. Lund Myhre and H. Tizek, *Vib. Spectrosc.*, 2004, **34**, 55.
30 H. Tizek, E. Knözinger and H. Grothe, *Phys. Chem. Chem. Phys.*, 2004, **6**, 972.
31 H. Grothe, C. E. Lund Myhre and C. J. Nielsen, *J. Phys. Chem. A*, 2006, **110**, 171.
32 H. Grothe, H. Tizek, D. Waller and D. J Stokes, *Phys. Chem. Chem. Phys.*, 2006, **8**, 2232.
33 R. Escribano, D. Fernández-Torre, V. J. Herrero, B. Martín-Llorente, B. Maté, I. K. Ortega and H. Grothe, *Vib. Spectrosc.*, 2007, **43**, 254.
34 R. Wagner, O. Möhler, H. Saathoff, O. Stetzer and U. Schurath, *J. Phys. Chem. A*, 2005, **109**, 2572.
35 B. Mate, I. K. Ortega, M. A. Moreno, V. J. Herrero and R. Escribano, *J. Phys. Chem. B*, 2006, **110**, 7396.
36 D. J. Stokes, J.-Y. Mugnier and C. J. Clarke, *J. Microsc.*, 2004, **213**, 198.
37 M. I. Mishchenko and L. D. Travis, *J. Quant. Spectrosc. Radiat. Transfer*, 1998, **60**, 309.
38 R. F. Niedziela, R. E. Miller and D. R. Worsnop, *J. Phys. Chem. A*, 1998, **102**, 6477.
39 M. Avrami, *J. Chem. Phys.*, 1939, **7**, 1103.
40 R. T. Tisdale, A. M. Middlebrook, A. J. Prenni and M. A. Tolbert, *J. Phys. Chem. A*, 1997, **101**, 2112.

PAPER

Is atmospheric aerosol an aerosol?—A look at sources and variability

Ruprecht Jaenicke

Received 24th January 2007, Accepted 16th April 2007
First published as an Advance Article on the web 16th July 2007
DOI: 10.1039/b701095h

Countless observations of the atmospheric aerosol reveal large concentration variabilities. Such variability is not characteristic for a colloid. It must have a certain degree of stability and that is not seen in the atmospheric aerosol. The question then is open if the atmospheric aerosol is really a colloid, an aerosol. On the other side, the atmospheric aerosol exhibits typical properties of colloids, which mainly stem from the large surface-to-volume ratio. This discrepancy is dissolved if the residence time of the aerosol particles is taken into account. This results in a more or less dynamic equilibrium. Because the residence time of the atmospheric aerosol is size dependent, sources (and sinks) have size dependent productions (removals). For the first time the paper presents data for selected atmospheric model aerosols, at which size ranges the majority of particles (whether mass or number) are produced (and removed). The results are also pointing toward cloud droplet residues as a supplementary aerosol particle source. In addition, the size dependent variability is asking for a change in measuring strategy and instrumental design. Short residence times require good time resolution. The paper shows that size ranges below .01 and above .1 μm are most susceptible to measurements with low time resolution. Present day measuring methods often lack the necessary time resolution.

Introduction

In the current discussion on the future of the climate, the atmospheric aerosol is the major difficulty alongside the uncertain social-economic behaviour of mankind.[1] For that reason, it is highly relevant to understand and investigate the atmospheric aerosol; however, are scientific studies focusing on the correct area? It is important to consider whether the atmospheric aerosol really is an aerosol, and if the investigations are truly objective or are tailored to current scientific thinking.

In 1929, it was stated for the first time[2] that the atmosphere should be regarded as a colloid, or aerosol, and the term atmospheric aerosol has been used ever since.

However, how is a colloid defined? The word comes from the Greek term for "glue" and was first used by Thomas Graham in 1861 to describe "pseudosolutions" prepared by Selmi. "The term emphasizes their low rate of diffusion and lack of crystallinity. Graham deduced that the low rate of diffusion implied that the particles were fairly large at least 1 nm in diameter in modern terms. On the other hand, the failure of particle sedimentation implied an upper size limit of 1 μm. Graham's

Institut für Physik der Atmosphäre, University, 55099, Mainz, Germany

definition of the range of particle sizes that characterize the colloidal domain is still widely used today".[3,4] Diffusion among polydisperse particles is faster than the diffusion among monodisperse particles, therefore the low rate of diffusion implies that the particles are rather monodisperse. As a result of this consideration, for the atmospheric aerosol to be a colloid its upper size should be limited to 1 μm and it should be monodisperse.

Countless atmospheric measurements confirm the omnipresence of particles in the size range 1 nm to several hundred micrometers, meaning that particles (greater then 1 μm) are present that, according to the definition, should not be there. In addition, the atmospheric aerosol experiences rather rapid coagulation (as a result of thermal diffusion) and pronounced settling to the Earth's surface, often summarized as ageing. Coagulation means that particles smaller than 0.1 μm disappear quite rapidly. Sedimentation means that particles larger than 1 μm disappear quickly as well. This results in a lack of stability of the atmospheric aerosol. The swiftness of settling results from the large size of some particles and the very small vertical air velocities in the atmosphere under standard conditions, apart from the area- and time-limited intense vertical motions in convective situations.

In addition, a colloid gets its special properties not only from the large surface-to-volume ratio, but also from interactions of the particles. While this is also certainly the case in the atmosphere, those interactions are often ignored, for instance in cloud physics and chemistry.

For a cloud droplet concentration of 100 cm^{-3}, the average separation of droplets is about 1 mm, about a factor of 100 larger compared to the droplet size of 10 μm. However, modellers assume that a droplet grows or shrinks independently of its neighbours, only driven by the external humidity field. Observations[5] show that 20% of the droplets are closer to each other than 100 radii, and 10% are closer than 60 radii. So water vapour interaction and chemical interaction is unavoidable. This means that droplets experience chemical reactions and grow or shrink in the humidity field of neighbouring droplets rather than in the macroscopic field.

In atmospheric optics, interactions are occasionally visible, such as an iridescent cloud. Some clouds are developing thin layer colours. Those clouds themselves are thin and contain an almost monodisperse cloud droplet† population. The absence of iridescent clouds under most circumstances indicates a polydisperse (although polydispersity in this case is still closer to monodispersity than in the atmospheric aerosol) droplet population of most clouds.

Multiple scattering, a process scattering radiative rays between cloud droplets and aerosol particles, has been well studied.[6] For chemical reactions, recently proposed methods utilising multiple scattering[7] estimate the path length extensions in clouds by such scattering, and suggest that it could enhance chemical reactions depending on concentration *via* photochemistry.

The arguments for the nature of the atmospheric aerosol are therefore both supported: on the one hand there are observations that curb seeing it as a colloid and on the other hand observations that support such a categorization. Definitely, the rapidly removed particles are replaced quickly, so the ensemble remains rather constant and could act as a colloid for certain properties. Such properties are total mass, total surface, optical properties, and acting as cloud condensation nuclei and ice nuclei.

In the following, the use of term aerosol for the atmospheric aerosol will be explored for the first time, and what the implications of the rapid change (due to coagulation and sedimentation) will be examined. In addition, the parameter "Differential Source Strength" (DSS) or "sink" will be introduced to identify certain particle size ranges where research is needed and adapted measuring methods need to be used.

---

† http://www.atoptics.co.uk/droplets/irid1.htm, 17 January 2007.

General considerations

If aerosol concentrations remain rather constant, a simple equation connects emission P (or removal S) and content M via residence time τ

$$P = \frac{M}{\tau} = S$$

P, the emission, could be expressed as mass of particles per unit volume and time or as mass of particles per globe and time. The content M could be mass per volume (mass concentration) or all suspended particle mass on the globe. τ is the residence time. In general, this simple equation is valid for number and other properties as well. In (dynamic) equilibrium, production P balances removal S.

The residence time of the atmospheric aerosol has been proposed[8] as a function of the particle radius r

$$\frac{1}{\tau} = \frac{1}{K}\left(\frac{r}{R}\right)^2 + \frac{1}{K}\left(\frac{r}{R}\right)^{-2} + \frac{1}{\tau_{wet}}$$

with the constant $K = 1.28 \times 10^8$ s, a normalizing radius $R = 0.3$ μm, and a wet removal time τ_{wet}. The wet removal time is estimated as eight days in the lower troposphere (because of removal by clouds and precipitation) and three weeks in the middle-to-upper troposphere.[9]

The aerosol mass density distribution usually is presented as $dM/d\log r$ in units of g cm^{-3}. The emission density p (in contrast to the total emission P)‡ would then be

$$p = \frac{1}{\tau}\frac{dM}{d\log r}$$

To scale that up for the world, a volume (a "box") has to be assumed. That volume multiplies the surface ("the box area") of the source and a suitable vertical extension. As the mass concentration of the aerosol gets smaller with altitude in the atmosphere, the vertical concentration profile in empirical equations is expressed as a function of scale height[9]

$$m_z = m_0\left\{\exp\left(\frac{-z}{|H_m|}\right) + \left(\frac{m_B}{m_0}\right)^{\frac{H_m}{|H_m|}}\right\}^{\frac{H_m}{|H_m|}} ; H_m \neq 0$$

With z the altitude and m_z the mass concentration at altitude z, m_B a background mass concentration, m_0 the surface mass concentration and H_m the scale height of the aerosol mass concentration. For all practical purposes the scale height could be taken for vertical extension in the above volume calculation. This way, differential as well as accumulated source strengths can be calculated. Differential source strength data have the advantage of clearly pointing to particle size ranges, in which particles are produced or removed.

Aerosol sources

Aerosol sources often are characterized in terms of mass emission. However, this property is only useful for larger particles; smaller particles have a tiny mass that is negligible compared to the total aerosol mass. For small particles, number production rates would be much better (see below).

---

‡ To state it clearly, P is the total emission in mass (per capacity and) per time, p is the differential emission in mass (per capacity and) per time and per incremental size interval. Capacity in that sense could be the earth as a whole or a certain volume of air. The terms in brackets are often ignored, for instance if the total earth in meant.

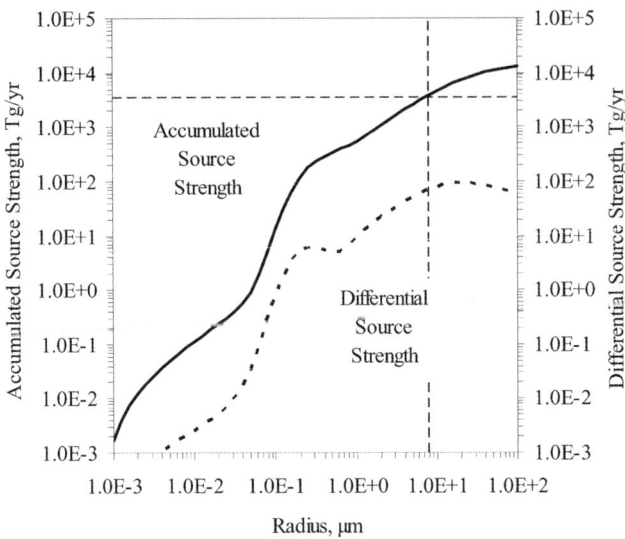

Fig. 1 Accumulated (left ordinate) and differential source strength (right ordinate) of the global oceans. The dashed line (left ordinate) indicates an independent estimate of global source strength.[1] The wet removal time (τ_{wet}) has been set to 8 d. Even if the units of accumulated and differential source strength seem to be equal, the differential source strength is normalized to d log r.

This paper will focus on some natural sources only. That does not mean that anthropogenic sources are less important, on the contrary, they are of importance in selected regions and for certain processes and are the key sources in some regions. However, on a global scale, anthropogenic sources have lost their past prominent position. In 1968, the ratio of man-made sources to natural sources was estimated at 0.22.[10] Today,[11] 0.05 is a more accepted value. This is not because the atmosphere became so much cleaner, it is only because the natural sources were severely underestimated 40 years ago.

Sources by particle mass

In terms of area, the major aerosol source on earth are the oceans, which possess a uniform surface. Because of this uniformity, oceans are a good case for validating the above idea. Particles become airborne through the bubble-bursting process and wind sheer. Model aerosol size distributions,[12] vertical profiles,[9] and source estimates[1] are available. Fig. 1 shows computed accumulated, as well as differential, source strength. Accumulated source strength includes all particles smaller than the given particle size. Differential source strength is an equivalent for source strength density in Tg per yr and per differential logarithmic radius interval. The figure also includes today's published accumulated source strength.[1]

Both accumulated strength estimates agree very nicely: For particles smaller than 8 μm radius the published[1] source strength is 3344 Tg per yr compared to 3503 Tg per yr from Fig. 1. In this paper, a wet residence time (τ_{wet}) of 8 d has been assumed. That value seems to be typical for the lower troposphere. This reflects the high probability of precipitation over the oceans, and thus the efficient removal of particles in a certain size range. In agreement to published estimates,[9] the scale height for this calculation has been set to 900 m. The agreement supports the different approaches and appears to support the values assumed[9] for the vertical profiles. The curves and the integrals are sensitive to the combination and selection of scale height and τ_{wet}, but the combination selected had been published independently.

There is further independent support[13] for the data obtained. If the sea salt vertical aerosol flux is integrated for wind velocities of 6 m s^{-1} (6.89 m s^{-1} is the average wind velocity over the oceans at 10 m for January 1958 to December 2001 using ERA40-reanalysis [1° × 1° resolution][14]), it results in 4 × 10^{-11} g cm^{-2} s^{-1}, a value in very good agreement with the above emission 3.95 × 10^{-11} g cm^{-2} s^{-1}.

Fig. 1 also shows the sensitivity of the source strength estimate to the upper size limit of the particles. Because of the rather short residence time and the large mass of giant particles, the largest value of differential source strength is around 30 µm in order to maintain the observed size distribution. This stresses the point that those particles have to be included in all investigations about the chemistry of the particles. This is not always followed in present day observations. The situation is very different for continents and deserts (see below).

Particles around 1 nm need only to be replenished with very small masses. Because of the shape of the size distribution, the method using the mass of source strength estimate is not well suited to these particles. While in technological processes ball milling producing nanoparticles is a standard process, there is no direct measured evidence that such small sea salt particles are produced (as part of a bulk-to-particle process). However, sea salt particles as small as 0.01 µm do exist.[15] There is also recent laboratory evidence[16] that the Coulomb instability of charged cloud droplets might produce jets, breaking into nanometre droplets. If such a droplet contains dissolved sea salt, then consequently nanometre sea salt particles remain after drying (see discussion about cloud residues below). Particles between 0.1 and 0.3 µm radius need a relatively strong differential source, even if their residence time is the longest.

Global source estimates[1,11] list mineral (soil) dust as a source, but usually model size distributions[12] differentiate between remote continental and desert dust storms. Both emit minerals, among biological particles. It turns out that for mineral particles on a global scale, the non-desert continents (excluding the deserts) are a small source compared to deserts only. Dust storms certainly do not represent the average conditions, but measurements of mineral aerosols have not been published for the complete size distribution under average desert conditions. Recent studies[17] still use the crude measurements made some 30 years ago.

In spring 2006, very comprehensive size distribution measurements were completed in the Moroccan Saharan desert[18] in which great care was taken in sampling the very large particles as well as the small ones. This was achieved by collecting those large particles without any intake loss and performing time consuming "eye ball" optical evaluation. Overall, particles between 0.009 and 130 µm were studied (although the remarks below about time resolution are noteworthy). Supplementary aircraft LIDAR probing revealed a very uniform dust layer up to 3000 m above ground[19,20] (scale height).

Using the measured average size distribution and assuming it suitable for all deserts of the world as an average, and applying the same procedure as above, Fig. 2 results. It shows the extreme sensitivity of the emission strength from the assumed upper size limit. A few micrometres more or less in radius and the accumulated source strength will vary by a factor of 10. Previously,[1] 2150 Tg per yr are assumed for particles smaller than 10 µm radius, while this calculation meets this source strength only for particles smaller than 3.5 µm. Therefore, great care must be taken in deciding what upper particles size is to be included in any observations and estimates. With this sensitivity in mind, these data are in good agreement.

Sources by particle number

Using the particle number in source strength estimates, a certain impression is formed about the effects of nucleation and possible cloud droplet evaporation as a particle source. For that purpose, the oceans are used with all the estimates given above but instead of estimating for the globe and year, computations have been made for number of particles cm^{-2} s^{-1} (Fig. 3).

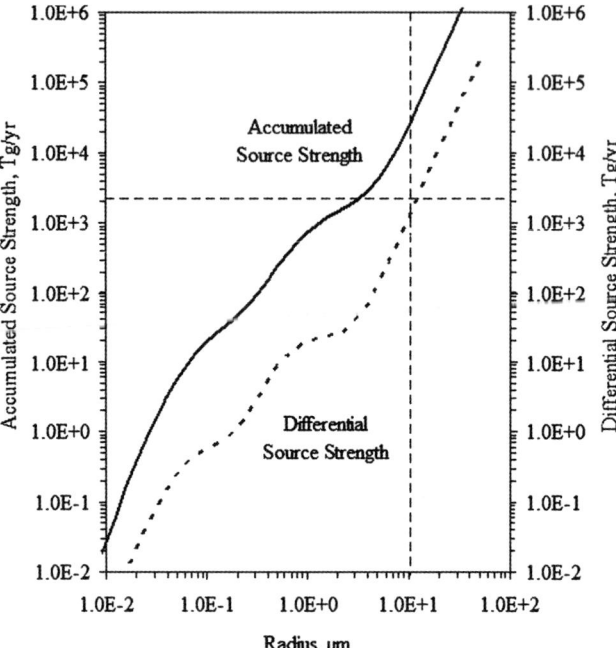

Fig. 2 Accumulated (left ordinate) and differential source strength (right ordinate) for global deserts. The dashed line (left ordinate) indicates an independent estimate.[1] The wet removal time has been set to three weeks, because removal by precipitation in deserts is very much reduced. Even if the units of accumulated and differential source strength seem to be equal, the differential source strength is normalized to d log r.

Clearly the differential production (nucleation) rate in terms of number is greatest for the smallest particles, which are removed rapidly by coagulation with larger ones and need to be replenished quickly in large numbers. Freshly formed nucleation particles are the smallest ones. Values of 0.01 and 10 particles cm^{-2} s^{-1} during nucleation events in a boreal forest have been observed.[21] No values for areas above oceans have been published so far. The values in Fig. 3 are averages for the global oceans to replenish the ocean-produced content of the whole troposphere. The large particles are those grown by coagulation from smaller ones and gaseous condensation on particle surfaces the instant before. Particle growth rates of 1 to 20 nm h^{-1} have been published.[21]

The secondary peak of differential particle production rate around 0.1 μm radius still needs understanding. This source is also indicated in Fig. 1 for particle mass. In that graph, the maximum is shifted to about 0.3 μm radius due to the folding with r^3. The source is definitely not nucleation and follow-up modification. Also, a growth out of smaller particles is not probable.

Input into any globe "box" could come from below—from the Earth's surface—as well as from above—from the "clouds". In addition, it seems that a process is needed that produces particles of a rather narrow size distribution. There has been speculation[22] over what happened to evaporating cloud droplets. Nine out of ten clouds evaporate before the last cloud finally precipitates. The residues could be considered as a "source" for particles, *i.e.*, material (particulates as well as transformed gases) that had been incorporated in forming the cloud. Consequently, the residues of nine evaporated clouds (this is in the order of 6000 Tg per yr on the globe) are left behind as aerosol particles. A rough calculation based on one cloud droplet having a size of ∼10 μm and an impurity content of 10^{-6} gives an evaporated particle residue of ∼0.1 μm in size, which is fine according to Fig. 3.

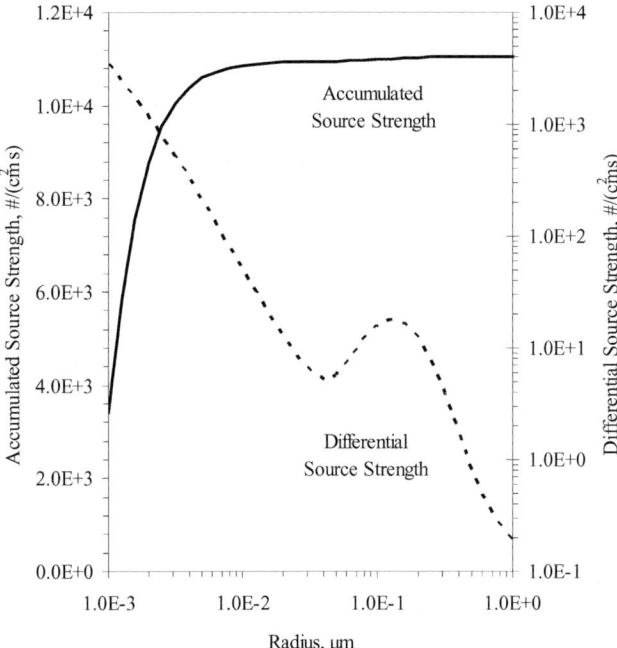

Fig. 3 Accumulated (left ordinate, linear) and differential source strength (right ordinate, logarithmic) in particles cm^{-2} s^{-1} for the global oceans. The wet removal time τ_{wet} has been adjusted to 8 d as above.

It could be argued that clouds are of equal power as sinks and sources. However, it is well known that clouds do chemically transform particulates and incorporated gases,[23] so at least the gaseous transformation is an additional source for particles. In this case, a smaller value of 6000 Tg per yr results. Regardless of the outcome, that figure must be compared to secondary particle production.

Consequence of the dynamic stability

Maintaining the continual presence of the atmospheric aerosol is only possible in a balance between emission and removal, between sources and sinks. Emission and removal (for instance, precipitation) show variations in time and location. As a function of particle size, emission must reach a maximum when removal peaks. Therefore, high variability (in mass or number) should be expected when the residence time is small.

This effect has been observed[24] recently. In 2003, measurements were carried out in Mainz, Germany, the results[25] of which are shown in Fig. 4.

The data of Fig. 4 have been obtained with an Electrical Aerosol Spectrometer.[26] The advantage of this instrument, as compared to other aerosol sizing instruments, is the exceptional good time resolution of below 1 min. In addition, all sizing channels are measured parallel in time, rather than sequentially as in other instruments. Sequential size channel measurements (usually spread over more than 10 min) could tend to fake an aerosol size distribution that might never have existed because of the size dependent variability.

These observations show very clearly that, only in a limited size range, the relative standard deviation (variability) is small. For larger (greater than 0.1 μm) as well as smaller particles (smaller than 0.01 μm), the standard deviation (variability) is larger. The particle size range with the smallest variability does not agree completely with the longest residence time, however, the measurements have been carried out at one

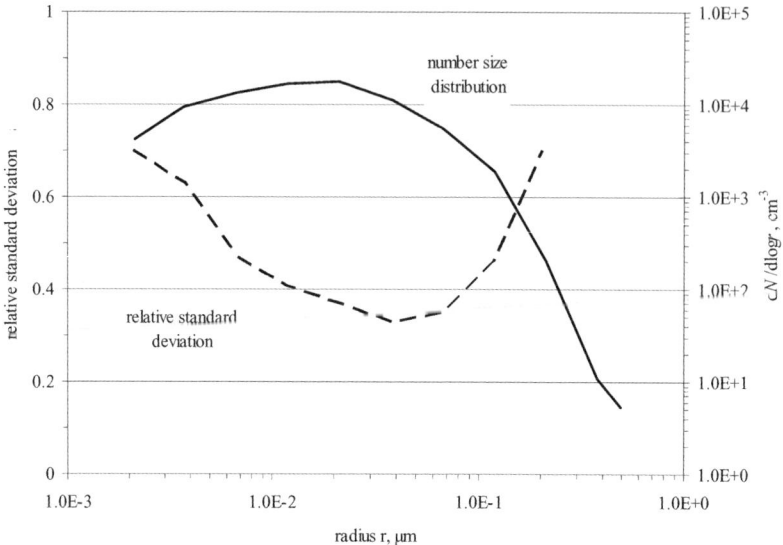

Fig. 4 Measurements of 29th July to 28th August 2003 in Mainz/Germany: Aerosol number distribution (right ordinate) and relative standard deviation (dashed, left ordinate). The time resolution through this sampling time period is 1 min.

location and one time only. More measurements on this topic would provide very useful data.

Unfortunately, to date, very large particles in particular are collected and measured with an extremely poor time resolution (because of their low number concentration), occasionally reaching 24 h.[20] Measurements with such a low time resolution can be tolerated as a first guess, for exploratory measurements only. Observations in this field need instrumental improvements in order to understand better the atmospheric aerosol.

Conclusions

Despite the comparatively short residence time of the atmospheric aerosol, a rather constant presence is observed. In this sense, the atmospheric aerosol can be regarded as exactly that—an aerosol or colloid. A constant concentration can only be maintained if constant differential emissions exist, which are largest for particles with a short residence time, which itself is short for the smallest and the largest particles. However, the discussion about the differential source strength indicates that additional sources besides nucleation and direct injection from the ground must be assumed. The evaporating clouds and their residues are potential candidates. As a consequence of these ever present emissions, a large variability in concentration (and mass) for certain particle size ranges is observed. These fluctuations are masked if investigations are performed with instruments having poor time resolution. Often, such low time resolution is preferred for keeping the amount of data to a manageable level or for newly emerging fields, like the study of primary biological particles or giant particles, as long as no adequate measuring devices are at hand. In such cases, low time resolution, exploratory studies can be tolerated.

With such data sets, however, the atmosphere and the atmospheric aerosol can never be clearly understood, and conclusions for future developments can't be drawn; further studies with acceptable time resolution are needed in addition. This could point to particle size regions where action is needed.

References

1 WMO/UNEP, IPCC Third Assessment Report 'Climate Change 2001'. Contributions of Working Groups I, II, and III to the Third Assessment Report of the Intergovernmental Panel on Climate Change, http://www.ipcc.ch, Accessed 29 November, 2006.
2 A. Schmauss and A. Wigand, *Die Atmosphäre als Kolloid.*, Vieweg, Braunschweig, 1929.
3 B. Wirsing, *What is a Colloid?*, http://www.mpikg.mpg.de/kc/what_is_a_colloid, Accessed 8 November, 2006.
4 D. F. Evans and H. Wennerström, *The Colloidal Domain. Where Physics, Chemistry, and Biology Meet*, Wiley-VCH, New York, 1999.
5 S. Borrmann, H. J. Vössing, E. M. Uhlig and R. Jaenicke, in *Dynamics and Chemistry of Hydrometeors*, ed. R. Jaenicke, Wiley-VCH, Weinheim, 2001, pp. 196–209.
6 D. Antoine and A. Morel, *Appl. Opt.*, 1998, **37**, 2245–2259.
7 O. Funk, PhD thesis, Ruprecht-Karls-Universität, 2000.
8 R. Jaenicke, *Ber. Bunsen-Ges. Phys. Chem.*, 1978, **82**, 1198–1202.
9 R. Jaenicke, in *Aerosol-Cloud-Climate Interactions*, ed. P. V. Hobbs, Academic Press, San Diego, 1993, pp. 1–31.
10 J. T. Peterson and C. E. Junge, in *Man's Impact on the Climate*, ed. W. H. Matthews, W. W. Kellogg and G. D. Robinson, MIT Press, Massachusetts, 1971, pp. 310–320.
11 R. Jaenicke, in *Observed Global Climate*, ed. M. Hantel, Springer, Heidelberg, 2005, pp. 8–17.
12 R. Jaenicke, in *Physical and Chemical Properties of Air*, ed. G. Fischer, Springer Verlag, Heidelberg, 1988, pp. 391–457.
13 C. W. Fairall, K. L. Davidson and G. E. Schacher, *Tellus*, 1983, **35B**, 31–39.
14 H. Wernli, personal communication, Mainz, 2007.
15 A. Mészáros and K. Vissy, *Aerosol Sci.*, 1974, **5**, 101–109.
16 D. Duft, T. Achtzehn, R. Müller, B. A. Huber and T. Leisner, *Nature*, 2003, **421**, 128.
17 K. E. Noll and O. Aluko, *J. Aerosol Sci.*, 2006, **37**, 1797–1808.
18 SAMUM, Saharan Mineral Dust Experiment, http://www.tropos.de/samum/publications/OVERVIEW.ppt#256,1,Folie 1, Accessed 7 December, 2006.
19 A. Petzold, personal communication, Oberpfaffenhofen, 2006.
20 K. Kandler, personal communication, Mainz, 2007.
21 M. Kulmala, *Science*, 2003, **302**, 1000–1001.
22 H. R. Pruppacher and R. Jaenicke, *Atmos. Res.*, 1995, **38**, 283–295.
23 Y. Yin, S. Wurzler, Z. Levin and T. G. Reisin, *J. Geophys. Res., [Atmos.]*, 2002, **107**, 4724–4738.
24 J. Williams, M. de Reus, R. Krejci, H. Fischer and J. Stroem, *Atmos. Chem. Phys.*, 2002, **2**, 133–145.
25 N. Prats Porta, Diploma Thesis, University Mainz, 2004.
26 H. Tammet, A. Mirme and E. Tamm, *Atmos. Res.*, 2002, **62**, 315–324.

PAPER   www.rsc.org/faraday_d | Faraday Discussions

Understanding hygroscopic growth and phase transformation of aerosols using single particle Raman spectroscopy in an electrodynamic balance†

Alex K. Y. Lee, T. Y. Ling and Chak K. Chan*

Received 26th March 2007, Accepted 16th April 2007
First published as an Advance Article on the web 30th July 2007
DOI: 10.1039/b704580h

Hygroscopic growth is one of the most fundamental properties of atmospheric aerosols. By absorbing or evaporating water, an aerosol particle changes its size, morphology, phase, chemical composition and reactivity and other parameters such as its refractive index. These changes affect the fate and the environmental impacts of atmospheric aerosols, including global climate change. The ElectroDynamic Balance (EDB) has been widely accepted as a unique tool for measuring hygroscopic properties and for investigating phase transformation of aerosols *via* single particle levitation. Coupled with Raman spectroscopy, an EDB/Raman system is a powerful tool that can be used to investigate both physical and chemical changes associated with the hygroscopic properties of individually levitated particles under controlled environments. In this paper, we report the use of an EDB/Raman system to investigate (1) contact ion pairs formation in supersaturated magnesium sulfate solutions; (2) phase transformation in ammonium nitrate/ammonium sulfate mixed particles; (3) hygroscopicity of organically coated inorganic aerosols; and (4) heterogeneous reactions altering the hygroscopicity of organic aerosols.

1. Introduction

Atmospheric aerosols have been shown to have profound direct and indirect impacts on the earth's radiation balance, air quality, cloud chemistry and physics and visibility degradation as well as human health.[1–3] Of central importance to these effects is the hygroscopic nature of atmospheric aerosols. By absorbing and evaporating water, atmospheric aerosols change their size and chemical compositions in response to changes in the Relative Humidity (RH) of the surrounding air. To assess the impacts that aerosols have, it is imperative to determine the physical and chemical properties of atmospheric aerosols under various humidities.

The ElectroDynamic Balance (EDB) has been widely accepted as a unique tool for measuring the hygroscopic growth of aerosols, including inorganic salts and organic compounds, as well as their mixtures.[4–8] Since levitated particles do not contact any

Department of Chemical Engineering, Hong Kong University of Science and Technology, Clear Water Bay, Kowloon, Hong Kong. E-mail: keckchan@ust.hk; Fax: (852) 2358-0054; Tel: (852) 2358-7124

† The HTML version of this article has been enhanced with additional colour images.

surfaces inside the EDB chamber, heterogeneous nucleation of the levitated particles can be suppressed, which enables the study of highly supersaturated solutions that would normally undergo rapid crystallization in bulk solutions. This is particularly important because supersaturated solution droplets likely exist in atmospheric environments. The recently-developed scanning EDB allows quick water activity measurements of droplets at different concentrations within an hour. This experimental technique facilitates hygroscopic measurements of semi-volatile organic species[7] and studies of the mass transfer effects of organic coatings on the hygroscopic growth of inorganic salts.[9,10]

Although the hygroscopicity and the phase transformation (deliquescence and crystallization) of levitated particles can be readily examined *via* the balancing voltage measurements using an EDB, some observed hygroscopic phenomena of complex aerosol particles are difficult to explain based on the measurements of particle mass changes alone. For example, Chan *et al.*[11] observed a significant reduction in the evaporation rate of highly concentrated magnesium sulfate ($MgSO_4$) droplets. They attributed this reduction to gel formation inside the droplets but they did not provide additional molecular-level information to support their hypothesis. Choi and Chan[7] investigated the effects of organic components on the hygroscopic growth of inorganic aerosols. They reported that both NaCl–malonic acid and $(NH_4)_2SO_4$–malonic acid mixtures absorb a significant amount of water before their respective Deliquescence RH (DRH), because malonic acid absorbs water reversibly without crystallization. Even though this suggestion was reasonable, the evolution of the phases of malonic acid and the inorganic components in response to the ambient RH was not experimentally verified. Additional information, such as the chemical composition, phase state and possible molecular interactions between solutes and between solutes and water in the particles, is essential in gaining a better understanding of the hygroscopic behavior of complex aerosol particles.

The stationary levitation of a particle under controlled conditions allows for laser spectroscopic analyses. Spontaneous Raman spectroscopy is one of the most common single particle spectroscopic techniques.[12] Some researchers have used single particle Raman spectroscopy to study morphological resonance structures in the inelastic scattering of droplets.[13,14] Other previous studies include mass transfer processes[15,16] and particle phase reactions.[17–19] Since the Raman spectrum of a particle is very sensitive to the phase states (dissolved or solid) and the molecular interactions of solute, it is particularly useful in studying phase transformations and molecular interactions inside levitated particles. For instance, Fung and Tang[20] investigated the solidification of ammonium sulfate and sodium nitrate aqueous droplets. Tang *et al.*[5] studied the phase transition and metastability of various inorganic particles. Recently, our group studied the formation of contact ion pairs in supersaturated droplets.[21–23]

In this paper, we present four typical examples to demonstrate how Raman spectroscopic analysis helps in understanding the hygroscopic behavior of levitated particles. They are: (1) contact ion pairs formation in affecting the hygroscopic growth of $MgSO_4$; (2) phase transformation in ammonium nitrate/ammonium sulfate mixed particles; (3) hygroscopicity of organically coated inorganic aerosols; (4) heterogeneous reactions altering the hygroscopicity of organic aerosols. The motivation for, and relevant experimental details of, each example are presented separately in their corresponding sections.

2. Experiment

2.1 Generation of solution droplets

Depending on its solubility, the solute was first dissolved in water or organic solvent. Inorganic salts and water-soluble organics were dissolved in water while unsaturated fatty acids used in heterogeneous reaction studies were dissolved in ethanol. A small amount of this solution was introduced into a piezoelectric particle generator (Uni-Photon Inc., NY., USA, Model 201). By applying electric pulses to the droplet generator, solution droplets with diameters of 20–40 microns (μm) were generated and charged by passing them through a metal induction plate before they were trapped in the ElectroDynamic Balance (EDB).

2.2 The electrodynamic balance for hygroscopic measurements

The principle of the EDB has been well documented[24] and therefore is not described in detail here. In brief, a charged solid particle or solution droplet is trapped at the null point of the cell with proper adjustments of the AC and DC electric fields surrounding the particle. Assuming that there is no loss of charge, the mass of the particle is proportional to the applied DC voltage. Hence, the relative mass change of the particle due to any physical (*e.g.*, evaporation or condensation of water in response to changes in the ambient RH) or chemical changes is determined by recording the DC voltage required to balance the weight of the particle.

When a droplet is equilibrated with the surrounding environment inside the EBD, the water activity (a_w) of the droplet is related to the RH as follows: $a_w = P_w/P_w^{Sat} = RH/100$, where P_w is the partial pressure of water in the ambient environment and P_w^{Sat} is the saturation water vapor pressure at the ambient temperature. The Kelvin effect to correct for vapor pressure dependence from the curvature of a droplet can be ignored, since the particles are larger than 10 microns in diameter. In our experiments, the molar Water-to-Solute Ratio (WSR) or the mass fraction of solute (*mfs*) of the droplets was changed by altering the ambient RH, and both were determined by the DC balancing voltage measurements. The experimental results can be also presented in the form of mass ratios (m/m_0) instead of *mfs*, as typically used in hygroscopic measurements in an EDB (*e.g.*, Chan *et al.*[25]) when the composition of the particles is highly uncertain and/or no bulk data can be used for identification of the reference state.[26] The RH inside the EDB was adjusted by mixing a stream of saturated air and another of dry air at controlled flow rates. The RH was determined by a dew-point hygrometer (EG&G DewPrime model 2000). The overall experimental error of the mass ratios obtained from the hygroscopic measurements in our experiments was within 1% for droplets, and the error in the determination of RH was estimated to be ±1% at RH = 40–80%. The air flow for controlling RH in the EDB was momentarily stopped when the balancing voltage was measured.

2.3 Raman spectroscopy of single levitated particles

We used a Raman spectroscopic system similar to that used in Zhang and Chan[21] (Fig. 1). It consisted of a 5 W argon ion laser (Coherent I90-5) and a 0.5 m monochromator (Acton SpectraPro 500) attached to a CCD (Princeton Instrument, TE/CCD-1100PFUV or Andor Technology, DV420-OE), which was integrated with the EDB system. The 514.5 nm line of an argon ion laser with output power between 25 and 50 mW was used as the source of excitation. A pair of lenses, which matched the *f*/7 optics of the monochromator, was used to focus the 90° scattering of the levitated droplet in the EDB onto the slit of the monochromator. A 514.5 nm Raman notch filter was placed between the two lenses to remove the strong Rayleigh scattering. A 300 g mm^{-1} grating of the monochromator was selected for our studies of organic coatings and heterogeneous reactions. Since our studies of the phase transformation and the formation of contact ion pairs required high spectral resolution for monitoring the changes in peak positions and full width at half height (fwhh), a 1200 g mm^{-1} grating was used in those studies. The resolution of the spectra obtained was about 6 cm^{-1} and 2.3 cm^{-1} for 300 g mm^{-1} and 1200 g mm^{-1}

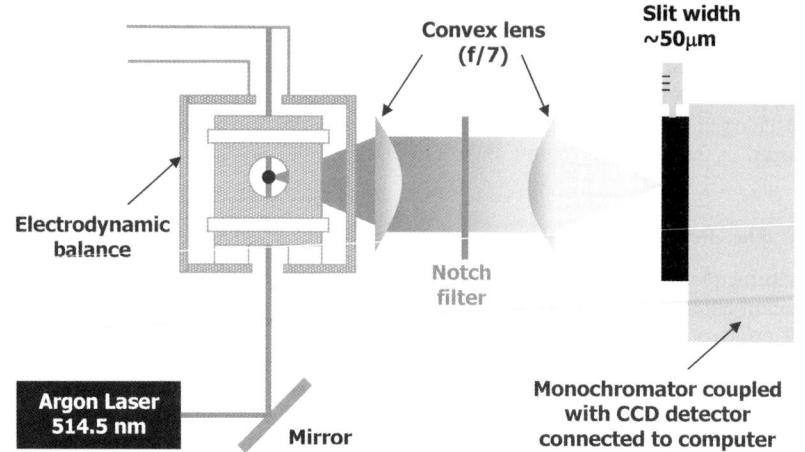

Fig. 1　A schematic diagram of the EDB/Raman system.

grating, respectively. All measurements were made at ambient temperatures of 22–24 °C.

3. Results and discussion

3.1 Contact ion pairs formation in affecting the hygroscopic growth of magnesium sulfate

Atmospheric aerosols consist of a large amount of inorganic ions such as sulfate and nitrate. Most of them are hydrophilic in nature and therefore they play a significant role in affecting the hygroscopicity of atmospheric aerosols. Although the measurements of water absorption using the EDB and other experimental approaches are useful in understanding the hygroscopicity of atmospheric aerosols, they do not provide information on the chemical interactions in the aqueous solutions, which are the molecular basis for water absorption. The chemical interactions between ions and between water and ions in concentrated aqueous aerosols, especially near efflorescence, are expected to be very different from those in diluted aqueous aerosols. For example, Chan et al.[11] reported that $MgSO_4$ droplets exhibit a significant reduction in the evaporation rate after a transition at about 1.2 hr (Fig. 2), suggesting that mass transfer limitations are important at high concentrations. They proposed that the $MgSO_4$ solution forms a gel at high concentrations. Although they did not study the molecular structures of the concentrated $MgSO_4$ solution, it is apparent that the concentrated $MgSO_4$ solution has different structural characteristics from those of diluted solutions.

Since levitated particles do not contact any surfaces inside the EDB chamber, the heterogeneous nucleation of the levitated particles can be suppressed, which allows studies of supersaturated aqueous droplets.[4–8] More importantly, Raman spectroscopy is very sensitive to the chemical interactions between ions/molecules. Monitoring both Raman peak positions and fwhh are particularly useful in determining the changes in chemical interactions inside aqueous droplets. These spectral features make the EDB/Raman system as an ideal tool in relating hygroscopicity to chemical interactions in both the diluted and supersaturated aqueous droplets. In this section, we describe the use of the EDB/Raman system to investigate the relationship between the hygroscopicity of, and the chemical interactions in, supersaturated $MgSO_4$ aqueous droplets. Our findings are important in understanding the peculiar evaporation characteristics of $MgSO_4$ droplets as observed by Chan et al.[11] The

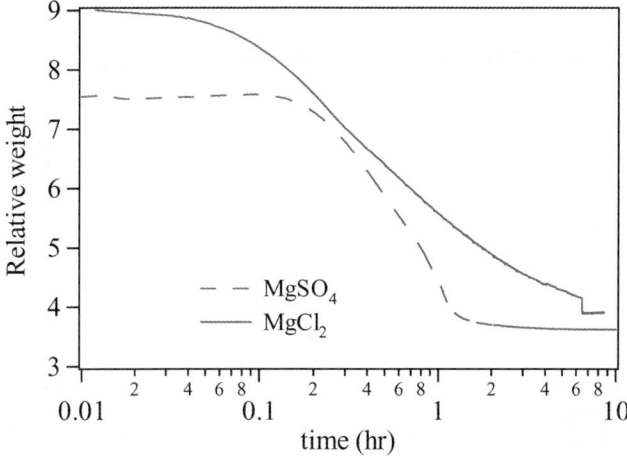

Fig. 2 Evaporation of water from supersaturated MgSO$_4$ and MgCl$_2$ droplets.

details of the experimental systems and working procedures were reported by Zhang and Chan.[21]

Fig. 3a and 3b show the Raman spectra of levitated droplets of MgSO$_4$ and (NH$_4$)$_2$SO$_4$ solutions, respectively, at various molar WSR. In diluted droplets, the ν_1-SO$_4^{2-}$ band of the (NH$_4$)$_2$SO$_4$ solution droplet at 981 cm^{-1} is symmetric (Fig. 3b), which is consistent with results in the literature for bulk solutions of similar concentrations.[27] In contrast, the ν_1-SO$_4^{2-}$ band of MgSO$_4$ solutions is asymmetric with a shoulder at about 995 cm^{-1} (Fig. 3a). This observation was also made for the spectrum of the bulk solution, as reported in the literature.[28] Obviously, the most distinct differences between the spectra of the two species are the changes in the peak position and the fwhh of the ν_1-SO$_4^{2-}$ band under highly supersaturated concentrations.

Fig. 4 shows the peak position and the fwhh of MgSO$_4$ and (NH$_4$)$_2$SO$_4$ particles as a function of the WSR. The strongest band of the MgSO$_4$ droplet, including the SO$_4^{2-}$ ion-related bands, had an overall blue-shift from 983 to 1007 cm^{-1} and an increase in its fwhh from 12 to 54 cm^{-1} when the WSR decreased from 17.29 to 1.54 (and RH decreased from 88.3% to 8.4%), which can be roughly divided into two steps. From the ratio of 17 to 8, the peak shifting and the changes in fwhh were not obvious even though the solution with a WSR smaller than 15.6 is already supersaturated. The spectral response is, however, a very sensitive function of the WSR between 8 and 1.54. The ν_1-SO$_4^{2-}$ band continued its blue-shift to 990 cm^{-1} and the fwhh was about 19 cm^{-1} when the ratio reached 5. As the ratio decreased further, both the peak position and the fwhh underwent abrupt changes. Finally, the ν_1-SO$_4^{2-}$ band appears at 1007 cm^{-1} as a rotund envelope at the WSR equal to 1.54. In contrast, the ν_1-SO$_4^{2-}$ band of (NH$_4$)$_2$SO$_4$ had few spectral changes as the WSR varied. The peak position had a little red-shift from 981 to 977 cm^{-1}, and its fwhh increased slightly from 9 to 12 cm^{-1} when the ratio was adjusted from 15.80 to 1.28. This band shifted to 973 cm^{-1} with an abrupt increase in intensity when the crystallization occurred to form an anhydrous particle. The fwhh also suddenly decreased to 8 cm^{-1}.

Raman spectra are sensitive not only to the strong interaction between adjoining atoms in a covalent bond but also to the weak intermolecular forces, including hydrogen bonding and the van der Waals force. Although there has been no agreement on the relationship between the contact ion pairs and the characteristics of the ν_1-SO$_4^{2-}$ band in diluted solutions, many researchers have used the shoulder peak at 995 cm^{-1} as an indicator of the presence of contact ion pairs in quantitative Raman studies.[27–31] In our experiments, the lowest WSR attained in the

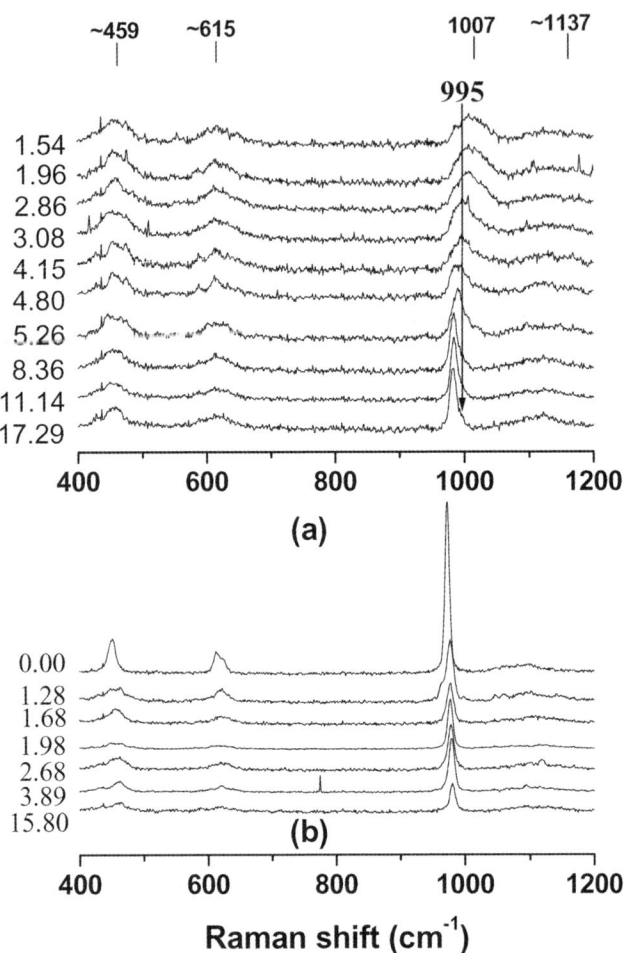

Fig. 3 The Raman spectra of droplets (a) MgSO₄ and (b) (NH₄)SO₄ at different WSR.

supersaturated droplets of MgSO$_4$ and (NH$_4$)$_2$SO$_4$ were 1.54 and 1.38, respectively. Under such conditions, the direct interaction between the ions is strong. The strong dependence of the peak position and the fwhh on the WSR in MgSO$_4$ solutions shown in Fig. 3 and 4 is an indication of the formation of contact ion pairs between Mg^{2+} ions and SO$_4^{2-}$ ions. Furthermore, the increase of the intensity of the high-frequency wing of the ν_1-SO$_4^{2-}$ band as the WSR decreases also indicates the formation of contact ion pairs. The broadening of the band results from the distribution of contact ion pairs with different structures. Some proposed structural units of contact pairs of SO$_4^{2-}$ and Mg^{2+} to form monodentate, bidentate, and bridge-bidentate complexes have been proposed by Zhang and Chan.[21] It is important to note that no spectral feature indicating the formation of contact ion pairs was observed when the WSR was higher than 8, at which point the solvent-separated ion pairs are the dominant components in the aqueous droplets.

On the basis of a comparison of the evolutions of the Raman spectra of MgSO$_4$ and (NH$_4$)$_2$SO$_4$ solutions, especially when the WSR is below 6, we believe that the shoulder usually appearing at 995 cm^{-1} in the diluted MgSO$_4$ solution is related to the formation of contact ion pairs. Our findings clearly show that the molecular interactions between ions and between water and ions in supersaturated aqueous droplets can be very different from those in diluted aqueous droplets. The formation

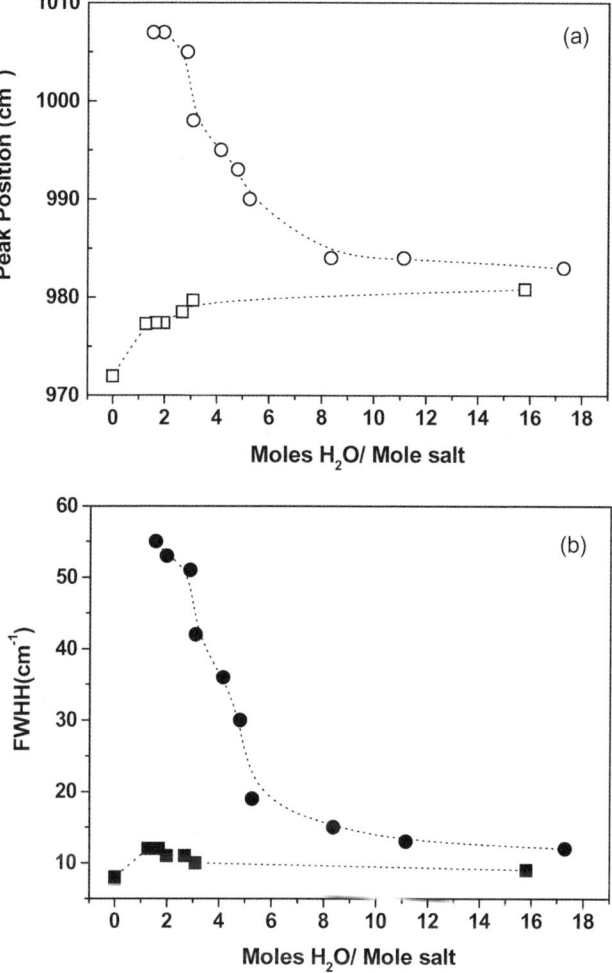

Fig. 4 The dependence of (a) the peak position and (b) fwhh of the ν_1-SO_4^{2-} band on the WSR. Open circles: peak positions of $MgSO_4$ spectra; solid circles: fwhh of $MgSO_4$ spectra; open squares: peak positions of $(NH_4)_2SO_4$ spectra; solid squares: fwhh of $(NH_4)SO_4$ spectra.

of a chain structure of the contact ion pairs can help us to explain the mass transfer limitation in the evaporation of concentrated $MgSO_4$ droplets observed by Chan et al.[11] Recently, we also used the same experimental approach in investigating the relationships between hygroscopic behaviors and the formation of contact ion pairs in different inorganic systems such as $MgNO_3$ and other metal sulfates.[22,23]

3.2 Phase transformation of ammonium nitrate/ammonium sulfate mixed particles

In hygroscopic and phase transformation measurements of aerosols, an EDB or a Tandem Differential Mobility Analyzer (TDMA) measures the overall particle mass/size changes only. Combining the hygroscopic and spectroscopic measurements provides further information on the physical state of each individual substance, and possibly on the chemical interactions present within the system during the phase transformation of an internal mixture of two or more solutes. An example is the use of TDMA and Fourier Transform InfraRed spectroscopy (FTIR) to study the phase transformation and hygroscopic growth of particles containing mixtures of humic acids and $(NH_4)_2SO_4$.[32] $(NH_4)_2SO_4$ is a deliquescent salt while humic acids exhibit continuous water uptake. Hence, at low RH, the mixed particles partially crystallize but retain water because of the presence of humic acids. Further evaporation of water takes place with decreasing RH after crystallization of $(NH_4)_2SO_4$. Based on the shifting of the NH_4^+ mode in the FTIR spectra as a result of the phase transformation, Badger et al.[32] measured the RH at which phase changes of $(NH_4)_2SO_4$ took place. In addition, they proposed a complex salt formation for the internally mixed aerosol based on the FTIR spectra. However, their hygroscopic and spectroscopic measurements were obtained in separate TDMA and FTIR aerosol flow tube experimental setups, respectively. In our study, we used the EDB/Raman system to measure both the particle mass change and Raman spectrum simultaneously as a function of RH.

The phase transformation and hygroscopic growth of mixed NH_4NO_3–$(NH_4)_2SO_4$ particles (NH_4NO_3 : $(NH_4)_2SO_4$ = 1 : 1 molar ratio) were investigated using the EDB/Raman system. The relative change in the fwhh of the Raman signal of each component (($NH_4)_2SO_4$ and NH_4NO_3) was used for phase (solid or aqueous) identification. $(NH_4)_2SO_4$ and NH_4NO_3 are major inorganic components in atmospheric aerosols and have been the subject of a number of hygroscopic studies (e.g., Chan et al.[25]). The NH_4NO_3–$(NH_4)_2SO_4$ mixture was chosen as a model system in our study because it has been well studied and modeled and the phases of its individual species (SO_4^{2-} and NO_3^- and their mixed salts) at a particular RH can be predicted by aerosol thermodynamic models. In the following, we compare our single particle Raman results with the predictions of the AIM-II model.[33]

Unlike FTIR analysis of $(NH_4)_2SO_4$, which has a sizable NH_4^+ peak shift of ~ 25 cm^{-1} as a result of phase transformation,[32] the shifts in the SO_4^{2-} and NO_3^- Raman peaks are relatively small, within a few wavenumbers only.[21,34] Hence, we did not use the peak shifts for the phase transformation analysis in our study. Because the Raman peaks of the solid phase are sharper, i.e., they have a smaller fwhh, than those of the corresponding aqueous phase, changes in the fwhh were used for phase analysis in the NH_4NO_3–$(NH_4)_2SO_4$ system. Instead of using the strongest peaks (i.e., the SO_4^{2-} peak at ~ 980 cm^{-1} and the NO_3^- peak at ~ 1050 cm^{-1}), we used the relatively weaker signals in the lower Raman shift region, which indicate more significant changes in the fwhh during the phase transformation. Fig. 5 shows the Raman spectra of the NH_4NO_3–$(NH_4)_2SO_4$ mixed particles in different phases. Peaks at ~ 450 cm^{-1} and ~ 620 cm^{-1} are signals from SO_4^{2-} and the peak at ~ 720 cm^{-1} is from NO_3^-. The apparent overlapping of two peaks at ~ 620 cm^{-1} would complicate our fwhh analysis. We thus used the SO_4^{2-} signal at ~ 450 cm^{-1} in our analysis. The changes in fwhh upon phase transformations are discussed below.

Fig. 6a shows the AIM-II predictions and the measurements of the hygroscopic growth of NH_4NO_3 : $(NH_4)_2SO_4$ = 1 : 1 (molar ratio) in terms of the mass ratio (relative to the dry mass of the particle) as a function of RH in both the evaporation and growth modes. The deliquescence curve was obtained using the normal mode of AIM-II, while the evaporation curve at supersaturation was obtained by the disabling solid formation in the model. AIM-II predicts DRH_{start} and $DRH_{complete}$ of 67% and 76%, respectively. DRH_{start} and $DRH_{complete}$ are the Deliquescence RH (DRH) at which water absorption starts to take place and at which deliquescence is completed, respectively. During evaporation, because of the lack of a foreign nucleation surface, the droplet was supersaturated before it crystallized at the crystallization RH (CRH = $\sim 40\%$). Solid inclusion was not expected prior to crystallization, as confirmed by the regular patterns observed in the elastic Mie scattering of the droplet (Fig. 6b).[35] In the growth measurements, the particles started absorbing water at RH = 67% and became completely deliquescent at RH = 79%. Complete deliquescence was confirmed by the observation of a regular Mie scattering pattern and the overlapping of the hygroscopic growth data with the evaporation data. It should be noted that the RH was changed in steps and the $DRH_{complete}$ falls between 76% (the last data point measured before deliquescence

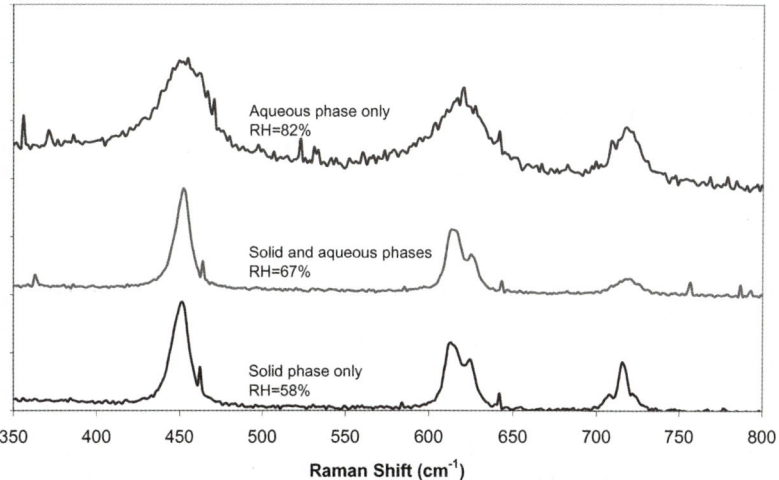

Fig. 5 Raman spectra of a NH_4NO_3–$(NH_4)_2SO_4$ (1 : 1 molar ratio) mixed particle at different RH. Spectra are normalized with the SO_4^{2-} peak at ~450 cm^{-1}. Decreasing intensity of the nitrate signal from RH = 58% to RH = 67% indicates the dissolution of NO_3^- while SO_4^{2-} remains in the solid phase. The spectrum at RH = 82%, when both species have completely dissolved, is scaled up for clarity.

was complete) and 79% (the first data point measured after deliquescence was completed). The results are generally consistent with the AIM-II predictions. Between DRH$_{start}$ and DRH$_{complete}$, AIM-II predicts the presence of a mixed phase particle in which both the solid and aqueous phases of $(NH_4)_2SO_4$ but only the aqueous phase of NH_4NO_3 are present. The amount of solid $(NH_4)_2SO_4$ decreased due to its dissolution as RH increased. We observed distorted Mie scattering patterns (Fig. 6c), which indicate the presence of solid inclusions in the particle during deliquescence.[35]

Fig. 7a and 7b show the fwhh of the SO_4^{2-} peak at ~450 cm^{-1} and the NO_3^- peak at ~720 cm^{-1} as a function of RH, respectively. During evaporation, the average fwhh of the evaporating droplet SO_4^{2-} peak ranged from 24 to 33 cm^{-1} but there was no obvious change in the trend before crystallization took place (Fig. 7a).

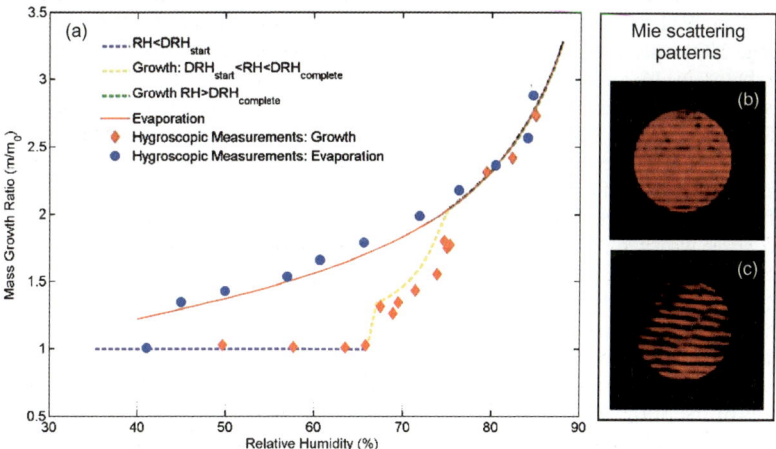

Fig. 6 (a) Hygroscopicity of NH_4NO_3–$(NH_4)_2SO_4$ (1 : 1 molar ratio) mixed particle predicted by AIM-II. The Mie scattering pattern of a particle (b) after complete deliquescence (RH = 85%) and (c) before complete deliquescence (RH = 75%).

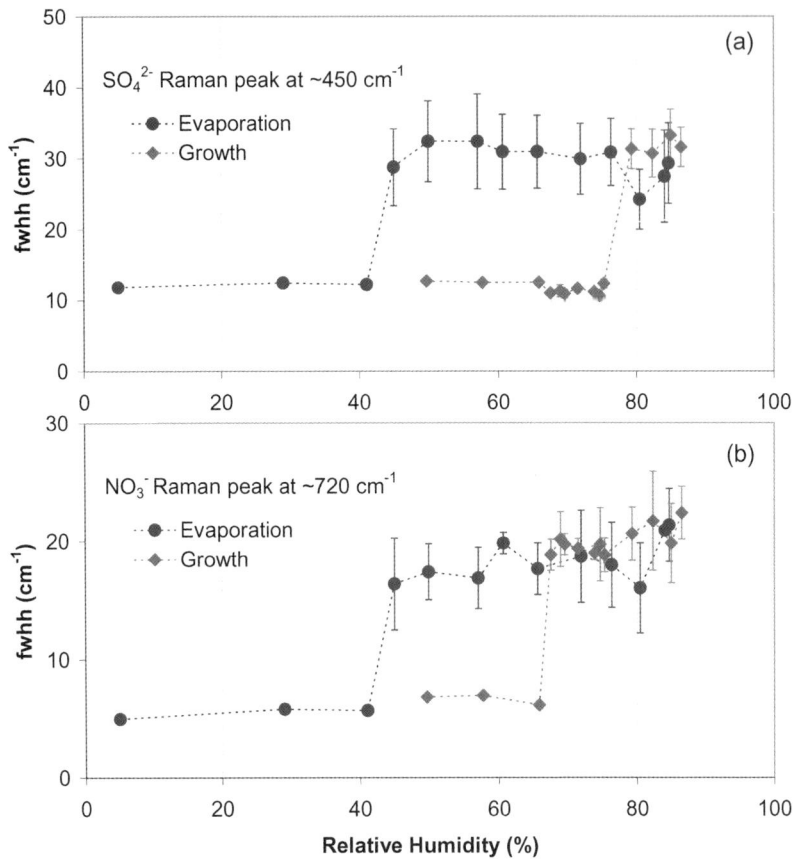

Fig. 7 Dependence of fwhh of (a) a SO_4^{2-} peak at 450 cm^{-1} on RH; (b) of a NO_3^- peak at 720 cm^{-1} on RH.

When the RH was further decreased to the CRH of 40%, the average fwhh of the SO_4^{2-} peak dropped to 12 cm^{-1}. The fwhh remained roughly constant at 12 cm^{-1} at lower RH of 29% and 5%. Hygroscopic measurements show no change in the water content at RH below the CRH of the particles. Similar observations were obtained for the NO_3^- peak at ~720 cm^{-1} (Fig. 7b). The average fwhh in the droplet phase was between 16 and 21 cm^{-1}, which decreased to between 5 and 6 cm^{-1} upon crystallization. We conclude that both NH_4NO_3 and $(NH_4)_2SO_4$ are in a solid state at RH below 40%. The data for particles at a droplet state show a larger standard deviation in repeated measurements because of their relatively weak signals when compared to those for solids.

The state of a particle undergoing deliquescence growth can be generally classified into three different regimes, according to the range of equilibrium RH: RH < DRH_{start}, DRH_{start} < RH < $DRH_{complete}$ and RH > $DRH_{complete}$. At RH < DRH_{start}, the particle remained in a solid form. In Fig. 7a and 7b, the averages of the fwhhs of the SO_4^{2-} peak and NO_3^- peak are about 12 cm^{-1} and 6 cm^{-1}, respectively, which are comparable to those of particles crystallized in the evaporation mode. At RH > $DRH_{complete}$, the particle became completely deliquescent. The averages of the fwhhs of the SO_4^{2-} peak and the NO_3^- peak are 32 cm^{-1} and 20 cm^{-1}, respectively, which are comparable to those for droplets undergoing evaporation.

In our experiments, the transformation from the solid particle to a completely deliquesced droplet took place when the RH increased from DRH_{start} to $DRH_{complete}$. The fwhh of the NO_3^- and SO_4^{2-} peaks responded differently to the

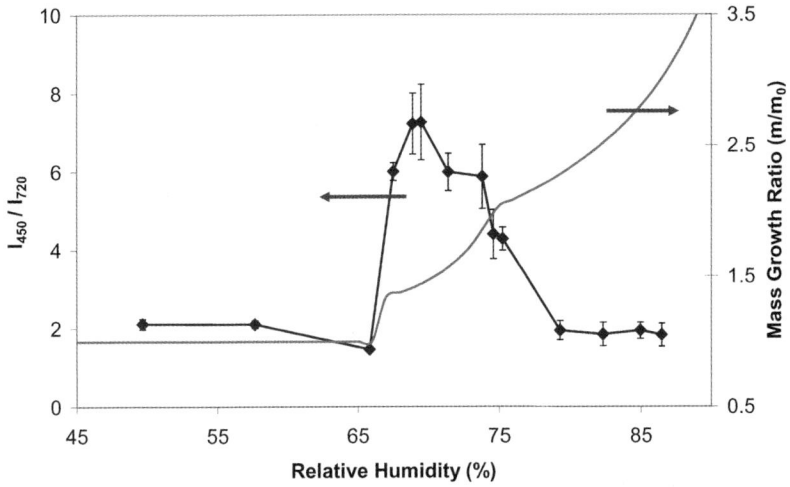

Fig. 8 The relative intensity ratio of the SO_4^{2-} peak at 450 cm^{-1} to the NO_3^- peak at 720 cm^{-1}.

increase in RH. A sharp increase in fwhh indicated the transformation from the solid phase to the aqueous phase, and this occurred at RH = 67% for NO_3^- and at RH = 79% for SO_4^{2-}. These results suggest that during deliquescence, the NH_4NO_3–$(NH_4)_2SO_4$ mixed solid particle started taking up water at RH = 67%, when the nitrate first dissolved but most of the sulfate remained in the solid phase. The sulfate gradually dissolved when RH reached DRH$_{complete}$. During deliquescence, as RH approached DRH$_{complete}$, more SO_4^{2-} was expected to dissolve into the aqueous phase. Fig. 8 shows the relative intensity ratio of the SO_4^{2-} peak at ∼450 cm^{-1} (in both solid and aqueous phases) to the NO_3^- peak (aqueous only) at ∼720 cm^{-1} as a function of RH during deliquescence. As RH increased, the ratio decreased because a larger fraction of SO_4^{2-} was in the aqueous phase, which gave weaker Raman peaks than in the solid phase.

Overall, we have shown that the relative changes in fwhh can be used to examine the phase of individual components when solid and aqueous phases co-exist within a particle, provided that the fwhh of fully deliquescent and completely solidified particles are available for reference. Since the relative change in fwhh during a phase transformation is a physical response that is independent of the chemical composition, we believe that fwhh analysis can be useful in understanding the deliquescence of other aerosol mixtures.

3.3. Hygroscopicity of organically coated inorganic aerosols

Organic compounds have been found on the surface of atmospheric particles as coatings,[36–42] and such coatings can alter the abilities of atmospheric particles to change their phases and sizes in response to changes in the ambient RH. Water-insoluble organic coatings have been found to retard the evaporation and condensation rates of water molecules from planar solution surfaces, water droplets, and aerosol particles.[10,43] However, the effects of water-soluble organic coatings on the aerosol hygroscopicity may be more complicated than those of water-insoluble coatings. Water-soluble organic coatings on the surface of inorganic particles may dissolve into the droplets once the inorganic core deliquesces, consequently altering the hygroscopicity of the particles due to chemical interactions between the dissolved organics and the inorganic solutes.[9] Since Raman spectra are sensitive to the phase state of the solute, they can be used to infer the phase information of a solute and to

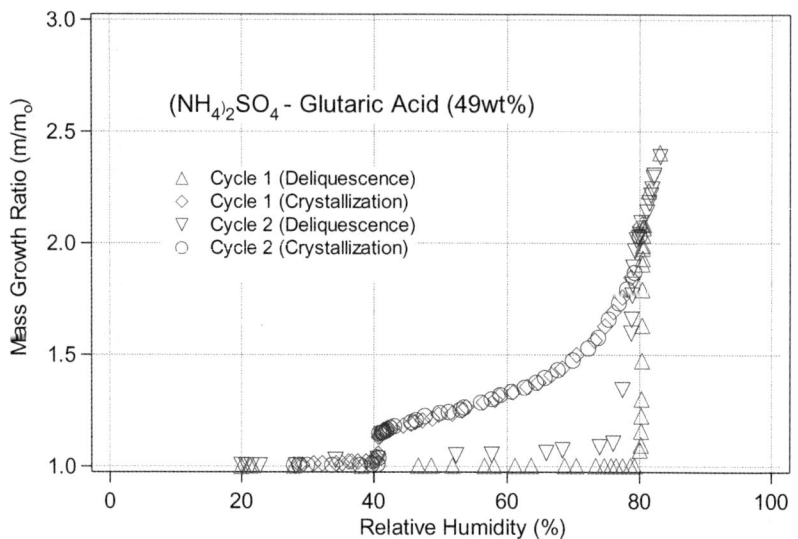

Fig. 9 The hygroscopicity of $(NH_4)_2SO_4$ particles coated with 49 wt% glutaric acid in two deliquescence and crystallization cycles. The deliquescence RH in the two cycles is different but the evaporative behaviors and crystallization RH in the two cycles are very similar.

provide additional information to support the observations in hygroscopic measurements of the coated particles upon repeated deliquescence and crystallization cycles.

In this section, we describe an experimental study in which the EDB/Raman system was used to examine the role of water-soluble organic coatings (glutaric acid) in the hygroscopicity of inorganic salt particles ($(NH_4)_2SO_4$). The hygroscopic measurements of the coated particles were conducted utilizing the scanning EDB method described by Chan et al.[9] Fig. 9 illustrates the hygroscopicity, in terms of the mass growth ratio, of $(NH_4)_2SO_4$ particles coated with 49 wt% glutaric acid in two deliquescence and crystallization cycles. The main difference in the water uptake behavior between the two cycles is that while freshly coated particles started to deliquesce at RH = 80.9%, which is close to the deliquescence RH (DRH) of $(NH_4)_2SO_4$ particles, the particles formed after the first measurement cycle absorbed a small amount of water at about RH = 50% before deliquescence and started to deliquesce at about RH = 74%. The different deliquescence characteristics in the two cycles can be explained by the formation of mixed $(NH_4)_2SO_4$–glutaric acid particles after the deliquescence in the first cycle. The formation of mixed $(NH_4)_2SO_4$–glutaric acid particles was confirmed by the Raman characterization and visual observations of particle morphology in another EDB experiment.

A pure solid $(NH_4)_2SO_4$ particle had a rough surface under white light illumination, as shown by the irregular bright spots in the image (Fig. 10a). After coating with glutaric acid, the 47 wt% coated particle was shaped like a snowflake (Fig. 10b and 10c), suggesting that the coating was formed via condensation/coagulation of small glutaric acid particles. Since the glutaric acid coating may be highly porous or it may have incompletely covered the surface of the $(NH_4)_2SO_4$ particle, water molecules could penetrate the organic coating and reach the $(NH_4)_2SO_4$ directly, making the freshly coated particle deliquesce close to the DRH of $(NH_4)_2SO_4$ particles. Our Raman measurements clearly show that the core region of the freshly coated particle mainly consists of solid $(NH_4)_2SO_4$ with small glutaric acid signals (Fig. 10b). When the position of the laser beam was adjusted to irradiate the coating, many strong Raman hydrocarbon signatures (~ 2900–3100 cm^{-1}) were obtained in the coating region (Fig. 10c), suggesting that the major component of the coating was glutaric acid. It is interesting to note that the Raman spectrum of the glutaric

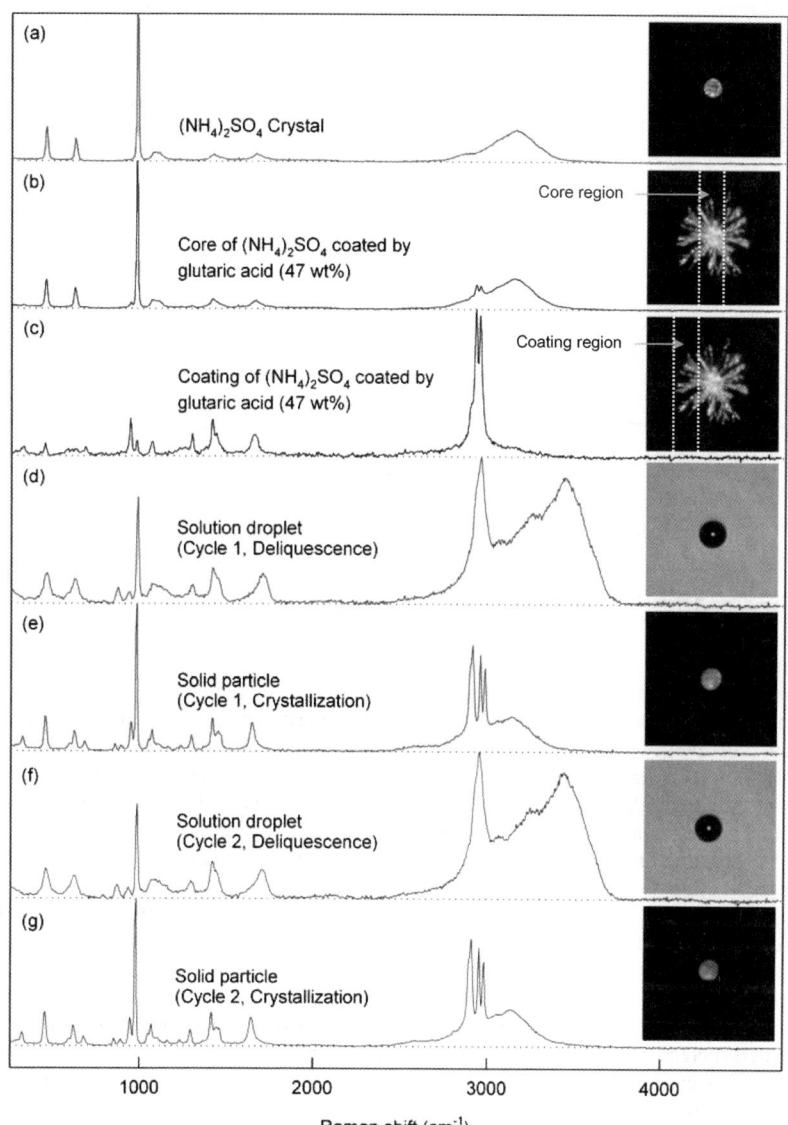

Fig. 10 The Raman spectra of $(NH_4)_2SO_4$ particles coated with 47 wt% glutaric acid in two deliquescence and crystallization cycles: (a) $(NH_4)_2SO_4$ particles; (b) solid $(NH_4)_2SO_4$ particles coated with glutaric acid (center); (c) solid $(NH_4)_2SO_4$ particles coated with glutaric acid (branch/coating); (d) solution droplets in the first deliquescence cycle; (e) solid particles in the first crystallization cycle; (f) solution droplets in the second deliquescence cycle; (g) solution droplets in the second crystallization cycle. The light source was positioned at the opposite side and at an angle of 90° with respect to the microscope to capture images of the solution droplets and the solid particle, respectively.

acid coating, especially between the region of 2900 and 3100 cm^{-1}, differs from that of a solid glutaric acid particle (Fig. 11c). The molecular structure of the coating formed from condensational/coagulational growth of glutaric acid might be different from that of solid glutaric acid particles formed from crystallization of glutaric acid solution droplets.

Fig. 10 depicts the evolution of the Raman spectra and particle morphologies of a $(NH_4)_2SO_4$ particle coated with 47 wt% glutaric acid in two cycles. Once the freshly

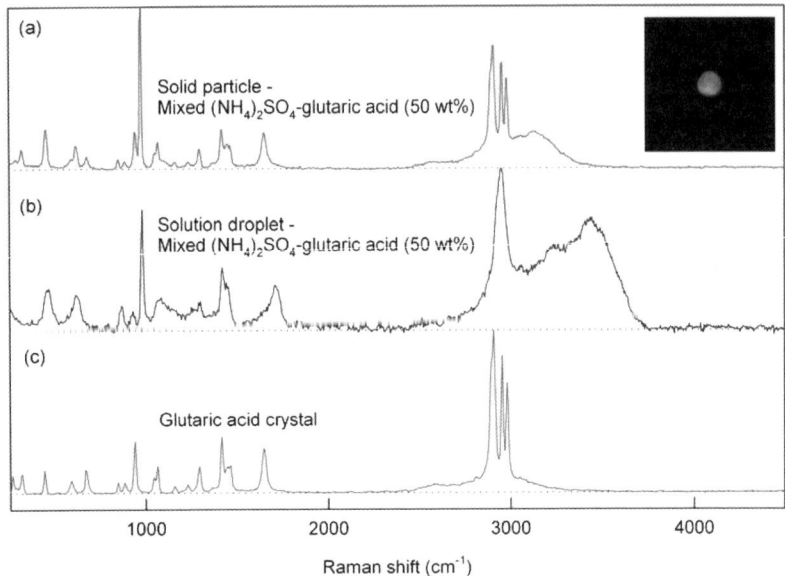

Fig. 11 The Raman spectra of mixed $(NH_4)_2SO_4$-glutaric acid particles (50 wt% glutaric acid): (a) solid particles; (b) solution droplets; and (c) the Raman spectra of solid glutaric acid particles.

coated particle deliquesced, the organic coating dissolved into the aqueous phase to form a droplet with a spherical shape, as shown by the black circle in the image (Fig. 10d). After crystallization in the first cycle, the solid particle had a spherical-like shape and a rough surface (Fig. 10e), which is totally different from the morphology of the freshly coated particle. The morphologies of the aqueous droplet and solid particle formed in the second cycle (Fig. 10f and 10g) were similar to those formed in the first cycle (Fig. 10d and 10e). More importantly, the Raman spectra of the aqueous droplets (Fig. 10d and 10f) and solid particles (Fig. 10e and 10g) formed after deliquescence of the glutaric acid coated solid $(NH_4)_2SO_4$ particles in the two cycles are almost identical to those of mixed $(NH_4)_2SO_4$–glutaric acid aqueous droplets and solid particles (Fig. 11a and 11b), which were generated by premixed solutions of a similar amount of glutaric acid (50 wt%). It is worth noting that the Raman spectra of freshly coated particles (Fig. 10b and 10c) are different from those of the solid particles formed after cycles of deliquescence and crystallization (Fig. 10e and 10g) and from solid mixed $(NH_4)_2SO_4$-glutaric acid particles (Fig. 11a).

To sum up, both the Raman measurements and the evolution of the particle morphologies support the suggestion that mixed $(NH_4)_2SO_4$–glutaric acid particles are formed after dissolution of the solid $(NH_4)_2SO_4$ core and the glutaric acid coating during deliquescence. This also explains the change in the DRH value from 80.9%RH to 74%RH in the second cycle and the observed similar crystallization characteristics of the particles in the two cycles. These results are consistent with previous studies of hygroscopic properties that used mixed $(NH_4)_2SO_4$–glutaric acid particles generated from a premixed solution of known $(NH_4)_2SO_4$ and glutaric acid concentrations.[7,44,45]

3.4 Heterogeneous reactions altering the hygroscopicity of organic aerosols

Organic compounds are recognized as ubiquitous components of atmospheric aerosols. While the hygroscopic growth of inorganic aerosols has been well studied in the last two decades, that of organic aerosols has been ignored until about a decade ago.[46,47] Non-polar hydrocarbons are hydrophobic and are not expected to absorb water, but oxygenated hydrocarbons, particularly water-soluble organic compounds, are likely to be hygroscopic. Dicarboxylic acids and their mixtures with inorganic salts (e.g., NaCl and $(NH_4)_2SO_4$) have been extensively studied.[2,48] A number of these acids have been found to be as hygroscopic as inorganic aerosols and vigorous modeling efforts have been undertaken to develop thermodynamic models that incorporate both inorganic and organic components with the consideration of interactions between these compounds.[49–51] Recent research has examined the effects of chemical reactions on hygroscopic properties of organic aerosols. In particular, heterogeneous reactions of organic aerosols with atmospheric oxidants, such as ozone OH and NO_3 radicals, are potentially important in enhancing the hygroscopicity and possibly the Cloud Condensation Nuclei (CCN) activity of atmospheric organic aerosols.[1,2] The oxidation products that remain in the particle phase are generally more oxygenated and hydrophilic than are their parent molecules.[52–54] However, our understanding of the heterogeneous reactions of organic aerosols and their atmospheric implications is rather limited.

Although Raman spectroscopy has been applied to a number of EDB or single particle investigations of the physical characteristics of particles, such as their light scattering or evaporation/condensation rates,[15,16,55] the use of Raman spectroscopy as a chemical tool for probing reactions of single particles is very limited.[17–19] Raman spectroscopy is particularly suitable to study organic reactions, since the Raman features of organic species are very rich and can provide information on the changes in the composition of reacting organic aerosols. Another distinct advantage of using the EDB/Raman system is that it allows long duration particle levitation (of the order of days) and thus long exposure of particles to gas phase oxidants or organics at concentrations relevant to atmospheric applications is possible. Furthermore, any changes in the hygroscopic properties of the levitated particles due to chemical reactions can be measured. In this section, we demonstrate the use of our EDB/Raman system to investigate the heterogeneous oxidation of unsaturated fatty acid particles, including oleic acid (C18:1), linoleic acid (C18:2) and linolenic acid (C18:3), with ozone under ambient temperatures (22–24 °C) and dry conditions (RH < 5%) during 20 hr of exposure. These unsaturated fatty acids have similar chemical structures as expected for their degree of unsaturation. The ozone concentration used was in the range of 200–280 ppb. The details of the experimental systems and working procedures were described by Lee and Chan.[56,57]

The Raman spectra of pure and ozone-processed oleic acid particles are shown in Fig. 12a. The spectral changes illustrate that oleic acid ozonolysis results in the consumption of carbon–carbon double bonds (C=C, peaks at $\sim$973, $\sim$1269, $\sim$1655 and $\sim$3008 cm^{-1}) of oleic acid molecules and leads to the formation of peroxidic products (O–O, peak at $\sim$850 cm^{-1}), carbonyl groups (C=O, peak at $\sim$1740 cm^{-1}) and hydroxyl groups (O–H, peak at $\sim$3450 cm^{-1}).[56,58,59] These changes in functional groups are in accordance with the predictions of the Criegee mechanisms[60] as well as the results reported in the recent literature.[61] In the cases of linoleic acid and linolenic acid, the consumption of C=C bonds (peaks at $\sim$973, $\sim$1269 and $\sim$3008 cm^{-1}) and similar product peak formations (O–O, C=O and O–H groups) were also observed, as shown in Fig. 12b and 12c, respectively. The close similarities of the functional group characteristics of the reaction products generated in ozone-processed oleic acid, linoleic acid and linolenic acid particles indicate that both linoleic acid and linolenic acid can also undergo direct ozonolysis within the ozone exposure period.[57]

In addition to the formation of O–O, C=O and O–H groups, other new product peaks were also observed in the Raman spectra of ozone-processed linoleic acid and linolenic acid particles. In Fig. 12b and 12c, the product peaks formed at $\sim$1590 cm^{-1} and $\sim$1640 cm^{-1} are assigned to the symmetric and asymmetric C=C stretching vibrations of the conjugated dienes (C=C–C=C), respectively.[58,59] However, it is important to note that conjugated dienes cannot be produced via the direct ozonolysis of linoleic acid and linolenic acid based on the Criegee

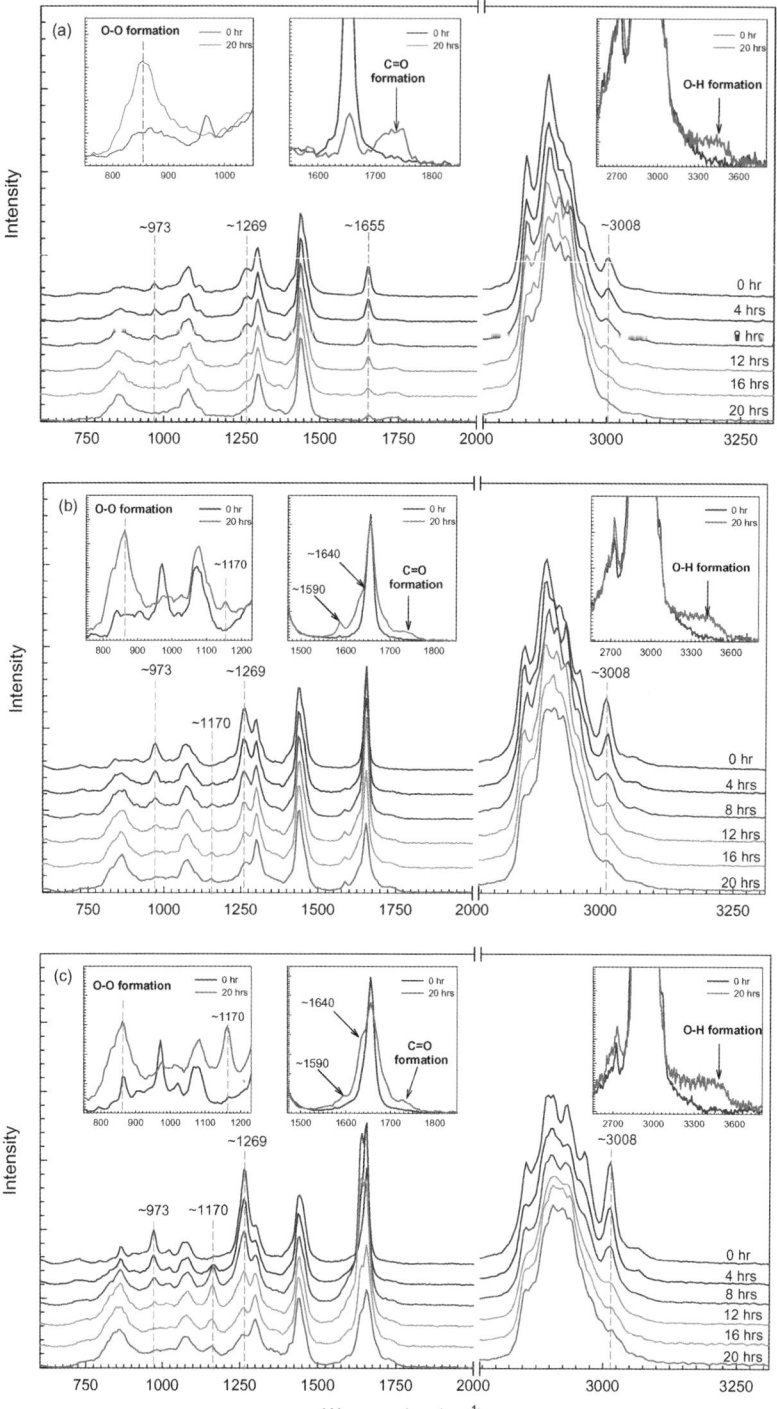

Fig. 12 The Raman spectra of (a) oleic acid, (b) linoleic acid and (c) linolenic acid particles at different ozone exposures. The Raman spectra are normalized to the peak located at 1443 cm^{-1}. The inserts highlight the difference of spectral features between pure and ozone-processed particles.

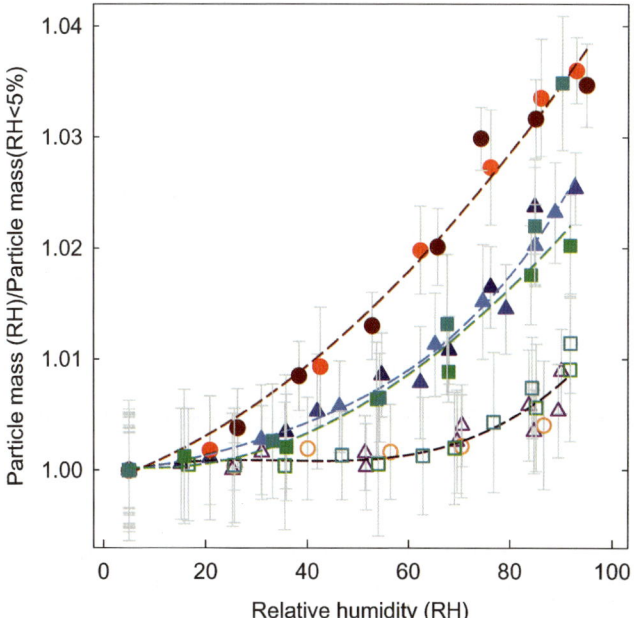

Fig. 13 The hygroscopicity of the pure unsaturated fatty acids (open squares: oleic acid, open triangles: linoleic acid, open circles: linolenic acid) and ozone-processed oleic acid (solid squares), linoleic acid (solid triangles) and linolenic acid (solid circles) particles. The lines included are to guide the eye and have no physical meaning.

mechanism. Instead, we attribute these spectral changes to the autoxidation (or peroxidation) process, which is a typical free radical reaction consisting of chain initiation, propagation, and termination steps.[62,63] Furthermore, a new band that peaks at ~1170 cm^{-1} is additional evidence of the occurrence of the autoxidation process, as this peak has been observed in studies of lipid or oil autoxidation.[64,65] The details of the reaction mechanisms of ozone-induced autoxidation of linoleic acid and linolenic acid particles were described by Lee and Chan,[57] and thus are not reported here. Overall, our Raman measurements show that ozone-induced autoxidation, in addition to direct ozonolysis, is a plausible pathway in the reactions between ozone and linoleic acid and linolenic acid particles.

Fig. 13 shows the mass ratios (the particle mass at a specific RH to the particle mass at RH < 5%) of oleic acid, linoleic acid and linolenic acid particles as a function of RH. The unreacted fatty acid particles are hydrophobic in nature and their mass ratios at RH = 85% only increased by approximately 0.5–1%, which is within experimental uncertainty. After 20 hr of ozone exposure, the mass ratios of ozone-processed oleic acid, linoleic acid and linolenic acid particles at RH = 85% increased by about 2–3%, 2–3% and 3–4%, respectively. The difference in the mass ratios between pure and ozone-processed oleic acid particles became more significant when the RH was higher than 85%. Although the changes in hygroscopic growth were small, our measurements clearly show that the oxidation products that remained in the particles are more hygroscopic than are their hydrophobic parent molecules, thus probably affecting the CCN activity of the particles.

In summary, *in situ* single particle Raman spectroscopic analysis can provide information on the changes in the chemical composition (functional groups) of levitated particles due to heterogeneous processes. Not only can spectroscopic information be used for investigating possible reaction pathways of heterogeneous processes, but it also helps to explain the hygroscopic behaviors of the reacted particles. In this work, we clearly illustrate that the enhancement in the hygroscopic growth of the fatty acid particles is in good agreement with the Raman results that the levitated particles were oxygen-enriched after the heterogeneous oxidative processes.

4. Conclusion

In this paper, we demonstrate the use of Raman spectroscopy as a complementary tool for studying hygroscopic growth and phase transformation of levitated particles in an EDB. Changes in Raman peak positions and full width at half height (fwhh) are useful indicators of the changes in molecular interactions between ions/molecules at high concentrations and phase transformation. An EDB/Raman system is also particularly suitable for investigating the roles of organic compounds in the hygroscopic growth of aerosols, both as coatings on inorganic aerosols and in heterogeneous reactions that altered the functional group composition of organic aerosols. Organic compounds are likely to remain the focus of much atmospheric research in the next decade. The EDB/Raman system is a valuable tool for enhancing our understanding of organic aerosols.

Acknowledgements

This work was supported by the Research Grants Council of the Hong Kong Special Administrative Region, China (Project No. HKUST600303 and HKUST610805).

References

1 Intergovernmental Panel on Climate Change (IPCC), *Climate Change: The Scientific Basis*, Cambridge University Press, UK, 2001.
2 M. Kanakidou, J. H. Seinfeld, S. N. Pandis, I. Barnes, F. J. Dentener, M. C. Facchini, R. Van Dingenen, B. Ervens, A. Nenes, C. J. Nielsen, E. Swietlicki, J. P. Putaud, Y. Balkanski, S. Fuzzi, J. Horth, G. K. Moortgat, R. Winterhalter, C. E. L. Myhre, K. Tsigaridis, E. Vignati, E. G. Stephanou and J. Wilson, *Atmos. Chem. Phys.*, 2005, **5**, 1053.
3 J. H. Seinfeld and S. N. Pandis, *Atmospheric Chemistry and Physics: from Air Pollution to Climate Change*, 2nd edn, John Wiley & Sons, New Jersey, 2006.
4 M. D. Cohen, R. C. Flagan and J. H. Seinfeld, *J. Phys. Chem.*, 1987, **91**, 4563.
5 I. N. Tang, K. H. Fung, D. G. Imre and H. R. Munkelwitz, *Aerosol Sci. Technol.*, 1995, **23**, 443–453.
6 I. N. Tang and K. H. Fung, *J. Chem. Phys.*, 1997, **106**, 1653.
7 M. Y. Choi and C. K. Chan, *Environ. Sci. Technol.*, 2002, **36**, 2422.
8 M. N. Chan and C. K. Chan, *Environ. Sci. Technol.*, 2003, **37**, 5109.
9 M. N. Chan, A. K. Y. Lee and C. K. Chan, *Environ. Sci. Technol.*, 2006, **40**, 6983.
10 M. N. Chan and C. K. Chan, *Atmos. Environ.*, 2007, **41**, 4423.
11 C. K. Chan, R. C. Flagan and J. H. Seinfeld, *J. Am. Ceram. Soc.*, 1998, **82**, 646.
12 E. J. Davis and G. Schweiger, *The Airborne Microparticle: Its Physics, Chemistry, Optics and Transport Phenomena*, Springer Verlag, Heidelberg, 2002.
13 R. E. Preston, T. R. Lettieri and H. G. Semerjian, *Langmuir*, 1985, **1**, 365.
14 R. Thurn and W. Kiefer, *Appl. Opt.*, 1985, **24**, 1515.
15 H. Moritz, R. Vehring and G. Schweiger, *J. Aerosol Sci.*, 1996, **27**, S517.
16 M. Trunk, J. Popp and W. Kiefer, *Chem. Phys. Lett.*, 1998, **284**, 377.
17 C. L. Aardahl and E. J. Davis, *Appl. Spectrosc.*, 1996, **50**, 71.
18 C. Esen, T. Kaiser and G. Schweiger, *Appl. Spectrosc.*, 1996, **50**, 823.
19 J. Musick and J. Popp, *Phys. Chem. Chem. Phys.*, 1999, **1**, 5497.
20 K. H. Fung and I. N. Tang, *Appl. Opt.*, 1988, **27**, 206.
21 Y. H. Zhang and C. K. Chan, *J. Phys. Chem. A*, 2000, **104**, 9191.
22 Y. H. Zhang and C. K. Chan, *J. Phys. Chem. A*, 2002, **106**, 285.
23 Y. H. Zhang, M. Y. Choi and C. K. Chan, *J. Phys. Chem. A*, 2004, **108**, 1712.
24 E. J. Davis, *Aerosol Sci. Technol.*, 1997, **26**, 212.
25 C. K. Chan, R. C. Flagan and J. H. Seinfeld, *Atmos. Environ.*, 1992, **26**, 1661.
26 Z. Ha and C. K. Chan, *Aerosol Sci. Technol.*, 1999, **31**, 154.
27 W. Rudolph, *Ber. Bunsen-Ges. Phys. Chem.*, 1998, **102**, 183.
28 A. R. Davis and B. G. Oliver, *J. Phys. Chem.*, 1973, **77**, 1315.
29 W. Rudolph and G. Irmer, *J. Solution Chem.*, 1994, **23**, 663.
30 C. C. Pye and W. W. Rudolph, *J. Phys. Chem. A*, 1998, **102**, 9933.

31 W. F. Rudolph and C. C. Pye, *J. Phys. Chem. B*, 1998, **102**, 3564.
32 C. L. Badger, I. George, P. T. Griffiths, C. F. Braban, R. A. Cox and J. P. D. Abbatt, *Atmos. Chem. Phys.*, 2006, **6**, 755.
33 S. L. Clegg, P. Brimblecombe and A. S. Wexler, *J. Phys. Chem. A*, 1998, **102**, 2137.
34 J. Musick, J. Popp and W. Kiefer, *J. Raman Spectrosc.*, 2000, **31**, 217.
35 A. P. Olsen, R. C. Flagan and J. A. Kornfield, *Rev. Sci. Instrum.*, 2006, **77**, 073901.
36 R. S. Gill, T. E. Graedel and C. J. Weschler, *Rev. Geophys. Space Phys.*, 1983, **22**, 903.
37 M. Posfai, H. Xu, J. R. Anderson and P. R. Buseck, *Geophys. Res. Lett.*, 1998, **25**, 1907.
38 M. Posfai, R. Simonics, J. Li, P. V. Hobbs and P. R. Buseck, *J. Geophys. Res., [Atmos.]*, 2003, **108**, 8483.
39 W. Seidl, *Atmos. Environ.*, 2000, **34**, 4917.
40 L. M. Russell, S. F. Maria and S. C. B. Myneni, *Geophys. Res. Lett.*, 2002, **29**, DOI: 10.1029/2002GL014874.
41 H. Tervahattu, J. Juhanoja and K. Kupiainen, *J. Geophys. Res., [Atmos.]*, 2002, **107**, DOI: 10.1029/2001JD001403.
42 H. Tervahattu, K. Hartonen, V. M. Kerminen, K. Kupiainen, P. Aarnio, T. Koskentalo, A. F. Tuck and V. Vaida, *J. Geophys. Res., [Atmos.]*, 2002, **107**, DOI: 10.1029/2000JD000282.
43 P. Y. Chuang, *J. Geophys. Res., [Atmos.]*, 2003, **108**, 4282.
44 A. J. Prenni, P. J. DeMott and S. M. Kreidenweis, *Atmos. Environ.*, 2003, **37**, 4243.
45 A. Pant, A. Fok, M. T. Parsons, J. Mak and A. K. Bertram, *Geophys. Res. Lett.*, 2004, **31**, L12111.
46 P. Saxena, L. M. Hildemann, P. H. McMurry and J. H. Seinfeld, *J. Geophys. Res., [Atmos.]*, 1995, **100**, 18755.
47 P. Saxena and L. M. Hildemann, *J. Atmos. Chem.*, 1996, **24**, 57.
48 C. Peng, M. N. Chan and C. K. Chan, *Environ. Sci. Technol.*, 2001, **35**, 4495.
49 D. O. Topping, G. B. McFiggans and H. Coe, *Atmos. Chem. Phys.*, 2005, **5**, 1223.
50 S. L. Clegg and J. H. Seinfeld, *J. Phys. Chem. A*, 2006, **110**, 5692.
51 S. L. Clegg and J. H. Seinfeld, *J. Phys. Chem. A*, 2006, **110**, 5718.
52 S. Decesari, M. C. Facchini, E. Matta, M. Mircea, S. Fuzzi, A. R. Chughtai and D. M. Smith, *Atmos. Environ.*, 2002, **36**, 1827.
53 Y. Rudich, *Chem. Rev.*, 2003, **103**, 5097.
54 A. Asad, B. T. Mmereki and D. J. Donaldson, *Atmos. Chem. Phys.*, 2004, **4**, 2083.
55 R. Symes, R. M. Sayer and J. P. Reid, *Phys. Chem. Chem. Phys.*, 2004, **6**, 474.
56 A. K. Y. Lee and C. K. Chan, *Atmos. Environ.*, 2007, **41**, 4611.
57 A. K. Y. Lee and C. K. Chan, *J. Phys. Chem. A*, 2007, **111**, 6285.
58 D. Lin-Vien, N. B. Colthup, W. G. Fateley and J. G. Grasselli, *The handbook of infrared and Raman characteristic frequencies of organic molecules*, Academic Press Inc., San Diego, USA, 1991.
59 G. Socrates, *Infrared and Raman Characteristic Group Frequencies. Tables and Charts*, 3rd edn, John Wiley & Sons, Chichester, 2001.
60 P. S. Bailey, *Ozonation In Organic Chemistry. Volume 1. Olefinic Compounds*, Academic Press, New York, 1978.
61 J. Zahardis and G. A. Petrucci, *Atmos. Chem. Phys.*, 2007, **7**, 1237.
62 S. P. Kochhar, in *Deterioration of edible oils, fats and foodstuffs, in Chapter 2 of Atmospheric oxidation and antioxidants, Volume 2*, ed. G. Scott, Elsevier Science Publishers B. V., Amsterdam, Netherlands, 1993.
63 N. A. Porter, S. E. Caldwell and K. A. Mills, *Lipids*, 1995, **30**, 277.
64 H. Tachikawa, M. Polaâsïek, J. Q. Huang, J. Leszczynski, A. K. Salahudeen and J. S. Kwiatkowski, *Appl. Spectrosc.*, 1998, **52**, 1479.
65 B. Muik, B. Lendl, A. Molina-Díaz and M. J. Ayora-Cañada, *Chem. Phys. Lipids*, 2005, **134**, 173.

PAPER

Deliquescence behaviour of single levitated ternary salt/carboxylic acid/water microdroplets†

L. Treuel,[a] S. Schulze,[b] Th. Leisner[b] and R. Zellner*[a]

Received 21st February 2007, Accepted 16th April 2007
First published as an Advance Article on the web 3rd August 2007
DOI: 10.1039/b702651j

The deliquescence relative humidities (DRH) of ammonium sulfate as well as ammonium sulfate/dicarboxylic acid (glutaric, maleic and tartaric) mixtures as a function of temperature and relative composition have been studied using an electrodynamic balance (EDB) in connection with optical microscopy and Mie scattering. The absolute DRH values for pure ammonium sulfate as well as their temperature dependence are consistent with literature data and with the AIM model of Clegg *et al.* The addition of either glutaric or maleic acid to ammonium sulfate leads to a decrease of the DRH value, with the temperature dependence either remaining constant (glutaric acid) or increasing (maleic acid) with increasing acid concentration. This difference is attributed to the higher acidity of maleic acid, which generates stronger ionic interactions with the ammonium sulfate system. In the case of tartaric acid, the deliquescence behaviour of ammonium sulfate is substantially influenced by the formation of insoluble ammonium tartrate.

1. Introduction

Atmospheric aerosols have significant effects on climate, atmospheric turbidity and air quality.[1,2] In addition to these effects, they also play an important role in many chemical processes occurring in the atmosphere.[3,4] The heterogeneous conversion of N_2O_5 to HNO_3 on aerosol surfaces presents an example of the importance of aerosol composition and phase on the rate of reaction. As a result, this reaction is now recognized as playing a crucial role in controlling the fate of nitrogen oxides in both the stratosphere and troposphere. The reactivity of the N_2O_5 on the aerosol surface strongly depends on the phase and water content of the aerosol particles.[5]

Furthermore, the protuberant role of atmospheric aerosol particles in the radiative transfer within the atmosphere by scattering and absorbing electromagnetic radiation[6] makes them an important parameter in modelling the Earth's climate.

Since chemical and radiative effects of atmospheric aerosols are size and phase related,[7,8] they are strongly influenced by the ambient relative humidity (RH) due to water absorbing hygroscopic components, changing both particle diameter and

[a] *Institute for Physical and Theoretical Chemistry, University of Duisburg-Essen, Germany. E-mail: reinhard.zellner@uni-due.de*
[b] *Institute for Environmental Physics, University of Heidelberg, Germany and Institute for Meteorology and Climate Research, Forschungszentrum Karlsruhe Germany*

† The HTML version of this article has been enhanced with colour images.

wavelength dependent refractive indices.[9–13] As a result, for a given atmospheric particle load the net effect on chemistry and/or climate will depend on the relative humidity, and will hence be modified by either temperature or water partial pressure.

Phase and water content of particles govern their mass and hence their surface area and reactivity. The hygroscopic properties and phase changes of atmospheric aerosols as well as the surface morphology (of solid particles) and chemical surface composition must be understood and represented accurately in order to improve aerosol-climate models.

Sulfate aerosols are widely abundant in the atmosphere and represent the largest anthropogenic mass source for the accumulation mode of atmospheric particles.[14] Being non-absorbing in the visible region of the electromagnetic spectrum, they provide the most significant anthropogenic cooling contribution to the global direct radiative forcing.[6,14–20]

Depending on factors like location, aerosols can contain various ratios of inorganic-to-organic material.[21] Results from field measurements indicate that organic material typically accounts for 10–50% of the fine particle mass,[1] with the organic material originating from both anthropogenic and natural sources.[3] Recent field data confirm these findings, indicating that indeed up to 50% or more organic material may be present in atmospheric aerosols.[22]

Dicarboxylic acids are amongst the chemical compounds found in atmospheric aerosol particles.[23,24] Like many other polar organic substances, they are predominantly present in condensed phases rather than in the gas phase,[25–27] resulting from their low vapour pressures. In aqueous aerosol particles, dicarboxylic acids are found to be major constituents of the water soluble organic compounds (WSOCs),[28] and composition measurements have shown that the organic material is internally mixed together with inorganic compounds in tropospheric particles.[22,29]

The physical behaviour of dicarboxylic acids is likely to be typical of many polar WSOCs in the atmosphere,[30] which puts them in the focus of many current laboratory studies.

Dicarboxylic acids are found in many different environments.[23] They have been abundant in aerosols from the urban,[31–34] marine,[35] polar[36] and tropical atmosphere.[37,38] The sources of organic aerosols include fossil-fuel combustion and biomass burning, whose global emission rates are estimated to be 28.5 and 44.6 Tg yr^{-1}, respectively.[39] From their field data, Mochida et al.[23] suggest that deposition is more important than chemical decomposition as a sink of diacids and that they are relatively stable end products in the atmosphere.

Whilst the behaviour towards varying relative humidity, as normally described by deliquescence and efflorescence, of pure ammonium sulfate (AS) particles is well established,[9,40] information about phase transitions and hygroscopic properties of organic and mixed organic/inorganic particles is not at a level comparable to inorganic particles.[41] Several groups have studied the deliquescence of pure organic systems[42–49] and these works have shown that the deliquescence strongly depends on the chemical nature of the organic substance. Moreover, most of the studies have been performed at room temperature only, although there is substantial interest in the temperature dependence of this property as well.

A number of groups have also studied the deliquescence and crystallisation of mixed organic/inorganic particles.[50–66] Studies with such internally mixed organic/AS particles[54] have shown that the organic component changes the deliquescence relative humidity (DRH) relative to pure ammonium sulfate. The water content of atmospheric aerosols is governed by the ambient relative humidity. Hence, high aqueous phase concentrations can be attained at low relative humidities. These result in large deviations from ideal solution behaviour and make the properties of the system difficult to predict.[67]

In the current work we present results from deliquescence measurements of AS and mixed dicarboxylic acid/ammonium sulfate crystals using an electrodynamic balance (EDB). The temperature dependence of the DRH of pure AS solutions,

solutions of dicarboxylic acids and water as well as of ternary mixtures containing AS/dicarboxylic acid/ water has not previously been studied in EDB experiments.

In the following sections of the paper, a short introduction of the experimental setup and the techniques involved will be given. The experimental results will be presented and discussed in the subsequent sections. A final chapter serves to summarise the results, give a conclusion of their significance and suggest further experimental and theoretical efforts in order to enhance the current level of comprehension of the processes involved.

2. Experimental

2.1 Electrodynamic balance

All deliquescence experiments described in this paper have been carried out with single levitated particles using an EDB.[68] The experimental setup used for this work was already described elsewhere in great detail[69–73] and only a brief introduction will be given here. However, specific aspects of the present setup, such as the height control and peripheral devices needed for temperature and humidity control, will be described more extensively.

The EDB is of the classical hyperbolical geometry as suggested by Paul.[74,75] It consists of a toroidal centre electrode connected to a liquid nitrogen cryostate, which is machined from gold-plated copper and serves as a climate chamber.

Six optical ports allow access to the trap, two of which accommodate the temperature and humidity sensors. Two hyperboloidal endcap electrodes seal the top and the bottom of the climate chamber. Here, an AC voltage (approximately 2 kV, 200 Hz) is applied to trap charged particles. A superimposed DC field compensates the gravitational force on the particle.

Fig. 1 shows the EDB in its opened aluminium climate chamber together with the relevant sensing devices.

2.2 Particle microscopy

The trap is integrated into a long working-distance microscope (Mitutoyo), which is equipped with a cooled CCD camera (PCO sensicam) and illuminated by an ultrafast spark flash lamp (HSP nanolight) in order to take microscopic still images of the particles at various stages of the efflorescence/deliquescence cycles (*cf.* Fig. 2).

2.3 Temperature control

In order to achieve a precise and reliable temperature control within the EDB, a copper cooling finger (CryoVac) is directly attached to the trap *via* two cooling rings to ensure a good thermal conductivity. To avoid condensation of ambient water on the trap, it is placed in an evacuated aluminium chamber.

Cooling is achieved by evaporating liquid nitrogen in a heat exchanger at the tip of the cooling finger. A membrane pump and a needle valve control the flow of gaseous nitrogen and allow adjusting the cooling power. A regulated heating system is applied to stabilize the trap at the desired temperature. The interplay of heating system and cooling setup allows it to achieve and retain constant temperatures over a wide temperature region with a precision of <0.5 K.

The EDB temperature is determined by a Pt-100 thermometer, which is integrated in the setup and has direct contact with the EDB body. A combined RH/Temperature sensor (Honeywell 3602, accuracy: ±2% RH, ±0.5 K) is also used to monitor the temperature within the trap.

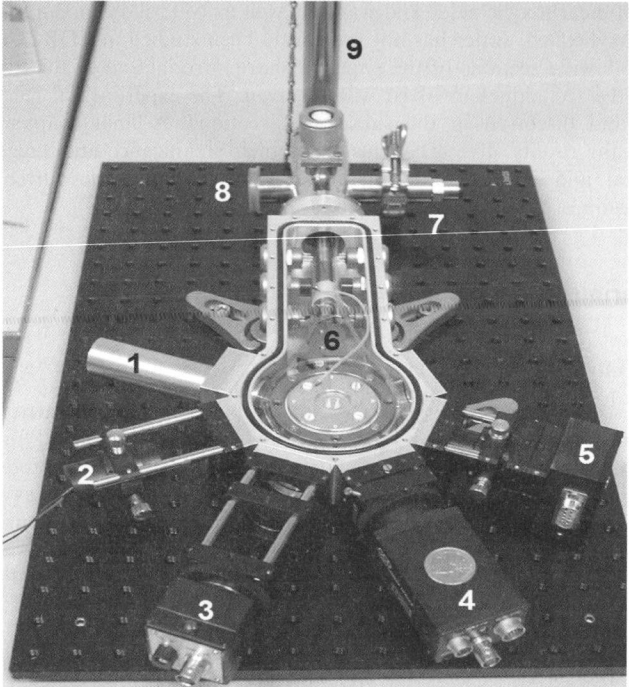

Fig. 1 Photograph of the experimental setup used. (1)—injector port, (2)—laser diode for droplet illumination, (3)—CCD camera for droplet imaging, (4)—CCD camera for Mie scattering detection, (5)—CCD column for automated height control, (6)—cooling finger, (7)—N_2 suction pipe of the cryostat, (8)—connection to vacuum pump, (9)—coupling for U-lifter.

2.4 Humidity control

An Ansyco (SycosHS) moisturiser has been used to regulate the RH inside the EDB. It relies on vapour diffusion through a semi-permeable membrane and allows to keep the gas flow through the EDB constant at all RH values. The RH is measured directly inside the EDB with the combined RH/T probe described in Section 2.3. An

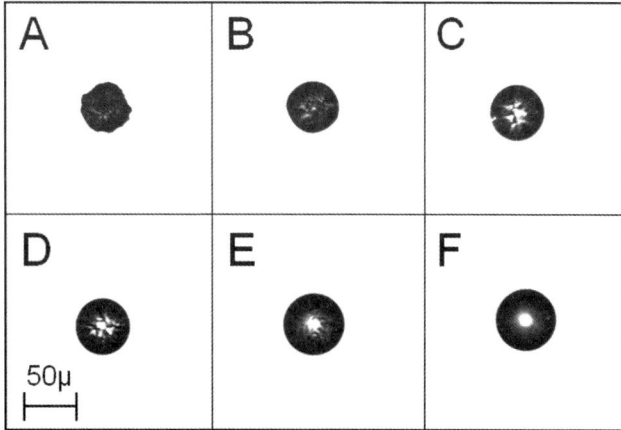

Fig. 2 A series of photographs taken before (A), during (B-E) and after (F) the deliquescence of a pure ammonium sulfate particle.

additional RH value is determined internally by the Asyco (SycosHS) in the outgoing moisturised gas flow.

The setup has an automated height control that keeps the trapped droplet at the EDB centre by changing the DC voltage accordingly. This allows following any changes in the mass/charge ratio of the droplet. Assuming that no loss of charge takes place during an experiment,[76] the height control data can be used to determine changes in the droplet mass. This allows the mass fraction of the solute (mfs = mass of solute on a dry basis/mass of solution) of a levitated particle as a function of ambient RH to be determined.

2.5 Experimental procedure

The droplet is inserted into the preconditioned EDB using a piezoelectric injector (Gesim mbH) and the equilibration process of the droplet can be observed. Subsequently, the RH can be changed in small steps until efflorescence of the droplet is observed. After a sufficient extent of drying, the RH is slowly raised until a sudden increase of the droplet mass indicates the deliquescence. Typical images of a droplet undergoing deliquescence are shown in Fig. 2.

The dry and crystalline particle (A) takes up water and loses all fine edges (B). Further water uptake (C–E) leads to the slow dissolution of the crystal and results ultimately in a completely liquid solution droplet (F).

Since water is a dielectric fluid, it seems plausible that electric fields might alter the structure of aqueous solutions and hence alter their phase transition behaviour. The influence of the EDB on phase transition behaviour of trapped particles has been probed in freezing experiments. Good agreement exists in literature, that electric fields will only be influential if field strengths of a couple of hundred kV m^{-1} are reached. Hobbs[77] gives a good overview of this problem. In the centre of the EDB used in the experiments, only a field strength in the order of a few V mm^{-1} is realised.[69] Krämer et al.[73,76] have probed the influence of the droplet charge on the nucleation rate and have not found any effect.

An influence of particle motion in the EDB on the phase transition can also be ruled out since work on similar setups[69,76] have found a good agreement of their data with literature data from other, non-EDB experiments. The data presented in this work also shows good agreement with literature values from different experimental techniques, thus underlining the reliability of this technique.

2.6 Chemicals

The purities and sources of the chemicals used are as follows: Ammonium Sulfate (AS), Fluka ≥ 99%, Glutaric Acid (GA), Acros organics, 99%, Maleic Acid (MA), Fluka, ≥ 99%, L-(+)-Tartaric Acid (TA), puriss. p.a. ≥ 99,5%. These chemicals were used without further purification. Water used for the preparation of solutions was purified by a Milipore apparatus.

3. Results and discussion

The deliquescence behaviour of mixed organic/inorganic solutions is a complex process. This paper primarily tries to contribute to a better understanding of the influence that dicarboxylic acids have on the deliquescence of AS. Two dicarboxylic acids, GA and MA have been chosen for their atmospheric relevance and chemical structure (Fig. 3). A third—TA—has solely been chosen for its chemical structure. It has no reported atmospheric relevance.

GA is a dicarboxylic acid containing a saturated carbon chain, whereas MA has a slightly shorter chain length and a double bond, introducing additional polarity into the molecule. Two hydroxyl-groups in the TA molecule add even more polarity and allow investigating the effect of higher polarity without a change in the carbon chain

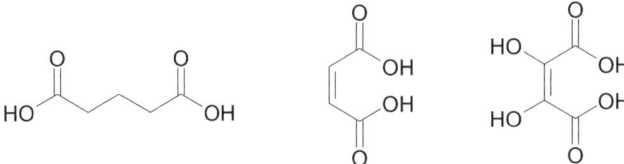

Fig. 3 Structures of glutaric-, maleic- and tartaric acid.

length. Before discussing the behaviour of mixed solutions, we will first present the results for the pure substances, starting with AS.

3.1 Pure Ammonium Sulfate (AS)

A typical result of one single deliquescence measurement of pure AS is shown in Fig. 4. For these results, a solution droplet has been injected into the trap and, after the complete efflorescence, the droplet was dried further down to ensure the complete loss of water from the crystalline particle. The relative humidity was increased again until deliquescence could be observed. The data in Fig. 4 shows the sudden rise in particle mass between 79 and 80% RH. Different experiments with AS showed a high degree of reproducibility.

The data points shown in Fig. 4 are taken directly from the height control signal and the RH sensor signal. A calibration correction for the RH signal has been applied and it has been normalised relative to a concentration of 1 mole SO_4^{2-}.

The deliquescence behaviour of the pure inorganics/H_2O system, as observed in the current work, has been compared with predictions based on the Aerosol Inorganics Model II (AIM) model. The AIM[78,79] is a multi-component mole-fraction-based model to represent aqueous phase activities, equilibrium partial pressures and saturation with respect to solid phases. Carslaw et al.[80] have developed a method using the mole-fraction-based equations of Pitzer, Simonson and Clegg,[81–84] which allows calculating activity coefficients for the system $HCl–HNO_3–H_2SO_4–H_2O$. Clegg et al.[78] applied this model to the system $H^+–NH_4^+–SO_4^{2-}–NO_3^-–H_2O$. Comparisons suggest that the model satisfactorily represents salt solubilities and water activities.[78] Mole fractions are calculated on the basis of the individual ionic and molecular species present. Equations for the activity

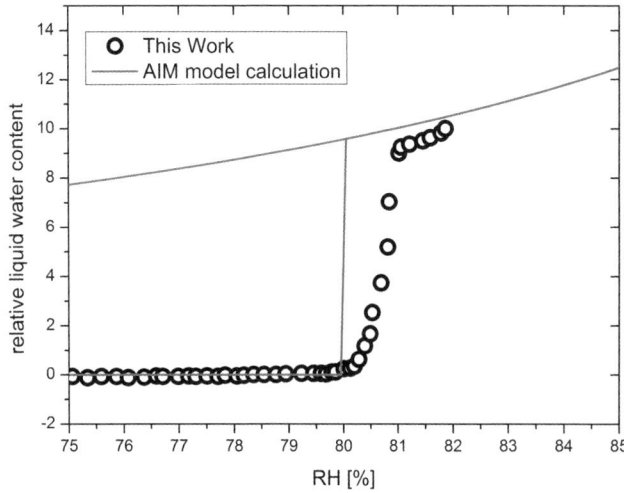

Fig. 4 Deliquescence of an AS droplet; comparison of experiment and AIM model.[78]

coefficients of neutral species, cations and anions in arbitrarily complex mixtures are derived from an equation for the excess Gibbs energy of a liquid mixture.[81–83] This equation accounts for three major contributions. First, an extended Debye–Hückel term accounting for long range forces between ions, a contribution most important in dilute solutions. Second, a higher order electrostatic term essentially modifying the Debye–Hückel contribution and arising from the unsymmetrical mixing of ions of the same sign but different charge. The third contribution is an expression for the short range forces between components that dominate in concentrated solutions. This term is based upon a Margules expansion[85] of terms in the mole fraction.

The equations generally contain parameters describing interactions in binary solutions and ternary mixtures whose values are determined by fitting to empirical data. The AIM model has been made available online at http://mae.ucdavis.edu/~sclegg/aim.html.

The comparison of the experimental results with respect to the DRH value with the predictions from the AIM model calculation shows that the experimental result is slightly higher than that predicted by the model, although this deviation is well within the margin of error of the experimental data. This data comparison, however, gives a good indication of the accuracy of the experimental results.

The temperature influence on the DRH of AS solutions is shown in Fig. 5. The graph compares our own experimental findings with data from the AIM.[78] In the observed temperature region, the AIM model yields a temperature dependence of -0.07% K^{-1} for pure AS. Included are also experimental results from other studies.[40,86]

Our experimental data covers only a relatively small temperature range, within which the temperature dependence is well reproduced by the temperature dependence as predicted by the AIM model calculations. In this model, the temperature dependence of the DRH for pure salts results exclusively from the change in water solubility of the salt with changing temperature.

In the temperature range presented, the dependence of DRH on temperature can be considered linear. Although the other experimental data available cover a somewhat larger temperature range, the scatter between individual data points is too large to imply a temperature variation other than linear. As a consequence all available data and theory must be considered consistent.

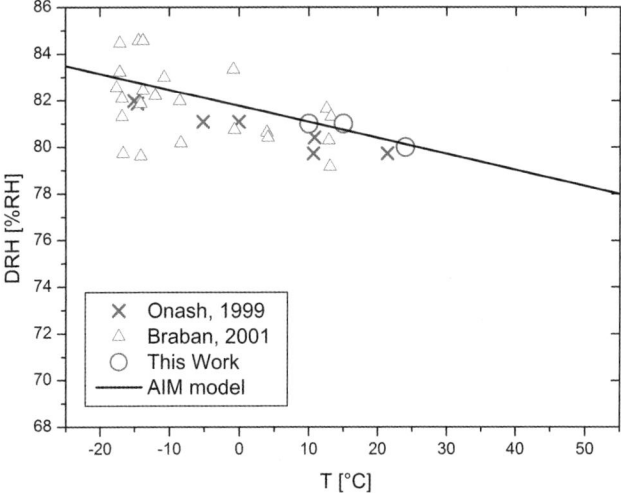

Fig. 5 Temperature dependence of the DRH of pure AS. A comparison of experimental and model data.

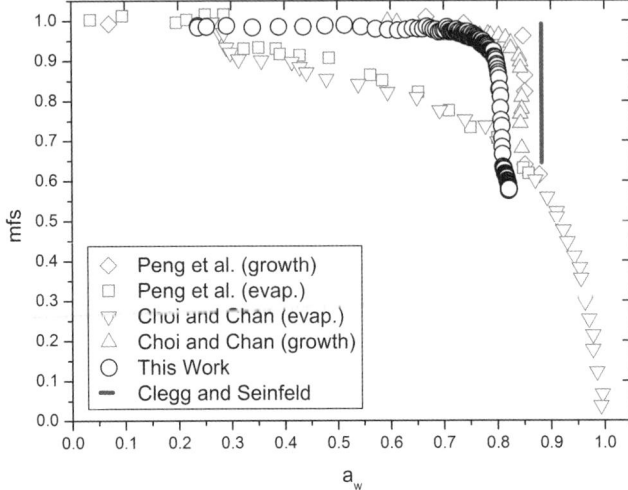

Fig. 6 Deliquescence data for pure GA.

3.2 Tartaric Acid (TA) and TA/AS mixtures

Tartaric acid has no significant atmospheric relevance. Nevertheless, it was chosen here as a dicarboxylic acid with a high polarity within the molecule, differing substantially from GA and MA.

In our experiments, however, we found no efflorescence for pure TA. Since the injection of liquid droplets into the trap is the initial step in our experiment, we rely on the efflorescence to produce a stably-trapped solid particle within our EDB. Hence, the absence of efflorescence makes it impossible to study the deliquescence process without modifications of the experimental setup.

For mixed TA/AS droplets, however, efflorescence has been observed and hence deliquescence could also be observed. Upon RH increase, the start of a deliquescence process has been recorded at 76 ± 2% RH. However, even after a time period of several hours, the Mie scattering indicated that the droplet still contained solid constituents which failed to resolve within all experiments. Furthermore, changing the concentration of tartaric acid within the solution had no effect on the starting point of this incomplete deliquescence. We attribute these effects to the formation of ammonium tartrate during the crystallisation process. Whilst tartaric acid has a fairly high solubility in water (1390 g L^{-1}), ammonium tartrate is poorly soluble (63 g L^{-1}). The remaining AS will undergo the normal deliquescence process influenced by the relatively small amount of tartaric acid present in solution whereas ammonium tartrate will remain undissolved.

The formation of undissolved ammonium tartrate is expected to exert a substantial influence on the fate of such a solution droplet, as the ammonium tartrate crystal will serve as a nucleus in any forthcoming crystallisation process. Hence, no hysteresis between efflorescence and deliquescence will be observed and crystallisation will always occur at the DRH, which is consistent with our experimental observations.

3.3 Glutaric Acid (GA) and GA/AS mixtures

Our experiments with pure GA (Fig. 6) yield a DRH of 83 ± 4%. The error takes into account the precision of the RH meter as well as the reproducibility of the individual measurements. Whereas measurements for AS as well as for the inorganic/organic mixtures showed a high degree of reproducibility, the results for GA and the other pure acids differed between the single measurements. Overall, in this work, deliquescence of pure GA at 298 K occurred in a range between 80 and 85% RH.

Previous deliquescence measurements by Peng et al.[46] reported a starting deliquescence at 83% and 85% (different particles) with a water uptake process over 12 h. They suggested the presence of mass transfer limitations that delayed the completion of deliquescence. The phenomenon of mass transfer limitation in the growth process has also been observed by Peng and Chan[49] for sodium pyruvate but not in other acids. Saxena and Hildeman[87] estimated a DRH range of 89–99% for pure GA. Cruz and Pandis[58] have reported the DRH of pure GA at 85 ± 5% using a TSMA System. Clegg and Seinfeld[30] calculated the saturation activity to 0.878 leading to a DRH of 87.8%.

Our own data are well in line with the data from both Peng et al.[46] and Cruz and Pandis.[58] However, as opposed to Peng et al., we did not experience strong mass transfer limitations and deliquescence was on the same timescale as for other organic and inorganic compounds (<1 h). In other measurements using an optical levitation method,[66] however, we have experienced growth delaying effects in internally-mixed GA/AS particles, which may be explained by mass transport limitations but needs further evaluation.

Within both margins of error, the data of Clegg and Seinfeld[30] is also in line with our findings. The temperature dependence has as yet not been studied for this system.

Studies of DRH of ternary GA/AS/water droplets have also been conducted. The aim of these experiments was to quantify the influence of the acid on the DRH of the AS. For this purpose, solutions with a concentration of 20 weight% have been used. The mass fraction of the acid in the ammonium sulfate was varied between 1 (pure AS) and 0.5 (1-to-1 ratio of AS and GA) in these mixtures.

We find a notable concentration dependence, which is displayed in Fig. 7. A decrease in the DRH for AS particles upon addition of GA can be found and compares well to the experimental findings of Brooks et al.[54]

The temperature dependence of the DRH of solutions of GA/AS and water has also been studied. The results obtained together with that for pure AS is shown in Fig. 8.

The graphical interpretation of the experimental data in Fig. 8 shows a clear temperature dependence, but no change of this temperature dependence upon addition of various amounts of GA. The dotted straight lines through the data for

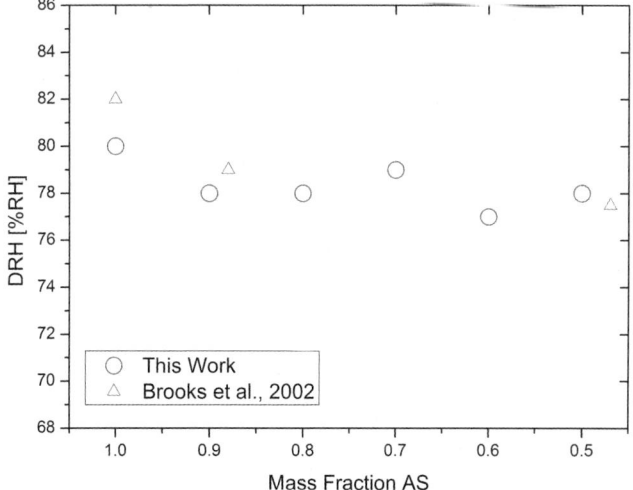

Fig. 7 Concentration dependence of the DRH for mixtures of AS and GA.[54]

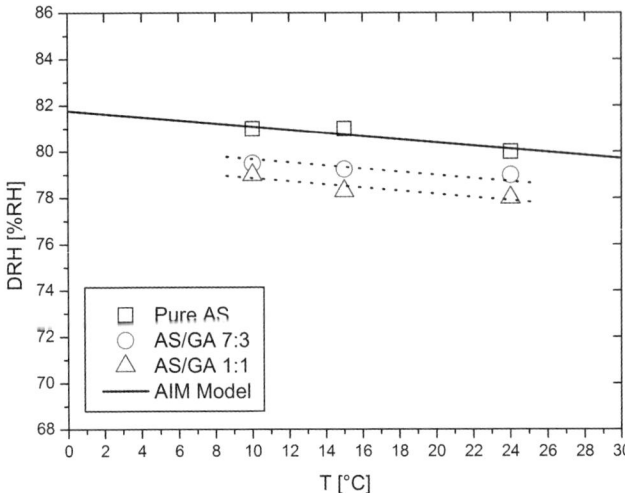

Fig. 8 Temperature dependence of DRH values of AS and mixed GA/AS solution droplets.

mixed particles are within error margins of the same gradient as the AIM model line for pure AS and indicate that no detectable difference is introduced by the GA. In terms of the nature of the physical effects driving the DRH temperature dependence, this observation may be interpreted as to indicate that the temperature dependence of the mixture remains to be dominated by the solubility change of the pure AS component. However, at all studied temperatures the mixed particles deliquesced at lower RH than the pure AS, as already noted above for the studies on the concentration dependence.

3.4 Maleic Acid (MA) and MA/AS mixtures

This section of paper focuses first on the deliquescence of the pure MA and we report a DRH of 83 ± 5%. Again, the reproducibility of the results is not as good as for pure inorganic or inorganic/organic acid solutions, which contributes to a larger margin of error. Fig. 9 compares our findings to the data of Choi and Chan,[57] and the calculations of Clegg and Seinfeld,[30] showing our data in good agreement with both data sets.

Choi and Chan determined their data using a similar EDB setup to the one used in this work. In the work of Clegg and Seinfeld,[30] the saturation activity, and hence the DRH of MA, is determined from solubility data and model calculations. They report a value of 0.885 for the saturation activity leading to a DRH of 88.5%.

With analogue solutions containing AS and MA, the concentration dependence of the DRH on the relative mass fraction of MA in AS solutions has also been probed. A clear influence of the MA on the deliquescence behaviour of the AS solution has been found, in agreement with previous experimental findings.[54] The comparison between both experimental data sets shows good agreement.

From the comparison between the data in Fig. 9 and Fig. 10, it is apparent that MA has a much stronger influence on the DRH of AS solutions than GA. This effect may be explained by increased polarity within the molecule as well as structural differences. Unfortunately, similar effects for tartaric acid, which were expected to be even more pronounced due to further increased polarity, could not be observed as a result of formation of ammonium tartrate.

Experimental results have also been obtained for the concentration dependence of the temperature influence on the DRH for internally mixed AS/MA droplets.

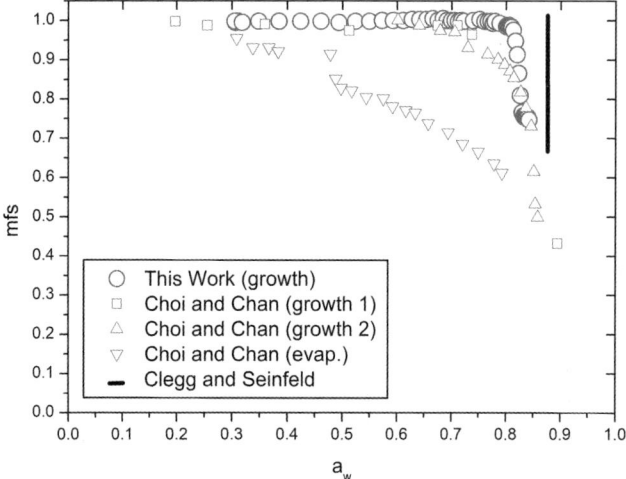

Fig. 9 Deliquescence behaviour of pure MA. Circles: own experimental data; triangles and squares: EDB data from Choi and Chan;[57] solid line: model calculations from Clegg and Seinfeld.[30]

Contrasting the results for GA, the temperature influence shows a clear dependence on the relative MA concentration (*cf.* Fig. 11).

Whilst the temperature dependence of the DRH of pure AS is reproduced well by the AIM model, the temperature dependence increases with increasing MA concentration. In the observed temperature region, the AIM model yields a temperature dependence of -0.07% K^{-1} for pure AS. At an AS/MA ratio of 7 : 3 (by weight), this dependence increases to -0.21% K^{-1}, and at a ratio of 1 : 1 (by weight) the temperature dependence increases even further to -0.37% K^{-1}.

This observation contrasts to that of GA/AS solutions, for which an enhancement of temperature dependence with increasing acid has not been observed. Further experimental data is needed over a broader range of temperatures to confirm and explain this effect. In addition, an extension of the AIM model to encompass ternary solutions including carboxylic acids is needed.

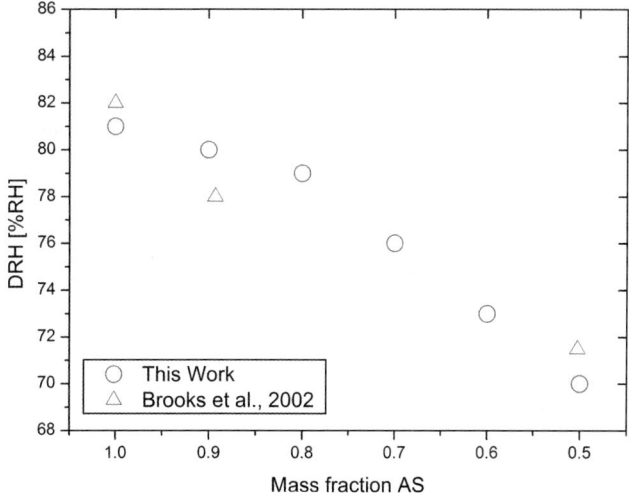

Fig. 10 Concentration dependence of the DRH for mixtures of AS and MA.

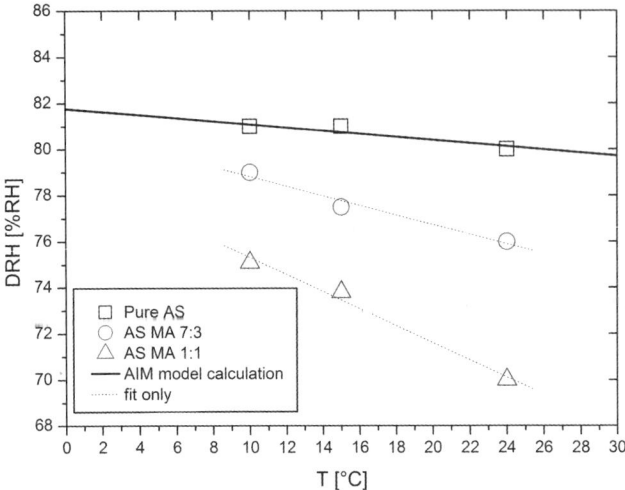

Fig. 11 Temperature dependence of DRH values for MA/AS mixtures.

4. Conclusions

The current work has provided deliquescence relative humidities (DRH) for ammonium sulfate (AS) as well as mixtures of AS with tartaric, glutaric and maleic acid (TA, GA and MA, respectively). In particular, concentration and temperature dependences of DRH values were derived. The major finding is that DRH values of AS decrease substantially in the presence of GA and MA. In terms of the atmospheric significance of such mixtures, these findings imply an earlier onset of hygroscopic growth of mixed particles as opposed to the pure AS system. It is worthy of note, however, that the radiative effects of such particles is not only a function of their size, i.e., depending on RH, but also of the refractive index of the mixture. Since, grossly speaking, the refractive indices of mixtures are additive, the effect of this quantity may be dominant.

A somewhat surprising finding has been the increase in temperature dependence for different relative compositions of AS/MA concentrations. This effect has not been observed for the mixtures of AS with GA and probably suggests much stronger interactions of MA with the inorganic system. That this may be the case is suggested from the dissociation constants (pK_a (MA) = 1.91,[88] pK_a (GA) = 3.77 [88]), which implies a much stronger dissociation of MA and hence the presence of larger concentrations of charged particles. In order to further quantify such effects, an extension of the experimentally accessible temperature range is also required.

Acknowledgements

The authors would like to thank René Müller and Daniel Rzesanke for their contribution of ideas and work force to the development and operation of the experimental setup. L. T. acknowledges support of the Max-Buchner-Foundation.

References

1 EPA, *Air Quality Criteria for Particulate Matter. EPA/600/P-95/001 aF-cF.3v.* 1996, Research Triangle Park, NC, National Center for Environmental Assessment–RTP office, Springfield, VA, National Technical Information Service.
2 D. O. De Haan and B. J. Finlayson-Pitts, *J. Phys. Chem. A*, 1997, **101**(51), 9993–9999.
3 J. H. Seinfeld S. N. Pandis, *Atmos. Chem. Phys.*, John Wiley & Sons Inc., New York, USA, 1998.
4 J. P. Reid and R. M. Sayer, *Sci. Prog.*, 2002, **85**(3), 263–296.

5   J. A. Thornton, C. F. Braban and J. P. D. Abbatt, *Phys. Chem. Chem. Phys.*, 2003, **5**(20), 4593–4603.
6   J. T. Houghton, Y. Ding, D. J. Griggs, M. Noguer, P. J. Linden, X. Dai, K. Maskell C. A. Johnson, *Climate Change 2001: The Scientific Basis.*, Cambridge University Press, Cambridge, 2001.
7   C. Mund and R. Zellner, *J. Mol. Struct.*, 2003, **661**, 491–500.
8   C. Mund and R. Zellner, *ChemPhysChem*, 2003, **4**(6), 630–638.
9   S. T. Martin, *Chem. Rev.*, 2000, **100**(9), 3403–3453.
10 J. P. Reid and L. Mitchem, *Annu. Rev. Phys. Chem.*, 2006, **57**(1), 245–271.
11 L. Mitchem, R. J. Hopkins, J. Buajarern, A. D. Ward and J. P. Reid, *Chem. Phys. Lett.*, 2006, **432**, 362–366.
12 L. Mitchem, J. Buajarern, R. J. Hopkins, A. D. Ward, R. J. J. Gilham, R. L. Johnston and J. P. Reid, *J. Phys. Chem. A*, 2006, **110**(26), 8116–8125.
13 J. P. Reid, H. Meresman, L. Mitchem and R. Symes, *Int. Rev. Phys. Chem.*, 2007, **26**, 139–192.
14 S. T. Martin, H. M. Hung, R. J. Park, D. J. Jacob, R. J. D. Spurr, K. V. Chance and M. Chin, *Atmos. Chem. Phys.*, 2004, **4**, 183–214.
15 R. J. Charlson, W. M. Porch, A. P. Waggoner and N. C. Ahlquist, *Tellus*, 1974, **26**, 345–360.
16 J. H. Seinfeld, National Research Council Panel on Aerosol Radiative Forcing Climate Change. A Plan for a research program on aerosol radiative forcing climate change, National Academy Press, Washington D.C., 1996.
17 J. Haywood and O. Boucher, *Rev. Geophys.*, 2000, **38**(4), 513–543.
18 J. M. Haywood, D. L. Roberts, A. Slingo, J. M. Edwards and K. P. Shine, *J. Clim.*, 1997, **10**, 1562–1577.
19 P. J. Adams, J. H. Seinfeld, D. Koch, L. Mickley and D. Jacob, *J. Geophys. Res., [Atmos.]*, 2001, **106**(D1), 1097–1111.
20 M. Z. Jacobson, *J. Geophys. Res., [Atmos.]*, 2001, **106**(D2), 1551–1568.
21 J. Heintzenberg, *Tellus*, 1989, **41B**, 149–160.
22 D. M. Murphy, D. S. Thomson and T. M. J. Mahoney, *Science*, 1998, **282**(5394), 1664–1669.
23 M. Mochida, K. Kawamura, N. Umemoto, M. Kobayashi, S. Matsunaga, H. J. Lim, B. J. Turpin, T. S. Bates and B. R. T. Simoneit, *J. Geophys. Res., [Atmos.]*, 2003, **108**(D23), 8638.
24 L. E. Yu, M. L. Shulman, R. Kopperud and L. M. Hildemann, *Environ. Sci. Technol.*, 2005, **39**(3), 707–715.
25 A. Limbeck, H. Puxbaum, L. Otter and M. C. Scholes, *Atmos. Environ.*, 2001, **35**(10), 1853–1862.
26 J. J. Schauer, M. J. Kleeman, G. R. Cass and B. R. T. Simoneit, *Environ. Sci. Technol.*, 1999, **33**(10), 1566–1577.
27 E. D. Baboukas, M. Kanakidou and N. Mihalopoulos, *J. Geophys. Res., [Atmos.]*, 2000, **105**(D11), 14459–14471.
28 P. Saxena and L. M. Hildemann, *J. Atmos. Chem.*, 1996, **24**(1), 57–109.
29 A. M. Middlebrook, D. M. Murphy and D. S. Thomson, *J. Geophys. Res., [Atmos.]*, 1998, **103**(D13), 16475–16483.
30 S. L. Clegg and J. H. Seinfeld, *J. Phys. Chem. A*, 2006, **110**(17), 5692–5717.
31 D. Grosjean, K. Van Cauwenberghe, J. P. Schmid, P. E. Kelley and J. N. Pitts, *Environ. Sci. Technol.*, 1978, **12**(3), 313–317.
32 K. Kawamura and I. R. Kaplan, *Environ. Sci. Technol.*, 1987, **21**(1), 105–110.
33 K. Kawamura and K. Ikushima, *Environ. Sci. Technol.*, 1993, **27**(10), 2227–2235.
34 R. Sempere and K. Kawamura, *Atmos. Environ.*, 1994, **28**(3), 449–459.
35 K. Kawamura and F. Sakaguchi, *J. Geophys. Res., [Atmos.]*, 1999, **104**(D3), 3501–3509.
36 K. Kawamura, R. Semere, Y. Imai, Y. Fujii and M. Hayashi, *J. Geophys. Res., [Atmos.]*, 1996, **101**(D13), 18721–18728.
37 M. Narukawa, K. Kawamura, N. Takeuchi and T. Nakajima, *Geophys. Res. Lett.*, 1999, **26**(20), 3101–3104.
38 S. Gao, D. A. Hegg, P. V. Hobbs, T. W. Kirchstetter, B. I. Magi and M. Sadilek, *J. Geophys. Res., [Atmos.]*, 2003, **108**(D13), DOI: 10.1029/2002JD002324.
39 C. Liousse, J. E. Penner, C. Chuang, J. J. Walton, H. Eddleman and H. Cachier, *J. Geophys. Res., [Atmos.]*, 1996, **101**(D14), 19411–19432.
40 T. B. Onasch, R. L. Siefert, S. D. Brooks, A. J. Prenni, B. Murray, M. A. Wilson and M. A. Tolbert, *J. Geophys. Res., [Atmos.]*, 1999, **104**(D17), 21317–21326.
41 M. C. Jacobson, H. C. Hansson, K. J. Noone and R. J. Charlson, *Rev. Geophys.*, 2000, **38**(2), 267–294.
42 M. T. Parsons, J. Mak, S. R. Lipetz and A. K. Bertram, *J. Geophys. Res., [Atmos.]*, 2004, **109**(D6), DOI: 10.1029/2003JD004075.

43 A. J. Prenni, P. J. DeMott, S. M. Kreidenweis, D. E. Sherman, L. M. Russell and Y. Ming, *J. Phys. Chem. A*, 2001, **105**(50), 11240–11248.
44 C. F. Braban, M. F. Carroll, S. A. Styler and J. P. D. Abbatt, *J. Phys. Chem. A*, 2003, **107**(34), 6594–6602.
45 M. Y. Choi and C. K. Chan, *J. Phys. Chem. A*, 2002, **106**(18), 4566–4572.
46 C. Peng, M. N. Chan and C. K. Chan, *Environ. Sci. Technol.*, 2001, **35**(22), 4495–4501.
47 E. Demou, H. Visram, D. J. Donaldson and P. A. Makar, *Atmos. Environ.*, 2003, **37**(25), 3529–3537.
48 A. R. Hansen and K. D. Beyer, *J. Phys. Chem. A*, 2004, **108**(16), 3457–3466.
49 C. Peng and C. K. Chan, *Atmos. Environ.*, 2001, **35**(7), 1183.
50 A. Pant, A. Fok, M. T. Parsons, J. Mak and A. K. Bertram, *Geophys. Res. Lett.*, 2004, **31**(12), L12111.
51 C. F. Braban and J. P. D. Abbatt, *Atmos. Chem. Phys.*, 2004, **4**, 1451–1459.
52 S. D. Brooks, P. J. DeMott and S. M. Kreidenweis, *Atmos. Environ.*, 2004, **38**(13), 1859–1868.
53 S. D. Brooks, R. M. Garland, M. E. Wise, A. J. Prenni, M. Cushing, E. Hewitt and M. A. Tolbert, *J. Geophys. Res., [Atmos.]*, 2003, **108**(D15), 4487.
54 S. D. Brooks, M. E. Wise, M. Cushing and M. A. Tolbert, *Geophys. Res. Lett.*, 2002, **29**(19), 23-1.
55 M. N. Chan and C. K. Chan, *Environ. Sci. Technol.*, 2003, **37**(22), 5109–5115.
56 Y. Y. Chen and W. M. G. Lee, *J. Environ. Sci. Health, Part A*, 2001, **36**(2), 229–242.
57 M. Y. Choi and C. K. Chan, *Environ. Sci. Technol.*, 2002, **36**(11), 2422–2428.
58 C. N. Cruz and S. N. Pandis, *Environ. Sci. Technol.*, 2000, **34**(20), 4313–4319.
59 K. Hameri, R. Charlson and H. C. Hansson, *AIChE J.*, 2002, **48**(6), 1309–1316.
60 H. C. Hansson, M. J. Rood, S. Koloutsou-Vakakis, K. Hameri, D. Orsini and A. Wiedensohler, *J. Atmos. Chem.*, 1998, **31**(3), 321–346.
61 J. M. Lightstone, T. B. Onasch, D. Imre and S. Oatis, *J. Phys. Chem. A*, 2000, **104**(41), 9337–9346.
62 E. Mikhailov, S. Vlasenko, R. Niessner and U. Poschl, *Atmos. Chem. Phys.*, 2004, **4**, 323–350.
63 A. J. Prenni, P. J. De Mott and S. M. Kreidenweis, *Atmos. Environ.*, 2003, **37**(30), 4243–4251.
64 M. E. Wise, J. D. Surratt, D. B. Curtis, J. E. Shilling and M. A. Tolbert, *J. Geophys. Res., [Atmos.]*, 2003, **108**(D20), 4487.
65 M. T. Parsons, D. A. Knopf and A. K. Bertram, *J. Phys. Chem. A*, 2004, **108**(52), 11600–11608.
66 N. Jordanov and R. Zellner, *Phys. Chem. Chem. Phys.*, 2006, **8**(23), 2759–2764.
67 S. L. Clegg, P. Brimblecombe and A. S. Wexler, *J. Phys. Chem. A*, 1998, **102**(12), 2155–2171.
68 E. J. Davis, *J. Aerosol Sci. Technol.*, 1997, **26**(3), 212–254.
69 P. Stöckel, PhD thesis, Department of Physics, FU Berlin, Berlin, 2001.
70 P. Stöckel, H. Vortisch, T. Leisner and H. Bäumgartel, *J. Mol. Liq.*, 2002, **96–7**, 153–175.
71 I. Weidinger, J. Klein, P. Stöckel, H. Bäumgartel and T. Leisner, *J. Phys. Chem. B*, 2003, **107**(15), 3636–3643.
72 P. Stöckel, I. M. Weidinger, H. Bäumgartel and T. Leisner, *J. Phys. Chem. A*, 2005, **109**(11), 2540–2546.
73 B. Krämer, O. Hubner, H. Vortisch, L. Wöste, T. Leisner, M. Schwell, E. Rühl and H. Bäumgartel, *J. Chem. Phys.*, 1999, **111**(14), 6521–6527.
74 W. Paul and M. Raether, Das elektrische Massenfilter, *Z. Phys.*, 1955, **140**(1), 262–273.
75 E. Fischer, *Z. Phys.*, 1959, **156**(1), 1–26.
76 B. Krämer, PhD thesis, FU Berlin, Berlin, 1998.
77 P. V. Hobbs, *Ice Physics*, Clarendon Press, Oxford, 1974.
78 S. L. Clegg, P. Brimblecombe and A. S. Wexler, *J. Phys. Chem. A*, 1998, **102**, 2137–2154.
79 A. S. Wexler and S. L. Clegg, *J. Geophys. Res., [Atmos.]*, 2002, **107**(D14), 4207.
80 K. S. Carslaw, S. L. Clegg and P. Brimblecombe, *J. Phys. Chem.*, 1995, **99**(29), 11557–11574.
81 S. L. Clegg, K. S. Pitzer and P. Brimblecombe, *J. Phys. Chem.*, 1992, **96**(23), 9470–9479.
82 S. L. Clegg, K. S. Pitzer and P. Brimblecome, *J. Phys. Chem.*, 1994, **98**(4), 1368–1368.
83 S. Clegg, K. Pitzer and P. Brimblecombe, *J. Phys. Chem.*, 1995, **99**(17), 6755–6755.
84 K. S. Pitzer and J. M. Simonson, *J. Phys. Chem.*, 1986, **90**(13), 3005–3009.
85 J. B. J. Thompson, *Thermodynamic Properties of Simple Solutions, Researches in Geochemistry*, ed. P. H. Abelson, Wiley, New York, 1967, pp. 340–361.
86 C. F. Braban, J. P. D. Abbatt and D. J. Cziczo, *Geophys. Res. Lett.*, 2001, **28**(20), 3879–3882.
87 P. Saxena and L. M. Hildemann, *Environ. Sci. Technol.*, 1997, **31**(11), 3318–3324.
88 J. A. Dean, *Lange's Handbook of Chemistry*, McGraw-Hill, New York, 15th edn, 1998, p. 1424.

PAPER

The complex refractive index of atmospheric and model humic-like substances (HULIS) retrieved by a cavity ring down aerosol spectrometer (CRD-AS)†

E. Dinar,[a] A. Abo Riziq,[a] C. Spindler,[a] C. Erlick,[b] G. Kiss[c] and Y. Rudich*[a]

Received 1st March 2007, Accepted 26th March 2007
First published as an Advance Article on the web 13th July 2007
DOI: 10.1039/b703111d

Atmospheric aerosols absorb and reflect solar radiation causing surface cooling and heating of the atmosphere. The interaction between aerosols and radiation depends on their complex index of refraction, which is related to the particles' chemical composition. The contribution of light absorbing organic compounds, such as HUmic-LIke Substances (HULIS) to aerosol scattering and absorption is among the largest uncertainties in assessing the direct effect of aerosols on climate. Using a Cavity Ring Down Aerosol Spectrometer (CRD-AS), the complex index of refraction of aerosols containing HULIS extracted from pollution, smoke, and rural continental aerosols, and molecular weight-fractionated fulvic acid was measured at 390 nm and 532 nm. The imaginary part of the refractive index (absorption) substantially increases towards the UV range with increasing molecular weight and aromaticity. At both wavelengths, HULIS extracted from pollution and smoke particles absorb more than HULIS from the rural aerosol. Sensitivity calculations for a pollution-type aerosol containing ammonium sulfate, organic carbon (HULIS), and soot suggests that accounting for absorption by HULIS leads in most cases to a significant decrease in the single scattering albedo and to a significant increase in aerosol radiative forcing efficiency, towards more atmospheric absorption and heating. This indicates that HULIS in biomass smoke and pollution aerosols, in addition to black carbon, can contribute significantly to light absorption in the ultraviolet and visible spectral regions.

1. Introduction

Trace gases, clouds, and atmospheric aerosols influence Earth's radiation balance. In contrast to the relatively robust understanding and low uncertainty associated with the effect of trace gases, large uncertainties remain regarding the contribution of

[a] Department of Environmental Sciences, Weizmann Institute, Rehovot 76100, Israel.
E-mail: Yinon.Rudich@weizmann.ac.il
[b] Department of Atmospheric Sciences, The Hebrew University of Jerusalem, Jerusalem 91904, Israel
[c] Air Chemistry Group of Hungarian Academy of Sciences, University of Pannonia, 8200, Veszprém, Hungary

† The HTML version of this article has been enhanced with colour images.

aerosols and clouds.[1] Aerosols affect Earth's radiation balance both directly and indirectly. The indirect effect is related to aerosol–water interactions when aerosols act as Cloud Condensation Nuclei (CCN), altering cloud microphysical properties and consequently precipitation and Earth's albedo.[2–5] The direct climatic effect of aerosols occurs through scattering and absorption of incoming solar radiation and absorption of outgoing longwave radiation. Absorption of radiation by aerosols heats the atmosphere and can cause cloud dissipation (the "semi-direct" climatic effect) and possible changes in atmospheric circulation.[6–9] Therefore, accurate measurements of aerosol scattering and absorption are crucial for estimating Earth's energy balance.

The interaction of radiation with particles through scattering and absorption leads to extinction of the incident light. This attenuation can be expressed as $\alpha_{ext} = \alpha_{sca} + \alpha_{abs}$, where α_{ext} is the extinction coefficient, and α_{sca} and α_{abs} are the scattering and absorption coefficients, respectively. Other important parameters in modeling the radiative impact of aerosols and in remote sensing are the Single Scattering Albedo (SSA) (the ratio of scattering to total light attenuation ($\varpi = \alpha_{sca}/\alpha_{sca} + \alpha_{abs}$)) and the mass absorption and scattering coefficients. To derive these parameters, knowledge of the complex index of refraction of the particles is needed.

Both the microphysical (indirect effect) and optical effects (direct and "semi-direct" effect) of aerosols depend on their chemical and physical properties. Traditionally, most climate models treat scattering by atmospheric aerosols composed mainly of inorganic salts (sea spray, sulfate aerosols) and minerals (dust). A few models include absorption by "black carbon" aerosols.[8,10] However, it is now recognized that carbonaceous species contribute significantly to atmospheric particle mass.[7,11] In contrast and in comparison to inorganic salts, mineral dust, and soot (C_{soot}), the optical properties of "brown carbon" (C_{brown}, which includes all Light Absorbing Carbonaceous (LAC) species excluding C_{soot}) have hardly been examined.[10,12–14] It was recently suggested that the assessment of absorption of solar radiation by C_{brown} may not be sufficient, since both measurements and modeling have not considered the strong absorption spectral dependence of C_{brown}, especially in the UV range.[12,14] Due to the high abundance of C_{brown} in continental aerosol and the sharp increase in its absorption towards shorter wavelengths, it may absorb significantly and influence local and global radiative forcing and photochemistry.[12,14]

Aerosols that contain C_{brown} are commonly classified on the basis of its water solubility. Water Soluble Organic Carbon (WSOC) dominates atmospheric organic matter.[15–17] Between 20 and 70 wt% of the WSOC fraction is composed of high molecular weight multifunctional compounds that contain aromatic, phenolic, and acidic functional groups.[18–24] These aerosol-associated compounds resemble Humic Substances (HS) from terrestrial and aquatic sources in many aspects and are therefore termed HUmic-LIke Substances (HULIS). HULIS probably comprise a major fraction of LAC or the "brown carbon" (C_{brown}). However, the optical properties of HULIS material are not well constrained, mostly due to difficulties in sampling and extraction of atmospheric HULIS for detailed laboratory measurements. To the best of our knowledge, only Hoffer *et al*.[13] have studied the optical properties of HULIS (extracted from biomass burning aerosols). The lack of atmospheric material raises the need to search for proxies, such as commercially available HS from aquatic and terrestrial sources. A model material often used is Suwannee River Fulvic Acid (SRFA).

In previous studies, we compared the CCN activity,[25] hygroscopicity,[26] and density[27] of molecular weight-fractionated SRFA samples to those of fulvic acid-type HULIS extracted from collected wood burning smoke and pollution particles. In this study, we focus on their optical properties in the visible (532 nm) and in the ultraviolet (390 nm). Using a dual-wavelength Cavity Ring Down Aerosol Spectrometer (CRD-AS), the extinction coefficients of laboratory-generated SRFA and HULIS-containing aerosols were measured. Using Mie scattering calculations, the complex Refractive Indices (RI) of the samples were retrieved, including the scattering (real part) and the absorption (imaginary part) indices. The complex refractive indices were further used to derive the particles' SSA and mass scattering and absorption extinction coefficients, and to estimate the possible radiative impacts of HULIS in mixed aerosols.

2. Experiment

2.1 HULIS from ambient particles

Aerosol particles (<10 μm, PM10) were collected on quartz filters using high volume samplers (flow rate 70 m^3 h^{-1}) on the roof of a four story building in an urban location (Weizmann Institute, Rehovot, Israel). The filters were pre-baked at 450 °C in open aluminium foil envelopes, sealed and stored at –18 °C.

Three samples were collected. Two were collected at the Weizmann Institute, Rehovot, Israel: (1) Fresh smoke particles (referred to hereafter as "Smoke") were sampled throughout the night (16–17 May 2006) during an extensive nation-wide wood burning event with average PM10 mass concentration near the sampling location of 200–300 μg m^{-3}; and (2) air pollution particles (referred to hereafter as "Pollution") were sampled from May 18 2006 until the end of June 2006 during daytime, with average PM10 mass concentrations of $\sim$40 μg cm^{-3}. The atmospheric conditions during May–June 2006 were relatively constant; the mass concentrations were determined by the Israeli Ministry of the Environment. The third HULIS sample was collected in a mixed forest clearing on the Great Hungarian Plain, about 80 km SE of Budapest (referred to hereafter as "K-puszta"). This sample is a representative of Central European rural background air. Measurements were performed at a height of 10 m above ground and with an impactor with a flow rate of $\sim$630 L min^{-1} with a cut-off value of 1.5 μm. Aerosol samples were collected on quartz filters from June 19 to June 26 and from June 26 to June 29, 2001. The filters were stored in a freezer until analysis.

HULIS of the fulvic acid type were extracted from the filters collected in Weizmann and separated from other aerosol components by an isolation procedure based on the scheme used by the International Humic Substances Society (IHSS) for separating aquatic fulvic acids (http://www.ihss.gatech.edu) and adapted for atmospheric particulate matter. Briefly, filters were subjected to consecutive water extractions and then filtered through a 0.45-μm filter (PAL, Supor-450) and acidified to pH = 2. The extracted HULIS were separated from other water-soluble organic and inorganic aerosol species by preparative chromatography using XAD-8 resin followed by elution in a basic solution. The eluant was cation exchanged on an H$^+$-saturated cation-exchange resin to produce protonated acids and then freeze-dried. The freeze-dried samples were stored at room temperature under vacuum and in the dark. Mill-Q water (18 MΩ) was used for all solutions throughout the study. The K-puszta sample was extracted by soaking filters in water for 18 h. The extracts were subsequently filtered through a Millex HV 0.45-μm membrane filter. The light yellow colored extract had a pH of 4.0, indicating the presence of acidic compounds. The extract was acidified to pH 2.1 and then loaded on Oasis HLB Solid Phase Extraction (SPE) columns. The columns were rinsed with water, dried, and the retained organic compounds were eluted with methanol. Detailed discussion of the SPE procedure, including column selection and performance, can be found in Varga et al.[24] Water-soluble organic matter isolated from aerosol samples by the above procedure was characterized by various analytical techniques (UV, fluorescence, and FTIR spectroscopy, elemental analysis, liquid chromatography–mass spectrometry), and a high degree of similarity was found between their properties and those of aquatic and terrestrial humic substances.[21,28]

2.2 Separation of Suwannee River Fulvic Acid by molecular weight

Suwannee River Fulvic Acid (SRFA, IHSS code 1R101F) of reference grade was used as a model for atmospheric HULIS. SRFA is commonly used in laboratory studies as representative of atmospheric HULIS.[25,29–32]

Separation of the SRFA bulk sample into fractions of different average molecular weight was performed using diafiltration, which yielded a coarse subdivision of the initially polydisperse material according to the effective size of the molecules in solution. The nominal pore sizes of the ultrafiltration membranes do not correspond to the actual molecular weights of the fractionated material, due to molecular conformation and molecule–membrane charge interactions.[33] Hence, further analysis was carried out. The number average molecular weight of the different SRFA fractions was estimated using correlations between UV absorbance, molecular weight, and aromaticity of different aquatic and terrestrial fulvic acid samples available in the literature.[33,34] It was explicitly assumed that these correlations hold also for the Smoke and Pollution HULIS and the K-puszta HULIS samples. The estimates for the K-puszta sample are in agreement with previous determinations of the molecular weight by vapor pressure osmometry and mass spectrometry.[28] The SRFA filtration yielded five fractions of water-soluble material with number average nominal molecular weight between 440 and 740 Da. In this study, only F2, F5, and the bulk sample were examined. A detailed description of the equipment and procedures used is given in Dinar et al.[25]

2.3 Aerosol generation and classification

A simplified illustration of the setup is shown in Fig. 1. Briefly, aqueous solutions of the HULIS samples (0.02–1 mg cm^{-3}, below the Critical Micelle Concentration (CMC)) were nebulized using a constant output atomizer (TSI-3076) with dry particle-free nitrogen. The droplets were dried in two silica gel column dryers (RH <3%) yielding a polydisperse particle flow. The aerosols were neutralized (TSI 3012A) to obtain an equilibrium charge distribution.

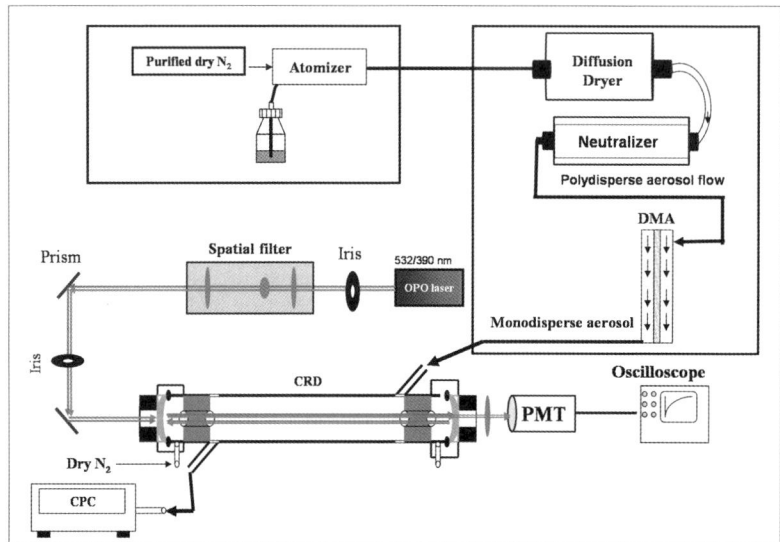

Fig. 1 The CRD-AS setup can be dividend into three main sections; (1) droplets are generated by nebulizing an aqueous solution; (2) aerosols are formed by drying the droplets, charged and size-selected in a DMA; (3) optical measurements of the aerosols. For simplicity, only one cavity is shown. By controlling particle size and concentration, the extinction efficiency of the aerosols is measured.

Size-selected monodisperse aerosol flow was generated using an electrostatic classifier (TSI Differential Mobility Analyzer (DMA)) operating with 5 SLM dry nitrogen sheath flow and fixed at an applied voltage. The size-selected monodisperse aerosols were directed through a dilution apparatus for precise control of particle number concentration and total flow, and then to the CRD cell (0.8 SLM). The resulting HULIS-containing particles are spherical, as was shown in previous studies.[13,27]

2.4 Cavity Ring Down Aerosol Spectrometer (CRD-AS)

The CRD-AS (shown in Fig. 1) is described in detail by Riziq et al.[35] However, this newly modified system consists of two identical cavities capable of measuring aerosol extinction in the visible (510–550 nm) and in the UV range (385–425 nm). Each cavity consists of two highly reflective plano-concave mirrors (curvature radii of 1 m and 99.995% reflectivity at 532 and at 405 nm (Los Gatos, USA)) mounted at the sides of a 90 cm 3/4" stainless steel tubes. A small purge flow of dry particle-free nitrogen (0.05 and 0.1 SLM) is introduced in front of each mirror and prevents mirror contamination by particle deposition. The aerosol flow enters the cavities through four tubes at 45° designed to ensure good mixing and even concentration of the particles. The particles exit the cavities with a flow rate of 0.95 SLM in a similar configuration. The length of the cavities occupied by particles is about 68 cm. The particle concentration is determined by a condensation particle counter (CPC, TSI 3022A), and it is measured before and at the exit of both CRD cells to ensure accurate concentration determination.

Both wavelengths (532 and 390 nm) are produced by an Optical Parametric Oscillator laser (OPO, EKSPLA, NT 342C/3/UVE, tunable 0.2–2 μm). The second harmonic of the Nd:YAG pump (10 Hz, 3–6 ns) is used for the 532-nm beam. The 390-nm beam is generated by doubling the 780-nm output of the OPO. The laser beam is directed through a spatial filter consisting of two lenses with focal length of 5 cm and 10 cm and a 100-μm pinhole in the common focus of the lenses. The beam diameter in front of the cavity is about 1 mm, with energy of 1–1.5 mJ per pulse.

The laser intensity emerging from the CRD cell is measured with a photomultiplier (Hamamatsu H6780-02 for visible and H6780-04 for the UV), and the resulting signal is fed into a digital oscilloscope (LeCroy, model 9361, 300 MHz). The digitized data for each laser pulse is transferred and stored in a computer.

The decay time for a cavity filled with particle-free dry nitrogen (τ_0) is 60 and 32 μsec for the visible and UV cavities, respectively. The decay times are shorter upon introduction of particles and with increasing concentration. The ability to precisely measure minimal differences in the ring down times between the empty cavity (τ_0) and a filled cavity (τ) provides a good estimate of the maximal sensitivity. The minimum detectable extinction coefficient (α_{min}) can be calculated using the following definition.[36]

$$\alpha_{min} = \frac{L\Delta\tau_{min}}{Cd\tau_0^2}, \qquad (1)$$

where $\Delta\tau_{min}$ is the minimal detectable change of the ring time ($\tau_0 - \tau$), obtained from the standard deviation of the decay time for a cavity filled with dry nitrogen. Taking $\tau_0 = 32$ and 60 μsec (for the UV and the visible, respectively) and $\Delta\tau_{min} = 0.2$ μsec for an average of 1000 shots, this amounts to $\alpha_{min} = 4.43 \times 10^{-9}$ cm^{-1} for the UV cavity and $\alpha_{min} = 1.26 \times 10^{-9}$ cm^{-1} for the visible cavity. Values of α_{min} are calculated for a cavity filled with nitrogen. In the case of aerosol measurement, the fluctuations in aerosol concentration inside the cavity may lead to larger errors in the decay time and to lower sensitivity compared to that calculated for the cavity filled with gas. In such a case, a statistical treatment may be used for estimating the error.[37] Such a calculation was not performed in the present study. However, previous studies suggest that the contribution of these fluctuations to the total error is small.[37]

2.5 Experimental procedure and data analysis

Extinction coefficients were measured by detecting the changes in decay time resulting from the presence of size-selected particles in the cavity. The extinction efficiency, which is the ratio of the Beer's law extinction cross section to the geometric area of the particle (Q_e), was obtained by varying the size and concentration of the particles in the cavity. For each sample, a plot of Q_e versus the size parameter, x (the ratio of the particle size (D) to the laser wavelength (λ), $x = \pi D/\lambda$), was obtained at both wavelengths.

The retrieval algorithm compares the measured extinction efficiency as a function of the size parameter, with the extinction efficiency calculated using a Mie scattering subroutine for homogeneous spheres,[38] while simultaneously varying the real and imaginary parts of the RI of the particles. The algorithm finds the complex RI by minimizing the "merit function" χ^2/N^2, where χ^2 is

$$\chi^2(n,k) = \sum_{i=1}^{N} \frac{(Q_{\text{ext measured}} - Q_{\text{ext calculated}}(n,k))_i^2}{\varepsilon_i^2}. \qquad (2)$$

The sum runs over N particle sizes, where n and k denote the real and imaginary parts of the RI, respectively. For every particle size, the measurement is performed subsequently with several different particle concentrations, so that ε_i, the measurement error for particle size i, given by the standard deviation of the individual extinction efficiencies, is also estimated. The built-in MATLAB function "fminsearch", which performs a direct search simplex algorithm in order to minimize the merit function, is used.[39] Starting with arbitrary initial values of the real and imaginary parts of the Refractive Index (RI), it builds a simplex consisting of three points in the (n, k)-space. At every optimization step, it generates a new point in the vicinity of the current simplex and adopts this point to the simplex if the resulting merit function is smaller. As a result, we obtain the complex RI of the substance at the minimal χ_0^2. To estimate the error of the retrieved RIs, we study the deviation of χ^2 in the vicinity of χ_0^2. Assuming the errors of the measurement ε_i^2 to be normally distributed, the values of χ^2 for different measurements follow a χ^2-distribution for the two degrees of freedom n and k. The quintile for the 68.3% (1 σ) confidence level for this parameter set is 2.298. Any measurement will fall within the 1σ error bound of the best fit of our measurement if $\chi^2 \leq \chi_0^2 + 2.298$. The values for n and k that fulfil this relation lie within contours around χ_0^2. Projections of the contour lines onto the n and k plane give the standard errors Δn and Δk, respectively.

3. Results

3.1 Validation of the system

The optical properties of PolyStyrene Latex spheres (PSL) are known.[37] Therefore, validation of the system was performed by measuring the extinction efficiency of 10 different PSL sizes at 532 nm, and a few sizes at 390 nm. The absorption (approaching zero) and the scattering indices of PSL do not change significantly between 532 and 390 nm (Fig. 2A and Table 1). The retrieved refractive indices were in good agreement with literature values.

The complex index of refraction of absorbing aerosols composed of nigrosin—a water soluble black dye, was also measured. The reported RI for nigrosin at 532 nm is $n = 1.67 + 0.26i$[40] and CRD-AS measurement by Lack et al.[41] support this value ($n = 1.7 + 0.31i$). Nigrosin is used for absorption validation because it is strongly absorbing at 532 nm and, furthermore, it forms spherical particles when generated from solution.[41] The extinction efficiency as a function of size parameter for nigrosin

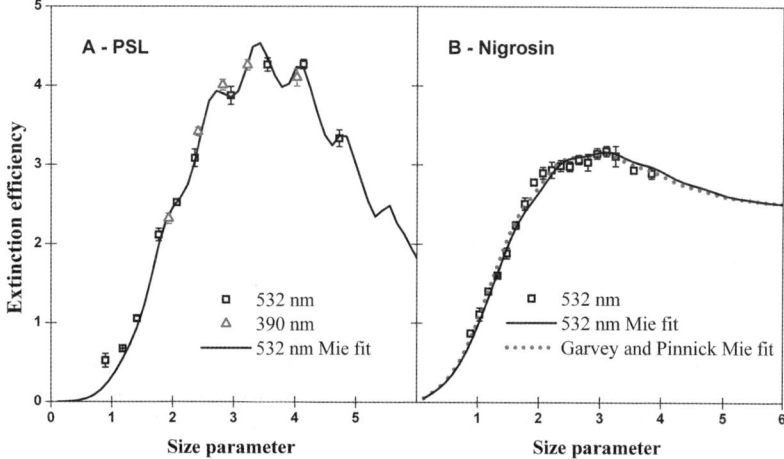

Fig. 2 The extinction efficiency (Q_{ext}) as a function of size parameter (x) of ten PSL sizes at 532 and 390 nm (A) and for nigrosin at 532 nm (B). The squares represent measurements points at 532 nm—□ and the triangles represent measurements points at 390 nm—△. The full lines are the Mie fit from which the RIs were derived (see Table 1 for values) and the dotted line (in B) represent the Mie curve of the RI measured by Garvey and Pinnick.

particles measured by the 532 nm CRD-AS is shown in Fig. 2B; the RI retrieved by our system is $n = 1.649 + 0.238i$, in agreement with the previous results (Table 1).

Finally, we measured the complex RI of Ammonium Sulfate (AS) at both 532 and 390 nm. At 532 nm, the index of refraction is known and the retrieved value agrees well with literature. As expected, at 390 nm, the value does not change significantly (Table 1).

3.2 The complex index of refraction of SRFA fractions

In this study, we use model fulvic acids to elucidate the relation between the chemical and physical properties of humic substances and their optical properties. Specifically, we study the optical properties of well characterized SRFA samples.[25–27] The chemiphysical properties of the samples are given in Table 2. Extinction efficiency as a function of size parameter for the three model samples is shown in Fig. 3A for 532 nm and Fig. 3B for 390 nm. The retrieved complex RI values are presented in Table 2. The real index (scattering) is lower and the imaginary index (absorption) is higher at 390 nm compared to their values at 532 nm for all SRFA samples. The absorption of SRFA increases with increasing molecular weight (M_N), aromaticity,

Table 1 Retrieved and published refractive indices for PSL, nigrosin, and ammonium sulfate aerosols

Sample	532 nm	390 nm	Literature values
PSL	$1.597 \pm 0.008 + i0.005 \pm 0.006$	$1.611 \pm 0.008 + i0.013 \pm 0.008$	This work
PSL	$1.59 + i0.0$	—	Manufacturer (589 nm)
PSL	$1.598 + i0.0$	—	Pettersson et al.[37] (532 nm)
Nigrosin	$1.649 \pm 0.007 + i0.238 \pm 0.008$	—	This work
Nigrosin	$1.67 + i0.26$	—	Garvey and Pinnick[40] (532 nm)
Nigrosin	$1.7 \pm 0.4 + i0.31 \pm .05$	—	Lack et al.[41] (532 nm)
Ammonium sulfate	$1.521 \pm 0.0026 + i0.002 \pm 0.002$	$1.525 \pm 0.0019 + i0.0001 \pm 0.0002$	This work (at 390 nm)

Table 2 Chemiphysical and optical parameters of the SRFA samples. The nominal molecular weight (M_N) and sample aromaticity were estimated by UV absorption correlations.[25] The effective density was measured directly.[27] The refractive indices were derived by CRD-AS (This work)

SRFA Sample	M_N/ Da	Aromaticity (%)	Effective density/ g cm^{-3}	532 nm	390 nm
F2	520	16	1.42 ± 0.01	1.613 ± 0.005 + i0.003 ± 0.005	1.597 ± 0.002 + i0.051 ± 0.002
Bulk	570	20	1.47 ± 0.02	1.634 ± 0.004 + i0.021 ± 0.005	1.602 ± 0.005 + i0.098 ± 0.005
F5	770	32	1.51 ± 0.01	1.650 ± 0.00 + i0.094 ± 0.009	1.644 ± 0.005 + i0.156 ± 0.005

and effective density in both the visible and UV. Up to a size parameter of ~1.5 (the smaller particles), the stronger absorbing species tend to have a higher extinction efficiency value compared to less absorbing species (Fig. 3).

3.3 HULIS samples

The extinction efficiencies as function of size parameter measured at 532 nm and at 390 nm for the Pollution, Smoke, and K-puszta samples are shown in Fig. 4A, 4B, and 4C, respectively. The chemo-physical parameters of the HULIS samples together with the retrieved RIs are summarized in Table 3. All samples absorb more at 390 nm compared to their absorbance at 532 nm. As for the SRFA, the general

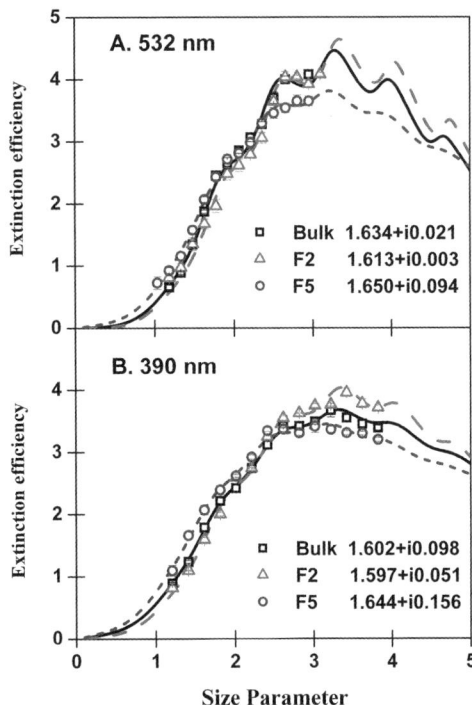

Fig. 3 The extinction efficiency (Q_{ext}) for SRFA as a function of size parameter (x) at 532 nm (A) and 390 nm (B). The symbols are the measurements taken for the model samples: SRFA (□), F2 (△) and F5 (○); the lines represent the best Mie curves fit from which the RI's were derived.

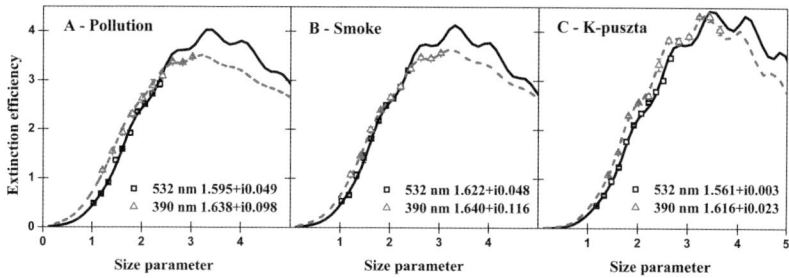

Fig. 4 The extinction efficiency (Q_{ext}) as a function of size parameter (x) as measured for HULIS samples at 532 nm (■) and 390 nm (△). The lines are the Mie curves using the retrieved complex refractive index.

trend is that the absorption of HULIS increases with increasing molecular weight (M_N), aromaticity, and density.

The K-puszta sample, which represents aerosols from a rural environment, has the lowest real index ($n = 1.561$, close to that of SRFA-F2) and the lowest imaginary index ($k = 0.003$) of all samples studied in this work for the visible measurements. In the UV, the K-puszta sample has the lowest real index ($n = 1.616$) of all HULIS samples and the lowest imaginary index ($k = 0.023$) of all HULIS and SRFA samples examined.

The optical properties of the Pollution and Smoke samples are close despite their different origins. In the visible, the Smoke sample has a higher real index ($n = 1.622$) compared to the Pollution sample ($n = 1.595$). Both have a similar imaginary index ($k = \sim 0.049$). At 390 nm, both samples have the same real index ($n = \sim 1.639$) with differences in the imaginary index; for Smoke, $k = 0.116$, while for the Pollution sample, $k = 0.098$.

4. Discussion

The complex RI was derived using a dual wavelength CRD-AS. The results indicate that CRD-AS can accurately derive aerosol scattering and absorption properties with high precision by measuring the extinction of the aerosol, without the need to collect high aerosol mass and extract various components in order to measure their individual properties. It also avoids the need to measure aerosol absorption and scattering separately.

Using a Mie code and the retrieved complex RIs of aerosols, it is possible to calculate the scattering and absorption efficiencies, Q_{sca} and Q_{abs}, which make up the total extinction efficiency, Q_{ext}. The total extinction and the scattering and absorption efficiencies for the three HULIS samples at 532 and 390 nm are shown in Fig. 5. The increase in absorbance in the UV region compared to the visible is clearly observed. There are significant differences between the optical properties of the Pollution and Smoke samples compared to those of the K-puszta sample. The

Table 3 Chemiphysical and optical parameters of the HULIS samples. The nominal molecular weight (M_N) and sample aromaticity are estimated based on UV absorption correlations.[25] The refractive indices were retrieved by CRD-AS (see experimental section)

HULIS Sample	M_N/ Da	Aromaticity (%)	532 nm	390 nm
Pollution	460	16	1.595 ± 0.004 + i0.049 ± 0.006	1.638 ± 0.006 + i0.098 ± 0.006
Smoke	420	15	1.622 ± 0.003 + i0.048 ± 0.009	1.640 ± 0.003 + i0.116 ± 0.005
K-puszta	340	11	1.561 ± 0.002 + i0.003 ± 0.004	1.616 ± 0.002 + i0.023 ± 0.002

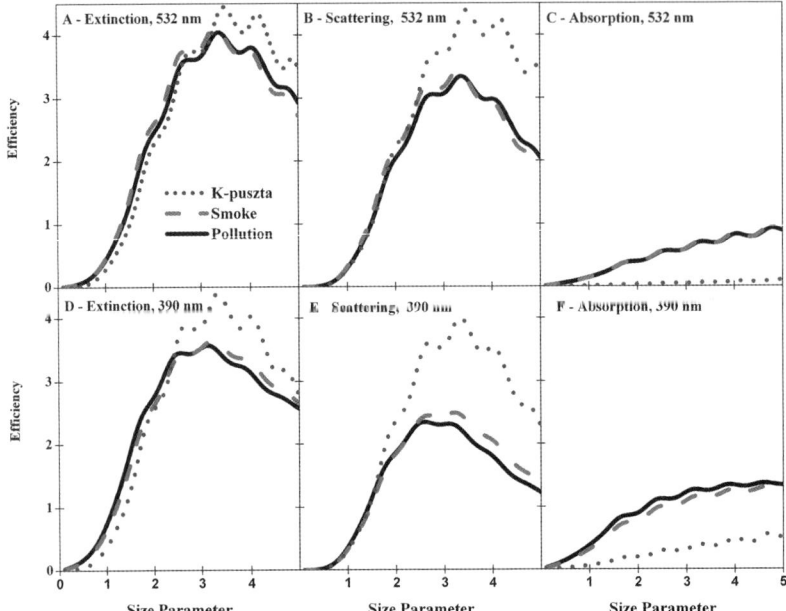

Fig. 5 Graphs A and D represent the extinction efficiencies of the three HULIS samples for 532 and 390 nm, respectively. Using a Mie code, the scattering and absorption components of the extinction can be derived (B, C, E and F).

increased absorbance of the Pollution and Smoke samples is associated with higher molecular weight and aromaticity compared to the K-puszta sample (Table 3).

The higher imaginary index measured for the Pollution and Smoke samples compared with the K-puszta sample may originate either from differences in the aerosol itself (lower aromaticity, lower molecular weight, Table 3) or from the sample preparation procedure, and deserves future investigation. The XAD resin used for Pollution and Smoke samples tends to concentrate more hydrophobic compounds, while the Oasis HLB Solid Phase Extraction (SPE) column applied to the K-puszta sample isolates more hydrophilic compounds, resulting in a minor deviation in chemical composition between the samples, as reflected in the lower aromaticity observed for the K-puszta sample, for example. This is supported by the fact that HULIS isolated from biomass-burning aerosol from Amazonia[12] using the HLB method had imaginary indices similar to that of K-puszta sample in the current study. The Smoke and Pollution samples, on the other hand, exhibit imaginary indices that are close to those estimated for organic carbon extracted by methanol from wood- and savannah-burning aerosols.[12]

Knowing the scattering and absorption efficiencies separately allows the calculation of further parameters that have direct applications in modeling the radiative impacts of aerosols and their possible climatic forcing. One of the most important parameters regarding aerosol optical properties is the SSA. As mentioned in the introduction, the SSA is defined as the fraction of scattering to total light attenuation. In terms of efficiencies:

$$\varpi(x) = \frac{Q_{\text{sca}}(x)}{Q_{\text{sca}}(x) + Q_{\text{abs}}(x)} \tag{3}$$

Using the scattering and absorption components of the extinction, the SSA of HULIS aerosol as a function of size was calculated at 532 and 390 nm (Fig. 6). The SSA is substantially lower in the UV than in the visible due to the increased

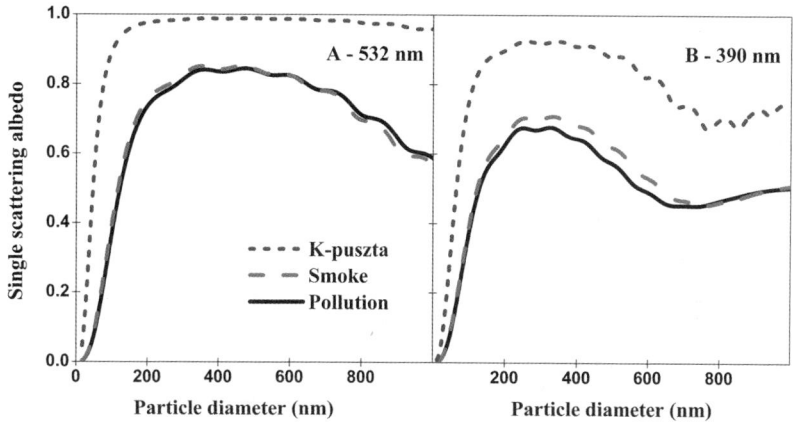

Fig. 6 Single scattering albedo for the HULIS samples at 532 nm (A) and 390 nm (B) as a function of size. The three samples absorb both in the visible and in the UV and could substantially influence absorbance in the atmosphere.

absorption at shorter wavelengths, as has been previously noted.[12,14] The SSA for the HULIS aerosols reflects the similarity between the Pollution and the Smoke samples, and points out the large difference between those and the rural sample, K-puszta. The K-puszta sample is mostly scattering while the other two samples represent HULIS which are more absorbing, and therefore can contribute more to atmospheric heating and to the semi-direct climatic effect. Note that, despite their lower absorbance at 532 nm, they also contribute to atmospheric absorbance in the visible range.

The dimensionless extinction efficiency Q_{ext} is independent of the particle mass. Other important parameters that can be derived by knowing Q_{ext}, Q_{sca}, Q_{abs} and the bulk density (ρ_P) are the mass efficiencies (Q_{ext}^{mass}, Q_{sca}^{mass}, and Q_{abs}^{mass}) of the aerosols in m^2 g^{-1}:

$$Q_i^{mass}(x) = \frac{3}{2\rho_P D_P} Q_{i\,calculated}(x) \qquad (4)$$

where the subscript i stands for extinction, scattering, and absorption, respectively. The mass efficiencies are, in principle, the extinction, scattering and absorption efficiencies of a particle normalized to its mass *via* the material density and the particle diameter (D_P). In this representation, the influence of particle mass on its extinction, scattering and absorption behavior as a function of particles size is demonstrated. The mass extinction, scattering, and absorption for the HULIS samples as a function of particle size are shown in Fig. 7. They were calculated based on the density of Pollution and Smoke samples measured by Dinar *et al.*[27] For the K-puszta sample, a density of 1.5 g cm^{-3} was assumed. Fig. 7 illustrates that in the UV range, smaller particles have a larger effect on both scattering and absorption of radiation compared to the visible. Differences between the rural (K-puszta) and the other HULIS samples are again observed. It is noted that, while for scattering there is a very sharp dependence on size, for particles 100–300 nm in diameter, absorption does not depend much on the size, and even small, atmospherically relevant aerosols (100–200 nm) absorb efficiently.

Our results indicate that the optical properties of HULIS may be associated with their origin, but may also depend on the extraction procedure, and further research should be done to clarify this. This study, as well as our previous work, indicates that SRFA can be used as surrogate for HULIS with care.

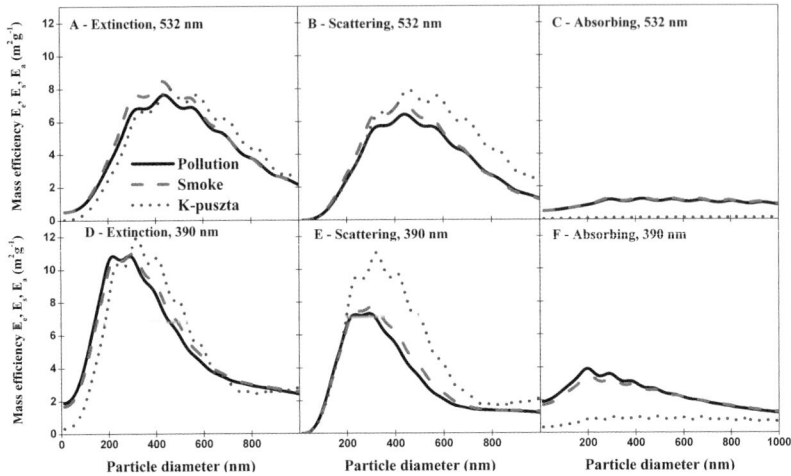

Fig. 7 Mass extinction (Q_{ext}^{mass}), scattering (Q_{sca}^{mass}) and absorbing (Q_{abs}^{mass}) efficiencies of the three HULIS samples at 532 nm (A–C) and 390 nm (D–F).

4.1 Implications for radiative forcing by HULIS

Evaluations of radiative forcing by organic aerosol components in the atmosphere have typically assumed that they behave similarly towards solar radiation as do sulfate type aerosols, *i.e.*, that they do not absorb in the visible part of the spectrum. This is true whether the organic aerosols are considered a separate entity, externally mixed with other aerosol types, as in the Geophysical Fluid Dynamics Laboratory (GFDL) AM2,[42] or whether they are modeled as internally mixed with elemental carbon and sulfates, as in the University of Michigan Radiative Transfer Model.[43] Such an assumption may influence estimates of the magnitude of the negative climate forcing (cooling) caused by fossil fuel, biomass burning, and secondary organic aerosols.[1] Although some studies have suggested that organic aerosol components may absorb visible radiation,[12,13,44–46] there has been limited quantification of the effect. Indeed, it has been suggested that a source of absorption is "missing" from theory as compared to recent measurements of biomass burning aerosol.[13,14] A light absorbing organic component such as the HULIS studied here may fill that gap.

In this section, we use the HULIS RI derived from CRD-AS measurements at 390 nm and 532 nm to estimate the SSA and Radiative Forcing Efficiency (the radiative forcing per unit optical depth; RFE) of pollution-type aerosols containing organic carbon. Specifically, we compare the results obtained using our retrieved HULIS RI with the results we would obtain assuming that the organic fraction has the same optical properties as sulfate, in order to assess the contribution of absorption by HULIS.

4.1.1 Methods. To simulate pollution-type aerosols, we assume an internal mixture of Ammonium Sulfate (AS), HULIS (as surrogate for the organic carbon), and soot, based on Bates *et al.*,[47] who found that sulfate, particulate organic matter, black carbon, and ammonium tend to exist as internal submicron mixtures. The "standard" aerosol particle size distribution that yielded the most reasonable results as compared with Bates *et al.*[47] is the water soluble component of the urban aerosol from Hess *et al.*,[48] Tables 1 and 4, which is loosely based on a similar internal mixture. We use this size distribution in conjunction with the retrieved HULIS RI, taking an RI of $n = 1.52 + i0.00$ for AS at both 390 and 532 nm (measured using our CRD-AS) and interpolating the RI for soot from d'Almeida,[49] Tables 4 and 3.

Table 4 Effective RI, SSA, and RFE at 390 nm and 532 nm for pollution aerosols both without and with HULIS. RFE is radiative forcing per unit optical depth. For reference, the values for pure AS are effective RI $n = 1.520 + i0.000$, SSA 1.000, and RFE -27.4 W m^{-2} at 390 nm, effective RI $n = 1.520 + i0.000$, SSA 1.000, and RFE -29.1 W m^{-2} at 532 nm, respectively

	Effective RI	SSA	Difference (%)	RFE/ W m^{-2}	Difference/ W m^{-2}
	390 nm, 78% AS, 20% HULIS, 2% soot				
Reference (98% AS, 2% soot)	1.525 + i0.009	0.947		−23.25	
Pollution	1.549 + i0.036	0.820	−13.4	−13.64	9.61
Smoke	1.549 + i0.031	0.840	−11.3	−15.22	8.03
K-puszta	1.544 + i0.013	0.925	−2.3	−21.91	1.34
	390 nm, 48% AS, 50% HULIS, 2% soot				
Reference (98% AS, 2% soot)	1.525 + i0.009	0.947		−23.25	
Pollution	1.584 + i0.078	0.697	−26.4	−4.61	18.64
Smoke	1.585 + i0.066	0.730	−22.9	−7.22	16.03
K-Puszta	1.573 + i0.020	0.895	−5.5	−20.19	3.06
	390 nm, 28% AS, 70% HULIS, 2% soot				
Reference (98% AS, 2% soot)	1.525 + i0.009	0.947		−23.25	
Pollution	1.608 + i0.107	0.641	−32.3	−0.59	22.66
Smoke	1.609 + i0.089	0.678	−28.4	−3.49	19.76
K-Puszta	1.592 + i0.024	0.877	−7.4	−19.18	4.07
	532 nm, 78% AS, 20% HULIS, 2% soot				
Reference (98% AS, 2% soot)	1.525 + i0.008	0.948		−25.04	
Pollution	1.540 + i0.018	0.897	−5.4	−21.21	3.83
Smoke	1.545 + i0.018	0.899	−5.2	−21.48	3.56
K-puszta	1.533 + i0.009	0.945	−0.3	−24.98	0.06
	532 nm, 48% AS, 50% HULIS, 2% soot				
Reference (98% AS, 2% soot)	1.525 + i0.008	0.948		−25.04	
Pollution	1.563 + i0.033	0.833	−12.1	−16.53	8.51
Smoke	1.576 + i0.032	0.839	−11.5	−17.20	7.84
K-puszta	1.546 + i0.010	0.941	−0.7	−24.90	0.14
	532 nm, 28% AS, 70% HULIS, 2% soot				
Reference (98% AS, 2% soot)	1.525 + i0.008	0.948		−25.04	
Pollution	1.578 + i0.043	0.798	−15.9	−13.93	11.11
Smoke	1.596 + i0.042	0.805	−15.1	−14.84	10.20
K-puszta	1.554 + i0.010	0.939	−1.0	−24.86	0.18

(We note that Bond and Bergstrom[10] recommend higher values for the soot RI. However, there is no other combination of size distribution and RI that gives satisfactory results when soot is involved. Bond and Bergstrom[10] also comment on this difficulty.) The effective RI of the internal mixtures is calculated using a two-inclusion implementation of the Maxwell–Garnett mixing rule (see, e.g., Sihvola,[50] pp. 61–63), where both HULIS and soot are considered inclusions within a matrix of AS. To test the sensitivity, we chose a volume fraction of 2% for soot and volume fractions of 20%, 50%, and 70% for HULIS. The single scattering parameters (SSA (ϖ) and asymmetry factor (g)) of the aerosol size distribution are then calculated using the Mie scattering subroutine of Bohren and Huffman[38] (in their Appendix A).

Since our emphasis is on differences between forcing values, in a similar fashion to Randles et al.,[51] we estimate the direct shortwave aerosol radiative forcing efficiency at the top of the atmosphere caused by a uniform, optically thin aerosol layer in the lower troposphere using:[52]

$$\frac{\Delta F}{\tau} = SD(1 - A_{\text{cld}})T_{\text{atm}}^2(1 - R_{\text{sfc}})^2\left[2R_{\text{sfc}}\frac{1-\varpi}{(1-R_{\text{sfc}})^2} - \beta\varpi\right], \quad (5)$$

where τ is the aerosol optical depth, S is the solar constant, set to 1370 Wm^{-2}, D is the fractional day length, set to 0.5, A_{cld} is the fractional cloud cover, set to 0.6, T_{atm} is the solar atmospheric transmittance, set to 0.76, and R_{sfc} is the surface albedo, set to 0.15 (appropriate for an urban area).[51,53] The parameter β is the average upscatter fraction (the fraction of scattered sunlight that is scattered into the upward hemisphere), which is a function of hemispheric backscatter fraction, b, defined as the ratio of backscattering efficiency to total scattering efficiency. While in principle β and b can be calculated using the Mie scattering subroutine, for simplicity we follow Anderson et al.[53] and use a simple functional relationship between β and b derived from the Henyey–Greenstein phase function:

$$\beta = 0.082 + 1.85b - 2.97b^2, \quad (6)$$

where in terms of g:[47,54]

$$b = \frac{1-g^2}{2g}\left[\frac{1}{\sqrt{1+g^2}} - \frac{1}{1+g}\right]. \quad (7)$$

ΔF is the change in net solar flux at the top of the atmosphere due to the presence of the aerosols (($F^\downarrow - F^\uparrow$)$_{\text{with aerosols}}$ − ($F^\downarrow - F^\uparrow$)$_{\text{without aerosols}}$). Since $F^\downarrow$, the incident solar flux, is a constant, a negative value of RFE indicates an overall increase in scattering by the atmosphere due to the presence of aerosols, while a positive value of RFE indicates an overall increase in absorption by the atmosphere. The less negative (or more positive) the RFE, the more the aerosol contributes to absorption by the atmosphere and to its subsequent heating.

Note that in all of the calculations, no water has been included. All values are effectively for 0% humidity.

4.1.2 Results. The effective RI, SSA, and RFE at 390 and 532 nm for pollution-type aerosols with different volume fractions of AS, HULIS, and soot are shown in Table 4. For reference, the first row of each section (labeled "reference") is the calculation for an aerosol comprised of the same volume fraction of soot but with the RI of AS in place of the RI of HULIS. The two columns labeled "Difference" contain the difference between the calculation with measured HULIS RI and the reference calculation. For the SSA, the difference is reported as the percent difference. For the RFE, the difference is the absolute difference in W m^{-2}. These differences express the magnitude of the effect of properly taking into account the HULIS optical properties, specifically absorption.

In all cases, when we include the measured HULIS RI, the Single Scattering Albedo (SSA), the ratio of scattering to total extinction, is lower than the reference SSA. Since the SSA decreases with increasing absorption, this clearly shows that absorption by HULIS can have a significant effect both in the visible and in the UV spectral ranges. As would be expected, as the volume fraction of HULIS increases, the SSA decreases even more with respect to the reference calculation. The difference between the SSA calculated with the retrieved HULIS RI and the reference calculation reaches 32.3% at 390 nm and 15.9% at 532 nm. Although estimated differently, the percentages are in agreement with Hoffer et al.,[13] who estimate that 35–50% of absorption in biomass burning aerosols at 300 nm may be attributed to HULIS. Similarly, in all cases, the RFE with retrieved HULIS RI is less negative than the reference RFE. As described in the previous section, the less negative or more positive the RFE, the more the aerosol contributes to atmospheric absorption and warming. The difference between the RFE calculated with the retrieved HULIS RI and the reference calculation reaches 22.66 W m^{-2} at 390 nm and 11.11 W m^{-2} at 532 nm. This sensitivity study emphasizes the increasing importance of the absorbance by HULIS at the UV range compared to the visible, although absorption in the visible should not be neglected.

Note that the calculations for the HULIS measured at K-puszta, the rural region in central Europe, exhibit a less substantial difference from the reference calculations than the other samples at both wavelengths. The reason is that the K-Puszta sample exhibits less scattering (a less refractory nature/lower real RI) than the other samples as well as less absorption (lower imaginary RI) than those samples. The coupling between scattering and absorption in the other samples, with their stronger refractory nature as well as stronger absorption, tends to increase atmospheric absorption more significantly as compared to the reference.

In summary, accounting for absorption by HULIS leads to a significant decrease in the SSA and a significant increase in the RFE (meaning more atmospheric absorption and atmospheric heating) of a pollution-type aerosol containing ammonium sulfate, organic carbon, and soot. However, if the HULIS do not have a high enough real RI as well as imaginary RI, the effect is less than with more refractory and absorbing HULIS. (Of course, HULIS always decreases the SSA and increases the RFE substantially in comparison to pure AS aerosols, but this is less relevant since pollution aerosol generally contains some soot; see the table caption for the pure AS values.) It should be noted that the magnitude of our results may vary if a different configuration of the aerosols is used, such as an external mixture or a core plus shell internal mixture. In addition, it is noted that we have not treated wet aerosol in this study. However, the tendency of our results should be robust. Finally, it should be noted that if the fraction of soot is increased, the effectiveness of the absorption by HULIS decreases because of masking by the soot. However, we still find a significant effect of HULIS with up to 25% soot by volume (results not shown). This study indicates that HULIS, in biomass smoke and pollution aerosols, in addition to black carbon, can contribute significantly to light absorption in the ultraviolet and visible spectral regions.

Acknowledgements

This work was partly funded by the German Israeli Science Foundation (GIF), Contract no. I-899-228.10/2005 and by the Minerva Foundation of the Max Planck Society, and by the Israel Science Foundation, Grant no. 1315/04.

References

1. *IPPC (2007) Climate Change 2007: The Physical Science Basis.* Summary for policy makers. Contribution of Working Group I to the Fourth Assessment Report of the Intergovernmental Panel on Climate Change, Cambridge University Press, Cambridge, 2007.
2. Y. J. Kaufman, I. Koren, L. A. Remer, D. Rosenfeld and Y. Rudich, *Proc. Natl. Acad. Sci. U. S. A.*, 2005, **102**, 11207–11212.
3. F. J. Nober, H. F. Graf and D. Rosenfeld, *Global Planet. Change*, 2003, **37**, 57–80.
4. V. Ramanathan, P. J. Crutzen, J. T. Kiehl and D. Rosenfeld, *Science*, 2001, **294**, 2119–2124.
5. D. Rosenfeld and A. Givati, *J. Clim. Appl. Meteorol.*, 2006, **45**, 893–911.
6. C. Erlick, V. Ramaswamy and L. M. Russell, *J. Geophys. Res., [Atmos.]*, 2006, **111**, 6204.
7. M. Z. Jacobson, *J. Phys. Chem. A*, 2006, **110**, 6860–6873.
8. S. Menon, J. Hansen, L. Nazarenko and Y. F. Luo, *Science*, 2002, **297**, 2250–2253.
9. Y. J. Kaufman and I. Koren, *Science*, 2006, **313**, 655–658.
10. T. C. Bond and R. W. Bergstrom, *Aerosol Sci. Technol.*, 2006, **40**, 27–67.
11. M. Kanakidou, J. H. Seinfeld, S. N. Pandis, I. Barnes, F. J. Dentener, M. C. Facchini, R. Van Dingenen, B. Ervens, A. Nenes, C. J. Nielsen, E. Swietlicki, J. P. Putaud, Y. Balkanski, S. Fuzzi, J. Horth, G. K. Moortgat, R. Winterhalter, C. E. L. Myhre,

K. Tsigaridis, E. Vignati, E. G. Stephanou and J. Wilson, *Atmos. Chem. Phys.*, 2005, **5**, 1053–1123.
12 T. W. Kirchstetter, T. Novakov and P. V. Hobbs, *J. Geophys. Res., [Atmos.]*, 2004, **109**, 21208.
13 A. Hoffer, A. Gelencsér, P. Guyon, G. Kiss, O. Schmid, G. Frank, P. Artaxo and M. O. Andreae, *Atmos. Chem. Phys.*, 2006, **6**, 3563–3570.
14 M. O. Andreae and A. Gelencser, *Atmos. Chem. Phys.*, 2006, **6**, 3131–3148.
15 M. C. Facchini, S. Fuzzi, S. Zappoli, A. Andracchio, A. Gelencser, G. Kiss, Z. Krivacsy, E. Meszaros, H. C. Hansson, T. Alsberg and Y. Zebuhr, *J. Geophys. Res., [Atmos.]*, 1999, **104**, 26821–26832.
16 S. Zappoli, A. Andracchio, S. Fuzzi, M. C. Facchini, A. Gelencser, G. Kiss, Z. Krivacsy, A. Molnar, E. Meszaros, H. C. Hansson, K. Rosman and Y. Zebuhr, *Atmos. Environ.*, 1999, **33**, 2733–2743.
17 M. Kanakidou, K. Tsigaridis, F. J. Dentener and P. J. Crutzen, *J. Geophys. Res., [Atmos.]*, 2000, **105**, 9243–9254.
18 E. R. Graber and Y. Rudich, *Atmos. Chem. Phys.*, 2006, **6**, 729–753.
19 S. Decesari, M. C. Facchini, E. Matta, F. Lettini, M. Mircea, S. Fuzzi, E. Tagliavini and J. P. Putaud, *Atmos. Environ.*, 2001, **35**, 3691–3699.
20 M. Gysel, E. Weingartner, S. Nyeki, D. Paulsen, U. Baltensperger, I. Galambos and G. Kiss, *Atmos. Chem. Phys.*, 2004, **4**, 35–50.
21 G. Kiss, B. Varga, I. Galambos and I. Ganszky, *J. Geophys. Res., [Atmos.]*, 2002, **107**, 8339.
22 Z. Krivacsy, A. Gelencser, G. Kiss, E. Meszaros, A. Molnar, A. Hoffer, T. Meszaros, Z. Sarvari, D. Temesi, B. Varga, U. Baltensperger, S. Nyeki and E. Weingartner, *J. Atmos. Chem.*, 2001, **39**, 235–259.
23 O. L. Mayol-Bracero, P. Guyon, B. Graham, G. Roberts, M. O. Andreae, S. Decesari, M. C. Facchini, S. Fuzzi and P. Artaxo, *J. Geophys. Res., [Atmos.]*, 2002, **107**, 8091.
24 B. Varga, G. Kiss, I. Ganszky, A. Gelencser and Z. Krivacsy, *Talanta*, 2001, **55**, 561–572.
25 E. Dinar, I. Taraniuk, E. R. Graber, S. Katsman, T. Moise, T. Anttila, T. F. Mentel and Y. Rudich, *Atmos. Chem. Phys.*, 2006, **6**, 2465–2481.
26 E. Dinar, I. Taraniuk, E. R. Graber, T. Anttila, Th. F. Mentel and Y. Rudich, *J. Geophys. Res., [Atmos.]*, 2007, **112**, DOI: 10.1029/2006JD007442.
27 E. Dinar, T. F. Mentel and Y. Rudich, *Atmos. Chem. Phys.*, 2006, **6**, 5213–5224.
28 G. Kiss, E. Tombacz, B. Varga, T. Alsberg and L. Persson, *Atmos. Environ.*, 2003, **37**, 3783–3794.
29 M. N. Chan and C. K. Chan, *Environ. Sci. Technol.*, 2003, **37**, 5109–5115.
30 G. Kiss, E. Tombacz and H. C. Hansson, *J. Atmos. Chem.*, 2005, **50**, 279–294.
31 A. Nenes, R. J. Charlson, M. C. Facchini, M. Kulmala, A. Laaksonen and J. H. Seinfeld, *Geophys. Res. Lett.*, 2002, **29**, 1848.
32 B. Svenningsson, J. Rissler, E. Swietlicki, M. Mircea, M. Bilde, M. C. Facchini, S. Decesari, S. Fuzzi, J. Zhou and J. Mønster, *Atmos. Chem. Phys. Discuss.*, 2005, **5**, 2833–2877.
33 A. I. Schafer, R. Mauch, T. D. Waite and A. G. Fane, *Environ. Sci. Technol.*, 2002, **36**, 2572–2580.
34 Y. P. Chin, G. Aiken and E. Oloughlin, *Environ. Sci. Technol.*, 1994, **28**, 1853–1858.
35 A. A. Riziq, C. Erlick, E. Dinar and Y. Rudich, *Atmos. Chem. Phys.*, 2007, **7**, 1523–1536.
36 S. S. Brown, H. Stark, S. J. Ciciora, R. J. McLaughlin and A. R. Ravishankara, *Rev. Sci. Instrum.*, 2002, **73**, 3291–3301.
37 A. Pettersson, E. R. Lovejoy, C. A. Brock, S. S. Brown and A. R. Ravishankara, *J. Aerosol Sci.*, 2004, **35**, 995–1011.
38 C. F. Bohern and D. R. Huffman, *Absorption and Scattering of Light by Small Particles*, Wiley, New York, 1983.
39 J. C. Lagarias, J. A. Reeds, M. H. Wright and P. E. Wright, *SIAM J. Control Optimization*, 1998, **9**, 112–147.
40 D. M. Garvey and R. G. Pinnick, *Aerosol Sci. Technol.*, 1983, **2**, 477–488.
41 D. A. Lack, E. R. Lovejoy, T. Baynard, A. Pettersson and A. R. Ravishankara, *Aerosol Sci. Technol.*, 2006, **40**, 697–708.
42 J. L. Anderson, V. Balaji, A. J. Broccoli, W. F. Cooke, T. L. Delworth, K. W. Dixon, L. J. Donner, K. A. Dunne, S. M. Freidenreich, S. T. Garner, R. G. Gudgel, C. T. Gordon, I. M. Held, R. S. Hemler, L. W. Horowitz, S. A. Klein, T. R. Knutson, P. J. Kushner, A. R. Langenhorst, N. C. Lau, Z. Liang, S. L. Malyshev, P. C. D. Milly, M. J. Nath, J. J. Ploshay, V. Ramaswamy, M. D. Schwarzkopf, E. Shevliakova, J. J. Sirutis, B. J. Soden, W. F. Stern, L. A. Thompson, R. J. Wilson, A. T. Wittenberg and B. L. Wyman, *J. Clim.*, 2004, **17**, 4641–4673.

43 J. Penner, M. Andreae, H. Annegam, L. Barrie, J. Feichter, D. Hegg, A. Jayaraman, R. Leaitch, D. Murphy, J. Nganga and G. Pitari, *Aerosols, Their Direct and Indirect Effects, in Climate Change 2001: The Scientific Basis*, Cambridge University Press, 2001.
44 W. F. Cooke, C. Liousse, H. Cachier and J. Feichter, *J. Geophys. Res., [Atmos.]*, 1999, **104**, 22137–22162.
45 J. E. Hansen, M. Sato, A. Lacis, R. Ruedy, I. Tegen and E. Matthews, *Proc. Natl. Acad. Sci. U. S. A.*, 1998, **95**, 12753–12758.
46 H. Flentje, R. Dubois, J. Heintzenberg and H. J. Karbach, *Geophys. Res. Lett.*, 1997, **24**, 2019–2022.
47 T. S. Bates, T. L. Anderson, T. Baynard, T. Bond, O. Boucher, G. Carmichael, A. Clarke, C. Erlick, H. Guo, L. Horowitz, S. Howell, S. Kulkarni, H. Maring, A. McComiskey, A. Middlebrook, K. Noone, C. D. O'Dowd, J. Ogren, J. Penner, P. K. Quinn, A. R. Ravishankara, D. L. Savoie, S. E. Schwartz, Y. Shinozuka, Y. Tang, R. J. Weber and Y. Wu, *Atmos. Chem. Phys.*, 2006, **6**, 1657–1732.
48 M. Hess, P. Koepke and I. Schult, *Bull. Am. Meteorol. Soc.*, 1998, **79**, 831–844.
49 G. A. d'Almeida, P. Koepke and E. P. Sheettle, *Atmospheric Aerosols: Global Climatology and Radiative Characteristics*, A. Deepak Publishing, Hampton, VA, 1991.
50 A. Sihvola, *Electomagnetic Mixing Formulas and Applications, IEEE*, 1996.
51 C. A. Randles, L. M. Russell and V. Ramaswamy, *Geophys. Res. Lett.*, 2004, **31**.
52 J. M. Haywood and K. P. Shine, *Geophys. Res. Lett.*, 1995, **22**, 603–606.
53 T. L. Anderson, D. S. Covert, J. D. Wheeler, J. M. Harris, K. D. Perry, B. E. Trost, D. J. Jaffe and J. A. Ogren, *J. Geophys. Res., [Atmos.]*, 1999, **104**, 26793–26807.
54 W. J. Wiscombe and G. W. Grams, *J. Atmos. Sci.*, 1976, **33**, 2440–2451.

General Discussion

Professor Signorell opened the discussion of Professor Choularton's paper:
(1) Related to the CPI measurements of ice particles: Is it possible to distinguish with this or other methods between different crystal structures or multiphase particles?
(2) What about very small (ice) particles?

Professor Choularton answered: As can be seen in Fig. 5 of the paper there is a very sharp and clear distinction between ice crystals and water droplets in the cloud from the CPI data and this distinction can be seen for particles larger than about 10 μm. For the ice crystals a wide range of habits can be seen. To identify very small ice crystals a device such as SID (Small Ice Detector) developed at the University of Hertfordshire would be ideal but one was not deployed during these experiments. However, our Forward Scattering Spectrometer Probe measures small droplets and ice crystals. A comparison with a separate instrument ADA suggested that the vast majority of these particles are liquid (this instrument is only sensitive to liquid droplets). A clear distinction was found between cloud particles, liquid or ice, typically 10 μm or more in size and unactivated aerosol particles (see also the comment by Dr Bower).

Dr Bower addressed Professor Signorell and Professor Choularton: There is a wide separation in size between aerosol particles that become activated into cloud hydrometeors and those that remain unactivated and interstitial to the cloud—so there is no problem differentiating between the two. Generally speaking there is no continuum in size between unactivated and activated particles—so the size distribution of activated cloud particles can be measured with confidence and by the instrumentation deployed in CLACE.

Dr Krieger asked. Can you exclude that the ice nuclei is something completely different from what you observe with the AMS (for example a mineral) but arrives along with the organics and sulfate?

Professor Choularton answered: It is not possible to totally exclude this. However, ice crystals were sampled through a counter flow virtual impactor and the resultant ice crystal residuals were analysed using ESEM by Mertes and colleagues.[1] These showed that there were many more ice crystals than mineral residues. Further many crystals showed carbon residuals with no mineral component so it is highly unlikely that mineral dust particles were the ice forming nuclei in many cases.

1   S. Mertes, B. Verheggen, S. Walter, P. Connolly, M. Ebert, J. Schneider, K. N. Bower, J. Cozic, S. Weinbruch, U. Baltensperger and E. Weingartner, *Aerosol Sci. Technol.*, 2007, **41**(9), 848–864.

Dr Grothe remarked: What is the impact of the oxygenated functional groups of soot on the nucleation process? Do you think that aged soot particles are better nuclei than fresh soot particles?

Professor Choularton replied: I really don't know whether aged soot particles are better as ice nuclei; chamber studies need to be performed to investigate this. However, the aged particles will incorporate inorganic hygroscopic material such as sulfate and this makes them a good cloud condensation nucleus for water activation. It is to be expected that solid residues from the soot once incorporated into a droplet in this way will act as a centre for the initiation of ice in a supercooled droplet.

Dr Cox commented: On page 16, lines 45-46. Here it is suggested that 'ice nucleation is initiated by oxidised organic aerosol coated with sulfate...' Such internally mixed aerosol particles are more likely to contain a sulfate core and an organic coating, considering the surfactant properties of sulfate (very low) and organics (could be highly surfactant).

Also, on p. 15, line 18, 'hydrophilic' should be 'hydrophobic'?

Professor Choularton answered: I agree with this comment once the particle has deliquesced, however, it is likely that a solid residue is left in the droplet from the soot. It is this solid residue, which is likely to be the ice nucleus within the liquid droplet formed from the hygroscopic components.

Concerning page 15 then indeed the initial seed particles are likely to be hydrophobic aliphatic hydrocarbons. This is what we find in the urban area. The oxidised secondary material added to these particles as they age makes them weakly hydrophilic.

Dr Reid remarked: The mass spectrometry measurements presented in Fig. 2 of your paper are interpreted as providing evidence for internal mixtures of organics, sulfate, nitrate and ammonium. For some components, particularly in the water time measurements, this is not entirely evident, particularly for sizes smaller than 200 nm. Please expand on the evidence for internal mixing.

Professor Choularton answered: The development of internally mixed particles is an ageing process and it is likely that the smaller particles will not have become fully internally mixed. The ageing process involves the hydrophobic aliphatic organic particles gaining oxidized secondary organic material as shown in Fig. 3. The particles then become weakly hygroscopic which favours then picking up sulfuric acid and other strongly hygroscopic inorganic species.

Another key piece of evidence for the particles being internally mixed after ageing is that when subject to low and high relative humidities in for example an HTDMA instrument then they show a single growth factor between the two humidities. This is commonly found in well-aged aerosol, including measurements during the CLACE experiments.

Professor Jaenicke asked: What is the ice particle concentration and temperature?

Professor Choularton replied: The ice particle concentration was typically around 50 per litre at a temperature of -10 °C. However, as shown in Fig. 4 it is highly variable, the ice crystal concentration was not closely related to local temperature at temperatures of -10 °C and below but was generally lower at higher temperatures.

Dr Murray asked: I've seen some presentations recently showing that ice particle shattering can sometimes give artificially large ice number densities. Could shattering be an issue in your experiments?

Professor Choularton responded: This is a controversial issue in cloud physics research and there is little doubt that some measurements are contaminated by fragments from the instrument inlet when flying in cloud at aircraft speeds (about 100 m s^{-1}). However, detailed laboratory studies have shown that this is not a problem in ground based experiments such as CLACE where sampling velocities are usually much less.

Dr Murray asked: NH_4^+ and H_3O^+ both have a mass of 18; how do you distinguish between them?

Professor Choularton answered: This question has been answered by Dr James Allan. The AMS calculates mass concentrations of ammonium and other species by inspecting multiple peaks in the mass spectra and where interferences occur, weighted subtractions are made based on known fragmentation patterns, which have been shown to be repeatable. In the case of water, the contributions to m/z 16 (O^+) and 17 (OH^+) are predicted based on the signal at m/z 18, which is almost entirely H_2O^+. The ammonium mass concentration is then calculated based on the remaining signals, which are due to NH_2^+ and NH_3^+. An interference from gas-phase oxygen is removed in a similar manner. This technique is described in more detail in Allan et al.[1]

1 A generalised method for the extraction of chemically resolved mass spectra from Aerodyne aerosol mass spectrometer data, J. D. Allan, H. Coe, K. N. Bower, M. R. Alfarra, A. E. Delia, J. L. Jimenez, A. M. Middlebrook, F. Drewnick, T. B. Onasch, M. R. Canagaratna, J. T. Jayne, and D. R. Worsnop, *J. Aerosol. Sci.*, 2004, **35**(7), 909–922.

Professor Signorell opened the discussion of Dr Grothe's paper: Is the crystallization kinetics influenced by the substrate? Do any equivalent measurements exist for (free) aerosol particles?

Dr Grothe replied: The substrate may have some influence on the nucleation process and crystal growth in very thin films, as demonstrated recently in our RAIR measurements (ref. 35 in our paper). However, thick films, as received by the quenching procedure in this work, are not so much affected, since the interface to bulk ratio of the overall sample is rather small.

There exists equivalent measurements by the group of M. A. Tolbert, who have used FTIR spectroscopy to investigate the crystallization kinetics of NAD aerosols.[1] They have also carried out film studies.[2] They received similar results to us.

1 R. S. Disselkamp, S. E. Anthony, A. J. Prenni, T. B. Onasch, M. A. Tolbert, *J. Phys. Chem.* 1996, **100**, 9127–9137.
2 R. T. Tisdale, A. M. Middlebrook, A. J. Prenni and M. A. Tolbert, *J. Phys. Chem. A*, 1997, **101**, 2112–2119.

Dr Huthwelker remarked: The NAD in your experiment has been produced by depositing aerosols to a very cold (80 K) surface, with subsequent heating until the NAD forms. In the atmosphere, aerosols are cooled down from higher temperatures, but never reach 80 K. Can NAD form when cooling down without reaching such low temperatures? Would not other hydrates (*e.g.* NAT) form first under real atmosphere conditions?

Dr Grothe answered: Our study is a laboratory model investigation aimed at understanding the kinetics of the phase transitions of metastable nitric acid hydrates. We didn't intend to follow exactly the atmospheric processes.

In an aerosol chamber (AIDA) and in a cryostat the nucleation of NAD aerosols have already been observed under conditions similar to the atmosphere (slow cooling below 196 K). The phases formed under real atmospheric conditions depend on the cooling rate, the respective degree of supersaturation, the temperature and the concentration. Our model experiments point out that α-NAD, β-NAD, α-NAT and β-NAT might be possible candidates to be formed.

Professor Rühl said: What is the relevance of the laboratory results in comparison with LIDAR work, where the depolarization of the back scattered light probes solid particles?

Dr Grothe replied: We may emphasize that it wasn't the intention of our laboratory model to simulate certain atmospheric processes but to prove which nitric acid hydrate phases might persist under stratospheric conditions, to record their spectroscopic signatures, their crystallization and transition kinetics and to image their particle morphologies.

LIDAR measurements have proven solid PSC particles consisting of nitric acid hydrates to exist between 180 K and 196 K. In Fig. 1 we present a phase diagram where our kinetic investigations have been summarized—the concentration is related to our particular model experiment. The temperatures mark the isokinetic data, where 90% conversion of the phase transition was observed after 60 min. The diagram unambiguously shows that beside thermodynamically stable β-NAT also metastable α-NAT, α-NAD and α-NAD might exist in the stratosphere, at least for a while.

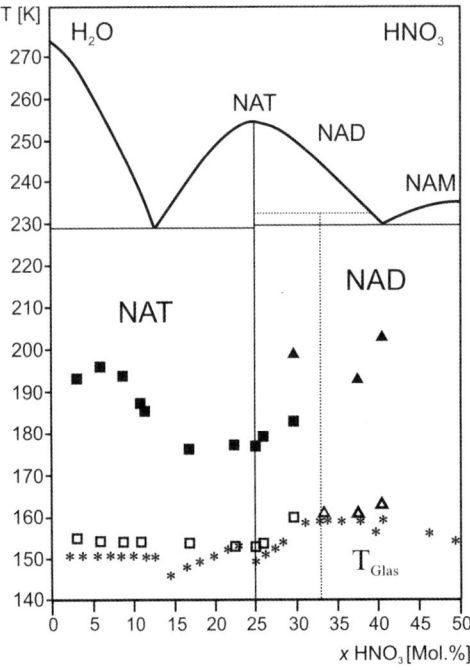

Fig. 1 Isokinetic data of nitric acid hydrates: squares refer to NAT and triangles refer to NAD. Open symbols indicate the crystallization temperatures of the α-phases and full symbols mark the phase transition temperatures from α- to β-phases. The stars denote glass transition points taken from the literature.

Dr Reid addressed Dr Grothe and Dr Murray: The paper presentation began with the statement that 'most particles are not in thermodynamic equilibrium'. Under what circumstances should we assume aerosols are at thermodynamic equilibrium and where, instead, should we really need to consider the kinetics of particle transformation?

Dr Murray answered: Metastable (non-equilibrium) phases may be important in many systems. A good example of this is the formation of ice from a gas supersaturated with water vapour. Huang and Bartell[1] performed an elegant experiment where they expanded a humid flow of gas and when it cooled to 200 K it became supersaturated. They found that water droplets homogeneously nucleated from the gas phase water vapour in a highly supercooled state rather than the more stable crystalline ice. They then observed these droplets to crystallise to the metastable cubic crystalline form of ice, rather than the stable hexagonal phase. In the water system the phase that forms most rapidly is the least stable and then the transformation to the more stable phases is governed by phase change kinetics. This is the Ostwald law of stages in action. In many atmospheric systems the cascade through the metastable phases is very rapid and we simply do not observe the metastable phases because they exist for very short periods of time, but this does not mean that they are not important in nucleation and crystallisation processes. However, where kinetics of phase changes are slower the metastable phases may persist for significant periods of time. Murray and Bertram[2] and Murray et al.[3] demonstrated that the cubic ice can form and persist for atmospherically significant periods of time in frozen water and solution droplets. In a similar way, Grothe has shown that metastable hydrates of nitric acid can form and will only very slowly transform to the thermodynamically stable phases. A further example of metastability in the atmosphere, which is very familiar, is that of supercooled water droplets which we know persist to a temperature of $\sim -38\ °C$, before finally crystallising to ice.

In answer to Dr Reid's first question (Under what circumstances should we assume aerosols are at thermodynamic equilibrium?): This question can only be answered for individual systems. Many phase changes occur through a sequence of steps where metastable phases initially form, due to faster kinetics (lower energy barrier, and only after some time can they transform to the thermodynamically favoured form. Others such as deliquescence of salt particles proceed with no kinetic barrier and probably do not form metastable phases. In short, metastable phases may form where there are kinetic barriers to the phase changes and therefore a competition between forming the more stable and less stable products. In answer to Dr Reid's second question (should [do] we really need to consider the kinetics of particle transformation?): Yes, in many circumstances we do need to consider kinetics. In some situations such as supercooled water in mixed phase or liquid clouds the atmospheric community are very happy about doing this, but in other situations, specifically in the low temperature upper troposphere and stratosphere, there is still a lot to be learnt in my opinion.

1   J. F. Huang and L. S. Bartell, *J. Phys. Chem.*, 1995, **99**, 3924.
2   Benjamin J. Murray and Allan K. Bertram, *Phys. Chem. Chem. Phys.*, 2006, **8**, 186.
3   B. J. Murray, D. A. Knopf and A. K. Bertram, *Nature*, 2005, **434**, 202.

Dr Grothe responded: Oswald's step rule states that under non-equilibrium conditions it is not the thermodynamically most stable crystalline phase which nucleates but the one with the lowest free energy barrier of formation. Its height determines whether and how fast the phase transformation occurs. In other words, nucleation as the introductory step in crystallization is controlled by kinetics rather than thermodynamics. The height of the nucleation barrier is indirectly proportional to supersaturation $(\ln S)^{-2}$ and is proportional to the third power of the interfacial energy (σ^3). If however the short range order of the surrounding supersaturated liquid incorporates already the same structural components as the later crystalline metastable phase does, then the respective interfacial energy is particularly low and also the nucleation barrier is significantly reduced. A good example is the crystallization of acetonitrile, where a metastable modification nucleates from a liquid containing the same anti-parallel dimers, see Tizek et al.[1]

In materials science crystallization of metastable phases is often reported for submicrometer particles. When the surfaces of polymorphs of the same material possess different interfacial energies, a change in phase stability can occur with decreasing particle size and in response to changes in the surface environment, *i.e.* the nature of the surrounding molecules, see H. Zhang et al.[2] and M. P. Finnegan et al.[3]

In two recent papers (Tizek et al.[4,5]) we have described the nucleation of nitric acid dihydrate (NAD) and nitric acid trihydrate (NAT). α-NAD, β-NAD and α-NAT are metastable phases, which irreversibly transform into β-NAT, which is thermodynamically stable. The reason why most field measurements (but not all) observe β-NAT as the prevalent phase in Polar Stratospheric Clouds (PSCs) is simply time. β-NAT is just the "end-product" of a complicated nucleation and phase transition process of these cloud particles. Another good example is the work of B. Murray and A. Bertram regarding the existence of cubic ice in the atmosphere.[6] In general, atmospheric physical chemistry has not paid enough attention to the existence of metastable modifications of solid aerosol particles and some are probably still unknown. However, the knowledge of the exact phase composition is essential in order to understand surface reactivity (uptake) and the optical properties (optical indices) of atmospheric aerosols. Therefore, we also have to know the time scale on which a metastable phase might persist.

It's just too simple to look at the phase diagram, which only exhibits the equilibrium phases, and to consider atmospheric conditions. Real atmospheric conditions are rather often non-equilibrium conditions, as it has been described in a recent paper by Th. Peter[7]—high supersaturation and surfactants have strong impact on ice nucleation and growth. The other way round, slow cooling, constant partial pressures, and low supersaturation would lead to slow crystallization under equilibrium conditions which forms particles being in thermodynamic equilibrium with its environment—this is often not the case in the atmosphere.

1   Heinz Tizek, Hinrich Grothe and Erich Knözinger, *Chem. Phys. Lett.*, 2004, **383**, 129–133.
2   H. Zhang, F. Huang, B. Gilbert and J. F. Banfield, *Nature*, 2003, **424**, 1025–1029.
3   M. P. Finnegan, H. Zhang and J. F. Banfield, *J. Phys. Chem. C*, 2007, **111**, 1962–1968.
4   Heinz Tizek, Erich Knözinger and Hinrich Grothe, *Phys. Chem. Chem. Phys.*, 2002, **4**, 5128–5134.
5   Heinz Tizek, Erich Knözinger and Hinrich Grothe, *Phys. Chem. Chem. Phys.*, 2004, **6**, 972–979.
6   Benjamin J. Murray and Allan K. Bertram, *Phys. Chem. Chem. Phys.*, 2006, **8**, 186–192.
7   Th. Peter, C. Marcolli, P. Spichtinger, T. Corti, M. B. Baker and Th. Koop, *Science*, 2006, **314**, 1399–1402.

Dr Reid commented: Is the equilibrium state an assumption made in your interpretation of the properties of ice clouds?

Professor Choularton replied: A truly mixed phase cloud is not in equilibrium as ice is thermodynamically favoured at temperatures below 0 °C over liquid water. Hence no such assumption is implicit in the analysis.

Dr Murray remarked: Can you identify all of the Bragg peaks in your XRD patterns of frozen HNO_3/H_2O solutions?

Dr Grothe responded: We have identified all Bragg peaks in our powder diffractograms by Rietveld refinement using the respective single crystal data. However, in the case of α-NAT this wasn't possible, since the crystal structure of α-NAT is unknown.

Professor Signorell asked: I do not believe that you can distinguish between prolate and oblate spheroids for NAD (with similar axis ratios) from infrared spectra (Fig. 6 of your paper) of an aerosol ensemble. The differences are too small if one considers uncertainties in refractive index data, size distribution, as well as a distribution of shape.

Dr Grothe replied: That's correct, in the paper we state: "It is interesting to observe that the calculated spectra for both prolate and oblate spheroids are relatively similar and resemble more closely the spectra of α-NAD aerosols reported by Wagner *et al.* (ref. 34 in our paper) from measurements in the AIDA facility. These authors concluded that the oblate shape provided a better fit to their data.

However, the ESEM results and the spectral simulations of the present work indicate that α-NAD aerosol particles might have a prolate shape." We are saying the same; we think that both spectra are rather similar, so you can't distinguish between both. However, taking in to account the ESEM results we can say that α-NAD aerosol particles should have a prolate shape.

Professor Mason asked: What is the reproducibility of your data from deposition to deposition? Is there any damage induced by X-ray analysis (*e.g.* reorientation effects of aerosols on surfaces)?

Dr Grothe responded: The reproducibility of the phase composition of the quenched samples is excellent. We have verified our data by four different techniques: XRD, FTIR, Raman and ESEM. We can exclude damage of the sample by X-rays, since the sample doesn't absorb in this frequency range.

Professor Rudich opened the discussion of Professor Jaenicke's paper:
Which atmospheric process (such as cloud formation, climate, health...) depend, or are affected by this high temporal variability?

Professor Jaenicke replied: Temporal Variability indicates that there are regions ("clouds", entities) with increased or decreased concentration. Increased concentrations could provoke cluster forming by faster coagulation. Decreased concentrations on the other hand could hinder coagulation and add to stability. Or think about clouds with droplets rather close together. The humidity field then is influencing each other droplet. You mention health. Health problems could start with a "puff" of increased concentrations. The average concentration resulting from long lasting measurements could shroud the actual load starting inflammation.

I would like to expand your question to clouds, which are aerosols as well. We have developed a holographic system to observe clouds.[1] The evaluation showed that droplets in clouds are not evenly distributed, rather 10% of the droplets with radii greater than 4.5 μm are closer to each other than 100 droplet radii. This could mean that the humidity field is modified with consequences for droplet grows (or shrinks), for chemistry and cloud cover. That remains to be studied.

1 E. V. Uhlig, S. Borrmann and R. Jaenicke, Holographic *in-situ* measurements of the spatial droplet distribution in stratiform clouds, *Tellus*, 1998, **50B**, 377–387.

Mr Malila asked: I would just like to point out that there are always stable cluster ions present. How does this fit to your picture that smallest and largest particles have the greatest variability?

Professor Jaenicke responded: You certainly are correct. But how is the variability of stable cluster ions on timescales I have indicated?

Dr Reid replied: A standard colloidal science textbook defines all colloidal dispersions as being thermodynamically unstable due to their high surface free energy. Their kinetic stability is dependant on the breadth in size distribution. Further, colloidal particle size distributions can indeed be very broad. Based on their thermodynamic instability, aerosols seem to be well classified as a dispersion colloid.

Professor Jaenicke answered: I have based my hypothetical and provocative question on the colloid definition in http://www.mpik.mpg.de/kc/what_is_a_colloid (8 Nov 2006).

Dr Bower remarked: Are the individual size distributions shown in the figure, individual real simultaneous (at all sizes per distribution) measurements which are then averaged to give the "average" measurement curve or are they theoretical distributions produced to show how such variability disappears when an average is taken?

Professor Jaenicke answered: The size distributions shown in the figure are real simultaneous distributions (all are taken at the same time). This is especially true for the EAS measurements. The other figure showed real simultaneous measurements based on a model, how to handle the simultaneously taken integral measurements.

Professor Choularton asked: What instrument was used to make the 10 s sample measurements? What is it's sample volume? Please can you comment on the contribution of sampling statistics to the fluctuations observed between the 10 s spectra ?

Professor Jaenicke responded: I have shown two graphs with aerosol size distributions. One was from Resolute, Canada and I estimated the measuring frequency in front of the auditorium to be 10 s. Those measurements are based on the evaluation of parallel measuring integral aerosol parameters with instruments, like 2 (with a different lower cut-off) of TSI LPC 3753, 2 (with a different lower cut-off) of LPS-C 501, CPC 3022. The instruments have typical time resolutions between 1 s and 60 s. For details about the evaluation method see Dreiling and Friederich.[1] Depending on the concentration, these instruments count individual particles or measure the whole volume. Individual counts have an error described with a Poisson distribution. Further details can be seen in the instruments' description. For the second graph we used an Electrical Aerosol Spectrometer (EAS). The whole work is described in Prats-Porta.[2]

1 V. Dreiling and B. Friederich, Spatial distribution of the arctic haze aerosol size distribution in western and eastern Arctic, *Atmos. Res.*, 1997, **44**, 133–152.
2 N. Prats-Porta, *Zeitliche Variabilität des atmosphärischen Aerosols* (*Untersuchung in einer semi-urban Umgebung*), Diploma-Thesis, Institut für Physik der Atmosphäre der Universität Mainz, University Mainz, 2004.

Dr Bower commented: Are the particles being measured charged (as in the atmosphere) or neutralised prior to measurement?

Professor Jaenicke answered: In measurements with the EAS, the particles are electrically charged. One fraction is diffusion, the other field charged. For details see the Diploma-Thesis mentioned previously.[1] Also see Tammet *et al.*[2] The data from Resolute have been obtained with instruments not using any charge.

1 N. Prats-Porta, *Zeitliche Variabilität des atmosphärischen Aerosols* (*Untersuchung in einer semi-urban Umgebung*), Diploma-Thesis, Institut für Physik der Atmosphäre der Universität Mainz, University Mainz, 2004.
2 H. Tammet, A. Mirme and E. Tamm, Electrical aerosol spectrometer of Tartu university, *Atmos. Res.* 2002, **62**, 315–324.

Professor Chan said: You presented results of temporal variations of size distributions. Spatial variation can also exist. I would like to know your comments on how these variations affect our understanding of atmospheric aerosols. Furthermore, wouldn't spatial variation be a more different problem to solve?

Professor Jaenicke replied: Spatial variations show as temporal variations, if the air moves by. So both variations are mixed for a stationary instrument. Temporal variations only can be measured, if the instrument drifts with the air.

The atmospheric aerosol is a highly variable subject. And only portions with a comparatively long residence time (between 0.1 and 1 μm) can be regarded as rather stable. This range influences the visibility in the atmosphere. So you can estimate by yourself the stability of that subject.

Dr Krieger asked: What is the difference of the rapid fluctuations you describe in aerosol size distributions to the concentration fluctuations in a gas phase species (like OH)?

Professor Jaenicke answered: I don't see a difference between those species (see Junge[1]). In my understanding, fluctuations and residence time are also connected in other subjects.

1 C. Junge, *Tellus*, 1974, **26**, 477.

Mr Homer remarked: You mentioned in the paper that a problem with these measurements was the low time resolution. Do you know of any techniques, or could you suggest any improvements to existing techniques that could improve this time resolution?

Professor Jaenicke replied: Many instruments today have a time resolving problem. Just think about the filter collection of particles, which is mostly influenced by the very large particles, and those have a very short residence time. But even for counting, those particles present a problem, because of their low concentration. So for those particles I hardly have a recommendation. For the smaller particles I have indicated techniques for counting.

Professor Rühl said: Can you specify what an 'acceptable time resolution' would be to probe inhomogeneously distributed particle size distributions in the atmosphere?

Professor Jaenicke responded: This depends on the residence time of that size. It is the understanding, that the residence time is the time constant of an exponential decay. To follow an exponential decay, it would be my guess that 5 measurements within one e-fold would be acceptable.

Miss Miles opened a general discussion of the papers by Professor Chan and Mr Treuel: What are the size of the error bars on your RH measurements? How confident are you that the RH your droplet experiences is the one that you measure, given that there are likely to be RH gradients across your cell between your RH probe region and the region occupied by your droplet?

Mr Treuel answered: The margin of error of our experimental data is given in the paper and differs for different substances dependent on the number of experiments and the degree of reproducibility. Since we changed the RH in small steps and give the system sufficient time to equilibrate we are confident that the droplet experiences the RH which we measure—within the applicable margin of error.

Professor Chan responded: The error of RH measurements is within 1%. Calibration experiments using compounds of known hygroscopic properties were conducted for the determination of RH as a function of time in the scanning experiments on glutaric acid/ammonium sulfate particles. For the hygroscopic measurements of the reaction products, RH was changed in discrete steps. More information of our control of RH in the scanning experiments can be found in Choi and Chan[1]

1 M. Y. Choi and C. K. Chan, Continuous Measurements of the Water Activities of Aqueous Droplets of Water-Soluble Organic Compounds, *J. Phys. Chem. A*, 2002, **106**(18), 4566–4572.

(277:116) **Dr Krieger** opened the discussion of Mr Treuel's paper: In Fig. 4 of your paper the mass uptake during deliquescence of a pure ammonium sulfate droplet is spread in relative humidity about a range of roughly 0.5% relative humidity. Is this due to kinetic limitations of your experimental setup (*e.g.* time response of the humidity sensor, equilibration time of the chamber) or due to the observation of an absorption film prior to deliquescence?

Mr Treuel replied: We attribute this effect to the response time of our RH probe. Kinetic limitations due to the equilibration time of the chamber are much faster than the sensor reaction time. We deduce this from monitoring of size changes of liquid droplets in our system which follows changes in the RH flow more rapidly than the sensor. However, by using a small step size when changing the RH and allowing enough time for the sensor reaction we have been able to reduce this effect to the roughly 0.5% RH shown in Fig. 4 which is well within the margin of error of our experiments.

Professor Ray asked Professor Chan and Mr Treuel: Do you alter the RH surrounding a particle in steps or in one shift? Are your measurements static or 'dynamic'? How do you create a gas stream of specific humidity? Are the acids used in the study hygroscopic or hydrophobic?

Mr Treuel responded: We alter the RH within our experimental chamber in small and well defined steps between which we allow sufficient time for an equilibration of the system. The gas stream is created by passing a controlled gas flow through a humidifier and then combining it with a controlled flow of dry nitrogen. Varying the mixing ratios of the two gas flows allows us to create a defined RH in the resulting gas flow.
    The dicarboxylic acids used in our experiments are all hydrophilic acids with hygroscopic properties.

Professor Chan responded: The RH was changed in steps for the reaction experiments and the hygroscopic measurements were "static". The glutaric acid/ammonium sulfate experiments were performed using the "dynamic" (scanning) method. Gas streams of specific RH were prepared by mixing a dry air stream and a humid air stream from a bubbler. The acids were hydrophobic and became more hygroscopic after reactions.

Mr Hunt addressed Professor Chan and Mr Treuel: As RH is varied by altering the proportion of dry/wet airflows, does the total flow rate remain constant? Does flow rate have an influence on the experimental outcomes?

Mr Treuel responded: We keep the total flow rate constant throughout the experiment to minimize its influence on the experimental error. However, for an accurate determination of the q/m ratio we momentarily interrupt the flow, as size and shape of the trapped particles have an influence on the aerodynamic behaviour. Therefore the experimental outcome should not be influenced by the gas flow in our chamber.

Professor Chan replied: For the scanning experiments that involve a single step change of RH and the use of a calibration RH (t) curve, the flow rate was held constant at the value used in the calibration experiments. For static experiments that involve multiple RH step changes, the equilibrium values were obtained and were not affected by the flow rate since the gas flow was turned off momentarily for balancing voltage measurements.

Professor Rühl asked Mr Treuel: Especially for temperature dependant experiments, it is useful to use instead of the relative humidity, the partial pressure of water (see Hamza *et al.*[1]). The reason is that the water saturation pressure is a temperature dependant property.

1   M. A. Hamza, B. Berge, W. Mikosch and E. Rühl, *Phys. Chem. Chem. Phys.*, 2004, **6**, 3484

Mr Treuel answered: We have indeed looked at our experimental data from that perspective and appreciate you comment.

Dr Murray addressed Professor Chan and Mr Treuel: Why is there no kinetic barrier to the deliquescence?

Mr Treuel responded: The kinetic limitation to the deliquescence is the gas phase diffusion of water molecules towards the particle.

Professor Chan replied: Our deliquescence experiments[1] do not allow us to examine any kinetic barrier beyond mass transfer effects. We have seen mass transfer effects in the deliquescence of ammonium sulfate particles coated with octanoic acid.

1   M. N. Chan and C. K. Chan, Mass Transfer Effects on the Hygroscopic Growth of Ammonium Sulfate Particles with a Water-Insoluble Coating, *Atmos. Environ.*, 2007, **41**, 4423–4433.

Dr Krieger commented: on why there is no nucleation barrier in deliquescence of a salt particle in contrast to efflorescence which is occurring at much drier conditions then expected from thermodynamics.

First, increasing the relative humidity up to the deliquescence relative humidity (DRH) leads to water absorption: at dry conditions the coverage will be sub-monolayer and then closer to the DRH it will evolve to multilayer coverage (*e. g.* Foster *et al.*[1]). At this stage ions will dissolve into the adsorption layer (*e. g.* Hücher *et al.*[2]). Therefore, the new, aqueous phase is already present at the deliquescence. If you try to melt a crystal not on its surface but within its bulk, you will indeed encounter a nucleation barrier.

Second, it helps that the new aqueous phase is less ordered than the crystalline phase.

1   M. Foster and G. E. Ewing, *J. Chem. Phys.*, 2000, **112**, 6817.
2   M. Hücher, A. Oberlin and R. Hocart, *Bull. Soc. Fr. Mineral. Cristallogr.*, 1967, **90**, 320.

Dr Murray replied: No further comment.

Professor Bain addressed Professor Chan and Mr Treuel: The particles in the EDB experiments are large (> 10 μm) so surface adsorption is relatively unimportant. For particles in the 100 nm range, thick adsorbed films or prewetting of the crystal surface by the saturated solution could significantly change the water uptake. How confidently can you extrapolate your results on large particles to the sub-μm range?

Mr Treuel replied: We would not extrapolate our results to particles in the sub-micron range.[1] We explore the nanosize effects on deliquescence of sodium chloride particles and the paper suggests an influence of the size on the DRH for nanoscale particles.

1   G. Biskos, A. Malinowski, L. M. Russell, P. R. Buseck and S. T. Martin, *Aerosol Sci. Technol.*, 2006, **40**, 97–106.

Professor Chan responded: EDB measurements are in general consistent with tandem differential mobility analyzer (TDMA) measurements which use submicron particles. However, it is possible that the prewetting of particles may result in a small water uptake (typically less than 5% mass change) prior to deliquescence and that depends on the morphology of the particles. Such a phenomenon has been widely observed in hygroscopic studies but the amount of water uptake is so small compared to that after deliquescence.

Dr Cox asked: In response to Dr Bain's question as to whether the deliquescence behaviour of mixed composition (organic + inorganic) electrolyte aerosols exhibited a size dependence, I refer to a paper by Badger *et al.*[1] which reported tandem DMA measurements of hygroscopic growth of mixed ammonium sulfate/humic acid aerosols. Fig 10 of that paper shows that growth curves for 100 nm and 50 nm particles were essentially the same. This was observed for several different compositions and it is concluded that there is no size dependence of hygroscopic growth in this range.

1   C. L Badger, I. George, P. T. Griffiths, C. F. Braban, R. A. Cox and J. P. D. Abbatt, *Atmos. Chem. Phys.*, 2006, **6**, 1–14.

Dr Grothe addressed Professor Chan: Why do your Raman spectra cut off at 400 cm^{-1}? Aren't lattice mode vibrations (below 400 cm^{-1}) better suited than FWHH and band positions of molecular vibrations in order to follow phase transformations?

Professor Chan answered: It is due to experimental limitations. I agree that lattice mode vibrations would be useful for solid characterization. Our results show that the FWHH analysis is sufficient to definitively show phase transformations for the systems we studied.

Dr Reid asked: Fig. 2 compares the evaporation rates of $MgSO_4$ and $MgCl_2$ droplets. Are these measurements performed at varying relative humidity or fixed humidity? Does the mass dependence of the sulfate droplet imply the absence of efflorescence and above what RH?

Professor Chan replied: These measurements were made at varying RH by introducing a dry stream to the EDB very slowly. The RH in the vicinity of the particle changes very slowly, in timescales of hours. Such experiments were designed to highlight the differences between the drying characteristics of $MgSO_4$ and $MgCl_2$ droplets. $MgSO_4$ did not effloresce, even at ∼0% RH. More information is available in ref. 1.

1   C. K. Chan, Z. Ha and M. Y. Choi, Study of Water Activities of Aerosols of Mixtures of Sodium and Magnesium Salts, *Atmos. Environ.*, 2000, **34**, 4795–4803.

Dr Ward addressed Professor Chang and Professor Chan: When obtaining Raman spectra from organic, or water based, droplets that are optically levitated it is common to encounter morphological resonances due to cavity effects. Could you comment on whether these resonances are present in your work.

If the resonances are weak, or not present, this may indicate that the droplet size is changing rapidly over the timescale of the spectral acquisition. Are the droplets in the EDB stable with respect to size changes?

Professor Chang responded: Both the fluorescence and Raman spectra contain MDRs which occur at different wavelengths from the resonances observed in the optical levitated experiments. It is best to compare the peaks in the Raman and fluorescence spectra with those of the elastic scattering peaks in the same wavelength-shifted region. The Raman and fluorescence peaks depend upon the resonances of the internal field parameters whereas the peaks in the levitation force experiments depend upon the scattering parameters that cannot exceed one. However, the resonances of the internal field parameters are also the resonances of the external scattered field parameters. The wavelength at which the resonances occur is the same for fluorescence, Raman, and elastic scattering. With respect to the next question, if the droplet sizes are changing during spectra acquisition time, then the resonances will be broadened. This broadening is called 'inhomogeneous broadening' and will give rise to confusion. However, in our experiments, the spectra are acquired in less than 10 ns and in this short time, we could probably safely assume that the size and shape of the droplets have not changed.

Professor Chan replied: Morphological dependent resonances (MDR) are commonly observed in elastic scattering. In Raman scattering, MDR can occur when the excitation radiation or the emission is in resonance (Chan et al., 1991[1]). In the former case, MDR appear in the Rayleigh scattering and radiation pressure during the Raman measurements of droplets and the Raman emissions at all wavelengths are enhanced. This type of MDR was frequently observed but it did not affect our interpretation of spectral results in terms of product identification or taking ratios of selected peaks. In the latter case, only enhancements at selected wavelengths in the Raman spectra were observed. The size of our droplets we used was large (20–60 microns) and the resolution of our Raman measurements was rather low, 3–6 cm^{-1}. Hence, only weak MDR in the broadband peak at around 3000–4000 cm^{-1} for water were observed. However, we did not focus on the water peaks in the experiments presented in this paper. In the Raman measurements, evaporation of droplets appeared and caused MDR effects on elastic scattering. However, the evaporation was insignificant in terms of mass change, as confirmed by the mass measurements before and after the Raman measurements.

1   C. K. Chan, R. C. Flagan and J. H. Seinfeld, Resonance Structures in Elastic and Raman Scattering for Microspheres, *Appl. Opt.*, 1991, **30**(4), 459–467.

Dr Mitchem remarked: In Fig. 8 the relative intensity ratio of the sulfate peak to the nitrate peak, I_{450}/I_{720}, as a function of relative humidity during deliquescence is shown. As the RH increases you state the ratio decreases because a larger fraction of the sulfate is in the aqueous phase giving a weaker Raman peak than in the solid phase. However the results in Fig. 5 suggest the variation in the intensity ratio between 58% RH and 82% RH is due to changes in the NO_3^- intensity not the SO_4^{2-}, as the intensity of the SO_4^{2-} does not appear to change. Can you please clarify this difference?

Professor Chan answered: The Raman spectra shown in Fig. 5 were normalized by the sulfate peak at about 450 cm^{-1}. Therefore, the relative peak intensities of this sulfate peak in each spectrum remains unchanged. Nevertheless, it is important to note that the Raman signal of solid particles is usually very intense and narrower than that of aqueous droplets for the same chemical species. In Fig. 5, the intensity of the sulfate peak at 82% RH was actually weaker than that at 58% and 67% RH. Furthermore, the intensity of the nitrate peak at 67% and 82% RH were comparable with each other and were less intense than that at 58% RH. Therefore, there is no conflict between these two figures.

Dr Pfrang asked: Professor Chan, did you try to obtain kinetic information from your experiments on heterogeneous reactions of organic aerosols with ozone? Are there any experimental difficulties or limitations of your equipment to be overcome to enable you to obtain this information?

Professor Chan replied: We could obtain kinetic information by following the changes in the Raman peak intensity of reactants and products inside the levitated particles. So, we can determine the "rate" information based on the appearance or disappearance of certain functional groups. However, this does not translate to the rate of formation of specific products and hence we did not focus on such calculations. Nevertheless, overall mass change as a function of time was determined. The kinetics of heterogeneous reactions depends on a number of factors including the size of the particles and mass transfer issues in addition to the intrinsic rate. Our particles are relatively large compared to atmospheric particles. In the reaction studies, we focus on the reaction mechanisms and the formation of specific functional groups in association with the change of hygroscopic properties of the particles.

Professor Bain asked: In Professor Chan's experiments on ozonolysis of the fatty acid droplets, what is the rate determining step in the oxidation?

Professor Chan responded: Our experiments took over 20 h for droplets of 40–70 microns in size. With such a long timescale, diffusion of ozone in the gas phase and within the droplets did not seem to be the rate limiting step. We also have observed initially an increasing trend and then a decreasing trend in the Raman signal of conjugated C=C–C=C systems for linoleic and linolenic acids, suggesting that there is a series of reactions involved. Hence it does not appear that the accommodation of ozone is rate determining either. The chemical reactions are likely to be the rate limiting step.

Dr Ammann stated: We have compared the hygroscopic properties of aerosol particles consisting of oleic acid (monounsaturated C_{18} fatty acid) and arachidonic acid (polyunsaturated C_{20} fatty acid) reacted with ozone at 200 ppb to 2 ppm over 5 min (Vesna *et al.*, manuscript in preparation). We have observed that the diameter growth factor does only change by about 1% in the case of oleic acid, consistent with the measurements reported by Lee and Chan of a few percent mass growth. In contrast, we show that the diameter growth factor can change up to 10% for arachidonic acid, which decomposes into hygroscopic C_3 and C_5 dicarboxylic acids. Furthermore, we observed a strong impact of humidity during the ozonolysis on the hygroscopicity of the products. This is correlated to the ratio of carboxylic to aliphatic functional groups in the product particles. This indicates that oleic acid is not a good model to demonstrate hygroscopic changes induced by oxidation by ozone, and that water can play an essential role in condensed phase alkene oxidation. The secondary processes such as autoxidation, oligomerisation *etc.* tie up carbon in non-soluble and non-hygroscopic compounds. Such processes are less likely to occur under atmospheric conditions, where water is present and where alkenes are not present in such high concentrations.

Professor Chan replied: We have also observed that ozone-processed linolenic acid particles ($C_{18:3}$) are more hygroscopic than ozone-processed oleic acid particles ($C_{18:1}$) (see references below), which is in general agreement with your hygroscopic measurement of oleic acid and arachidonic acid. We agree that water is potentially important in affecting the reaction mechanism of alkene oxidation in the condensed phase. In our EDB/Raman study, we observed strong fluorescence signals emitted from the ozone-processed linolenic acid particle under high RH but not under low RH condition. This indicates that water molecules can play a significant role in the condensed phase oxidation for certain organic species. We agree that reacted oleic acid particles are only slightly hygroscopic and reacted particles of polyunsaturated organic compounds are more hygroscopic. Nevertheless, oleic acid is used as a model unsaturated organic compound for investigating the role of ozonolysis of organic compounds. It is not easy to interpret the role of ozonolysis on hygroscopic properties of aerosols when the reaction is carried out under high RH. (See also our responses to Prof. Cox's comments.)

1 A. K. Y Lee and C. K. Chan, Single particle Raman spectroscopy for investigating atmospheric heterogeneous reactions of organic aerosols, *Atmos. Environ.*, 2007, **41**(22), 4611–4621.

2   A. K. Y. Lee, and C. K. Chan, Heterogeneous Reactions of Linoleic Acid and Linolenic Acid Particles with Ozone: Reaction Pathways and Changes in Particle Mass, Hygroscopicity, and Morphology, *J. Phys. Chem. A*, 2007, **111**(28), 6285–6295.

Professor Rudich asked: Can you attribute the hygroscopicity changes to chemical changes in the particles? What is the role of water in the reaction in your opinion? We observed the same effects in the reaction of ozone with HULIS.

Dr Ammann answered: We have correlated the hygroscopicity changes with changes in functional group composition determined by Proton Nuclear Magnetic Resonance spectroscopy. This indicates that more carboxylic acid groups were generated under more humid conditions. We suspect that ozone reacts with the Criegee intermediates to form hydroxyhydroperoxides that could decompose into water and an acid. Thus water would prevent to some degree the formation of larger molecular weight and insoluble diperoxides, alkoxyperoxide and similar. It is not clear to me, whether such processes could be relevant for the ozonolysis of HULIS, they might rather be important in the formation of HULIS from their precursors.

Dr Cox addressed Professor Chan and Dr King: Fig. 13 in Prof. Chan's paper shows a clear increase in hygroscopicity in the aerosol particles containing unsaturated organics following reaction with ozone, giving rise to particle growth at high humidity. On the other hand, Dr King, in his paper on O_3 reaction with organic aerosols, saw no increase in particle size and hence hygroscopicity following reaction. Is there a conflict here?

Professor Chan replied: In our previous publication, we have shown that the particle mass, which depends on the molecular structure of the chemical species, can be modified upon ozone exposure.[1,2] In the case of oleic acid ozonolysis, the particle mass decreased by about 3–4% after 20 h ozone exposure. The mass loss of the ozone-processed oleic acid particles is mainly due to the evaporation of volatile and semi-volatile organic products. On the other hand, reaction products of polyunsaturated acids (linoleic and linolenic acids) show mass increase.

Since fumarate has a relatively small molecular weight and benzoate is a highly unsaturated organic compound, it is possible that some volatile and semi-volatile organic products were formed during ozone exposure and consequently evaporated into the gas-phase in the King's study. If this is the case, the hygroscopic growth of the ozone-processed particles, which are expected to be more oxygenated than their parent particles, may be compensated by the evaporative loss of organics and thus cannot be easily observed. In other words, they reported the overall changes in particle size which are due to both water uptake and reactions. In contrast, the reference state used in our hygroscopic measurements was the mass of reacted particles at low RH (*e.g.*, RH < 5%).

1   A. K. Y. Lee and C. K. Chan, Single Particle Raman spectroscopy for Investigating Atmospheric Heterogeneous Reactions of Organic Aerosols, *Atmos. Environ.*, 2007, **41**(22), 4611–4621.
2   A. K. Y. Lee and C. K. Chan, Heterogeneous Reactions of Linoleic acid and Linolenic acid Particles with Ozone: Reaction Pathways and Changes in Particle Mass, Hygroscopicity, and Morphology, *J. Phys. Chem. A*, 2007, **111**(28), 6285–6295.

Dr King answered: There is no conflict here. Professor Chan's work should be compared to our previously published laser-tweezer work on the reaction between ozone and oleic acid/seawater droplets, (King *et al.*[1]) where we clearly demonstrate an increase in particle size during the oxidation of oleic acid with ozone. Professor Chan's work is thus consistent with our previous work. However in the work presented here the reaction between ozone and α-pinene in an organic droplet demonstrated no increase in hygroscopicity and this may be interpreted as evidence for oligiomeration of the reaction products.

1   M. D. King, K. C. Thompson and A. D. Ward, *J. Am. Chem. Soc.*, 2004, **126**(51), 16710–16711.

Dr Reid addressed Professor Chan and Mr Treuel: As an example, Fig. 9 in Treuel *et al.*[1] is considered to show good agreement between theory and experiment. What counts as 'good' agreement and how accurately are the model calculations able to predict the experimental size?

1   L. Treuel, S. Schulze, Th. Leisner and R. Zellner, *Faraday Discuss.*, 2008, **137**, DOI: 10.1039/b702651j

Mr Treuel replied: The work of Choi and Chan,[1] as well as our own data (Fig. 9 in our paper) show that the deliquescence of maleic acid is not a sharp step. This and the degree of reproducibility lead to a large error bar on our own experimental results ($\pm 5\%$). Considering this, the data of Clegg and Seinfeld is within our margin of error and hence shows good agreement with our experimental results.

1   M. Y. Choi and C. K. Chan, *Environ. Sci. Technol.*, 2002, **36**, 2422–2428.

Professor Chan answered: Thermodynamic models are typically based on bulk solution data for subsaturated concentrations and EDB results at supersaturated concentrations. Hence errors in RH in EDB experiments, which are about 1%, are embedded in the model predictions. We would consider 1-2% in RH differences acceptable. In real atmospheric applications, the uncertainty in chemical compositions is probably much more significant than errors in these laboratory experiments in predicting hygroscopic growth of atmospheric aerosols.

Mr J. Butler communicated: Regarding your relative humidity control:
—What is the total gas flow rate that you use?
—Have you experimented with other rates of flow, and if so why did you select the one you now use?
—Does the gas flow directly toward your trapped particle?
—Have you noted or do you suspect any increased evaporation from mass transport effects above that from relative humidity changes?
—And finally, what is the volume of your experimental chamber?

Mr Treuel communicated in reply: We have used a gas flow rate through our cell of 55 sccm throughout the experiments. This flow rate was selected because lower flow rates lead to increased equilibration times of the system as the RH is changed whereas higher flow rates decrease the stability of the particle which makes it harder to accurately detect q/m changes.

During the experiments the gas flows directly towards the particle. In order to observe an increase in the evaporation from mass transfer effects we would require a higher accuracy in the RH determination to reliably distinguish between the two effects.

The volume of our experimental chamber is approximately 1 cm^3.

Professor Chan communicated in reply:
—The total gas flow rate is about 200–300 ml min^{-1}. For the equilibrium measurements, we can use any combination of flow rate and exposure time in principle. The chosen flow rate is a compromise between the time required to change the RH and the risk in off-balancing the particle because of the flow. The flow is directly toward the levitated particle.

—We have not conducted a detailed kinetic study of the rate of evaporation. Typically it is controlled by how fast we change the RH in the vicinity of the particle but not the equilibration time of the particle, which is very short compared to that of the change of RH in the EDB.

—The volume of the chamber is about 60 cm^3.

—More information of our control of RH can be found in ref. 1 and 2.

1 M. Y. Choi and C. K. Chan, Continuous measurements of the water activities of aqueous droplets of water-soluble organic compounds, *J. Phys. Chem. A*, 2002, **106**(18), 4566–4572.
2 Z. Liang and C. K. Chan, A fast technique for measuring water activity of atmospheric aerosols, *Aerosol Sci. Technol.*, 1997, **26**(3), 255–268.

Professor Chan also replied: We have performed experiments using 10 ppm O_3 on linoleic acid and linoleic acid particles. Reacted particles are not spherical and their images indicate the presence of solids whereas reacted particles at 200 ppb O_3 remained as droplets. This suggests the importance of conducting reaction experiments at atmospherically relevant ozone concentrations. More detailed information is available in Lee *et al.*, (2007).[1]

1 Alex K. Y. Lee and Chak K. Chan, Heterogeneous reactions of linoleic acid and linolenic acid particles with ozone: reaction pathways and changes in particle mass, hygroscopicity, and morphology, *J. Phys. Chem. A*, 2007, **111**(28), 6285–6295.

Timothy J. A. Butler, Daniel Mellon and Andrew Orr-Ewing† opened a discussion of Professor Rudich's paper: As is demonstrated clearly by the paper of Rudich and coworkers, cavity ring-down spectroscopy (CRDS) is a powerful tool for the study of optical properties of aerosol particles. We have used the technique of optical feedback (OF) CRDS[1] to develop new methods to obtain quantitative information on aerosol particle extinction. Our test system involves study of monodisperse polymer beads, the set-up for which is shown in Fig. 2. The main advantages of the OF-CRDS technique are the very fast ($\sim$1.2 kHz) repetition rate, and the use of relatively inexpensive diode lasers. The sub-ms temporal resolution and the rapid accumulation of data for statistical analysis are both exploited in our experiments.

Fig. 3 shows how the extinction coefficient changes with time (over a few tens of milliseconds) as a single, 4-μm diameter spherical particle traverses the laser beam. The extinction increases and reaches a maximum when the particle is in the centre of the laser beam where the light intensity is highest. We proposed a model to describe such observations of time-dependent extinction coefficients, $\alpha(t)$ from which absolute extinction cross sections (σ) can be derived for single particles:[2]

$$\alpha(t) = \frac{2\sigma \exp(-2b^2/w^2)}{\pi w^2 l} \exp(-2(vt \sin\theta/w)^2) \tag{1}$$

Here, l is the cavity length, b is the distance of closest approach of the particle trajectory to the central axis of cylindrical symmetry of the intra-cavity laser beam, which has a beam waist w, v is the speed of the particle and θ is its angle of travel relative to the laser beam axis. The data clearly demonstrate that the measured extinction depends on the position of the particle within the intra-cavity laser beam.

We have used the same experimental set-up to study samples containing large numbers of smaller ($\sim$400 nm diameter) monodisperse particles. We made numerous measurements of extinction coefficients and their variance for different number densities (and thus total extinctions) of aerosol particles. Representative results are

† Also Professor Andrew Orr-Ewing, University of Bristol, UK.

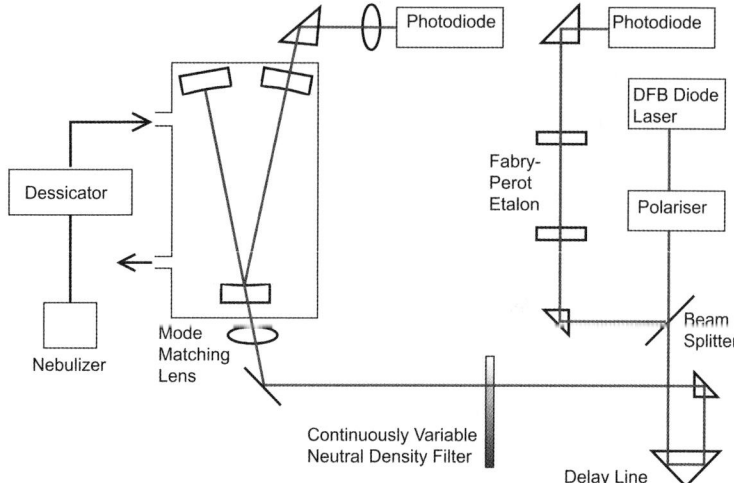

Fig. 2 A schematic diagram of the OF-CRDS experimental apparatus.

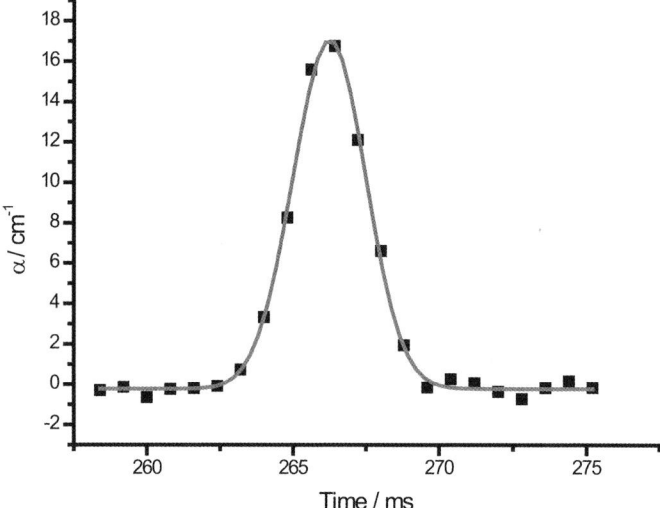

Fig. 3 A plot of the extinction peak caused by a single 4-μm diameter melamine sphere and a fit to a Gaussian function of the form of eqn (1).

plotted in Fig. 4. The variance has a predicted quadratic dependence on α, as determined by:

$$\mathrm{Var}(\alpha) = \frac{4\mathrm{Var}(d)}{d_0^2}\alpha^2 + \frac{\sigma}{V}\alpha + \alpha_{\mathrm{min}}^2 \qquad (2)$$

In eqn (2), V is the volume of the laser beam within the cavity, d is the diameter of the polymer beads, with average value d_0, and α_{min} is the baseline noise level of the apparatus. The analysis accounts for the Gaussian laser beam profile and the distribution of particle positions. The term that is linear in α arises from Poisson statistical fluctuations in the number of particles in the laser beam at any one time,[3,4] and the quadratic term derives from the size distribution of nominally monodisperse particles. Quantitative retrieval of particle properties appears to be feasible from

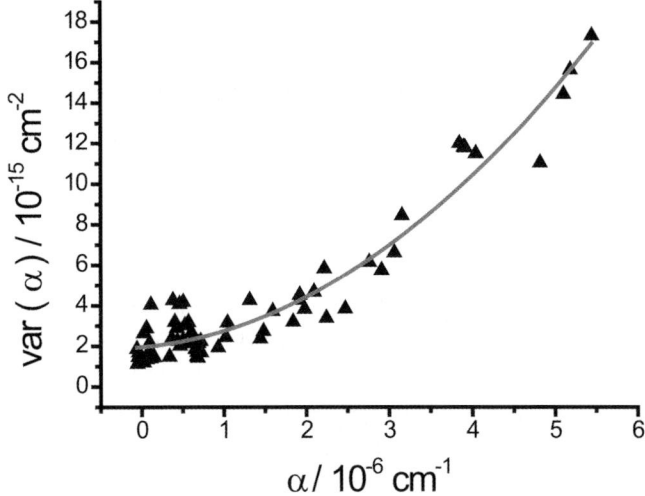

Fig. 4 A plot of variance in the extinction coefficient against extinction coefficient for different number densities of particles. The solid line is a quadratic fit to the data.

such measurements (see Table 1), and we are thus extending our work to the study of atmospheric aerosols.

Table 1 Comparison between the experimentally determined parameters derived from the fit in Fig. 4 based on eqn (2) and those calculated from the known conditions of the experiment.

Parameter	Measured value	Calculated value
$4\mathrm{Var}(d)/d_0^2$	$(4.4 \pm 0.6) \times 10^{-4}$	8.5×10^{-4}
σ/V	$(3.6 \pm 2.3) \times 10^{-10}$	3.8×10^{-10}
$\alpha_{\min}^2$	$(2.0 \pm 0.2) \times 10^{-15}$	1.8×10^{-15}

1   J. Morville, D. Romanini, A. A. Kachanov and M. Chenevier, *Appl. Phys. B*, 2004, **78**, 465.
2   T. J. A. Butler, J. L. Miller and A. J. Orr-Ewing, *J. Chem. Phys.*, 2007, **126**, 174302.
3   A. Pettersson, E. R. Lovejoy, C. A. Brock, S. S. Brown and A. R. Ravishankara, *J. Aerosol Sci.*, 2004, **35**, 995.
4   V. Bulatov, Y. H. Chen, A. Khalmanov and I. Schechter, *Anal. Bioanal. Chem.*, 2006, **384**, 155.

Dr Rudić stated: In the above paper Prof. Rudich showed how from a comparison between measured and calculated extinction efficiency as a function of size parameter α, ($\alpha < 10$) the real and imaginary part of the refractive index of the particles can be derived, *i.e.* the ability of particles to scatter and absorb light can be estimated. We would like to take this opportunity to emphasize similar measurements of extinction efficiency for particles that are characterized by much larger values of the size parameter, α element of (170, 340).[1] Pure water droplets were generated using a vibrating orifice aerosol generator, and were probed by 560 nm wavelength laser light employing the technique of cavity ringdown spectroscopy. By continuously varying the droplet size in the cavity while recording the CRD time, a periodic modulation of $\pm 1.5\%$ in the value of the extinction coefficient was clearly observed. The modulation is caused by the oscillatory nature of the extinction efficiency, which can be inferred from the measured data. In order to obtain a good agreement between theoretically simulated and experimentally derived values of the extinction coefficient, the effects of the strong forward scattering by such large water droplets must be included in the modelling. Indeed, the importance of forward scattering decreases as the particle size parameter decreases and thus may be considered negligible for particles characterized by a very small size parameter.

1 S. Rudić, R. E. H. Miles, A. J. Orr-Ewing and J. P. Reid, *Appl. Opt.*, 2007, **46**(24), 6142–6150.

Professor Chang commented: I am aware of the sensitivity advantages of the cavity ring-down technique for gases. However, I am not familiar with the advantages and complications of using cavity ring-down techniques for particle diagnostics. Wouldn't the forward and backward scattering enter into the Fabry–Perot and cause difficulty in interpretation?

Professor Rudich replied: The high directionality of the CRD avoids this problem for small particles. For large droplets (10-20 μm) the forward scattering has to be taken into account, as was shown in this meeting by Rudić *et al.*

Professor Orr-Ewing† said: Dr King asked about problems associated with particles causing sufficient losses of intracavity light to extinguish the ring-down delay. Such problems will be avoided in the work of Rudich because size-selection of particles of $\leq$ 1μm diameter ensures low cavity losses per particle. The DMA used for size selection has a transfer function that results in a (narrow) distribution of particle sizes and thus size parameters. What is the effect of this spread of sizes on the precision of the determination of the real and complex parts of the refractive index?

Professor Rudich replied: The dispersion from the "monodisperse" aerosol population wasn't taken into account in performing the Mie fitting curve; however, this dispersion is very small (5–7%) for small particles in the range of 150–650 nm. Since we only worked with such small particles, we did not take this into account in our calculations of the Mie fitting curve and the reported precision is a result of the fit. The number of points and the resonances in the Mie curve leads to better fits. The good agreement with literature values for various aerosol types (absorbing and non-absorbing aerosol) also suggests that this source of error is not large.

Recently we also conducted experiments where the size distribution was implicitly considered. Using several distributions, we can get a very good agreement with literature and CRD measurements. This paper is now under review (Spindler, Abo-Riziq and Rudich, submitted).

Professor Signorell asked: I would like to know which parameters limit the accuracy of the refractive index data determined?

Professor Rudich replied: The number of points and the amount of resonances limit the precision. However the validations we performed with substances with known refractive indices (PSL, ammonium sulfate, NaCl, nigrosin) is within a few percent.

Ms Kahan asked: Photobleaching and other reactions of HULIS can lead to reduced or otherwise changed absorption spectra, which might lead to aged HULIS having different effects on radiative forcing than fresh HULIS. Is anything known about the effects of aging on HULIS optical properties, and is this likely to be important?

Professor Rudich responded: We are not aware of such studies at the moment. Our results with various proxies suggests that HULIS aging may affect the molecular weight and aromaticity. This affects CCN activity, surface tension and density. Hence I expect that it will affect also optical properties. You may want to consult our papers:

1 E. Dinar, I. Taraniuk, E. R. Graber, S. Katsman, T. Moise, T. Anttila, Th. F. Mentel and Y. Rudich, Cloud condensation nuclei properties of model and atmospheric HULIS, *Atmos. Chem. Phys.*, 2006, **6**, 2465–2482.

2 E. Dinar, Th. F Mentel, Y. Rudich, Density of atmospheric Humic Like Substances (HULIS) and fractionated Fulvic Acids, *Atmos. Chem. Phys.*, 2006, **6**, 5213–5224.
3 E. Dinar, I. Taraniuk, E. R. Graber, T. Anttila, Th. F. Mentel and Y. Rudich, Hygroscopic growth of model and atmospheric HULIS, *J. Geophys. Res., [Atmos.]*, 2007, **112**, D05211.
4 I. Taraniuk, E. R. Graber, A. Kostinski and Y. Rudich, Surfactant properties of atmospheric and model humic-like substances (HULIS), *Geophys. Res. Lett.*, 2007, **34**, L16807.

Dr Grothe remarked: You found different optical properties of HULIS from rural and urban areas, is that due to changes in aromaticity or can heterogeneous reactions with pollutants affect the properties?

Professor Rudich responded: We have seen changes in properties between fresh and aged HULIS samples. After aging, those HULIS usually have lower molecular mass and less aromaticity (ref. 1–5). The "rural" HULIS we used are "aged" and have a lower molecular weight and hence we expect that these differences can explain the observed difference in the refractive index.

1 E. Dinar, Th. F Mentel, Y. Rudich, Density of atmospheric Humic Like Substances (HULIS) and fractionated Fulvic Acids, *Atmos. Chem. Phys.*, 2006, **6**, 5213-5224.
2 E. Dinar, I. Taraniuk, E.R. Graber, T. Anttila, Th.F. Mentel and Y. Rudich, Hygroscopic growth of model and atmospheric HULIS, *J. Geophys. Res., [Atmos.]*, 2007, **112**, D05211.
3 A. Abo Riziq, C. Erlick, E. Dinar and Y. Rudich, Optical properties of absorbing and non-absorbing aerosols retrieved by cavity ring down (CRD) spectroscopy, *Atmos. Chem. Phys.*, 2007, **7**, 1523–1536.
4 E. Dinar, I. Taraniuk, E. R. Graber, S. Katsman, T. Moise, T. Anttila, Th. F. Mentel and Y. Rudich, Cloud condensation nuclei properties of model and atmospheric HULIS, *Atmos. Chem. Phys.*, 2006, **6**, 2465–2482.
5 I. Taraniuk, E. R. Graber, A. Kostinski and Y. Rudich, Surfactant properties of atmospheric and model humic-like substances (HULIS), *Geophys. Res. Lett.*, 2007, **34**, L16807.

Dr Ammann remarked: The method that is presented in this study uses the water soluble fraction of organic aerosol samples. I wonder to what degree the significance of the results is affected by the fact that the insoluble fraction may (and certainly does) contain light absorbing material. In addition, especially aromatics may tie up in the insoluble fraction, which may artificially lead to the relatively low aromaticity of the investigated HULIS samples.

Professor Rudich answered: I agree with this comment. We do have some samples of the insoluble part. It is a smaller fraction of the total mass, but can contribute differently to the absorption. We intend to look at this fraction in the future.

Dr Murray asked: How representative of real atmospheric HULIS are the reference samples from the International Humic Substance Society, such as the fulvic acid from the Suwannee River?

Professor Rudich replied: Our previous work suggests that they differ from the properties of the HULIS we extracted from our samples. Perhaps the smaller fractions are more similar in their CCN properties but less in their absorption.

Dr Murray remarked: Can you recommend a more realistic proxy for atmospheric HULIS (other than that from IHSS)?

Professor Rudich responded: There is no simple proxy in my opinion. We think that using fractions with low molecular weight may be better proxies for hygroscopic purposes. For optical properties, the high variability we observe seem to preclude and proxy. You can look at more detailed comparisons that we performed in the following papers:

1   E. Dinar, I. Taraniuk, E. R. Graber, S. Katsman, T. Moise, T. Anttila, Th. F. Mentel and Y. Rudich, Cloud condensation nuclei properties of model and atmospheric HULIS, *Atmos. Chem. Phys.*, 2006, **6**, 2465–2482.
2   E. Dinar, Th. F Mentel, Y. Rudich, Density of atmospheric Humic Like Substances (HULIS) and fractionated Fulvic Acids, *Atmos. Chem. Phys.*, 2006, **6**, 5213–5224.
3   E. Dinar, I. Taraniuk, E. R. Graber, T. Anttila, Th. F. Mentel and Y. Rudich, Hygroscopic growth of model and atmospheric HULIS, *J. Geophys. Res., [Atmos.]*, 2007, **112**, D05211.
4   I. Taraniuk, E. R. Graber, A. Kostinski and Y. Rudich, Surfactant properties of atmospheric and model humic-like substances (HULIS), *Geophys. Res. Lett.*, 2007, **34**, L16807.

Dr Hopkins asked: Do you have any plans or do you think it would be of interest to measure the complex refractive index of HULIS extracted from aerosols generated during controlled burning of different biomass material?

For example, aerosols generated from burning different types of plant fuels representative of mid-latitude forests were examined in the FLAME project which took place in July 2006/2007 at the Fire Service Laboratory (FSL, Missoula, MT, USA). The chemical, physical and optical properties of the combustion products generated from the different plant fuels were found to vary substantially.[1]

1   R. J. Hopkins, K. Lewis, Y. Desyaterik, Z. Wang, A. V. Tivanski, W. P. Arnott, A. Laskin and M. K. Gilles, Correlations between Optical, Chemical and Physical Properties of Biomass Burn Aerosols, *Geophys. Res. Lett.*, 2007, DOI: 10.1029/2007GLO30502.

Professor Rudich replied: Currently we do not have such plans.

Dr Reid said: Your consideration of the impact of the mixing state on the optical properties of the aerosol is based on application of the Maxwell–Garnett mixing rule. How sensitive are your predictions to different approaches for treating the mixing state?

Professor Rudich answered: I cannot answer at this point. We are conducting more work on this very topic.

Dr Pfrang communicated: What would you consider to be a useful proxy for HULIS in laboratory studies of its reactions with initiators of atmospheric oxidation? How valuable is kinetic information about the behaviour of HULIS model structures (*e.g.* aromatic rings bearing substituted aliphatic chains with –COOH, –COCH$_3$ or –CH$_2$OH functionalities) from your perspective?

Professor Rudich communicated in reply: It depends on what one wants to study. We have worked with the Suwannee River in order to understand reaction mechanisms and possible products from reactions with ozone. I believe that kinetic information can be obtained from fractions, such as the ones we made in the past (ref. 1–5). The correlation between the kinetic information and chemical information about those fractions can provide valuable information.

1   E. Dinar, Th. F Mentel, Y. Rudich, Density of atmospheric Humic Like Substances (HULIS) and fractionated Fulvic Acids, *Atmos. Chem. Phys.*, 2006, **6**, 5213–5224.
2   E. Dinar, I. Taraniuk, E. R. Graber, T. Anttila, Th. F. Mentel and Y. Rudich, Hygroscopic growth of model and atmospheric HULIS, *J. Geophys. Res., [Atmos.]*, 2007, **112**, D05211.
3   A. Abo Riziq, C. Erlick, E. Dinar and Y. Rudich, Optical properties of absorbing and non-absorbing aerosols retrieved by cavity ring down (CRD) spectroscopy, *Atmos. Chem. Phys.*, 2007, **7**, 1523–1536.
4   E. Dinar, I. Taraniuk, E. R. Graber, S. Katsman, T. Moise, T. Anttila, Th. F. Mentel and Y. Rudich, Cloud condensation nuclei properties of model and atmospheric HULIS, *Atmos. Chem. Phys.*, 2006, **6**, 2465–2482.
5   I. Taraniuk, E. R. Graber, A. Kostinski and Y. Rudich, Surfactant properties of atmospheric and model humic-like substances (HULIS), *Geophys. Res. Lett.*, 2007, **34**, L16807.

PAPER

Characterization of microparticles with driven optical tweezers†

Tiffany A. Wood, G. Seth Roberts, Sarayoot Eaimkhong and Paul Bartlett*

Received 16th March 2007, Accepted 3rd April 2007
First published as an Advance Article on the web 16th July 2007
DOI: 10.1039/b703994h

We discuss how actively-driven optical tweezers may be used to characterize Brownian microparticles. Two experiments are described in detail. We follow the thermal fluctuations of a charged particle in an oscillatory electric field and demonstrate that charges as low as a few elementary charges can be measured accurately and reproducibly. Secondly, we measure the orientational dynamics of a trapped rotating droplet and use circular polarimetry within optical tweezers to determine *in situ* birefringence.

1. Introduction

One of the distinctive features of many micro-particulate systems is their diversity—no two particles have precisely the same size, charge or optical properties. Recent advances in soft matter that allow the interactions between individual pairs of microparticles to be studied have spawned an increased interest in characterizing the optical and electrical properties of single microparticles. Single particle detection and imaging offers the promise of new insights into many fundamental physical processes and a better understanding of the functionality of soft matter systems. This information is frequently obscured by ensemble measurements that provide only average material properties.

In this paper, we detail a general approach that utilizes the change in the random walk of a submicron particle following the application of a periodic perturbation to probe the optical and electrical properties of a single microparticle. The way in which the random Brownian motion of a small sphere immersed in a fluid is modified by an external force lies at the core of many diverse problems in physics, chemistry and biology. Examples include the operation of molecular motors,[1] the phenomenon of stochastic resonance[2] and the driven transport of colloids through a potential landscape.[3] We employ a focused laser beam to confine a microparticle in an optical trap and then track the Brownian fluctuations of the trapped particles with nanometer resolution.[4] In the harmonic potential well, generated by the laser beam, the microparticle oscillates around its equilibrium position as a consequence of the random thermal motion of the solvent molecules. By measuring the subsequent changes in these fluctuations produced by external periodic forces, the optical and electrical properties of single microparticles are derived. We analyze the effects of

School of Chemistry, University of Bristol, Bristol, UK BS8 1TS. E-mail: p.bartlett@bristol.ac.uk

† The HTML version of this article has been enhanced with colour images.

two classes of external perturbations: an *oscillatory linear force*, produced by applying an alternating electric field to a charged particle, and a *constant optical torque*, generated by illuminating a spherical birefringent microparticle with circularly-polarized light. The absolute value of the charge and the birefringence of a single microparticle are found from the autocorrelation of the particle's coordinates. We present a detailed analysis of the Brownian fluctuations of the trapped particle in each of these two situations and show how our measurements provide new quantitative tools for characterization of the optical and electrical properties of micrometric systems.

The paper is organised as follows: Section 2 describes the basis of the optical tweezers technique, in Section 3 we place a trapped particle in an oscillatory electric field and use the response to determine the charge on the particle, and in Section 4 we apply a constant angular torque to a trapped particle and record its orientational dynamics.

2. Optical tweezers

Optical tweezers are formed by tightly focusing a laser beam with a high numerical aperture objective lens.[5] The operation of the trap is based on the fact that a dielectric material subject to an external electric field $\mathbf{E}$ is polarized and generates an induced dipole moment per unit volume, $\mathbf{P}$, given by $\mathbf{P} = \chi\mathbf{E}$, where χ is the electric susceptibility.[6] For a uniform dielectric sphere of radius a, χ is isotropic and the polarization $\mathbf{P}$ is parallel to the electric intensity $\mathbf{E}(\mathbf{r}, t)$ and equal to‡,

$$\mathbf{P}(\mathbf{r}, t) = 3n_{\mathrm{m}}^2\varepsilon_0 \left(\frac{m^2 - 1}{m^2 + 2}\right)\mathbf{E}(\mathbf{r}, t), \tag{1}$$

where n_{m} is the index of refraction of the medium and m is the ratio of the index of refraction of the particle to the medium ($n_{\mathrm{p}}/n_{\mathrm{m}}$). There is no net force on the dipole in the field $\mathbf{E}$ unless the field is non-uniform such that one end of the dipole is subjected to a greater force than the other. In the *non-uniform* electric field generated at the focus of a laser beam there is an instantaneous gradient force per unit volume on the dielectric of,

$$\mathbf{F}_{\mathrm{g}}(\mathbf{r}, t) = [\mathbf{P}(\mathbf{r}, t) \cdot \nabla]\mathbf{E}(\mathbf{r}, t) = 3n_{\mathrm{m}}^2\varepsilon_0 \left(\frac{m^2 - 1}{m^2 + 2}\right)\frac{1}{2}\nabla\mathbf{E}^2(\mathbf{r}, t). \tag{2}$$

The time-averaged gradient force that the particle experiences

$$\mathbf{f}_{\mathrm{g}} = \int \mathrm{d}\mathbf{r}\langle[\mathbf{P}(\mathbf{r}, t) \cdot \nabla]\mathbf{E}(\mathbf{r}, t)\rangle, \tag{3}$$

is proportional to the intensity gradient and is directed towards the intensity maximum for $m > 1$. As a result, a highly-focused light beam will tend to confine a high refractive index microparticle within its focal volume. The strength of the isotropic optical forces is controlled by the spatial profile of the beam and the size and refractive index of the particle.[7]

Angular trapping occurs in optical tweezers with spherical particles made from materials such as nematic Liquid Crystals (LCs) that are birefringent. In this case, the susceptibility is no longer isotropic and the expression for the polarization $\mathbf{P}$ must be generalized to

$$\mathbf{P} = \chi_i E_i\hat{\mathbf{i}} + \chi_j E_j\hat{\mathbf{j}} + \chi_k E_k\hat{\mathbf{k}} \tag{4}$$

where $\hat{\mathbf{i}}$, $\hat{\mathbf{j}}$ and $\hat{\mathbf{k}}$ are unit vectors along the principal axes of the liquid crystal and χ_i, χ_j and χ_k are the corresponding susceptibilities. A uniaxial birefringent material, such

---

‡ Here, by way of illustration, we have assumed that the trapped sphere is much smaller than the wavelength of the trapping laser, *i.e.* $a \ll \lambda$. Particle shape effects are neglected. More complete theories are required when the dimension of the trapped particle is comparable to the wavelength of the trapping laser.

as a nematic LC, can be described by two susceptibilities: an *ordinary* susceptibility, χ_o, for electric fields normal to the optic axis, and an *extraordinary* susceptibility, χ_e, for fields parallel to the optic axis. Choosing the unit vector $\hat{\imath}$ parallel to the optic axis gives, $\chi_i = \chi_e$ and $\chi_j = \chi_k = \chi_o$. Since in general a birefringent nematic LC is more polarizable along the director (the extraordinary axis) than the ordinary axes, the electric polarization vector **P** is not parallel to the external electric field **E** but is tilted towards the extraordinary axis. The misalignment of **P** and **E** means that in a *uniform* electric field, a spherical birefringent particle experiences a torque

$$\tau = \int d\mathbf{r} \mathbf{P}(\mathbf{r}, t) \times \mathbf{E}(\mathbf{r}, t) \tag{5}$$

that tends to align the ordinary axes of the nematic LC drop with the electric field direction. The polarization state of the trapping beam controls the time-dependence of the resulting torque imparted to the trapped particle. In linearly-polarized light, the particle is angularly trapped and aligned along a particular orientation, and a particular particle is always aligned along the same direction whenever trapped. By contrast, in circularly-polarized light, where the plane of polarization rotates continually, a birefringent particle experiences a constant torque and, if free, rotates with a fixed frequency and angular speed.

2.1 Optical configuration

Fig. 1 shows a schematic of our optical tweezers system. A linearly-polarized TEM_{00} mode of a Ytterbium fibre-coupled laser of wavelength λ = 1064 nm (IPG Photonics, Germany) was focussed to a diffraction-limited beam spot with a diameter of approximately of 0.9 µm by an oil-immersion microscope objective (Plan-Neofluar, ×100, N.A. 1.3, Zeiss) mounted in an inverted microscope (Axiovert S100, Zeiss). The intensity of the laser beam was varied using a combination of a $\lambda/2$ waveplate and a polarizing beam-splitter cube placed in the beam path. For the optical torque experiments an additional $\lambda/4$ plate was placed immediately before the objective to convert the linearly polarized light of the laser into the circularly-polarized beam required to rotate the birefringent particles. The transmitted laser beam and light scattered by the trapped microparticle was collected by a high numerical aperture oil-immersion objective and projected onto a polarizing beam splitting cube, before being sent to two orthogonal quadrant photodetectors (model QD50-4X, Centronics, UK).

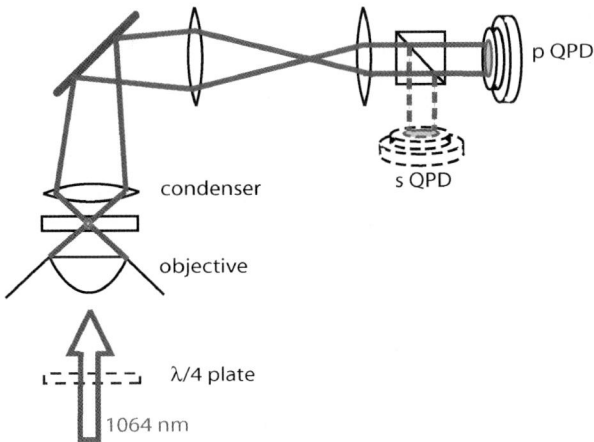

Fig. 1 Schematic representation of the optical tweezer apparatus. Components shown dashed were added for measurements under applied optical torque.

The signals from the two orthogonal detectors were analyzed in different ways depending on the optical experiment. In the linear force experiments detailed in Section 3, the in-plane position of the trapped particle was tracked with the resolution of a few nanometers by back-plane optical interferometry.[8] The quadrant photodetectors were aligned so that they recorded the intensity profile of the pattern generated by the interference of scattered and transmitted light in a plane conjugate to the focal plane. Appropriate combinations of the four quadrant signals give a voltage proportional to the instantaneous in-plane x and y displacement of the fluctuating particle, relative to the axis of the laser beam. Positions measured in voltages were converted into displacements in nanometers by recording the time-dependent mean-square voltage $\langle \Delta V^2(\tau) \rangle$ of five particles from the same batch of particles, immediately before each set of measurements. Assuming the voltage recorded is proportional to the displacement, $\langle \Delta V^2(\tau) \rangle$ was fitted to the theoretical expression for the mean-squared displacement $\langle \Delta x^2(\tau) \rangle$ of a Brownian sphere in a harmonic potential, to yield the detector calibration and the optical trap stiffness k. In the optical torque experiments detailed in Sec. 4, the voltages from the four sections of each detector were summed together. Each of the two orthogonal detectors were separately calibrated so that they provided the power of each polarization component, in units of the power at the trap focus. The transmitted powers recorded by the x and y polarization detectors (P_x and P_y) were combined (see Section 4.1) to analyze the polarization change of the laser beam after it had passed through the nematic LC drop. Data was acquired by a Labview programme (National Instruments, Austin, Texas, USA) and digitized using a high-speed data acquisition card (National Instruments model PCI-MI0-16E-1).

3. Measurement of the charge on a single microparticle

Many key properties of colloidal suspensions are directly or indirectly controlled by the electrical charge carried by their constituent particles. Consequently, methods to measure the electrophoretic mobilities of colloidal particles have received a lot of attention during the last couple of decades. Most techniques available today, however, rely on ensemble averages and are based on monitoring the linear motion of particles in an electric field. Laser Doppler Velocimetry (LDV) is widely used[9] to investigate the light scattered by particles suspended in an electric field and to infer their charge, but is limited to large samples of monodisperse materials. Phase Analysis Light Scattering (PALS) is commonly used to measure very low particle charges in either non-polar suspensions[10] or near the isoelectric point but suffers from similar practical limitations to LDV. These methods, although fast and accurate, provide little or no information on the details of the charge distribution. In many cases, however, knowledge of the charge distribution is of vital importance. Colloidal suspensions have, for example, proved to be valuable model systems in condensed matter physics. There is strong evidence that the width of the charge distribution (the charge polydispersity) significantly influences the glass transition and at high levels may suppress the freezing transition in these systems.[11] It is therefore important that techniques are available that provide an accurate characterization of the charge distribution of microparticles.

Here, we report the development of an ultra-sensitive technique for the measurement of the charge carried by a *single* colloidal particle in a non-polar suspension. Our approach is to use the well-known phenomenon of the resonance of a harmonic oscillator driven by a oscillatory linear force. A linearly-polarized laser beam is used to trap a single microparticle. The laser and trapped particle constitute a harmonic oscillator with a stiffness that is proportional to the laser power. The modulation of the Brownian motion generated by an alternating electric field is measured with nanometer sensitivity using an interferometric position detector and the charge is obtained directly from the power spectrum of the thermal fluctuations. Charges as low as a few elementary charges can be measured with an uncertainty of about 0.25 e.[12] This is significantly better than existing techniques and opens up new possibilities for the study of non-polar suspensions.

3.1 Experimental details

We use a model colloidal system of sterically-stabilized poly(methyl methacrylate) spheres [PMMA] of radius a = 610 ± 30 nm.[13] Electron microscopy revealed the particles were highly uniform in size with a radius polydispersity (root mean square variation/mean radius) of 0.046. Single PMMA spheres were suspended in dry dodecane and trapped in three dimensions using a linearly-polarized laser beam in a purpose-built microelectrophoresis cell. The cell contained two flat platinum foil electrodes separated by 128 μm ± 2 μm in a cylindrical glass sample chamber. Voltages of up to 10 V were applied to the electrodes. The electric field was estimated from the expression $E = \lambda V/d$, where V is the applied voltage, d is the plate separation and, λ is a factor correcting for the finite size of the electrodes. Finite element simulation of the electric field gave λ = 0.93. The confined microparticle was held about 70 μm above the base of the cell and 2^{18} data points were collected at 10 kHz. The Brownian motion of the particle was followed by back-plane optical interferometry using data collected from just one quadrant photodetector. The duration of each measurement was ≈ 26 s.

3.2 Brownian motion of an oscillating microparticle

The total force on a *charged* microparticle in an optical trap has four elements: a harmonic force, $-kx(t)$, arising from the optical trap, where k is the force constant of the potential well; a viscous drag force $-\xi_t \dot{x}(t)$, where from Stokes law ξ_t = $6\pi\eta a$ for a sphere of radius a; an oscillatory driving force $f_D(t) = A \sin \Omega t$ defined by the frequency Ω and amplitude A of the applied electric field; and Brownian forces $f_B(t)$ characterizing the random fluctuating forces due to collisions with molecules of the solvent. The Langevin force equation for the Brownian motion of a sphere of mass m is,[14]

$$m\ddot{x} = -kx - \xi_t \dot{x} + A \sin \Omega t + f_B(t) \qquad (6)$$

where $f_B(t)$ is a white-noise Gaussian random process with time-averaged correlations $\langle f_B(t) f_B(t') \rangle = 2\xi_t k_B T \delta(t - t')$. To simplify the problem, we neglect all inertial terms that are negligible at frequencies $\omega \ll 10^6$ rad s^{-1}. Defining the amplitude of the oscillatory linear force as $A = ZeE$, where Z is the charge on the microparticle and E is the amplitude of the electric field, then eqn (6) reduces to

$$\xi_t \dot{x} + kx - ZeE \sin \Omega t = f_B(t). \qquad (7)$$

Since the Brownian and periodic forces are uncorrelated, the general solution of eqn (7) is the superposition,

$$x(t) = x_B(t) + x_D(t), \qquad (8)$$

where x_B is the solution in the presence of random Brownian forces *only* (i.e. $A = 0$) and x_D is the solution where only driven forces act (i.e. $T = 0$). The factorization evident in eqn (8) is universal, so, for instance, the spectral density $S_D(\omega)$ of the particle's positional fluctuations is simply a combination of the spectral densities for purely Brownian and driven motion,

$$S_D(\omega) = \frac{k_B T}{\pi \xi_t} \frac{1}{\omega^2 + \Omega^2} + \frac{k_B T \gamma^2}{2k} [\delta(\omega - \Omega) + \delta(\omega + \Omega)]. \qquad (9)$$

Here, we have introduced γ^2, the ratio of the mean-square driven and Brownian forces,

$$\gamma^2 = \frac{\langle f_D^2 \rangle / \langle f_B^2 \rangle}{1 + (\Omega/\omega_c)^2} \quad (10)$$

and the corner frequency of the potential well, $\omega_c = k/\xi_t$. The mean square amplitude of the driven forces is $\langle f_D^2 \rangle = Z^2 e^2 E^2/2$ while the mean square modulus of the fluctuating Brownian forces in a harmonic oscillator is $\langle f_B^2 \rangle = k_B T k$. The force ratio γ appears naturally in the theory and, as we show below, is also the quantity most readily extracted from experiment. In the weak-field limit, where $\gamma \ll 1$, the motion of the trapped microparticle is dominated by random Brownian forces. By contrast, in the strong-field limit where $\gamma \gg 1$, the thermal forces are only a relatively small perturbation and the oscillatory electrical forces dominate.

The experimentally measured spectral density of the Brownian fluctuations are plotted in Fig. 2 at different driving frequencies. The measured spectra are particularly simple, being a superposition of a Lorentzian spectrum (shown by the solid line) reflecting diffusive motion in a harmonic potential and a single δ-peak (arrowed) at the fundamental electrode drive frequency Ω. The data contains no higher harmonics of Ω, confirming the linear response of particle and field. To extract the power in the periodic signal, the data around the δ-spike was masked and the diffusive spectrum fitted to the first term of eqn (9), by adjusting the unknown corner frequency ω_c. Subtracting the fitted Lorentzian from the measured power spectra yielded the signal spectrum. The mean-square displacement P_{sig} of the microparticle in the electric field was obtained by integrating the signal spectrum around the peak at $\omega = \Omega$. From eqn (9),

$$P_{\text{sig}} = \frac{k_B T}{k} \gamma^2. \quad (11)$$

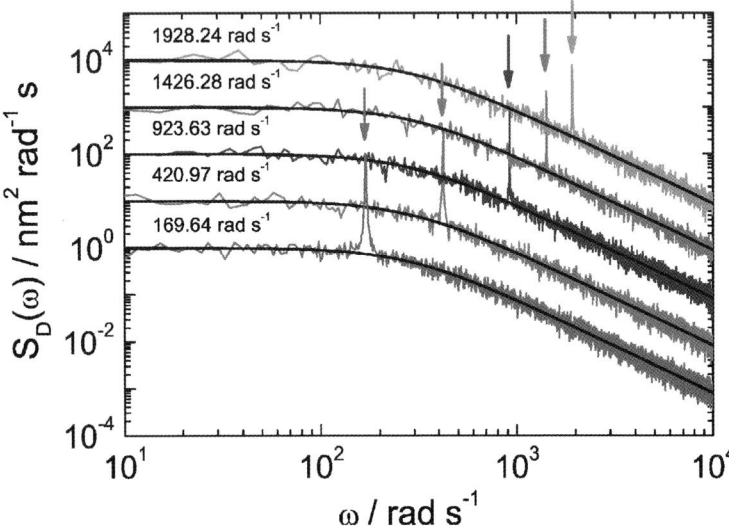

Fig. 2 The frequency dependence of the power spectral density $S_D(\omega)$. The sample was a 610 nm PMMA particle suspended in dodecane with 0.035 wt% added PHSA-g-PMMA copolymer. The amplitude of the electric field was fixed at $E = 72.8$ kV m^{-1} and the field frequency Ω varied. The arrows indicate the oscillatory fluctuations generated by the applied field.

so a measurement of P_{sig} yields directly an estimate of the force ratio γ. An expression for the charge on the trapped microparticle follows immediately from eqn (10) and (11) as,

$$(Ze)^2 = \left[\frac{2\xi_t^2(\Omega^2 + \omega_c^2)}{E^2}\right] P_{sig}. \quad (12)$$

At low driving frequencies $\Omega \ll \omega_c$ this result simplifies to

$$(Ze)^2 = 2(k/E)^2 P_{sig}. \quad (13)$$

3.3 Applications

The optical tweezers technique, detailed above, for the measurement of charge on individual microparticles has several advantages over existing methods. First, the approach is capable of very high sensitivity. Second, the measurements can be made rapidly, and third, the measurements are accurate and reproducible. We illustrate these benefits with two examples.

Fig. 3 show the effect of adding the surfactant sodium bis(2-ethylhexyl) sulfosuccinate [Na-AOT] to dilute PMMA suspensions in dry dodecane. At low concentrations c_s of Na-AOT ($c_s < 0.1$ mM), the particles have a very small residual charge. Attempts to measure the level of charge with a commercial PALS instrument were unsuccessful—the charge was below the detectable limit for this technique and the particle was incorrectly identified as having no charge. The low particle mobilities, however, presented no problems for the optical tweezers measurements, as the data in Fig. 3 confirms. The mean charge was measured as -2.9 ± 0.2 e. Increasing the concentration of surfactant results in the particle developing a large negative charge of $\langle Z \rangle \sim -50$ e. While such micelle-mediated charging has been known for at least the last 60 years,[15] the exact mechanism of charging in non-polar media has remained problematic and is not well understood. Recent work has highlighted the growing importance of charge in non-polar suspensions[16] and its technological significance.[17] The data of Fig. 3 reveals that our method is capable of accurate and reliable measurements of the extremely low level of particle charges characteristic of non-polar suspensions. Such improvements in quantitative characterization are an

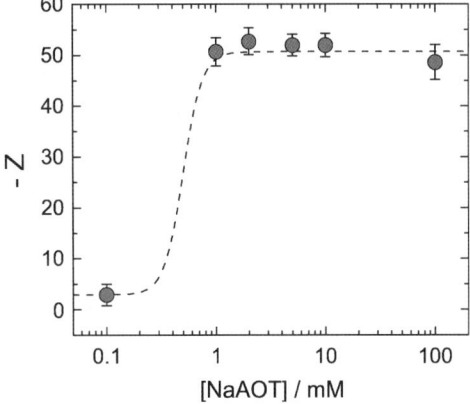

Fig. 3 The charge Z (in units of e) on undyed PMMA microspheres in dry dodecane as a function of AOT concentration. The error bars depict the variation in the charge of different particles under the same conditions. The uncertainty in Z from measurements on a single particle is significantly smaller.

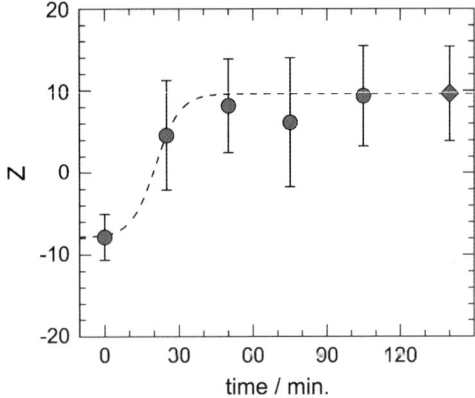

Fig. 4 Charge inversion caused by adsorption of water. The charge Z (in units of e) of dyed PMMA spheres in dry dodecane is plotted as a function of the exposure time (circles). The diamond indicates the limiting charge where the dispersion was left overnight in contact with excess water.

essential first step to developing a more detailed understanding of the role of charge in non-polar suspensions.

The small particle charges frequently found in non-polar suspensions means the level of charge is very sensitive to low concentrations of any surface-sensitive impurities. So, for instance, there have been several, and sometimes contradictory, reports in the literature[15] of the role of small amounts of water on the particle charge level. A key requirement for any technique used to investigate such an effect is rapidity of measurement. Fig. 4 reveals the speed capabilities of the optical tweezers technique, where reproducible measurements on single microspheres can be achieved within ~ 30 s. Poly(methyl methacrylate) spheres containing the cationic dye $DiIC_{18}$ were suspended in dodecane, which had been carefully dried with activated molecular sieves (Acros, size 4A) and stored under dry nitrogen. The charge on the particle was followed as the dispersion was exposed for progressively longer periods to a damp atmosphere. The data in Fig. 4 shows the results. While initially the dyed particles are negative with a charge of about -8 e, the charge starts to reduce immediately on initial exposure to water, reversing its sign within 30 min and saturating at $+10$ e after ~ 50 min.

4. Measurement of optical birefringence

The transfer of angular momentum from a light beam to a micron-sized particle by optical tweezers has become an important tool for the manipulation and characterization of objects in nanoscience.[18] The major application of this technique has been the study of microscopic biological systems. Microspheres have been used as a handle to rotate biological structures and to measure the associated torque. Other uses have emerged in the field of microfluidics, where optically-driven particles have been used as miniaturized pumps and stirrers[19] and as quantitative probes of rheological properties in microscopic volumes.[20] Quantitative applications depend on an accurate knowledge of the optical properties of the microparticles used. Here, we report on the birefringence of liquid crystal droplets dispersed in water. Previous work has shown that the internal molecular alignment of a liquid crystal is very sensitive to the local environment of the droplet.[21,22] So, for instance, the alignment of nematic LC drops may be changed from bipolar to radial by adding a surfactant. The coupling between rotating particle and fluid is, to date, little understood and difficult to probe. We demonstrate an optical technique that can accurately measure the *in situ* birefringence of a rotating probe particle. These measurements allow the effect of the local environment on the molecular alignment to be quantified directly.

Two nematic liquid crystals were used in our experiments. MLC-6815 from Merck is a positive uniaxial birefringent material, with $n_o = \sqrt{1+\chi_o} = 1.444$ and $n_e = \sqrt{1+\chi_e} = 1.495$ while MDA-00-1444 has $n_o = 1.484$ and $n_e = 1.661$ at $\lambda = 1064$ nm. Both liquid crystals are hydrophobic so when dispersed in deionised water they naturally forms spherical drops with a dipolar alignment. The droplets have a distribution of radii a of between 0.5–5 µm. The sample was prepared by encasing the dispersion within a 10 mm wide hole in a 500 µm thick glass slide between two glass coverslips. Captured in a circularly-polarized light beam a droplet spins continuously. To measure the optical anisotropy of the birefringent droplet the fluctuations in the intensity of the elliptically polarized outgoing beam exiting the rotating drop were measured using the two orthogonal quadrant photodetectors, shown in Fig. 1. In order to determine the birefringence of the spinning LC droplet, we have to analyze the change in the polarization of the laser trap illumination after it has passed through the droplet.

4.1 Angular momentum transfer

The oscillating electric field **E** of a monochromatic plane wave may be described in terms of the superposition of two orthogonal components,

$$\mathbf{E} = (E_x \hat{\mathbf{x}} + E_y \hat{\mathbf{y}}) \exp(ikz - i\omega t) \quad (14)$$

where the wave is propagating in the z-direction and ω, k are the angular frequency and wavevector of the beam, respectively. Since in general an incident beam will be elliptically polarized, that is it contains both circularly polarized and linearly polarized components, the amplitudes E_x and E_y will be complex. In the special case of circularly polarized light, where $E_y = \pm iE_x$, the total electric field has a constant magnitude and a field direction that rotates around the z-axis with the optical frequency ω. The sign depends on whether the beam is left (+) or right (−) circularly polarized. The spin angular momentum of left and right circularly polarized photons is $\pm\hbar$,[23] so that the flux of angular momentum in a circularly polarized light beam of power P is $\pm P/\omega$.[24,25]

The spin angular momentum carried by a general light beam may be readily calculated from the amplitudes of the equivalent left and right circular components of the electric field,

$$E_L = \frac{1}{\sqrt{2}}(E_x - iE_y)$$
$$E_R = \frac{1}{\sqrt{2}}(E_x + iE_y). \quad (15)$$

The polarization coefficient,

$$\sigma = \frac{E_L^* E_L - E_R^* E_R}{E_L^* E_L + E_R^* E_R} = (P_L - P_R)/P \quad (16)$$

summarizes the state of polarization of a general beam, where P_L and P_R are the powers of the two orthogonal circularly-polarized components and P is the total power. A left circularly polarized beam has $\sigma = 1$. The spin angular momentum flux of the beam is accordingly,[25]

$$L = \sigma P/\omega. \quad (17)$$

In general, if the trapped material is birefringent then the polarization state of the illuminating laser beam will change after it has passed through the droplet and the angular momentum of the emerging beam will differ from that of the illuminating beam. To conserve the total angular momentum of the system, momentum must be transferred to the birefringent particle and a torque on the particle will result. In the particular case where the illuminating beam is purely left circularly-polarized ($\sigma_{in} = 1$) and the emerging beam has the polarization σ_{out}, then the reaction torque on the particle will equal the loss of the angular momentum flux of the light beam,

$$\tau = (1 - \sigma_{out})\frac{P}{\omega}\hat{z}. \quad (18)$$

The amplitude of the reaction torque is a maximum when the birefringent microsphere acts as a $\lambda/2$-plate, so that the handedness of the emerging light beam is totally reversed and $\sigma_{out} = -1$.

The polarization coefficient of the emerging trapping beam σ_{out} was determined directly from the total voltage recorded by the two orthogonal quadrant photodetectors. The power of the left and right hand circularly-polarized components of the emerging beam are $P_L = (1 + \sigma_{out})P/2$ and $P_R = (1 - \sigma_{out})P/2$, respectively. Rewriting these components in terms of the electric fields along the lab-fixed x- and y-axes it is straightforward to show that the transmitted power measured, after the polarized beam splitter, by the two orthogonal quadrant detectors is

$$P_{x,y}(t) = \left(1 \pm \sqrt{1 - \sigma_{out}^2}\sin 2\theta\right)\frac{P}{2}, \quad (19)$$

where the positive sign is associated with the x-detector and θ is the angle between the optic axis of the liquid crystal and the lab-fixed x-axis. The reaction torque applied to the particle may be determined directly from the difference between the powers recorded on the two detectors scaled by the total beam power,

$$\mathcal{R}(t) = \frac{P_x(t) - P_y(t)}{P_x(t) + P_y(t)} = \sqrt{1 - \sigma_{out}^2}\sin 2\theta. \quad (20)$$

As the droplet rotates, the ratio $\mathcal{R}(t)$ alternates between maximum and minimum values of $\pm\sqrt{1 - \sigma_{out}^2}$, corresponding to orientations where the angle between the optic axis and the x-direction is $\theta = n\pi \pm \pi/4$. In between, where the optic axis of the birefringent droplet is parallel to the x-axis, $\mathcal{R}(t) = 0$. An example dataset of the $P_x(t)$ and $P_y(t)$ detector signals from a rotating LC droplet is shown in Fig. 5a and the corresponding ratio, $\mathcal{R}(t)$, in Fig. 5b. The extrema of $\mathcal{R}(t)$ following eqn (20) provide strictly only a measurement of the degree of circular polarization in the emerging beam $|\sigma_{out}|$ and not its direction (or sign). However, the sign of σ_{out} may be readily determined by inspection.

Measurement of the outgoing polarization σ_{out} and the beam power P gives an absolute measurement of the constant torque τ applied to the birefringent droplet, from eqn (18). The resulting droplet rotation rate Ω is determined from the Fourier

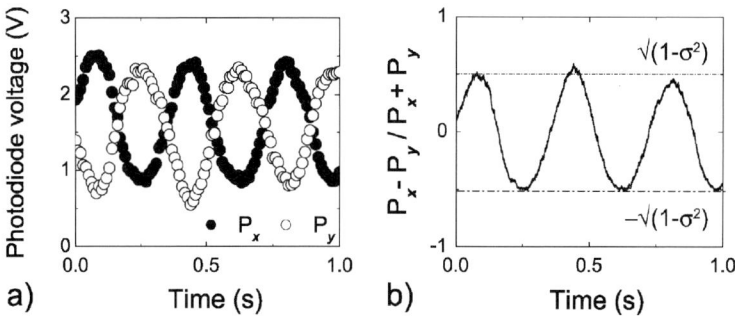

Fig. 5 Representative dataset showing (a) unnormalised P_x and P_y photodiode signals and, (b) the ratio $\mathcal{R}(t)$.

analysis detailed in Section 4.2. To establish the optical birefringence of the drop, we assume that the liquid crystal acts as an ideal wave plate with a thickness equal to the diameter of the droplet, upon which a circularly polarized beam impinges at normal incidence. The polarization coefficient of the emerging beam is then $\sigma_{\text{out}} = \cos \Delta$, where the phase shift Δ between electric field components travelling parallel and normal to the optic axis of the liquid crystal is equal to,

$$\Delta = \frac{4\pi a \Delta n_{\text{eff}}}{\lambda} \tag{21}$$

with $\Delta n_{\text{eff}} = n_e - n_o$. The radius a of the LC droplet was measured by video microscopy and used together with the experimental value of σ_{out} to calculate the (effective) birefringence Δn_{eff} of the droplet.

4.2 Brownian motion of a rotating droplet

The normalized detector signal $\mathcal{R}(t)$ is sensitive to the instantaneous angular coordinate θ of the droplet. In this section, we exploit this dependence to study the rotational diffusion of a sphere driven by a constant torque.

The dynamics of the birefringent droplet is controlled by a balance of three contributions: a constant driving torque τ that arises from the transfer of angular momentum from the circularly polarized beam; a viscous drag torque $\xi_R \dot{\theta}$; and a white noise thermal torque $\tau_B(t)$ that represents the random torques imparted by thermal collision with the solvent molecules. The fluctuation dynamics of the droplet are governed by the Langevin torque equation,

$$I\ddot{\theta} = \tau - \xi_R \dot{\theta} + \tau_B(t), \tag{22}$$

where I is the moment of inertia of the droplet and ξ_R is the rotational friction coefficient, which for a sphere of radius a is, from Stokes law, $\xi_r = 8\pi\eta a^3$. The Brownian torques have a time-averaged mean value of zero and are delta correlated,

$$\langle \tau_B(t) \rangle = 0; \quad \langle \tau_B(t)\tau_B(t') \rangle = 2\xi_R k_B T \delta(t' - t). \tag{23}$$

The spectral density of the random torque $C_B(\omega)$ is frequency independent,

$$C_B(\omega) = \frac{1}{2\pi} \int_{-\infty}^{+\infty} \langle \tau_B(t')\tau_B(t' + t) \rangle \exp(-i\omega t) \, dt = \frac{\xi_R k_B T}{\pi}. \tag{24}$$

The effect of a constant driving torque τ is to produce a uniform rotation of the droplet which modulates the correlation functions and leads to a subsequent resonance peak in the power spectrum. To understand these changes we consider first the rotational diffusion in the absence of a torque. In the limit of ordinary diffusion ($\tau = 0$), the solution of eqn (22) is obtained readily by Fourier techniques.[14] The Fourier transform of the angular correlation function, $C_{\theta\theta}(t - t') = \langle \theta(t)\theta(t') \rangle$, is

$$C_{\theta\theta}(\omega) = \frac{k_B T \xi_R}{\pi\omega^2[I^2\omega^2 + \xi_R^2]} \tag{25}$$

where we have used eqn (24). In the low frequency limit $\omega \ll \xi_R/I$, appropriate here since ξ_R/I is of order 10^7 rad s^{-1}, this expression reduces to the well-known inverse square spectrum:

$$C_{\theta\theta}(\omega) = \frac{k_B T}{\pi\omega^2 \xi_R}. \tag{26}$$

In the presence of a constant torque, the optic axis of the droplet is set spinning at a constant average rate. Neglecting the inertial terms, which are negligible at frequencies $\omega \ll \xi_R/I$ and averaging eqn (22) over the fluctuating Brownian torques reveals

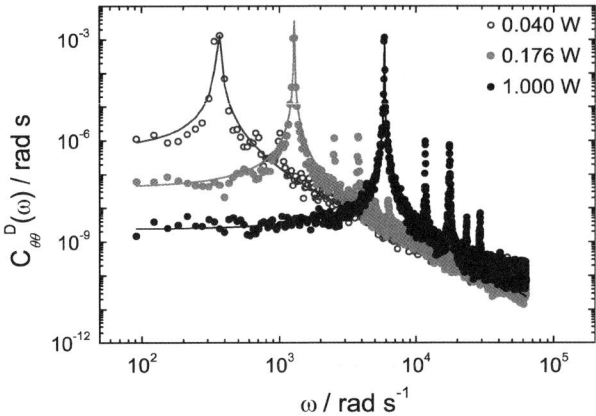

Fig. 6 The spectral density of the orientational fluctuations $C^D_{\theta\theta}(\omega)$ measured on a single spinning birefringent LC droplet (MDA-00-1444) of radius $a = 1.2$ μm with retardation $\Delta \sim 0.5$. The solid lines show fits of the measured spectral densities to eqn (29).

that the average rotation rate of the droplet is,

$$\langle \dot{\theta} \rangle = \frac{\tau}{\xi_R} = \Omega. \tag{27}$$

Although the average rotation rate is fixed at Ω, there remains fluctuations in the rotation rate due to collisions between the trapped drop and molecules of the solvent, which result in the application of additional random torques. Provided the time for rotation Ω^{-1} is longer than the characteristic relaxation time of the angular correlations of the droplet, then the effect of the continuous (non-Brownian) rotation is simply to modulate the purely diffusive angular correlation function $C_{\theta\theta}(t)$. The driven correlations $C^D_{\theta\theta}(t)$ will therefore, to a first approximation, have the form[26]

$$C^D_{\theta\theta}(t) = C_{\theta\theta}(t)\cos 2\Omega t, \tag{28}$$

where the periodic term arises from the non-random rotation and the factor of 2 is because the rotating linear component of light emerging from the droplet is parallel to each polarizer twice during a complete rotation of the droplet. The associated power spectrum exhibits a resonance at the frequency characteristic of the driving torque. Using the Fourier shift theorem and eqn (26)–(28), the spectral density of the angular fluctuations in the continuously driven droplet is,

$$C^D_{\theta\theta}(\omega) = \frac{k_B T}{\pi \xi_R (\omega - 2\Omega)^2} \tag{29}$$

In Fig. 6, we fit the power spectra measured from a single rotating drop at three different torques to eqn (29) and there is excellent agreement. At low frequencies, $\omega \ll \Omega$, eqn (29) saturates at a constant value, $C^D_{\theta\theta}(\omega)_{\text{sat}} = k_B T/4\pi\xi_R\Omega^2$, which is inversely proportional to the square of the driving frequency, as evident in Fig. 6. At high frequencies, $\omega \gg \Omega$, eqn (29) simplifies to $C^D_{\theta\theta}(\omega)_{\text{sat}} = k_B T/\pi\xi_R\omega^2$, indicating that diffusion dominates.

4.3 Applications

Detailed analysis of the orientational fluctuations show that the spinning liquid crystal droplets behave as mechanically rigid spheres. Fig. 7 shows that the rotation frequency Ω, obtained from the peak in the power spectra $C^D_{\theta\theta}(\omega)$, is a linear function of the laser beam power as expected from eqn (18) and (27). Similarly, the

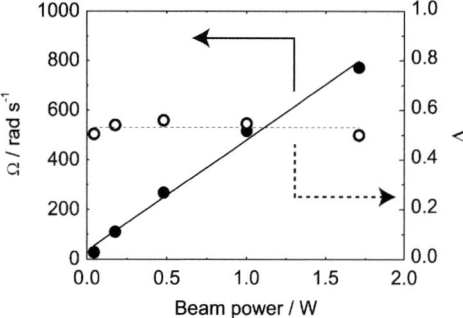

Fig. 7 The rotation frequency Ω (filled points) and phase shift Δ (open points) as a function of beam power (a = 1.2 μm, MDA-00-1444).

retardation Δ is constant, independent of the laser power, signifying that there is no change in internal molecular ordering or flow within the droplets for the powers typical of our experiments.

Accounting for the measured phase shift of the optical trapping beam as it passes through a nematic liquid crystalline droplet in terms of the birefringence of the material is not trivial. Within each droplet there is a range of molecular alignments. Whilst molecules near the centre of the droplet will be free to adopt the preferred dipolar orientation, surface energetics will favour parallel anchoring at the water interface. The competition between elastic deformation and anchoring energies define a continuous director field that is a function of the radial distance from the centre of the drop. As a result, the birefringence determined from the measured phase shift Δ (eqn (21)) will be an average over the complete director field within the drop and is likely to depend on the drop radius a. Fig. 8 summarizes the experimental data on the size dependence of the effective birefringence Δn_{eff} for droplets of MDA-00-1444. For comparison, the bulk birefringence of the liquid crystal is Δn = 0.177. Fig. 8 reveals that in large drops, $a > 2$ μm, Δn_{eff} saturates at a near size-independent limit of ~ 0.12, approximately two-thirds of the bulk birefringence, whilst in small drops, $a < 2$ μm, the effective birefringence is significantly smaller, with $\Delta n_{\text{eff}}/\Delta n \approx 0.34$. The size dependent birefringence indicates that the central region where the director alignment is uniform is significantly reduced in small droplets.

The effective birefringence of a droplet provides a clear indication of detrimental environmental factors. This is demonstrated by the data of Fig. 9, where light absorption causes progressive heating in an optical trap and the liquid crystal to switch from the nematic to the isotropic state. At the transition, a sharp drop in the

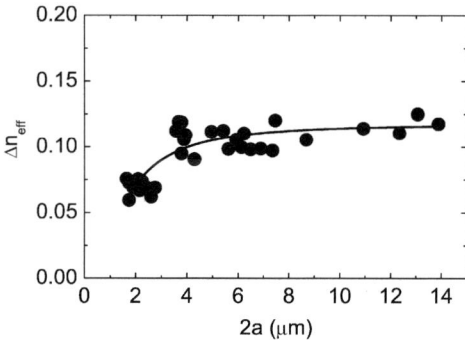

Fig. 8 The effective birefringence Δn_{eff} of nematic droplets of MDA-00-1444 as a function of diameter. The bulk birefringence is Δn = 0.177.

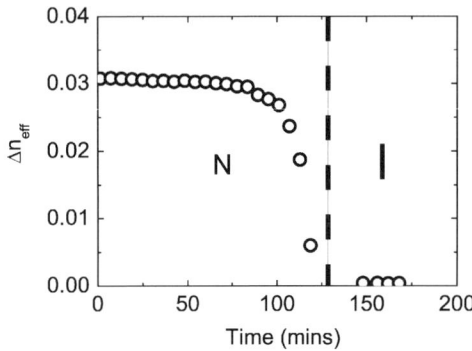

Fig. 9 Heating on prolonged irradiation. The birefringence Δn_{eff} reveals that, in a 4 W laser beam, heating destroys the nematic state after $\sim$2 h ($a = 2.6$ μm, MLC-6815).

effective birefringence of the droplet is observed, even though the particle still remains optically trapped. We can get an indication of the degree of heating by noting that MLC-6815 has a nematic-to-isotropic transition temperature of 67 °C. It is interesting to note that Fig. 9 reveals that in the nematic phase, $\Delta n_{\text{eff}} \sim 0.031$, which is approximately 60% of the bulk birefringence of MLC-6815, is in reasonably good agreement with the value obtained for the more birefringent MDA-00-1444. This consistency suggests that the effective birefringence of a spherical liquid crystal depends on the geometry of the system rather than any material property, and presents an interesting topic for further study.

5. Conclusions

The study of Brownian fluctuations in actively-driven situations provides new tools for the characterization of microparticles. In this manuscript, we have described two such examples. By using optical tweezers in combination with an external electric field, we have developed an ultra-sensitive method to measure the charge carried by an individual microparticle. The technique is fast, accurate and reproducible. Secondly, we have detailed how the application of a constant torque may be used to measure quantitatively the optical anisotropy of a spherical particle on a micronscale. We anticipate that optical tweezers in combination with external fields will lead to further new techniques for particle characterization. Work in this direction is in progress.

Acknowledgements

We acknowledge financial support from the Engineering and Physical Sciences Research Council and Unilever PLC.

References

1. P. Riemann and P. Hanggi, *Appl. Phys. A*, 2002, **75**, 169.
2. L. Gammaitoni, P. Hänggi, P. Jung and F. Marchesoni, *Rev. Mod. Phys.*, 1998, **70**, 223.
3. Y. Roichman, V. Wong and D. G. Grier, *Phys. Rev. E*, 2007, **75**, 011407.
4. W. Denk and W. W. Webb, *Appl. Opt.*, 1990, **29**, 2382.
5. A. Ashkin, J. M. Driedzic, J. E. Bjorkholm and S. Chu, *Opt. Lett.*, 1986, **11**, 288.
6. P. Lorrain and D. Corson, *Electromagnetic Fields and Waves*, W. H. Freeman, San Francisco, 2nd edn, 1970.
7. P. Bartlett and D. Henderson, *J. Phys.: Condens. Matter*, 2002, **14**, 7757.
8. F. Gittes and C. F. Schmidt, *Opt. Lett.*, 1998, **23**, 7.
9. E. E. Uzgiris, *Adv. Colloid Interface Sci.*, 1981, **14**, 751.
10. J. F. Miller, K. Schatzel and B. Vincent, *J. Colloid Interface Sci.*, 1991, **143**, 532.
11. B. V. R. Tata and A. K. Arora, *J. Phys.: Condens. Matter*, 1995, **7**, 3817.

12 G. S. Roberts, T. A. Wood, W. J. Frith and P. Bartlett, *J. Chem. Phys.*, 2007, **126**, 194503.
13 L. Antl, J. W. Goodwin, R. D. Hill, R. H. Ottewill, S. M. Owens, S. Papworth and J. A. Waters, *Colloids Surf.*, 1986, **17**, 67.
14 R. Kubo, M. Toda and N. Hashitsume, *Statistical Physics II. Nonequilibrium statistical mechanics*, Springer Series in Solid-State Sciences, Springer-Verlag, Berlin, 1978.
15 I. D. Morrison, *Colloids Surf.*, 1993, **71**, 1.
16 M. E. Leunissen, C. G. Christova, A.-P. Hynninen, C. P. Royall, A. I. Campbell, A. Imhof, M. Dijkstra, R. v. Roij and A. v. Blaaderen, *Nature*, 2005, **437**, 235.
17 B. Comiskey, J. D. Albert, H. Yoshizawa and J. Jacobson, *Nature*, 1998, **394**, 253.
18 J. E. Molloy and M. J. Padgett, *Contemp. Phys.*, 2002, **43**, 241.
19 J. Leach, H. Mushfique, R. di Leonardo, M. Padgett and J. Cooper, *Lab Chip*, 2006, **6**, 735.
20 A. I. Bishop, T. A. Nieminen, N. R. Heckenberg and H. Rubinsztein-Dunlop, *Phys. Rev. Lett.*, 2004, **92**, 198104.
21 N. Murazawa, S. Juodkazis, S. Matsuo and H. Misawa, *Small*, 2005, **1**, 656.
22 N. Murazawa, S. Juodkazis and H. Misawa, *J. Phys. D*, 2005, **38**, 2923.
23 R. A. Beth, *Phys. Rev.*, 1936, **50**, 115.
24 M. E. J. Friese, T. A. Nieminen, N. R. Heckenberg and H. Rubinsztein-Dunlop, *Nature*, 1998, **394**, 348.
25 T. A. Nieminen, N. R. Heckenberg and H. Rubinsztein-Dunlop, *J. Mod. Opt.*, 2001, **48**, 405.
26 B. J. Berne and R. Pecora, *Dynamic Light Scattering with applications to chemistry, biology and physics*, John Wiley and Sons, New York, 1976.

Optical manipulation of airborne particles: techniques and applications†

David McGloin,*[ab] Daniel R. Burnham,[ab] Michael D. Summers,[a] Daniel Rudd,[ab] Neil Dewar[a] and Suman Anand[c]

Received 12th February 2007, Accepted 17th April 2007
First published as an Advance Article on the web 30th July 2007
DOI: 10.1039/b702153d

In the following paper, we discuss new methods to trap and manipulate airborne liquid aerosol droplets. We discuss the single gradient force trapping of water aerosols in the 2–14 micron diameter range using both 532 nm and 1064 nm light, as well as the holographic optical trapping of arrays of aerosols. Using this holographic technique, we are able to show controlled aerosol coagulation. We also discuss two techniques based on the radiation pressure trapping of aerosols, namely the dual beam fibre trap and the controlled guiding of aerosols using Bessel beams. We conclude with a discussion of new topics for study based upon these techniques and some possible applications.

Introduction

The ability of light to exert forces on microscopic bodies is encapsulated in the technology known as optical tweezers.[1] Such devices make use of a tightly focussed laser beam to trap and manipulate particles and have, over the last two decades, enabled a wide range of studies in the physical, chemical and biological sciences to be carried out.[2–5] Of particular note is that optical tweezers offer a non-contact method to manipulate particles in the size range of a few microns, which typically includes particles such as industrially-relevant colloids, biological cells and, as will be discussed in this paper, aerosols of particular interest to atmospheric scientists.

The beauty of the optical tweezers technique is its relative simplicity, with little more than a laser and a simple microscope system required for basic studies. This has led to a large number of labs throughout the world making use of the unique opportunities to probe particle behaviour, dynamics and composition within the trap. It is fair to say, however, that the vast majority of the experiments carried out to date involve particles immersed in a liquid. Little work has been carried out on airborne particles, primarily due to the relative difficulties associated with airborne trapping, and perhaps the field not being fully appreciated by those working on the dynamics and characterisation of airborne particles.

---

[a] *SUPA, School of Physics and Astronomy, University of St. Andrews, St. Andrews, Fife KY16 9SS*
[b] *Electronic Engineering and Physics Division, University of Dundee, Dundee DD1 4HN*
[c] *Polymeric & Soft Material Section, National Physical Laboratory, Dr. K. S. Krishanan Road, New Delhi, India*

† The HTML version of this article has been enhanced with colour images.

In recent years, however, a small number of pilot studies have been carried out looking at how airborne particles, in particular liquid aerosols, interact with optical fields and the type of processes that can be studied using such trapped particles. The first study of using optical tweezers for airborne particles was carried out by Omori et al.,[6] which looked at trapping solid particles in air (and would appear to be the only such study to date) using a non-inverted trapping geometry. All subsequent aerosol studies have used an inverted geometry that uses gravity to aid the trapping process. Six years later, Magome et al.[7] demonstrated the first use of optical tweezers to trap an aerosol particle (a water droplet produced from a supersaturated vapour of water using a NH_4Cl microparticle as a nucleation centre). In 2004, two further papers appeared, both carried out in collaboration with Andy Ward's group at the Rutherford Appleton Laboratory (RAL). Hopkins et al.[8] demonstrated the controlled coagulation of optically trapped aerosol droplets using a dual beam tweezers system and measured the droplet volumes very accurately using Cavity Enhanced Raman Spectroscopy (CERS), while King et al.[9] examined how seawater and oleic acid droplets reacted with ozone. Jonathan Reid's group have followed up their initial paper with further studies looking at the CERS signals from growing and evaporating droplets,[10] methods for examining how immiscible aerosols mix when coagulated,[11] the characterisation of such coagulation using optical tweezers[12] and a method to compare the growth of more than one aerosol droplet in parallel. Furthermore, our group has carried out studies on making use of holographic optical tweezers to trap many particles simultaneously and coagulate them,[13] and using radiation pressure to guide droplets over large distances.[14] Thus, while there is a rather more extended literature dealing with optical levitation, the experimental work carried out on airborne particles using optical tweezers is very limited, and as such should prove to be an area ripe for study.

This paper is intended to highlight some of the technologies that are perhaps well established within the optical tweezers community that can be utilised to trap aerosol particles. We will also discuss some of the practical aspects associated with such techniques and discuss ways in which the technologies could develop, as well as outlining some of the interesting effects we have noticed in carrying out these studies, which run, perhaps, against the grain of conventional thinking with regard to optical tweezers. We will discuss four main experimental ideas: (1) a study of the trapping efficiency of various aerosol particles at both infrared (1064 nm) and visible wavelengths, (2) the use of holographic optical tweezers techniques to trap, manipulate and coagulate aerosol droplets, (3) the use of a dual beam fibre trap to trap aerosols and (4) the use of laser beams to levitate and guide aerosols over large distances. We will begin with a brief introduction to optical forces.

Physics of optical manipulation

In the early days of optical manipulation, before the development of a single beam optical tweezers, much use was made of radiation pressure, the force exerted by light through the momentum transfer of photons hitting the surface of a body. Using this technique, one can simply trap objects by balancing them against gravity, and in this early work there were a number of studies looking at aerosol particles.[15,16] This force can be written as:

$$F_{scatt} = \frac{n_m}{c} \frac{128\pi^5 r^6}{3\lambda^3} \left(\frac{m^2-1}{m^2+2}\right)^2 I_0 \qquad (1)$$

Where n_m is the refractive index of the medium, c is the speed of light, r is the particle radius, λ is the laser wavelength, I_0 the laser intensity and $m = n_p/n_m$, with n_p the refractive index of the particle (m is referred to as the relative refractive index).

In 1986, Ashkin et al.[1] published a landmark paper demonstrating the use of a single beam to trap particles by making use of the optical gradient force, which is proportional to the intensity gradient of the trapping light (typically a Gaussian laser beam). This force can be written as:

$$F_{\text{grad}} = \frac{2\pi r^3}{c} \left(\frac{m^2 - 1}{m^2 + 2}\right) \nabla I_0 \qquad (2)$$

The force is strong enough to overcome the radiation pressure force when the light is focussed down to an extremely small spot, which is typically achieved by making use of a high NA oil immersion microscope objective. The gradient force acts to pull the particles towards the focal point of the beam (both laterally and axially) provided the refractive index of the particle is higher than the surrounding medium, *i.e.*, $m > 1$. Once trapped, the particle can be considered as residing in a harmonic potential and as such the particle dynamics can be described by the following equation of motion:

$$\gamma_0 \dot{x}(t) + k x(t) = \xi(t) \qquad (3)$$

Where γ_0 is the viscous drag coefficient and k is the trap stiffness. ξ is the stochastic force due to thermal fluctuations. Typically, the motion of the particle is heavily overdamped and the inertial forces can be neglected, as in eqn (3). However, this issue is slightly less clear cut when dealing with particles trapped in air, as the damping of the medium is reduced. Although a typical airborne particle is still in the overdamped regime, it can be coerced towards the underdamped regime.[17] This fact coupled with the slightly higher-than-normal refractive index difference between the particle and the medium (compared to fluid immersed particles) is one of the likely explanations for the unusual behaviour observed when trapping airborne particles, some of which are discussed below.

Optical trapping technique for aerosols

Optical manipulation of aerosols poses a few challenges when one compares it with trapping in liquid. The primary challenge is that the process becomes passive as opposed to active. In water-based samples, one can move the laser beam around until a particle is found, *i.e.*, we can actively seek a particle out. In air this is not possible due to the motion of the aerosols. We must passively wait until an aerosol falls into our trap. As such the process can be rather laborious.

Another consideration is that we must control the environment to a greater degree than when dealing with fluid, by taking care of the relative humidity and the vapour pressure of the droplet.[8] Typically, the humidity can be controlled by placing some fluid into the sealed sample chamber and the vapour pressure of water can be controlled by adding sodium chloride. This has the effect of reducing the vapour pressure and enables the droplets to quickly reach an equilibrium size, allowing stable trapping times of many hours.

Our studies are motivated by the emerging need for tools that can hold and manipulate micrometer, and hopefully smaller, diameter aerosols[18] so as to study the particle composition, particle size, morphology, phase and mixing state, which all contribute to the chemical and physical state of the aerosol. Such properties include light scattering, toxicity, chemical activity and particle diffusion rates. These in turn are important in a wide range of fields such as atmospheric chemistry and physics, combustion science, dusty plasmas and health science. By probing such droplets we can measure properties such as nucleation rates, mass and heat transfer between the droplet and its surrounding medium and coagulation dynamics.

The development of a controllable optical platform provides an excellent basis for such studies and it is the development of such a platform that we are primarily motivated by here. Principally we will make use of established fluid (*i.e.*, water)-based techniques and apply them to airborne particles. These techniques are mainly optical gradient force techniques, that is, based on optical tweezers, but we also develop some new techniques based on radiation pressure, which are more widely developed, having received some initial attention by Ashkin in the 1970s.[15,16]

Aerosol trapping using optical tweezers

The ability of green laser light (532 nm) to trap and hold indefinitely single aerosol droplets is now well established.[9,10,13] Typically, we can measure axial (in the z-direction) Q values (which are a measure of the efficiency of the tweezers, with $Q = 1$ being the maximum efficiency) of $Q \sim 0.22$, which compares favourably with conventional fluid immersed particles, where Qs are typically <0.2, and often are an order of magnitude lower. The low absorption of water at this wavelength indicates that heating should not be a significant problem and that trapping should be robust and long lasting, which indeed we find to be true.[13]

However, the first paper on the optical trapping of aerosols using the gradient force[7] used infrared (IR) wavelengths and the suggestion was that while aerosols could be trapped, there was associated heating and the trapping was short lived, indeed the experiment was to investigate this heating effect. In ref. 7, the water droplets were not created using a nebuliser as in subsequent work, but by reacting ammonia and hydrochloric acid to create nucleation sites of ammonium chloride, around which droplets form. It has also been suggested to us[19] that long term trapping of aerosols using IR light was not possible and that this would limit the usefulness of aerosol tweezers. However the absorption spectra suggest that heating should not be an issue at IR wavelengths and we set out to explore optical trapping at this wavelength to try to refute this suggestion. We made use of a standard inverted tweezers setup, Fig. 1. The advantage of carrying out the experiments at IR wavelengths is that the cost per watt is much lower than with visible lasers and may be more beneficial for studies involving, for example, bioaerosols, where visible wavelengths may damage the specimens.

We make use of a 1064 nm fibre laser with a maximum power of 5 W (IPG Photonics). The power delivered to the trap is typically a few milliwatts, with power control by way of a half-wave plate and a polarising beam cube. We used a 100x oil immersion objective (Nikon CFI E Plan Achromat 100X oil, N.A. 1.25) to focus the beam into the sample chamber. This was identical to the objective used in our 532 nm work to allow a direct comparison. The aerosol droplets (for this and all the experiments detailed below) are created using an ultrasonic nebuliser (Omron U22(NE-U22-E)), which produces aerosols with a Mass Median Aerodynamic Diameter (MMAD) between approximately 3 and 5 microns.[20] Custom glassware is attached to the nebuliser, allowing the droplets to be fed accurately into a sealed glass cube through an inlet hole, which acts as our trapping chamber. It is worth pointing out that the length and diameter of the glass piece used to couple the aerosols from the nebuliser into the trapping chamber plays a critical role in producing a flow of aerosols at the correct velocity and

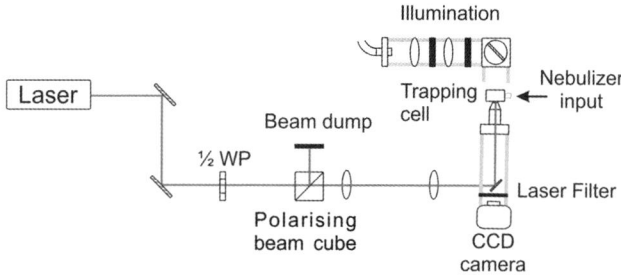

Fig. 1 Standard aerosol tweezers setup.

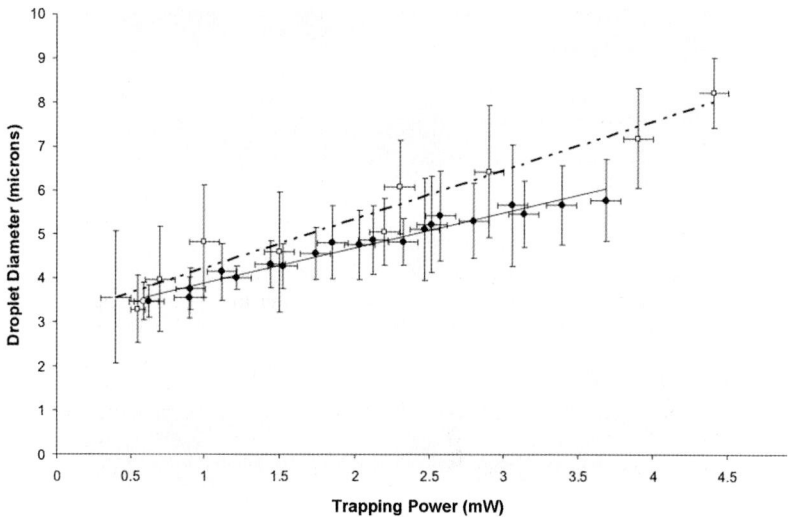

Fig. 2 Graph of droplet diameter against trap power for 2% w/v salt water solution at 532 nm (lower, solid trace) and 1064 nm wavelengths (upper, dashed trace). The points represent the mean diameter of droplets trapped over a range of experimental runs. The error bars give the standard error. The error bar in the y-direction gives an indication of the range of droplet sizes that can be trapped for a given power.

size into the chamber to allow trapping to occur. We have, however, not quantified this process and, while it is important, we can only state that for our nebuliser we use a glass tube of 5 cm in length with a 1.5 cm diameter, which starts to taper down to a diameter of 2 mm after 2.5 cm. The droplets that we find in the trapping chamber that we typically trap tend to be in the size regime 2–15 microns in diameter, the larger size being produced by coagulation. We measure the droplet diameter *via* calibrated video microscopy, which has shown to be a reliable method for estimating droplet size.[12]

We measured the power required to trap aerosols of a given size and also the Q values. The Q values are measured by trapping a droplet and then reducing the trapping power until it falls out. We can then use eqn (4) to find the Q:

$$Q = \frac{c}{n_m} \frac{F}{P} \qquad (4)$$

Where F is the force applied by the laser beam at power P. For a typical 6 micron diameter droplet, we find $Q \sim 0.25$, which is in keeping with our 532 nm work, but is lower than previously reported work.[7] The smallest droplets of water that we have been able to trap are ~ 3 μm in diameter and need ~ 0.5 mW minimum trapping power to be held. A full droplet size (water droplets with 2% w/v sodium chloride) *versus* trapping power plot is shown in Fig. 2, comparing 1064 nm and 532 nm. We would expect slightly larger droplets to be trapped for a given power using the green light, but this is not observed, although within experimental error the graphs are very similar. The reason for this slight discrepancy is probably due to slight differences in the optics, trapping chambers and overall experimental setups.

Importantly, by controlling the sample relative humidity within the chamber we can observe stable long term trapping over periods of hours using infrared beams.

We anticipate that other liquids such as ethanol and dodecane should also be straightforward to trap using infrared light and this will be reported in a future publication.

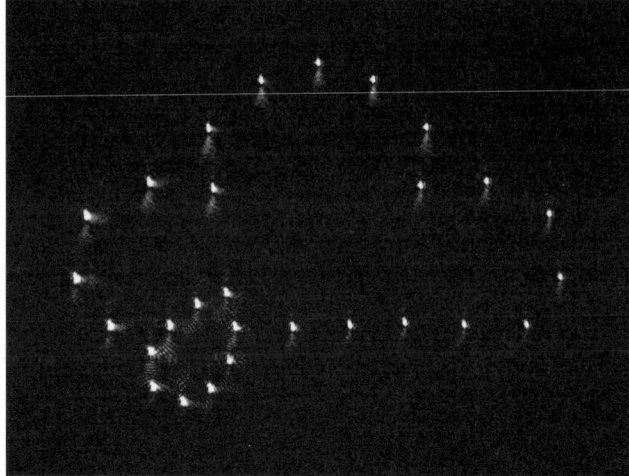

Fig. 3 Illustration of optical trap sites produced by a HOT. Note, the image has had some contrast enhancement to improve visibility.

Holographic optical trapping

A single beam optical tweezers system is a very powerful tool that can be used for many studies, but it is particularly useful in airborne studies to have the ability to carry out coagulation of droplets, and also to be able to carry out comparative studies on droplets simultaneously. There are many multibeam techniques that are able to trap and hold many particles simultaneously. The simplest is the dual beam optical tweezers,[21] which was first used in the context of airborne particles by Hopkins *et al.*[8] to fuse two water droplets. Other techniques include time-sharing traps using an Acousto-Optical Deflector (AOD) to rapidly move a beam between trapping sites.[22] The beam must be moved rapidly enough to ensure that the droplets do not diffuse away or fall too far due to gravity in the time taken to scan the beam between them.[12] Another approach, which we outline below, is to make use of Holographic Optical Tweezers (HOTs).

HOTs simply modify the phase of the initial laser that produces the trap and alter it into a desired intensity pattern. The devices used to modulate the phase are typically liquid crystal Spatial Light Modulators (SLMs).[23–25] The patterns that can be generated can be as simple as two spots or be a complicated array as in Fig. 3. For most trapping work, these arrays of spots are sufficient, but continuous patterns can be generated,[26] as can other beam types, such as Laguerre–Gaussian beams,[27] which are discussed below. The patterns can be produced using various algorithms,[28] which vary in their complexity depending on the amount of information that is known about the desired phase and whether spots or continuous patterns are required. Moreover, we can control optical aberrations of the traps by changing the applied holograms to correct for aberrations in the optical train, Fig. 4.

We recently demonstrated the first use of HOTs applied to aerosols,[13] in which we showed that one could trap multiple aerosol water droplets simultaneously and then control their coagulation, illustrated in Fig. 5.

Our system makes use of a Holoeye LC-R 2500 SLM to modulate the phase of 532 nm continuous wave light from a Laser Quantum Finesse laser (maximum output 4 W). We first rotate the laser's plane of polarisation to optimize the efficiency of the phase modulation, then, having expanded the beam using a telescope, the laser is incident on, and covers, the SLM. Two 4f imaging systems are placed directly after the SLM to reduce the beam size to slightly overfill the back aperture of the microscope objective (Nikon CFI E Plan Achromat 100X oil, N.A.

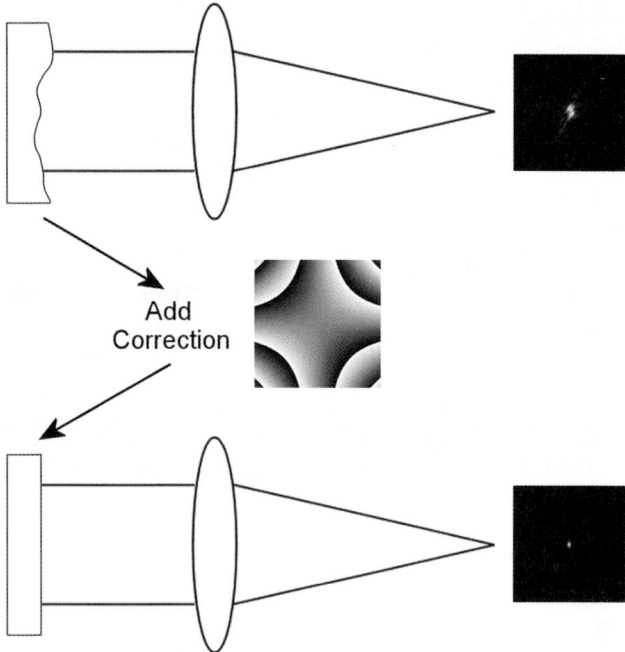

Fig. 4 Illustration of the way in which the SLM can be used to correct optical aberrations. The upper figure shows an uncorrected spot at the focal plane of the lens formed by an incoming aberrated waveform. The correcting hologram is shown in the central inset and the corrected spot in the bottom right.

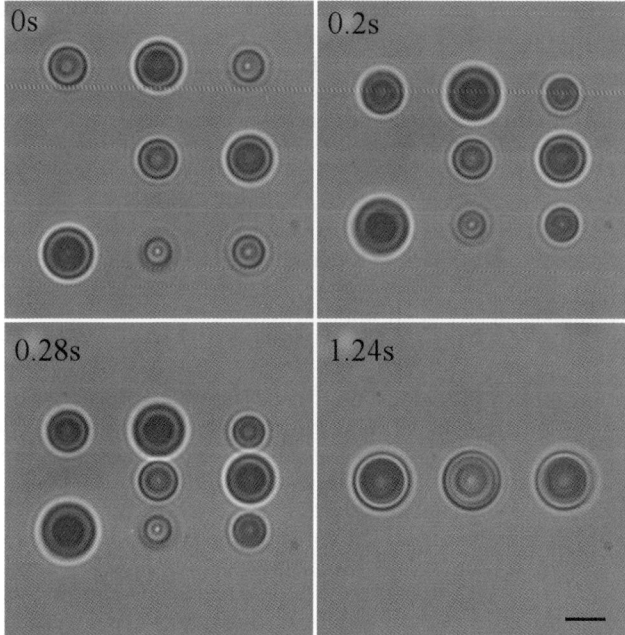

Fig. 5 Coagulation of water droplets using HOTs. The eight droplets trapped in the first image are coagulated into three droplets in around 1 s. The scale bar is 5 microns.

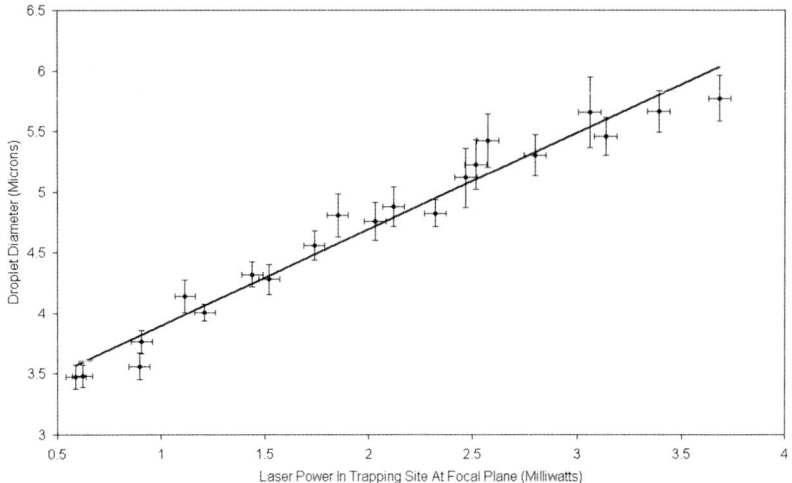

Fig. 6 Trapped droplet size *versus* power in a 532 nm optical tweezers. The points represent the mean diameter of droplets trapped over a range of experimental runs. The error bars give the standard error. The error bar in the *y*-direction gives an indication of the range of droplet sizes that can be trapped for a given power.

1.25). The 4f imaging systems also make the SLM and back aperture planes conjugate. Having focused the light into the trapping plane, the same objective is used for imaging. We generate our holograms making use of a custom LabVIEW program implementing an adaptive-additive algorithm that treats the SLM as a secondary microdisplay.

One of the key results of this work was a measurement of the trapping parameters of a single droplet held in an optical trap. The findings are illustrated in Fig. 6. The error bars in the *y*-direction are an indication of the range of sizes that can be trapped at a given power. The points on the graph are the mean droplet diameter and the error bars the standard error calculated over a number of experimental runs. The graphs indicate that at higher powers it is harder to trap smaller droplets. We do not have experimental evidence as to why this is but we believe it is due to the random trajectories that the droplets move in on their way to the trap. At higher powers the droplets feel a radiation pressure force far from the trap which makes it hard for them to get into the region where the gradient force dominates. Only droplets moving near the plane in which the focal region of the laser resides have a high probability of making it into the trapping region. As we increase the power, it becomes harder for the droplets to avoid the increasing radiation pressure zones around the trap focus. This effect is also evident from the graph in Fig. 2 above, showing the trapping ability of IR optical tweezers.

A second phenomenon that is observed is that if a droplet is trapped at a low power, then turning up the power leads to rapid particle oscillations and eventually the particle falls out of the trap. This is not behaviour normally associated with particles that are trapped in liquid. Why this should be the case is not yet fully clear, but is likely to be associated with a change in the dynamical regime in which it is trapped (*i.e.* a shift from overdamped to underdamped).[17]

The use of holograms can also be used to explore the transfer of angular momentum to airborne particles. To do so we make use of Laguerre–Gaussian beams, which carry Orbital Angular Momentum[29] (OAM) due to the helical nature of their phase fronts. Such beams can be characterised by an azimuthal winding number *l* and they carry an orbital angular momentum of $l\hbar$ per photon,[30] and this can be transferred to a particle trapped within the beam. We have shown rapid rotation rates are possible using $l = 80$ beams in air, illustrated in Fig. 7.

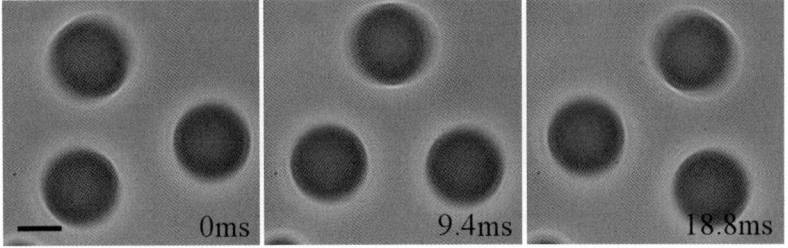

Fig. 7 Rotation of water aerosol using Laguerre-Gaussian beams. The three droplets are trapped in a single beam. The rotation is clearly seen as a function of time. Full rotation of complete circuits is possible but not shown. Scale bar is 5 µm.

Dual beam fibre trapping of aerosols

The idea of a dual beam trap goes back to the early days of optical trapping,[15] where two focussed Gaussian beams were used to trap particles. This technique was developed by Constable *et al.*[31] to make use of dual beams emerging from two aligned optical fibres. The trapping mechanism is due to radiation pressure and the advantage over conventional optical tweezers is that the beams are not delivered by high NA optics and can, if necessary, trap much larger particles than can be normally accessed by optical tweezers. Work by Kishan Dholakia's group in St. Andrews has shown optical manipulation of 0.1 mm diameter objects, for example.[32]

In the context of airborne trapping, we envisage the advantages of dual beam trapping to be the ability to trap a range of particle sizes not accessible to optical tweezers, increased probability of trap loading (*i.e.*, better particle sampling capabilities) and also providing a more robust platform to trap particles in hostile environments (namely environments outside the safe, well controlled optical lab). This might allow us to build an engineered optical trapping and sampling device capable of being used in the field.

To this end, we have examined the possibility of trapping water aerosols using a dual beam fibre trap (which have far lower trapping power requirements than a recent dual beam trap making use of focussed laser beams[33]).

Coarsely, we can categorise optical fibres into two types, multimode and single mode. We envisage single mode fibre to be the better choice for this type of trap, but to simplify laser coupling, power requirements and cost, we restrict ourselves to multimode fibre here.

Our experimental setup consists of a 532 nm diode-pumped solid state laser (a 4 W maximum power Laser Quantum Finesse) which is coupled into two multimode optical fibres (0.22 NA multimode Vis-IR step index fibre) (item#Afs50/125Y Thorlabs) with a 50 micron diameter core. We use a half-wave plate placed before a polarising beam cube to divide the power equally into the two arms of the system, where each beam is then coupled, *via* a fibre coupler, into a fibre. The half waveplate also allows us to control the relative power going into each arm of the device. A schematic of the experimental system is shown in Fig. 8.

The ends of the fibres are connectorised using ST connectors. We initially tried to make use of non-connectorised fibres but we found that they were very brittle, and even gentle air currents or knocks to the system could move the fibres out of alignment. Also, as we don't directly protect the ends of the fibres from the aerosols, it meant that we needed to repeatedly re-cleave the ends of the fibres after a given experimental run. Using the connectorised fibres makes them easier to clean, keeps them in more rigid alignment (by using standard optical mechanical mounts to hold them) and generally makes the system more robust.

The fibre ends are placed approximately 300 microns apart and are enclosed by a custom-made glass sheath that covers both the fibre ends and has inlet hole in the side to allow the aerosols to enter. A model of the trapping region is shown in Fig. 9.

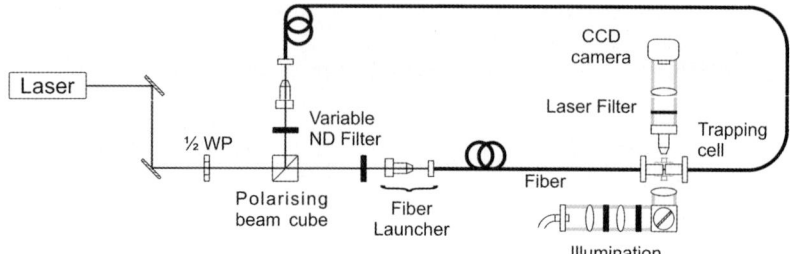

Fig. 8 Schematic of the dual beam fibre trap experimental setup.

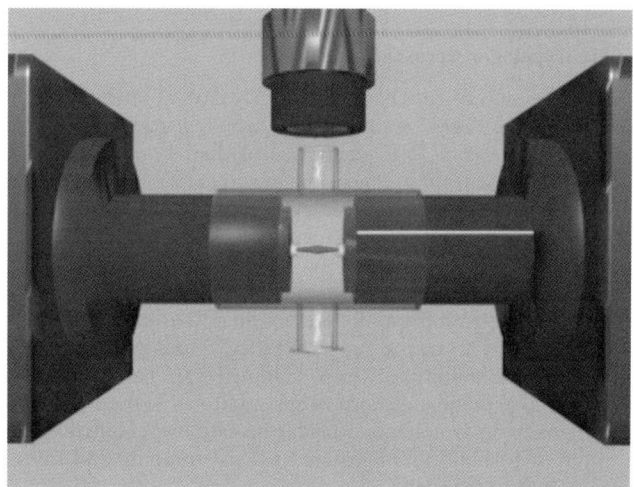

Fig. 9 Model of the dual beam fibre trap and the trapping region. The fibre ends (just visible as white insets into the transparent chamber) enter the transparent chamber, which acts to cut out air currents. The long working distance microscope objective is shown above the chamber. The chamber projections towards and directly opposite the objective are flattened windows to allow imaging into the chamber.

We find that the air currents are a significant factor in trying to trap the droplets. This is also true for the optically tweezed droplets, but here the geometry of the system coupled with the multimode nature of the light coming out of the fibre makes trapping slightly less robust. We hope improved engineering of the device will improve the stability—indeed, the work we have already done to make the device more rigid has improved our ability to trap significantly.

We make use of a long working distance microscope objective (Mitutoyo M Plan NIR 50x, NA 0.42) to observe the droplets in the trapping region due to the difficulty of using standard microscope objectives with short working distances to image beyond the ends of the connectorised fibres. An image of a trapped droplet is shown in Fig. 10.

We find that with using the multimode fibre, we are able to trap only a narrow range of droplet sizes as a function of power, in the range 3–7 microns, with an average size of approximately 5 microns. The gradient of the graph, Fig. 11, is low while the error bars are high, indicating that we do not gain much by increasing the power. To trap these droplets, we need a minimum of around 100 mW in each arm of the trap. We have observed trapping with 40 mW in each arm but this is less stable. While such powers are significantly higher than the optical tweezers discussed above, they are much lower than in a previously reported dual beam trap used to trap aerosols using focussed laser beams.[33] While the amount of power does not seem to

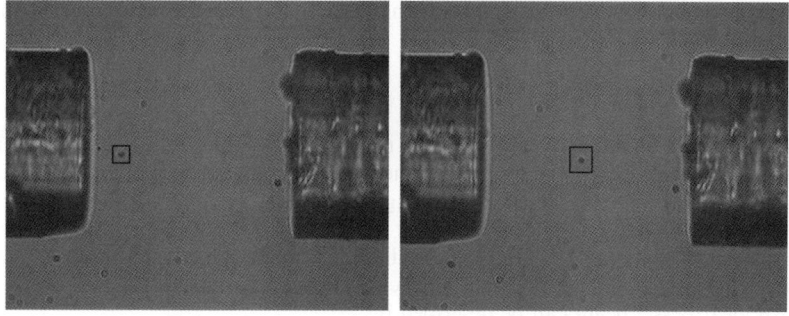

Fig. 10 Images of droplet trapped in fibre trap. The droplets in the trap are surrounded by a square box. The two images also show the droplet being guided with the initial trapping position on the left and then a new trapping position (right) about 50 microns from the initial position is reached by changing the power in one arm of the trap.

affect the droplet in any way and does not pose a constraint on the system, we anticipate that lower powers will be required to avoid heating and evaporation in certain circumstances. Lower trapping powers should be possible by altering the fibre separation and by using single mode fibres.

The use of multimode fibre makes a theoretical analysis of the system difficult due to the presence of numerous 'hotspots' in the intensity profile, resulting in the observation of ill-defined equilibrium positions in contrast to what one would expect in a single mode fibre trap.[34] Thus, single droplets are found to be trapped at nearly any position between the fibre ends. We can, however, move the droplets between the fibres by changing the relative power provided to each fibre. For example, we can move the droplets up to 50 microns, as shown in Fig. 10.

We find that multiple droplets can be trapped within the trapping region due to the presence of multiple hot spots. This process is largely uncontrollable and does not constitute any form of optical binding,[35] which we expect to see when using single mode fibre. The use of hot spots does allow multiple droplets to be trapped in different horizontal planes, and it may be that in future work we can optimise this process to allow controllable trapping of multiple droplets, allowing parallel

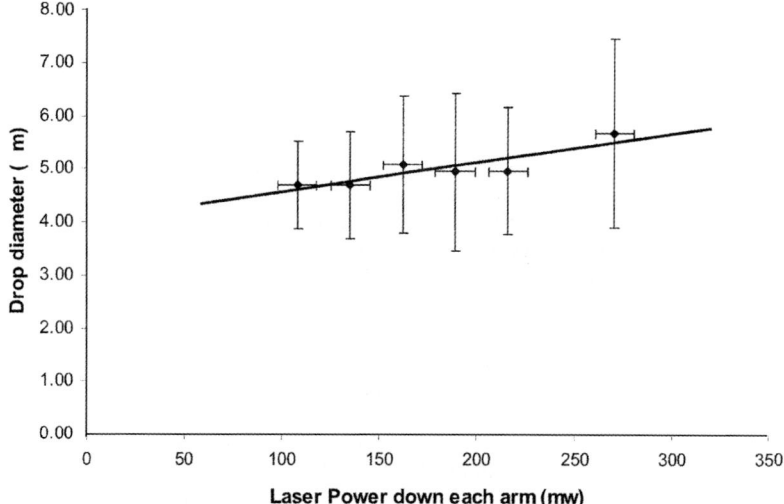

Fig. 11 Plot of trapped droplet size *versus* laser power in each arm for the fibre trap. The error bars indicate the standard error from the mean calculated over a number of experimental runs.

processing of droplet composition, morphology and other chemical and thermodynamic processes.

Optical guiding of aerosols

We conclude our look at novel aerosol trapping techniques and technologies by considering a variant of the well established radiation pressure trapping technique, using a single beam to hold the particle against gravity. Here, we demonstrate that we can controllably guide a particle over 2.5 mm by making use of a novel beam known as a Bessel beam.[36] A Bessel beam has the property that, when compared to a similar Gaussian beam, it does not diffract over an appreciable distance.

An ideal Bessel beam can be described by:

$$F(r, \phi, z) = A_0 \exp(ik_z z) J_n (k_r r) \exp(\pm in\phi) \quad (5)$$

where J_n is a nth-order Bessel function, k_z and k_r are the longitudinal and radial wavevectors, with $k = \sqrt{k_z^2 + k_r^2} = 2p/l$ (λ being the wavelength of the electromagnetic radiation making up the Bessel beam) and r, ϕ and z are the radial, azimuthal and longitudinal components, respectively. The intensity structure for a zeroth order Bessel beam is shown in Fig. 12.

For the Bessel beams considered here, $n = 0$. Our Bessel beams are created using an axicon,[37] or conical lens element. The opening angle of the axicon is given by:

$$q = (n - 1)g \quad (6)$$

Where n is the refractive index of the axicon material and g is the opening angle of the axicon. As we cannot physically have a beam that is truly non-diffracting, the beam we generate can be considered quasi-non-diffracting over a given *propagation distance*. For the axicon-generated beam, we have a propagation distance given by:

$$z_{max} = \frac{k}{k_r} w_0 \gg \frac{w_0}{q}. \quad (7)$$

In the following, we compare the guiding of water droplets using both a Bessel beam and a Gaussian beam. Our Gaussian beam has a beam diameter of 6 μm and the Bessel beam has a central core diameter of 4 μm and a propagation of 4 mm. The core was found to expand to 25 μm at the end of the 4 mm propagation distance, at which the 2nd ring became indistinct and the central spot became irregular.

A 2 W CW ytterbium fibre laser at 1064 nm (IPG Photonics) was used to produce the Gaussian and Bessel beams. For the Gaussian beam experiment the beam is collimated with an appropriate beam waist so as to form the desired spot size after being focused with a final $f = 25$ mm lens. The Bessel beam is made using an axicon

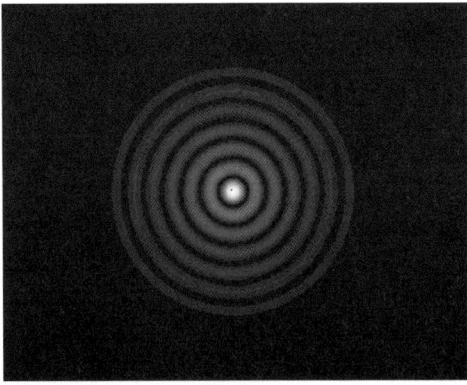

Fig. 12 Bessel beam intensity profile.

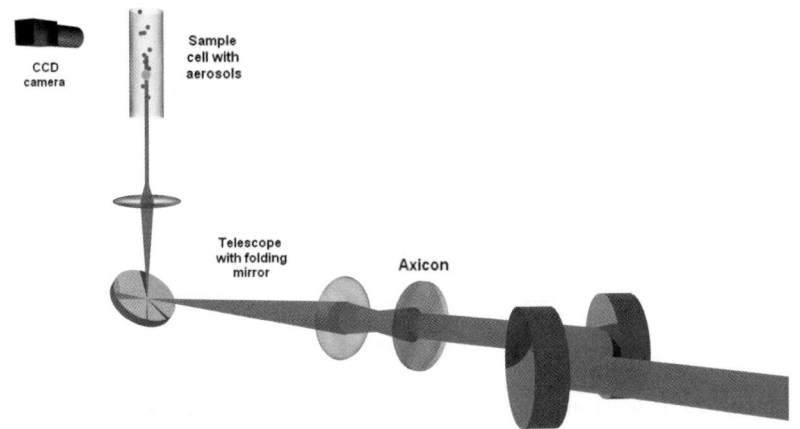

Fig. 13 Experimental setup for optical levitation using a Bessel beam.

in a double telescope arrangement.[38] The first telescope expands the beam to allow adjustment of the propagation distance of the beam, while the second telescope is used to adjust the central core size of the beam (to match the spot size of the Gaussian beam). The aerosols are generated using a nebuliser attached to a nitrogen tank. The nitrogen flow into the nebuliser is controlled by a mass flow controller. The experimental arrangement is shown in Fig. 13 for the Bessel beam.

In order to trap the droplets we had to start with a high initial power (~ 1 W for the Gaussian beam, 200 mW in the Bessel beam core). Low powers (100 mW for the Gaussian, >10 mW for the Bessel beam core) could be used to hold particles once they were captured, but were insufficient to trap the droplets moving under gravity and in the air currents in the sample chamber, which in this experimental arrangement were more of an issue than when using the optical tweezers to trap. Thus, we could trap a droplet with a high power and then subsequently guide it up and down by varying the power.

Using the Gaussian beam, water was stably trapped and could be made to move up and down with slow adjustment of the output power. The maximum guiding distances was found to be 250 μm for water (Fig. 14).

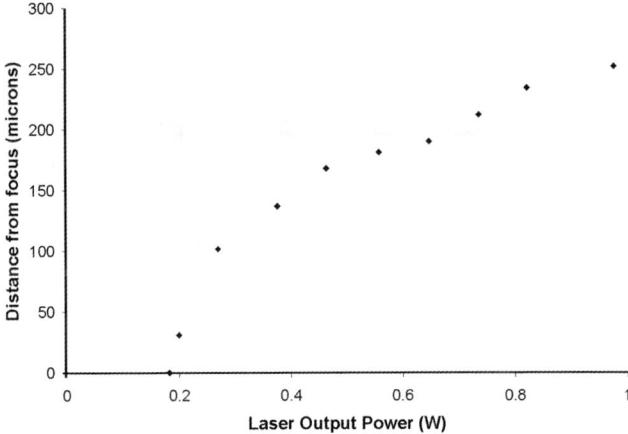

Fig. 14 Water droplet height against laser power (W) for water levitated in a Gaussian laser beam for a constant flow rate of nitrogen into the nebuliser of 0.58 litres min^{-1}. The particle displacement is relative to the position for maximum power achieved at that flow rate.

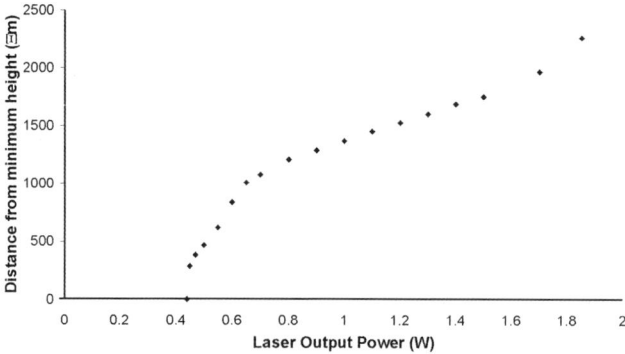

Fig. 15 Particle height against laser power (W) for water at a constant flow rate in a Bessel beam. The particle displacement is relative to the position for maximum power achieved at that flow rate.

We then performed similar experiments with the Bessel beam. The problem of trapping multiple droplets was a greater issue for the Bessel beam trap. Isolating single droplets was particularly difficult and the flow rate had to be carefully selected to avoid trapping multiple droplets. In some cases, droplets were trapped over a significant portion of the beam, forming extended arrays of particles. This was possible due to the relatively large vertical trapping region presented by the core over the large propagation distance. Another factor may have been the documented self-healing effect of Bessel beams, which we have shown to form similar arrays in colloidal solutions.[39]

For water droplets, the Bessel beam allowed guiding to occur over 2.5 mm. This is approximately ten times the distance achieved in the Gaussian beam, illustrated in Fig. 15.

Further results for guiding ethanol and dodecane can be found in ref. 14

Technologies and outlook

Above, we have outlined a number of optical techniques that can be used to trap liquid aerosol droplets. Each of the techniques is relatively new and the full parameter space in which they operate has yet to be explored. However, it is clear that from a physics and technology development point of view that there are a number of interesting things to explore. Perhaps most importantly is the type of material that can be trapped optically, and the means to deliver that material into the trap. Most of the work discussed above made use of water droplets and the primary reason for this is that water is perhaps the most interesting liquid from an atmospheric point of view. However, we are largely limited to water and aqueous solutions due to our delivery method of an ultrasonic nebuliser, and organic materials, for example, damage the seals within our devices and cannot be regularly and controllably aerosolised. That is not to say that with alternative techniques these limitations cannot be overcome, and we are currently assessing alternative ultrasonic nebulisers that hold the liquids in a more compatible way.

Another question, and one motivation behind the fibre trap development, is whether we can trap non-liquid or non-transparent particles in such optical traps. This is of importance as many aerosols are not aqueous, black carbon being an important example. Clearly, we do not expect optical tweezers to be able to trap such particles but we are cautiously optimistic about the possibility of using a dual beam radiation pressure trap to trap objects that cannot be trapped in tweezers, either due to their optical properties or their size. One can, for example, trap much larger objects in a dual beam trap than in conventional fluid-based tweezers, as noted above.[32]

To carry out more quantitative experiments on aerosols in optical traps we need to develop new, more controllable, ways to load the trap sites. Various options exist for loading techniques. We envisage a system design that makes use of techniques in atom optics where atoms are trapped in a 'dirty' chamber before being transferred through some mechanism to the 'science' chamber where the actual measurements are carried out. We believe that this can be achieved with aerosols either through sorting techniques,[40] *via* guiding, as illustrated with the Bessel beams above, or *via* guiding through a physical object such as hollow core optical fibre. We are currently exploring these options to see if any will improve the general robustness of trapping or the specificity in what we can actually trap, *i.e.* improving the monodispersity of the aerosols in the science chamber.

More generally, optical manipulation techniques seem to offer a stable platform for the manipulation of small, micron-sized droplets, which are difficult to trap using other methods such as electrostatic (electrodynamic balances) or sonic traps. Damping would also seem to be better in optical traps, allowing the particles to be better localised. The optical techniques also lend themselves very well to integration with spectroscopic analyses, and with holographic techniques offer the ability to manipulate multiple particles simultaneously, which cannot be done in a controlled manner using other methods.

We conclude, then, with the optimistic comment that optical manipulation techniques can be put to good use in probing single and small numbers of aerosol droplets. There is, however, still much work to be done to understand the specific processes involved in trapping things in air using laser beams and in designing and engineering robust systems that can have a wide range of uses both in laboratory and field environments.

Acknowledgements

We would like to thank the Royal Society, EPSRC and NERC for funding the above work. DM is a Royal Society University Research Fellow, DRB and DR acknowledge EPSRC for supporting their PhDs, MDS is supported by an RSC/EPSRC Analytical Trust studentship. JB and ND thank the Nuffield Trust for support and SA was supported by a Royal Society of Edinburgh International Visitor award. We thank Jonathan Reid and Laura Mitchem for helpful discussions and advice.

References

1. A. Ashkin, J. M. Dziedzic, J. E. Bjorkholm and S. Chu, *Opt. Lett.*, 1986, **11**, 288.
2. K. Dholakia and P. Reece, *Nano Today*, **1**, 18.
3. F. Ritort, *J. Phys.: Condens. Matter*, 2006, **18**, R531.
4. D. G. Grier, *Nature*, 2003, **424**, 810.
5. J. E. Molloy and M. J. Padgett, *Contemp. Phys.*, 2002, **43**, 241.
6. R. Omori, T. Kobayashi and A. Suzuki, *Opt. Lett.*, 1997, **22**, 816.
7. N. Magome, M. I. Kohira, E. Hayata, S. Mukai and K. Yoshikawa, *J. Phys. Chem. B*, 2003, **107**, 3988.
8. R. J. Hopkins, L. Mitchem, A. D. Ward and J. P. Reid, *Phys. Chem. Chem. Phys.*, 2004, **6**, 4924.
9. M. D. King, K. C. Thompson and A. D. Ward, *J. Am. Chem. Soc.*, 2004, **126**, 16710.
10. L. Mitchem, J. Buajarern, R. J. Hopkins, A. D. Ward, R. J. J. Gilham, R. L. Johnston and J. P. Reid, *J. Phys. Chem. A*, 2006, **110**, 8116.
11. L. Mitchem, J. Buajarern, A. D. Ward and J. P. Reid, *J. Phys. Chem. B*, 2006, **110**, 13700.
12. J. Buajarern, L. Mitchem, A. D. Ward, N. H. Nahler, D. McGloin and J. P. Reid, *J. Chem. Phys.*, 2006, **125**, 114506.
13. D. R. Burnham and D. McGloin, *Opt. Express*, 2006, **14**, 4175.
14. M. D. Summers, J. P. Reid and D. McGloin, *Opt. Express*, 2006, **14**, 6373.
15. A. Ashkin, *Proc. Natl. Acad. Sci. U. S. A.*, 1997, **94**, 4853.
16. D. McGloin, *Philos. Trans. R. Soc. London, Ser. A*, 2006, **364**, 3521.
17. R. Di Leonardo, G. Ruocco, J. Leach, M.J. Padgett, A. Wright, J. Girkin, D. Burnham and D. McGloin, submitted (cond-mat/0702557).

18 J. P. Reid and L. Mitchem, *Annu. Rev. Phys. Chem.*, 2006, **57**, 245.
19 This was communicated in referee's comments to a grant proposal we submitted.
20 J. H. Dennis, C. A. Pieron and K. Asai, *J. Aerosol Med.*, 2003, **16**, 213.
21 E. Fallman and O. Axner, *Appl. Opt.*, 1997, **36**, 2107.
22 A. Resnick, *Rev. Sci. Instrum.*, 2001, **72**, 4059.
23 J. E. Curtis and D. G. Grier, *Opt. Commun.*, 2002, **207**, 169.
24 H. Melville, G. F. Milne, G. C. Spalding, W. Sibbett, K. Dholakia and D. McGloin, *Opt. Express*, 2003, **11**, 3562.
25 G. Gibson, L. Barron, F. Beck, G. Whyte and M. Padgett, *New J. Phys.*, 2007, **9**, 1367.
26 D. McGloin, G. C. Spalding, H. Melville, W. Sibbett and K. Dholakia, *Opt. Express*, 2003, **11**, 158.
27 R. M. Lorenz, J. S. Edgar, G. D. M. Jeffries, Y. Zhao, D. McGloin and D. T. Chiu, *Anal. Chem.*, 2007, **79**, 224.
28 G. Sinclair, P. Jordan, J. Courtial, J. Cooper and Z. J. Laczik, *Opt. Express*, 2004, **12**, 6475.
29 L. Allen, M. J. Padgett and M. Babiker, *Prog. Opt.*, 1999, **31**, 291.
30 L. Allen, M. W. Beijersbergen, R. J. C. Spreeuw and J. P. Woerdman, *Phys. Rev. A*, 1992, **45**, 8185.
31 A. Constable, J. Kim, J. Mervis, F. Zarinetchi and M. Prentiss, *Opt. Lett.*, 1993, **18**, 1867.
32 P.R.T. Jess, private communication.
33 K. Taji, M. Tachikawa and K. Nagashima, *Appl. Phys. Lett.*, 2006, **88**, 141111.
34 P. R. T. Jess, V. Garcés-Chávez, D. Smith, M. Mazilu, L. Paterson, A. Riches, C. S. Herrington, W. Sibbett and K. Dholakia, *Opt. Express*, 2006, **14**, 5779.
35 N. K. Metzger, K. Dholakia and E. M. Wright, *Phys. Rev. Lett.*, 2006, **96**, 068102.
36 D. McGloin and K. Dholakia, *Contemp. Phys.*, 2005, **46**, 15.
37 J. H. MacLeod, *J. Opt. Soc. Am.*, 1954, **44**, 592.
38 J. Arlt, V. Garces-Chavez, W. Sibbett and K. Dholakia, *Opt. Commun.*, 2001, **197**, 239.
39 V. Garcés-Chávez, D. Roskey, M. D. Summers, H. Melville, D. McGloin, E. M. Wright and K. Dholakia, *Appl. Phys. Lett.*, 2004, **85**, 4001.
40 M. P. MacDonald, G. C. Splading and K. Dholakia, *Nature*, 2003, **426**, 421.

PAPER

In situ comparative measurements of the properties of aerosol droplets of different chemical composition

Jason R. Butler,[a] Laura Mitchem,[a] Kate L. Hanford,[a] Lennart Treuel[b] and Jonathan P. Reid*[a]

Received 4th May 2007, Accepted 15th May 2007
First published as an Advance Article on the web 3rd August 2007
DOI: 10.1039/b706770b

Aerosol optical tweezers can be used to manipulate multiple aerosol particles simultaneously. When coupled with spontaneous and stimulated Raman scattering, the composition, size and phase partitioning of different chemical components within a liquid droplet can be investigated. In combination, these two techniques suggest the possibility of a new strategy for characterising the thermodynamic behaviour of aerosols and the kinetics of mass transfer between the gas and condensed phases. We demonstrate here that two droplets can be characterised simultaneously, examining specifically the variation in wet particle size with relative humidity, recording the changes in size with nanometre accuracy. In a further demonstration, we use the size of a sodium chloride droplet to determine the relative humidity of the gas phase, allowing the variation in hygroscopicity of a second aqueous glutaric acid/sodium chloride droplet to be studied. We suggest that such a comparative approach can provide new insights into aerosol dynamics.

1. Introduction

Liquid surfaces are highly dynamic with rapid transport of molecules towards and across the interfacial region.[1–3] If it is assumed that the liquid and gas phase compositions are at equilibrium, the collision rate of gas phase water molecules with a water surface can be estimated to be $\sim 10^{23}$ collisions cm^{-2} s^{-1}. This is equivalent to an incoming gas phase molecule striking an area on the surface at a rate of one water molecule every 10 ns. Fluctuations in the surface structure occur on picosecond timescales, with molecules at the surface interchanging with molecules in the bulk.[2] Further, molecules approaching the surface can undergo elastic scattering, returning directly to the gas phase.[1] Thermal accommodation can lead to an incoming molecule adopting a state of transient existence at the surface followed by desorption. A molecule that adsorbs to the surface and becomes fully solvated diffusing into the bulk, can be described as undergoing mass accommodation.[3] Chemical reactions may occur rapidly at the surface or may proceed following diffusion into the bulk phase.

[a] School of Chemistry, University of Bristol, Bristol, UK BS8 1TS
[b] Physik. und Theoretische Chemie, Universität Duisburg-Essen, Universitätsstraße 5, D-45141, Essen, Germany

From such a molecular picture, it is unsurprising that the fundamental nature of intermolecular interactions in the interfacial region is important in governing the kinetics of mass transfer and the process of mass accommodation. To fully understand the interfacial transport, it is important that the heterogeneity of the surface be characterised. For example, the presence of a surface active organic component (a surfactant) can lead to an aqueous surface that is enriched in the organic component and the formation of an ordered monomolecular film.[4,5] It has been suggested that the existence of such a film on the surface of aqueous atmospheric aerosol may influence the mass accommodation of water, retarding droplet growth or evaporation: an incoming water molecule will be presented with a hydrophobic rather than a hydrophilic surface for accommodation.[6,7] Further, recent experiments have suggested that an organic surface coating can inhibit the accommodation and reaction of trace gases such as N_2O_5.[8,9]

Not only can the character of the surface play a significant role in governing the kinetics of mass transport and heterogeneous reactions in aerosols, but the thermodynamic properties of particles may be modified by changes in surface and bulk composition.[10] The partitioning of water between the condensed and gaseous phases has a crucial impact on the size distribution of aerosol in the atmosphere.[11] In most cases, it is assumed that the partitioning attains an equilibrium state and is governed by the thermodynamic properties of the aerosol. Although the behaviour of inorganic/aqueous aerosol is well described by Köhler theory, the properties of mixed organic/inorganic/aqueous aerosol remain poorly characterised.[12] The enrichment of a droplet surface with an involatile surfactant can lead to a reduction in the vapour pressure of the aqueous component, displacing water from the surface of the aerosol droplet.[13] Further, the impact of the curvature of the droplet surface must be considered when estimating the vapour pressure of droplets < 100 nm in size, and is treated by including the Kelvin equation in the framework for the thermodynamic analysis.[10,11,14] The surface curvature term is dependent on surface tension and, consequently, on the composition of the surface.[15]

Conventional measurements have demonstrated that it is possible to explore the kinetics of aerosol transformation through an examination of the changes in gas phase composition.[3,16] Although not providing information directly on the changes in condensed phase or surface composition, this strategy can be used to explore mass accommodation and reactive uptake. Measurements of changes in particle mobility by electrostatic classification have allowed the change in particle size to be explored. For example, with two differential mobility analysers operate in tandem it is possible to explore the response of particles of a selected size to changes in the Relative Humidity (RH) of the surrounding gas phase.[17] Such measurements allow an interrogation of the thermodynamic properties of aerosol, and in some cases allow the kinetics of mass transfer to be investigated. More recent advances in aerosol analysis allow the direct interrogation of aerosol composition through the development of aerosol mass spectrometry.[18] Further, single particle measurements on electrostatically trapped particles have been used to interrogate the phase behaviour and variation in wet particle mass with RH.[19] However, it remains challenging to correlate directly and unambiguously variations in the compositions of the bulk condensed phase, the surface and the surrounding gas phase with rates of mass transfer and thermodynamic behaviour. For example, discrepancies exist between thermodynamic measurements made on aerosol ensembles and single particles.[20]

We describe in this publication a novel strategy for examining the coupling between gas phase, surface and bulk condensed phase composition in aerosols. Spontaneous and stimulated Raman scattering can be used to interrogate and discriminate between the bulk and surface composition of a single aerosol droplet.[21] The requirements for an accurate probe of the gas phase composition are demanding. Ideally the probe should be able to examine the composition close to the surface of the trapped droplet on the micron length scale, minimising artefacts from spatial variability in gas composition. Further, the probe should exhibit rapid time response and should have a dynamic range appropriate for the conditions experienced by the droplet of interest. We will demonstrate here that comparative measurements can be made on two aerosol droplets simultaneously controlled within two optical traps and separated by a few droplet diameters. One of the two droplets can be used as a highly accurate and responsive probe of gas phase composition, allowing the RH to be determined with an accuracy of better than ±0.09% with 1 s time-resolution. This can then allow the behaviour of the second droplet to be interrogated accurately by resolving nanometre size changes, approaching the molecular scale.

We describe briefly the experimental technique, before exploring the accuracy with which the hygroscopic properties of a single droplet can be investigated and the problems associated with such measurements. This will provide an illustrative example of the difficulties associated with characterising condensed phase and gas phase composition. We will then report benchmark comparative measurements with two aerosol droplets containing sodium chloride before demonstrating that the variations in wet particle size with time of two droplets of different composition can be compared directly.

2. Aerosol optical tweezers

The strategy of combining a single-beam gradient force optical trap, commonly referred to as optical tweezers, with Raman spectroscopy has previously been described in detail, and only a brief summary will be reported here. A schematic of the experimental layout is shown in Fig. 1. In recent work, we have demonstrated that two aerosol droplets of different chemical components can be manipulated and steered to coagulation.[22,23] Brightfield microscopy and Raman spectroscopy can be used to interrogate the mixing state of the two components and determine the composition.[24] Non-linear Raman scattering can provide a signature of the existence of surface films. Further, we have demonstrated that the variation in wet particle size with RH can be interrogated for a single droplet, with measurements possible over timescales of days.[25]

Light of wavelength 532 nm produced from a Nd:YVO$_4$ laser (Coherent VERDI V5) is focussed by a 100× oil immersion objective (NA of 1.25) with a working distance of ~130 μm. This produces a gradient force optical trap in which the scattering forces imparted to the trapped particle are dominated by the gradient forces at the focal point.[26] Light of wavelength 532 nm is chosen as this is close to the

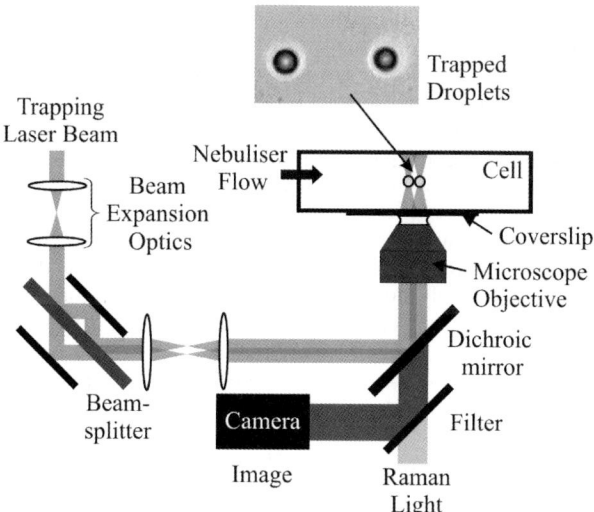

Fig. 1 Experimental schematic.

minimum in the complex refractive index of water, minimising light absorption and droplet heating.[23,25] For this reason, we have observed no significant variation in droplet size and evaporation rate with laser power, and can assume negligible droplet heating from the trapping laser beam.

An aqueous solution of the inorganic salt of interest (aqueous sodium chloride) is nebulised and passes into a custom designed and fabricated cell in which the optical trap is formed from the tightly focussed laser beam. The RH within the cell is regulated by flowing a humidified stream of nitrogen gas through the cell, produced by bubbling nitrogen through a water bubbler at flow rates of between 0 and 0.2 L min^{-1}. The RH and temperature within the cell are recorded using a capacitance probe (capacitive sensor, HIH-3602-A, Honeywell Sensing and Control) with an accuracy of ±2% below 90% RH and a time-response of 50 s. The trapped droplet is imaged by conventional brightfield microscopy using a blue LED for illumination. This allows convenient separation of the light for imaging, the laser light for trapping and the Raman scattered light, which occur in the blue, green and red parts of the visible spectrum, respectively. The backscattered Raman signal from the droplet is collected by the objective and imaged onto the entrance slit of a spectrograph of focal length 50 cm equipped with a diffraction grating blazed at 500 nm and with 1200 grooves mm^{-1}. The spectra are recorded by a CCD camera, with a spectral dispersion of 0.037 nm per CCD pixel.

In measurements involving two optical traps, which allow simultaneous control over two droplets, the optical design is adapted to incorporate a beam splitter at the conjugate plane of the back-aperture of the microscope objective, as discussed by Fallman and Axner.[27] The two optical traps that are formed can be translated independently in the imaging plane over distances greater than 50 μm. This allows each droplet to be translated into the region where backscattered Raman scattered light can be collected, allowing the independent spectral characterisation of composition and size for each droplet in turn. In such measurements, the interchange of the two droplets can be achieved in ~2 s, and droplet translation speeds are typically tens of μm per second. The droplet initially centred on the Raman collection optics is first moved out of the Raman active area and the second droplet is then translated into the active area.

3. Characterising the equilibrium size behaviour of an aerosol droplet

We first describe measurements of evolving wet droplet size with change in RH made on a single trapped sodium chloride droplet. This will provide an illustration of the problems associated with measurements in which the droplet size and composition are characterised accurately but the gas phase is not. An aerosol flow is introduced into the trapping cell and a single droplet is trapped by the optical tweezers. The RH is measured with the capacitance probe and the evolving size is determined from the Raman fingerprint. The droplet is retained in the trap for many hours, allowing measurements of the equilibrium wet droplet size over a wide range of RHs.

Sodium chloride particles deliquesce to become aqueous sodium chloride droplets at a RH of 75%.[28] With further increases in RH, the increasing activity of water in the gas phase leads to an increase in the activity of water in the condensed phase, an increase in the number of moles of water partitioned to the droplet, and a corresponding growth in droplet size. The number of moles of water partitioned to the condensed phase can be quantified by Köhler theory, which accounts for the influence of the Kelvin effect and the solute effect on the vapour pressure of the droplet.[10,11,28] The Kelvin effect describes the increase in vapour pressure of the droplet with diminishing size: the vapour pressure of a curved aqueous surface is greater than that of a flat aqueous surface. The solute effect describes the decrease in vapour pressure of an aqueous droplet with diminishing size as the concentration of an involatile solute increases. When the vapour pressure of the droplet resulting from the interplay of these two effects is equal to the surrounding RH, the droplet is at equilibrium with the surroundings and maintains a constant size. The condition for the droplet size to be in equilibrium with the surroundings at a given RH can be written as:

$$RH = a_w \exp\left(\frac{2V_w\sigma}{RTR_p}\right) \quad (1)$$

where a_w is the activity of water in an equivalent bulk solution with the same solute molality as the droplet, V_w is the partial molar volume of water, σ is the surface tension of the solution, and R_p is the radius of the liquid droplet. The activity of water can be written in terms of the activity coefficient, γ_w, and the mole fraction of water, x_w.

$$a_w = \gamma_w x_w \quad (2)$$

Thus, Köhler theory allows the calculation of the variation in the wet diameter with RH. This variation is dependent on the number of moles of involatile solute within the droplet, often expressed in terms of the dry particle diameter. Although the number of moles of solute in the wet droplet is invariant with time, the concentration of the solute changes depending on the number of moles of water partitioned to the condensed phase from the surrounding vapour.

An example of the evolving Raman fingerprint recorded for a trapped aqueous sodium chloride droplet over a period of approximately 2 hr with 1 s time resolution is shown in Fig. 2. A time of 0 s corresponds to the time at which the droplet was first captured. Spontaneous Raman scattering is observed at wavelengths around 650 nm, a Stokes shift of ~ 3400 cm^{-1}, corresponding to excitation of the O–H stretching vibrations of water. The information that can be gained from analysing this Raman scattering band has been described in detail in a previous publication.[29] Of central importance to the work presented here is the appearance of a resonance structure superimposed on the broad underlying spontaneous Raman band. The resonance structure appears at wavelengths commensurate with Whispering Gallery Modes (WGMs).[30] At these discrete wavelengths, the droplet acts as a low-loss optical cavity, with an integer number of wavelengths forming a standing wave around the droplet circumference, and the threshold intensity is surpassed for

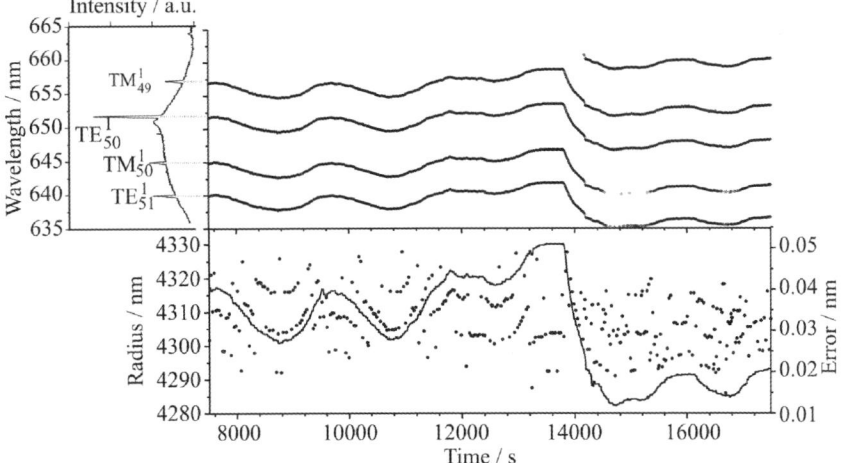

Fig. 2 An example of the evolving Raman fingerprint and radius of an aqueous sodium chloride droplet with time. The first spectrum with the assignment of modes is shown in the left panel, with the evolving wavelengths of the WGMs in the right panel. The corresponding variation in droplet size is shown in the bottom panel (solid line), along with the mean wavelength error per resonance arising in the spectral fit.

stimulated Raman scattering to occur. At WGM wavelengths, the light undergoes total internal reflection each time it encounters the interface. Mie scattering calculations can be compared with the unique pattern of WGM wavelengths, allowing the determination of droplet size with nanometre accuracy.[25,31] The principle sources of error in this size determination are the uncertainty in refractive index and the error in determining the wavelengths of the WGMs. These will be discussed below. Each WGM can be assigned a mode number, which reflects the integer number of wavelengths forming the standing wave within the droplet. Further, the mode order specifies the number of maxima in the radial distribution of the light intensity within the droplet. For all droplet sizes considered in this study, only first order modes appear in the spectra.[29] Finally, the mode may exhibit no radial dependence in the electric field of the light (a Transverse Electric mode, TE) or no radial dependence in the magnetic field of the light (a Transverse Magnetic mode, TM). The spectrum and mode assignment are shown for the first spectrum considered in the time variation in Fig. 2. TE or TM is used to designate the polarisation state of the mode. The mode number is specified by the subscript and the mode order by the superscript.

The variation in the WGM wavelengths is shown in Fig. 2, with slow and periodic drifts of a few nanometres occurring over a time period of ~ 2 hr, before a more abrupt change is observed. The drifts in wavelength arise from fluctuations in the wet particle radius of approximately 12 nm, a consequence of the periodic fluctuations in temperature arising from the air conditioning within the laboratory. At a time of $\sim 14\,000$ s, the RH decreases abruptly as the humidified nitrogen gas flow is temporarily switched off, leading to a significant reduction in the wet particle size. The error reported in Fig. 2 corresponds to the mean error in the calculated WGM wavelengths when compared with the experimental measurements. This is calculated from the differences between the experimentally recorded WGM wavelengths (the wavelength at a pixel centre) and the corresponding wavelengths calculated from the Mie calculations. 70% of the droplet sizes calculated have a simulated average wavelength error lower than the spectral dispersion of 0.037 nm per CCD pixel. Thus, in these cases each reported WGM wavelength is reproduced by the calculation within 1 full pixel width of its recorded wavelength.

Representative data for the variation in wet particle size with RH is shown in Fig. 3 between $\sim 22\,000$ and $32\,000$ s following initial trapping. The correlation between changes in size and RH is good, although the size change appears to occur in advance of the change in RH. We will use part of this period to discuss the errors arising in the RH measurements and errors in the size determination from uncertainties in refractive index. To assist in this, we reproduce the time variation in the

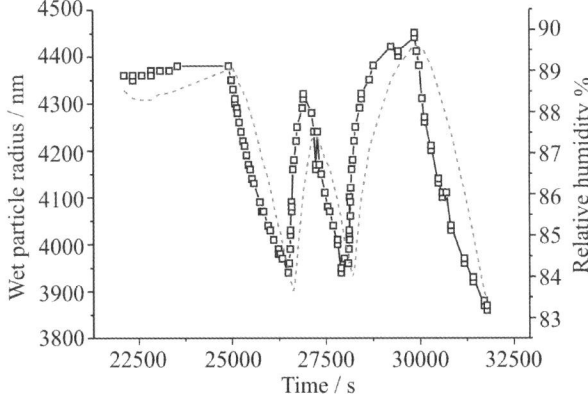

Fig. 3 Variation in wet particle size with time (symbols and solid line). The variation in RH measured by the capacitance probe is shown by the dotted grey line.

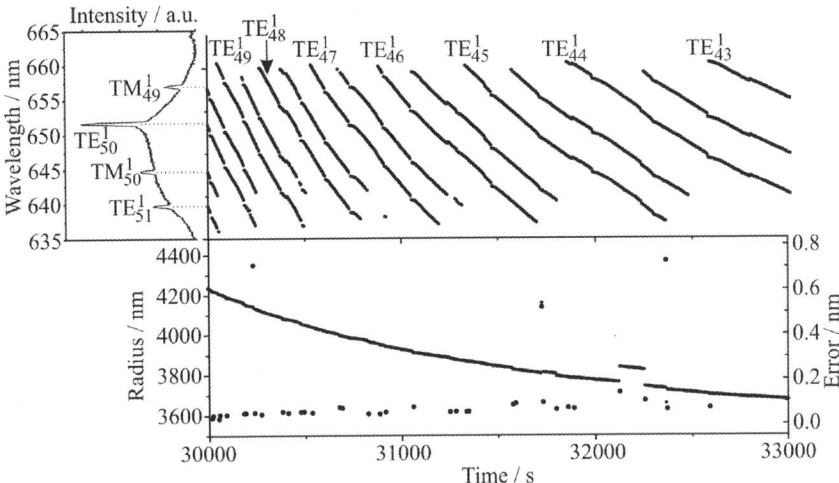

Fig. 4 The evolving Raman fingerprint and droplet radius between 30 000 and 33 000 s following trapping showing significant size variation. The first spectrum with the assignment of modes is shown in the left panel, with the evolving wavelengths of the WGMs and TE mode assignments in the right panel. The corresponding variation in droplet size is shown in the bottom panel (black line), along with the mean wavelength error per resonance arising in the spectral fit (filled circles).

Raman fingerprint and size over the time period 30 000 to 33 000 seconds in Fig. 4. The size change over this period is $\sim$ 500 nm, arising from a decrease in the RH from $\sim$ 90 to $\sim$ 83%, once again achieved by reducing the flow of humidified nitrogen into the aerosol cell. In contrast to the time period shown in Fig. 2, the size change is so significant that WGMs are observed to shift to the blue by more that the entire bandwidth of the Raman band and are replaced by modes of lower mode number at the red side of the band; a smaller droplet circumference requires a lower number of wavelengths to form the standing wave around the droplet circumference. Again, the error in reproducing the WGM wavelengths is acceptable and comparable to the spectral dispersion on the CCD for most of the sizes determined. However, at approximately 32 200 s a sudden step in size of $\sim$ 70 nm is observed. When fitting such spectra, local minima in the fit can be close in error. In the case shown here, the TE modes appearing in the spectrum prior to the step are assigned mode numbers 45 and 44. After the step up in size, these same two modes are assigned to mode numbers 46 and 45, respectively. These two pairs of modes have a similar spacing in wavelength and the mean errors in the wavelength fitting are comparable. However, considering the variation in radius as a whole over the entire time history of the droplet, the variation in size over the entire experiment that corresponds to the minimum cumulative error is clear. All subsequent experiments are analysed to minimise the cumulative error of the fit over the entire measurement, preventing the occasional incorrect mode assignment.

The treatment of refractive index in the Mie scattering calculations must account for changes in refractive index accompanying the variation in salt concentration arising from changes in droplet size, and must also explicitly include dispersion. The refractive index formulation used is that of Millard et al.,[32] which allows the calculation of refractive index at any sodium chloride concentration, wavelength and temperature. Failing to account for the size dependence of the concentration can lead to significant errors, particularly when the droplet size varies significantly from its starting value and concentration. The decrease in volume by a factor of $\sim$ 1.8 observed in Fig. 4 leads almost to a doubling in the concentration of the sodium chloride. Failing to account for the accompanying change in refractive index leads to an error in size of 20 nm later in this 3000 s window.

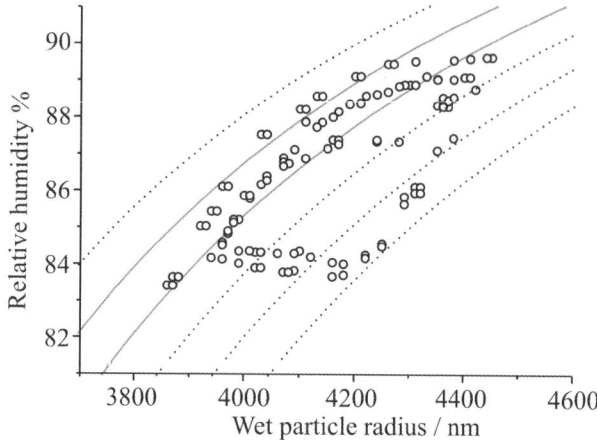

Fig. 5 A comparison of the wet particle sizes from Fig. 3 with Köhler curves calculated for dry sodium chloride particles of size 1750, 1800, 1850, 1900, 1950 and 2000 nm (dotted and grey solid lines from left to right, respectively). Reasonable agreement between measurements and theory is observed with predictions from dry particle sizes of 1800 and 1850 nm (shown by the two solid grey lines).

By taking the RH measurements in parallel to the variation in wet particle size, a Köhler curve can be plotted directly; this is shown in Fig. 5. The initial wet particle size immediately following trapping can be used to estimate the number of moles of solute within the droplet and, thus, the dry particle size representing the number of moles of involatile solute contained within the droplet. The data are broadly consistent with Köhler predictions when the RH is approximately constant for a long period of time.[25,28] The method for calculating the Köhler curves will be discussed in detail in Section 4. However, considerable discrepancies occur at early times immediately after trapping and over time windows in which the RH is varying rapidly. Indeed, two different pathways in RH/size are followed with decreasing and increasing size. These data indicate that the measurements of RH can be significantly different from the RH experienced by the droplet in the trapping region, arising from the existence of RH gradients within the cell, the slow time response of the capacitance probe to changes in RH and the error associated with the RH measurement of ±2%.

It is clear from the analysis of the measurements made on a single particle that the accuracy of Köhler curve measurements made on a single droplet can be compromised by errors in measurement of the gas phase composition (RH measurement) and errors in refractive index, which lead to errors in size. In the following section, we describe a more accurate way of determining the RH at which wet particle sizes are recorded. Further, the variations in refractive index with wavelength and droplet size (salt concentration) are explicitly included in all calculations.

4. Comparative measurements of the equilibrium size behaviour of two aqueous sodium chloride aerosol droplets

A significant possibility arises from the discussion in Section 3. The equilibrium size of a droplet can be used to determine the RH with high accuracy and fast time-response, allowing accurate measurements of gas phase composition. On referring to eqn (1), the activity of water for an equivalent bulk solution with a given solute concentration must be known, as must the partial molar volume of water and the surface tension of the solution to estimate the RH from the wet particle size. The online Aerosol Inorganics Model (AIM) provides the number of moles of water partitioned to the liquid phase for a given number of moles of a specified salt—

sodium chloride in this case—and for a given RH.[33,34] From the number of moles of water and salt partitioned in the liquid phase, the mole fraction of the salt (x_s) and water (x_w) can be determined. AIM also provides the appropriate activity coefficients and, thus, activities, with the activity of water required for eqn (1). The mass fraction of the solute (X_s) can then be calculated using the appropriate molar masses (M) of water and the salt and eqn (3).

$$X_s = \frac{M_s x_s}{M_s x_s + M_w x_w} \quad (3)$$

The density of NaCl–H$_2$O solutions at various mass fractions is well established and a second order polynomial fits this data.[35] From a knowledge of the solution density and mass fraction, it is then possible to calculate the partial molar volume of water, V_w, from eqn (4).[28]

$$V_w = \frac{M_w}{\rho}\left(1 + \frac{100 X_s}{\rho}\frac{d\rho}{dX_s}\right) \quad (4)$$

Further, the wet particle size can be determined through knowledge of the density and mass. Finally, the dependence of the surface tension on the mass fraction of solute has been characterised experimentally.[35] With knowledge of all of these parameters and for a fixed number of moles of solute, the wet particle size and the RH at which this size is in equilibrium with the surrounding vapour can be calculated. Thus, a unique Köhler curve results for a droplet containing a fixed number of moles of involatile salt allowing the accurate determination of RH for the experimental measurements.

The experimental strategy is as follows: An aqueous sodium chloride droplet is used as a control droplet to determine the RH within the trapping region. A spectrum of the control droplet is acquired immediately after trapping to determine its size. It is assumed that the initial composition is established by the concentration of salt in the nebulised solution, allowing the mass of solute and the dry particle size to be estimated. This assumption is supported by measurements of refractive index of the solution prior to nebulising and a sample collected from the delivered aerosol, which are measured to be identical within an accuracy of ±0.0001. This suggests that the maximum deviation in salt concentration between the nebulised salt solution and the trapped droplet is ±3%, corresponding to an error of <±1% in dry particle size. An appropriate Köhler curve for this solute loading is then calculated and the variation in wet particle size is used to directly and accurately monitor the variation in RH within the trapping region. An error in the dry particle size of ±1% would lead to an error in the RH determination of ±0.09% for RHs greater than 90%. In all cases, these errors represent the maximum possible.

By using a control droplet as an accurate probe of RH, the Köhler curve of a second droplet can then be measured, referred to as the droplet of interest. The experimental procedure is as follows. One of the two optical traps is loaded with a droplet of interest. The size immediately following trapping is used to determine the mass loading of solutes within the droplet. The second trap is then loaded with a control droplet. Loading of the second trap is often accompanied by coagulation between the control aerosol flowing through the cell and the droplet of interest. Recording the size change of the droplet of interest allows the change in composition to be determined. Spectra are acquired for each droplet sequentially by translating the droplets in and out of the active imaging area of the Raman spectroscopy, as illustrated in Fig. 6. Spectra of the control droplet are recorded to measure the RH. The control droplet is then replaced by the droplet of interest and spectra acquired. In principle, it is possible to collect spectra from the control droplet and droplet of interest simultaneously, although this requires that the two droplets be separated by a distance of less than ∼20 µm. In this work, we maintain a droplet separation that is a few droplet diameters, often greater than ∼30 µm, such that each droplet can be

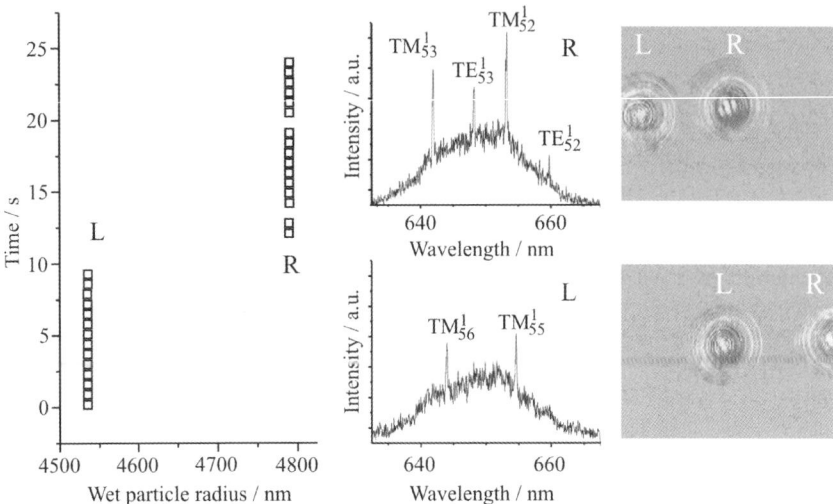

Fig. 6 Illustration of the interchange procedure for monitoring two droplets. Spectra of the left droplet are first recorded and the droplets are then interchanged and spectra recorded of the right droplet. Droplet sizes are calculated for each droplet independently. The droplets are interchanged every few seconds, depending on the rate of change of the RH and the rate of change of size.

assumed to behave independently.[36] The droplets can be interchanged within 1–2 s, minimising the delay between recording the two spectra.

An example of the variation in size of two aqueous sodium chloride droplets is shown in Fig. 7(a), along with the time dependence of the RH determined from the wet particle size of the control droplet. Over time, the RH declines and the two droplets decrease in size. One droplet is marginally larger in size, reflecting that it contains a higher mass loading of salt. In this experiment, the droplets were interchanged only infrequently with the RH displaying a continuous fall. By fitting the time-dependent variation in RH estimated from the control droplet, the RH variation with experimental time can be found. Thus, the Köhler curve for the droplet of interest can be obtained directly, as shown in Fig. 7(b). Based on the initial size of the droplet of interest, it is anticipated that the mass loading of sodium chloride should be equivalent to a dry particle radius of 1269 nm. Indeed, the measurements for the droplet of interest fall close to the Köhler curve for a dry particle size of 1270 nm. This measurement provides a benchmark demonstration of the accuracy with which Köhler curves can be determined. It is also interesting to note that in this size regime, neglecting the curvature term in the predictions of wet particle size increases the equilibrium size by 19 nm. Although this is a minor correction, the measurement of such a size discrepancy is possible with this technique. However, other uncertainties in the theoretical predictions are anticipated to represent a larger source of error, precluding such accurate measurements.

5. Comparative measurements of the equilibrium size behaviour of two aerosol droplets of different composition

In a final demonstration, we compare the variation in wet particle size for two droplets of different composition. The control droplet is again chosen to be an aqueous sodium chloride droplet. The droplet of interest is chosen to contain a water soluble organic compound, Glutaric Acid (GA), as well as sodium chloride. Field measurements indicate that 20–70% of the condensed phase organic carbon in the

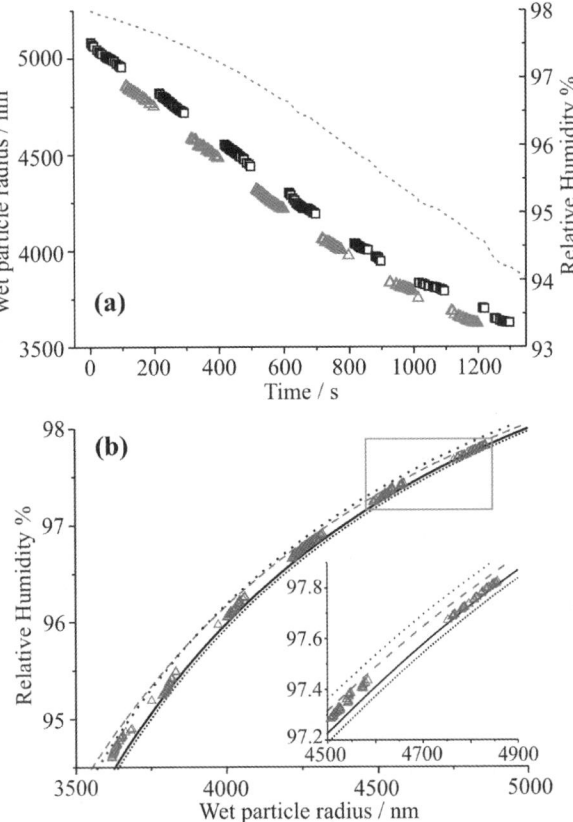

Fig. 7 (a) Variation in wet size of the two trapped sodium chloride droplets and variation in RH (dashed line, left axis) as determined from the control droplet (represented by black squares). (b) Comparison of the Köhler curve measured for the droplet "of interest", represented by grey triangles in (a), with theoretical predictions. A full Köhler analysis is shown for sodium chloride droplets of dry particle size 1250 and 1270 nm by the black dotted line above the experimental data and the solid black line respectively. The expected dry particle size is 1269 nm. Predictions neglecting the Kelvin term are shown by the fine dotted black line, below the solid black line. The dashed line corresponds to the growth factor formalism discussed in Section 5.

atmosphere is water soluble.[37] GA ($C_5H_8O_4$) is representative of organics components partitioned to the condensed aerosol phase in the atmosphere and is a dicarboxylic acid.[10] Dicarboxylic acids have low vapour pressures and are abundant throughout the atmosphere. GA has a vapour pressure of 7×10^{-4} Pa at 298 K[38] and a solubility equivalent to 61.5 wt% or 1.6 kg per kg of water.[12]

The impact of GA on the hygroscopicity of sodium chloride aerosol is well documented and can be treated within the Zdanovskii, Stokes and Robinson (ZSR) assumption.[39] Thus, the organic and inorganic components can be assumed to take up water independently. This suggests that the number of moles of water associated with the number of moles of the inorganic solute can be estimated at any RH, and similarly the number of moles of water associated with the organic component. The size of the composite internally mixed droplet containing GA, sodium chloride and water can then be estimated assuming that the organic and inorganic components can be treated independently.

Specifically, we use the parameterisation of the hygroscopic growth presented by Prenni *et al.*,[39] Kreidenweis[14] *et al.* and Koehler *et al.*[12] derived from laboratory measurements. The radius of the wet aerosol at a specified RH can be related to the radius of the dry particle by the growth factor, G, which is a function of RH.

$$G(\text{RH}) = \frac{R_{\text{wet}}(\text{RH})}{R_{\text{dry}}} \tag{5}$$

This provides a convenient and straightforward parameterisation, which allows the estimation of the wet particle size at any RH and for any amount of solute within the particle, determined by the dry particle radius, R_{dry}. For the sodium chloride and GA components, the growth factors can be expressed as:

$$G(\text{RH}) = \left[1 + (a + b\text{RH} + c\text{RH}^2)\frac{\text{RH}}{1 - \text{RH}}\right]^{1/3} \tag{6}$$

where RH is the relative humidity expressed as a fraction. The constants a, b and c are (4.83257, −6.92329, 3.27805) and (0.99597, −2.34929, 1.58981) for sodium chloride and GA, respectively. From the estimated mass loading of the solutes in the initially trapped wet droplet (estimated from the droplet size and solute concentrations), we can estimate the appropriate dry particle sizes for each component, given that the densities of sodium chloride and GA are 2.17 g cm^{-3} and 1.424 g cm^{-3}, respectively. The independent growth factors for each component can then be calculated, allowing equivalent volumes of two externally mixed droplets to be calculated. Then, the combined volume of a droplet in which the two components are internally mixed can be calculated. This final step assumes that there is no apparent change in the solution density associated with each component on mixing, a further approximation inherent to the ZSR treatment and a possible source of error. This growth factor analysis is compared in Fig. 7(b) to the more accurate theoretical determinations of the Köhler curves for the single component sodium chloride droplets. Although showing an increasing deviation from the exact treatment with decreasing water content and increasing solute mass fraction (consistent with the literature), the agreement remains satisfactory for this study. At 98% RH, the growth factor parameterisation underestimates the size by 40 nm (<1%). This rises to 70 nm (∼2%) at 95% RH.

The experimental procedure is as follows. The control sodium chloride droplet is the second droplet to be loaded into an optical trap (solution of sodium chloride 20 g L^{-1}), following loading of a mixed GA/sodium chloride droplet into the first optical trap (solution containing GA and sodium chloride each at 15 g L^{-1}). The loading of the control droplet unavoidably leads to dosing of the droplet of interest with further sodium chloride mass. To enable the size of the mixed organic/inorganic droplet to be estimated accurately, the concentration dependence of the refractive index has been measured directly at 589 nm and can be represented by the linear functional form {(1.3330 ± 3E−5) + (2.903E−4 ± 0.040E−4) C} for a solution containing a 1 : 1 mixture of each solute by mass, where C is the aqueous concentration of one of the solutes in units of g L^{-1}. The wavelength dependence can also be approximated as linear in this limited wavelength range, with a gradient that is correlated to the refractive index at 589 nm. Thus, the refractive index of the mixed component droplet can be estimated with an accuracy of <0.1%, leading to an uncertainty in size of <0.1% or <5 nm for the droplet sizes considered here.

The time variation in the sizes of the two droplets is shown in Fig. 8(a). The droplets were interchanged approximately every 5 s, providing an almost continual history of RH and wet particle size. The size of the original GA/sodium chloride droplet was determined within 5 s of trapping and has a radius of 3.861 μm containing an estimated 3.62 pg of each solute. All wet droplets sizes should be assumed to have an error <5 nm at the time at which the size can be determined. However, this does not account for any rapid changes in size that occur immediately on trapping prior to the first sizeable Raman fingerprint. Given that the droplets are at equilibrium with the surrounding gas flow on nebulisation, an initial rapid change

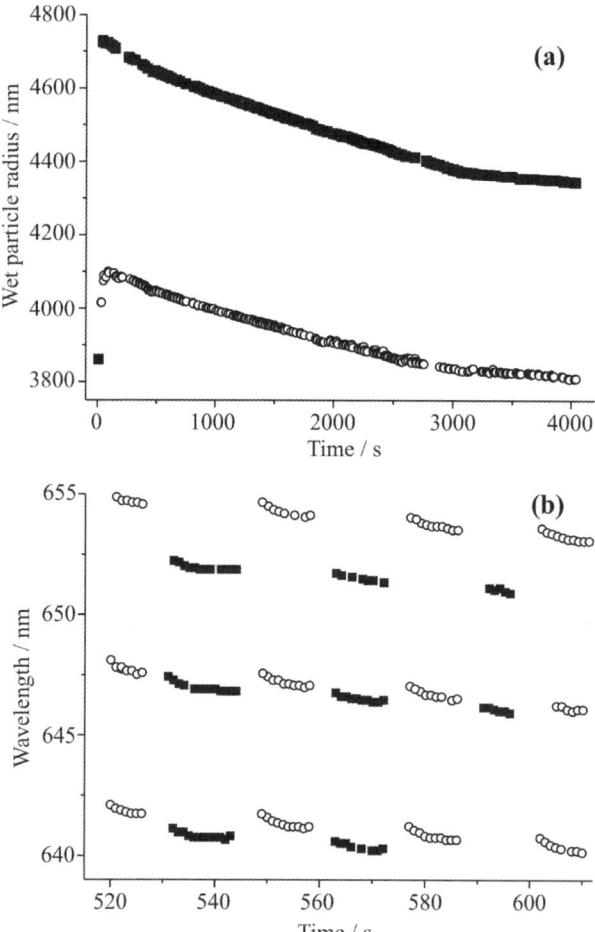

Fig. 8 (a) Variation in wet particle sizes with time for the aqueous sodium chloride droplet (open circles) and the mixed aqueous GA/sodium chloride droplet (filled squares). (b) Comparison of the time dependence of the WGM wavelengths over a narrow time window. The symbols are the same as in part (a).

in size is not expected but cannot be routinely investigated. An error of 5 nm in droplet size corresponds to an error in dry particle mass of 0.5% or 0.015 pg for the droplets considered here. On loading of the control droplet, the GA/sodium chloride droplet increased in size to 4.728 μm through coagulation with sodium chloride aerosol. This size is corrected for change in refractive index. Thus, coagulation led to an increase in the sodium chloride loading in this droplet to 7.65 pg. The equivalent dry particle radii for each of the GA and sodium chloride components in the mixed droplet are 0.846 and 0.944 μm, respectively. Each dry particle size can be assumed to have an error of <2 nm, based on the error in the wet particle size. However, the change in solute loading accompanying coagulation cannot be estimated until the first sizeable Raman fingerprint is recorded, leading to an unquantifiable error in the solute loading. From the analysis of sodium chloride aerosol, this is not expected to lead to a significant error.

The control sodium chloride droplet was initially 4.016 μm in radius, containing 5.43 pg of solute and equivalent to a dry particle radius of 0.842 μm. The GA/sodium chloride droplet was larger at all times, containing a larger solute mass. Once the control dry particle size is known, the RH can be estimated by considering the

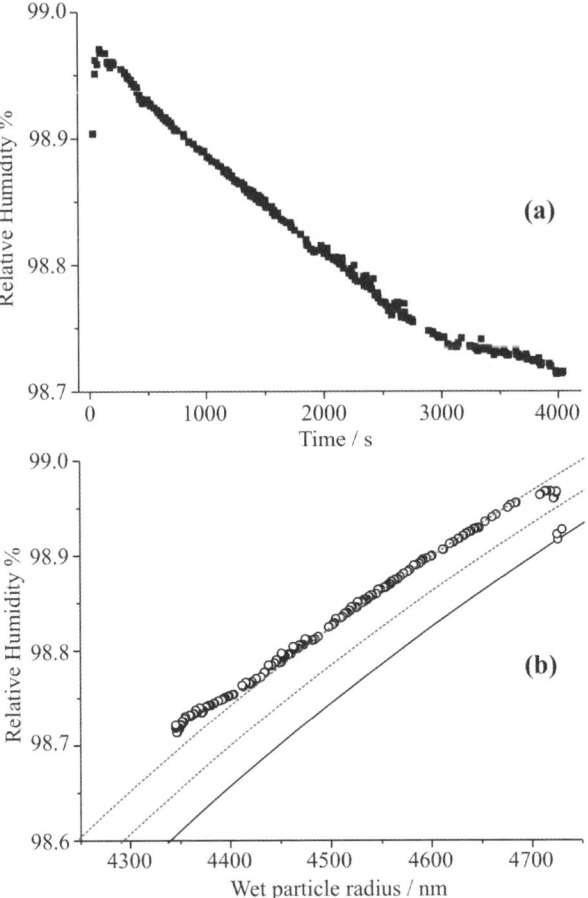

Fig. 9 (a) Time variation in the RH determined from the control droplet size. The overall variation in RH is <0.3%, measurements that would be impossible with the conventional RH probe described earlier. (b) Comparison of the measured and predicted Köhler curves. The solid black line is for the composition stated in the text corresponding to dry particle sizes of 0.846 and 0.944 μm for GA and sodium chloride, respectively. The dashed lines are for compositions with dry particle sizes of 0.826 and 0.924 μm (upper) and 0.836 and 0.934 μm (lower).

relationship between wet and dry particle sizes defined by eqn (5) and (6). The variation in RH with time is shown in Fig. 9(a). This then allows the evaluation of the mixed component droplet data and comparison with predictions of wet particle size from eqn (5) and (6).

It is evident in the time dependence of size that the control droplet and mixed droplet show different time dependencies for the first ∼100 s, with the latter decreasing in size while the former continues to grow. It is unclear why this is the case but it could be a signature of mass transfer limited equilibration, either limited by mass transfer through the gas phase (diffusion) or mass transfer across the interface (accommodation). It is widely assumed that sodium chloride aerosol equilibrates on a sub-10 s timescale, although it has been suggested that the timescales for the equilibration of organic droplets could be prolonged.[20] A further signature of non-equilibrium behaviour is evident from the time dependence of the WGMs on interchanging the droplets, an example of which is shown in Fig. 8(b). Both the mixed component and control droplets show initial rapid decreases in size as they are moved into the immediate atmosphere left by the preceding droplet before equilibrating. This could indicate the slow response of the gas phase composition over these micron length scales to small changes arising from the displacement of the droplets. The rapid decline and levelling of the resonant modes observed in Fig. 8(b) corresponds to size changes of <5 nm.

Good agreement is observed between the predicted equilibrium size dependence for the GA/sodium chloride droplet and the measured variation shown in Fig. 9(b), although the recorded curve is systematically higher than prediction. Considering the error of ±0.09% in RH possible in these measurements and the rather coarse treatment provided by the theory, the agreement is excellent. Typical errors associated with the measurements used to provide the parameterisation of eqn (5) and (6) are ±1.5% in RH and only size changes of greater than ~4% can be resolved.[39] Thus, the systematic difference in RH of ~0.1% observed from these measurements and the difference in size of ~100 nm, an error of ~2.2%, compare well with the errors associated with conventional growth measurements. Indeed, this difference is comparable to the magnitude of the difference between the different theoretical treatments for sodium chloride aerosol shown in Fig. 7(b). From a single measurement it is not possible to determine if the systematic error arises from the coarse nature of the original published growth measurements, from the errors inherent to the ZSR approach, or from the errors in the experimental measurements presented here. Further detailed measurements are needed to resolve the origin of this systematic difference. However, this study provides a clear example of the ability to perform comparative measurements on droplets of different composition using optical tweezers.

6. Concluding remarks

Optical tweezers and Raman spectroscopy can be used to interrogate the thermodynamic properties of aerosols and directly compare the behaviour of aerosol droplets of different composition. Specifically, we have shown that a control droplet can be used to probe the gas-phase composition with high accuracy and fast time-response, allowing determination of the RH with an accuracy better than ±0.09%, even within the challenging RH regime close to saturation. We have illustrated that this can allow direct measurements of the Köhler curve of a second droplet of different composition for which nanometre changes in size can be explored.

More generally, we suggest here that this approach can be used to not only examine the thermodynamic behaviour of aerosols but the kinetic factors governing mass transfer and interfacial transport. In a simple extension of this work, hysteresis in the time-dependence in size of one droplet relative to another during humidifying/dehumidifying cycles would provide direct evidence of mass transfer limited equilibration. Further, this approach could be extended to examine the mass accommodation of species other than water, allowing comparison of the uptake of trace species on droplets of different surface composition and properties. Thus, the ability to size droplets with nanometre accuracy and with high time resolution, and to compare two different droplets directly, could provide a flexible new strategy for characterising the thermodynamic properties and kinetic behaviour of aerosols.

Acknowledgements

JPR, LM, JRB and KLH acknowledge support from the EPSRC and NERC through support of an Advanced Research Fellowship, a Post-Doctoral Research Fellowship, and studentship support. LT acknowledges the support of the ESF through the INTROP programme, who provided a travel grant to support this work. The authors also wish to acknowledge Dr G. McFiggans, Dr D. Topping and Dr S. Clegg for valuable discussions.

References

1. G. M. Nathanson, *Annu. Rev. Phys. Chem.*, 2004, **55**, 231–255.
2. B. C. Garrett, G. K. Schenter and A. Morita, *Chem. Rev.*, 2006, **106**, 1355–1374.
3. P. Davidovits, C. E. Kolb, L. R. Williams, J. T. Jayne and D. R. Worsnop, *Chem. Rev.*, 2006, **106**, 1323–1354.
4. D. J. Donaldson and V. Vaida, *Chem. Rev.*, 2006, **106**, 1445–1461.
5. Y. Rudich, *Chem. Rev.*, 2003, **103**, 5097–5124.
6. G. B. Ellison, A. F. Tuck and V. Vaida, *J. Geophys. Res., [Atmos.]*, 1999, **104**, 11633.
7. P. S. Gill, T. E. Graedel and C. J. Weschler, *Rev. Geophys. Space Phys.*, 1983, **21**, 903–920.
8. J. A. Thornton and J. P. D. Abbatt, *J. Phys. Chem. A*, 2005, **109**, 10004–10012.
9. V. F. McNeill, J. Patterson, G. M. Wolfe and J. A. Thornton, *Atmos. Chem. Phys.*, 2006, **6**, 1635–1644.
10. J. H. Seinfeld and S. N. Pandis, *Atmospheric Chemistry and Physics: From Air Pollution to Climate Change*, John Wiley & Sons, New York, 1998.
11. G. McFiggans, P. Artaxo, U. Baltensperger, H. Coe, M. C. Facchini, G. Feingold, S. Fuzzi, M. Gysel, A. Laaksonen, U. Lohmann, T. F. Mentel, D. M. Murphy, C. D. O'Dowd, J. R. Snider and E. Weingartner, *Atmos. Chem. Phys.*, 2006, **6**, 2593–2649.
12. K. A. Koehler, S. M. Kreidenweis, P. J. DeMott, A. J. Prenni, C. M. Carrico, B. Ervens and G. Feingold, *Atmos. Chem. Phys.*, 2006, **6**, 795–809.
13. R. Sorjamaa, B. Svenningsson, T. Raatikainen, S. Henning, M. Bilde and A. Laaksonen, *Atmos. Chem. Phys.*, 2004, **4**, 2107–2117.
14. S. M. Kreidenweis, K. Koehler, P. J. DeMott, A. J. Prenni, C. Carrico and B. Ervens, *Atmos. Chem. Phys.*, 2005, **5**, 1357–1370.
15. Z. Li, A. L. Williams and M. J. Rood, *J. Atmos. Sci.*, 1998, **55**, 1859–1866.
16. G. M. Nathanson, P. Davidovits, D. R. Worsnop and C. E. Kolb, *J. Phys. Chem.*, 1996, **100**, 13007–13020.
17. C. L. Badger, I. George, P. T. Griffiths, C. F. Braban, R. A. Cox and J. P. D. Abbatt, *Atmos. Chem. Phys.*, 2006, **6**, 755–768.
18. M. P. Tolocka, T. D. Saul and M. V. Johnston, *J. Phys. Chem. A*, 2004, **108**, 2659–2665.
19. C. A. Colberg, U. K. Krieger and T. Peter, *J. Phys. Chem. A*, 2004, **108**, 2700–2709.
20. M. N. Chan and C. K. Chan, *Atmos. Chem. Phys.*, 2005, **5**, 2703–2712.
21. J. P. Reid and L. Mitchem, *Annu. Rev. Phys. Chem.*, 2006, **57**, 245–271.
22. J. Buajarern, L. Mitchem, A. D. Ward, N. H. Nahler, D. McGloin and J. P. Reid, *J. Chem. Phys.*, 2006, 114506.
23. R. J. Hopkins, L. Mitchem, A. D. Ward and J. P. Reid, *Phys. Chem. Chem. Phys.*, 2004, **6**, 4924–4927.
24. L. Mitchem, J. Buajarern, A. D. Ward and J. P. Reid, *J. Phys. Chem. B*, 2006, **110**, 13700–13703.
25. L. Mitchem, J. Buajarern, R. J. Hopkins, A. D. Ward, R. J. J. Gilham, R. L. Johnston and J. P. Reid, *J. Phys. Chem. A*, 2006, **110**, 8116–8125.
26. K. J. Knox, J. P. Reid, K. L. Hanford, A. J. Hudson and L. Mitchem, *J. Opt. A: Pure Appl. Opt.*, 2007, **9**, S180–S188.
27. E. Fallman and O. Axner, *Appl. Opt.*, 1997, **36**, 2107–2113.
28. D. O. Topping, G. B. McFiggans and H. Coe, *Atmos. Chem. Phys.*, 2005, **5**, 1205–1222.
29. J. P. Reid, H. Meresman, L. Mitchem and R. Symes, *Int. Rev. Phys. Chem.*, 2007, **26**, 139–192.
30. S. C. Hill and R. E. Benner, in *Optical Effects Associated with Small Particles*, ed. P. W. Barber and R. K. Chang, World Scientific, Singapore, 1988, vol. 1, p. 3.
31. J. D. Eversole, H. B. Lin, A. L. Huston, A. J. Campillo, P. T. Leung, S. Y. Liu and K. Young, *J. Opt. Soc. Am. B*, 1993, **10**, 1955–1968.
32. R. C. Millard and G. Seaver, *Deep-Sea Res., Part A*, 1990, **37**, 1909–1926.
33. S. L. Clegg and A. S. Wexler, *On-line Aerosol Inorganics Model*, http://www.aim.env.uea.ac.uk/aim/aim.html.
34. S. L. Clegg, P. Brimblecombe and A. S. Wexler, *J. Phys. Chem. A*, 1998, **102**, 2155–2171.
35. *Handbook of Chemistry and Physics*, 87th edn, CRC Press LLC, 2006–2007.
36. R. Vehring, C. L. Aardahl, E. J. Davis, G. Schweiger and D. S. Covert, *Rev. Sci. Instrum.*, 1997, **68**, 70–78.
37. P. Saxena and L. M. Hildemann, *J. Atmos. Chem.*, 1996, **24**, 57–109.
38. M. Bilde and S. N. Pandis, *Environ. Sci. Technol.*, 2001, **35**, 3344–3349.
39. A. J. Prenni and P. J. De Mott, and S. M. Kreidenweis, *Atmos. Environ.*, 2003, **37**, 4243–4251.

PAPER

The spectroscopy and chemical dynamics of microparticles explored using an ultrasonic trap

N. J. Mason,*[a] E. A. Drage,[a] S. M. Webb,[a] A. Dawes,[a] R. McPheat[b] and G. Hayes[b]

Received 21st February 2007, Accepted 29th March 2007
First published as an Advance Article on the web 30th July 2007
DOI: 10.1039/b702726p

Microsized particles play an important role in many diverse areas of science and technology, for example, surface reactions of micron-sized particles play a key role in astrochemistry, plasma reactors and atmospheric chemistry. To date much of our knowledge of such surface chemistry is derived from 'traditional' surface science-based research. However, the large surface area and morphology of surface material commonly used in such surface science techniques may not necessarily mimic that on the surface of micron/nano scale particles. Hence, a new generation of experiments in which the spectroscopy (e.g., albedo) and chemical reactivity of micron-sized particles can be studied directly must be developed. One, as yet underexploited, non-invasive technique is the use of ultrasonic levitation. In this article, we describe the operation of an 'ultrasonic trap' to store and study the physical and chemical properties of microparticles.

Introduction

Microparticles play a crucial role in many diverse areas of science and technology. Many of the complex molecules (e.g., methanol, formic acid and acetic acid) recently discovered in the InterStellar Medium (ISM) are now believed to be formed by surface chemistry on ice-covered micron-sized dust grains[1–3] (Fig. 1). The low densities in the ISM limit the probability of binary collisions, such that chemical reactivity is often restricted to the surfaces of the interstellar dust (carbonaceous and/or silicate) where, through slow accretion, icy mantles of simpler molecules (e.g., CO, CO_2, H_2O and NH_3) formed in ion–molecule reactions may be irradiated by energetic radiation (UV and cosmic rays).[4] Radicals released in the ice under irradiation may then react to form larger polyatomics, which may eventually be released through desorption or stay trapped in the ice, acting as a molecular reservoir for planetary and star formation.[5] Indeed, such microparticle chemistry may ultimately be the origin of life itself, providing the vital prebiotic material that can 'seed' a developing planet such as our own Earth.[6]

Microparticles, more commonly classified as dust, also play a key role in industrial processes, being an inevitable product of combustion, through the formation of soot. One of the major problems confronting the modern semiconductor electronics industry as it seeks to develop ever smaller (sub 40 nm) components is the formation of dust in the etching plasmas. Here, the dust is composed of micron-sized silicon

[a] Department of Physics and Astronomy, The Open University, Milton Keynes, UK MK7 6AA
[b] Molecular Spectroscopy Facility, CCLRC Rutherford Appleton Laboratory, Chilton, Didcot, Oxon, UK OX11 0QX

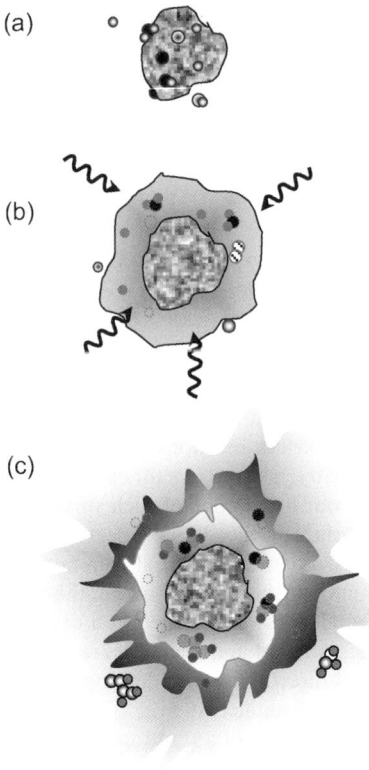

Fig. 1 Schematic of the possible method for synthesis of complex molecules on dust grains in the interstellar medium. (a) Ices are formed by accretion on the surface of the cosmic dust grain. (b) The icy mantle is bombarded with cosmic rays and solar UV. Chemical modification occurs. (c) Newly formed molecules desorb back into the gas phase.

particles that distort the electrical properties of the etching plasma (by acting as multiple charge carriers) and form deposits on the etched surface.[7,8]

However, perhaps the most important role of microparticles is in atmospheric chemistry. Specifically, in the case of the stratospheric ozone hole, heterogeneous chemistry on ice/water particles comprising Polar Stratospheric Clouds (PSCs) are responsible for the repartitioning of chlorine reservoir species into photochemically active species capable of catalytically destroying ozone.[9,10] At lower altitudes in the troposphere, photolysis and subsequent chemical reactions of biogenic compounds and Volatile Organic Compounds (VOCs) produces aerosol particulates, a product often known as 'Photochemical Smog' (PS). A well known natural PS is the 'blue haze' over the forested mountains of West Virginia, USA—the so-called 'blue ridge mountains'—where biogenic emissions from trees (terpenes) and reactions with ozone and nitric oxides (produced by solar photolysis) leads to the formation of biogenic aerosols that subsequently scatter sunlight leading to the observation of blue haze. Similar processes are induced when anthropogenic VOC emissions (from *e.g.*, petroleum products such as gasoline, turpentine or kerosene) react with urban-generated ozone to create PAN (peroxyacyl nitrate), a process that was responsible for the urban smogs in many US cities in the 1950s and 1960s and is still prevalent in the cities of many developing countries. Industrial emissions also include solid and liquid particulates released, for example, through flue emissions. The role of such aerosols in human health is a much discussed topic, with several studies reporting correlations between aerosol production and bronchial complaints (*e.g.* asthma).

The recent IPCC report stated that determining the role of aerosols on the global radiation budget is critical to the evaluation of global climate change but commented that, at present, *our knowledge of the radiative properties and chemical transformation of atmospheric aerosols is rudimentary and requires detailed research.* The terrestrial aerosol budget is therefore growing with as-yet-unpredictable consequences for global climate regulation. Depending on their altitude, size distribution and chemical composition, aerosols may either warm (hence, enhance global warming) or cool (moderate global warming).[11–14] It is therefore necessary to develop a scientific research programme to quantify the physical and chemical properties of such atmospheric aerosols. Such a programme requires both remote sensing observations of the terrestrial aerosol and 'field' studies (*e.g.*, through use of research aircraft). In turn, such data can only be interpreted if compared with laboratory experiments in which microparticles/aerosols are studied under controlled experimental conditions.

Trapping microparticles

Several techniques have been developed to study the physical and chemical properties of microparticles, many of which are derived from surface science-based research. However, the study of physical and chemical processes on the large areas of surface material commonly used in surface science techniques may not necessarily mimic that on the smaller surfaces founds on aerosols (typically micron-sized). Indeed, the chemistry of ozone depletion of the PSC particulates is now known to be significantly different from that monitored on laboratory ice surfaces prepared in many of the early experimental investigations of PSC chemistry.[15,16]

For example, the latent heat released by deposition of molecular species onto a large bulk ice has minimal consequences whilst the cumulative effect of successive depositions on a micron-sized grain may lead to surface melting, in turn changing the morphology of the ice. When probing bulk ice structures, localized regions of crystalline and amorphous ice may be important while the propagation of such features on the micron scale may have little or no meaning. Diffusion and mobility of chemical species (*e.g.*, photolysis-induced radicals) may have only a localized effect in the bulk but may dominate on the micron scale. Hence, it has long been surmised that the actual physical and chemical properties of aerosol species may be quite different from the bulk. Thus, extrapolation of chemical and physical data from bulk/large ice surfaces to aerosol models may be dangerous; however, in the absence of any other data, the bulk data has been widely used in atmospheric and climate models.[17,18]

Recently, several techniques have been developed in which particles/aerosols are trapped for experimental investigation of their physical and chemical properties. Two trapping techniques that have been explored are adopted from the atom trapping community, namely electrostatic traps[19] and laser trapping.[20,21] However, such methods may be expensive and involve rather complex designs.

One as yet underexploited, non-invasive technique is the use of *ultrasonic levitation*. Ultrasonic levitation is an example of the more general 'acoustic levitation' in which the pressure changes induced in a medium by generation of sonic fields are used to counteract the force of gravity. Acoustic levitation[22] has been used in many diverse areas of science and technology to suspend particles in liquid media,[23] for material processing[24] and underpins the study of the intriguing phenomena of sonoluminescence.[25]

When a particle is suspended in a sound field, the fluid necessary to transport the sound wave exerts hydrodynamical forces on the particle. These forces are greatly enhanced if the sound field is generated within a resonant cavity in which a standing wave is generated. To levitate a particle, the sound pressure level must be greater

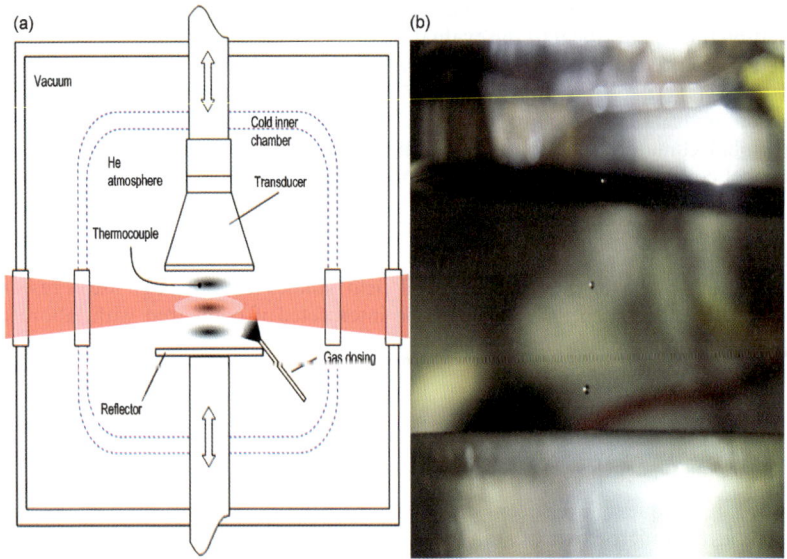

Fig. 2 (a) Schematic of the ultrasonic trap design illustrating the nodal points between transducer and reflector. (b) Illustration of three water droplets each separately suspended in a nodal region of the ultrasonic trap.

than the gravitational force,[26] but if the pressure is too high the particle may disintegrate (*e.g.*, a suspended liquid droplet will break up into small droplets).

Using this methodology, we have developed a system in which microparticles are trapped within a standing ultrasonic field established within an inert gas (nitrogen or helium) atmosphere. Microparticles placed within an ultrasonic field feel a force towards the acoustic pressure nodes of the stationary ultrasonic field. By ensuring that this force is strong enough to overcome other forces on the particulate, such as gravity, the microparticles can be suspended (levitated) at the nodal points of the pressure standing wave.

In our system, we have generated an ultrasonic (27 kHz) sound field (the frequency being chosen so as to be above the threshold for human hearing) capable of trapping micron-sized particles in the anti-node of a standing wave, thus forming an *ultrasonic particulate trap* (Fig. 2 and 3). The particles may be easily viewed using a simple He:Ne laser.

The microparticles can be held in the trap for long periods (dependent upon the strength of the driving field and turbulence in the surrounding air). Indeed, if the sonic field is strong enough even small metal spheres can be suspended. However, when working with smaller particles in any trap system (optical, electrostatic and/or ultrasound), coagulation becomes a major problem. In an ultrasonic trap the probability of coagulation is a function of the frequency and, unfortunately, for 0.1 → 10 microns this frequency is close to that of the 27 kHz resonant frequency of the transducers. However, by increasing the pressure we can shift the optimum coagulation frequency. Therefore, to study 0.1 → 10 microns we simply alter the pressure in the trap to shift this optimum coagulation frequency away from the resonant frequency of the transducers. A pressure of 2 atm will shift the optimum coagulation frequency to ∼15 kHz, sufficiently far from the 27 kHz resonant frequency to remove the probability of coagulation.

Furthermore, if the air surrounding the sonic field is humid, water is found to freeze out in these same nodes onto the particulate surface, forming an ice coating that may mimic ice covered particles in the terrestrial atmosphere.[27,28] Increasing the gas pressure also increases the temperature flow to the nodes from the reflector,

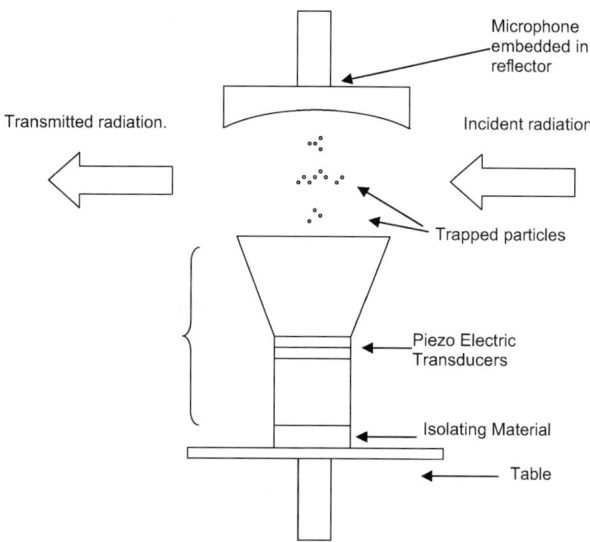

Fig. 3 Schematic of the ultrasonic trap design illustrating the trap set up to measure the spectra of suspended microparticles.

allowing us to control the temperature in the nodes more efficiently. We have therefore been able to readily form ice particles in the trap nodes from a flow of humid air into the ultrasound chamber.

Results

Spectroscopic studies of microsopic particles

Having demonstrated that the ultrasonic trap is capable of storing microparticles, we have performed an initial series of experiments to explore the scattering and absorption properties of such particles.

Some of the largest uncertainties in our understanding of the physico-chemical processes affecting the Earth's radiative balance are linked to our knowledge of the scattering and absorption properties of atmospheric aerosol particles. Aerosol scattering and photo-absorption processes reduce the solar irradiation reaching the terrestrial surface, leading to local surface cooling—hence aerosols are generally thought to mitigate the effects of global warming by anthropogenic gaseous emissions. However, photo-absorption by the aerosol may also warm the atmosphere, altering the local temperature and relative humidity profiles, leading to changes in the chemical composition and physical properties of the nascent aerosol. Hence, in any scientific programme that aims to quantify the role of the atmospheric aerosol in climate change, it is necessary to gain a detailed understanding of the radiative properties of the myriad of microparticles that comprise the terrestrial aerosol. Considerable variability is expected in scattering and absorption from different aerosols. Thus, different radiative responses are expected for different aerosol models.

Using the Bruker IFS 66 Fourier Transform Infrared Spectrometer (FTIR) located in the NERC Molecular Spectroscopy Facility at the Rutherford Appleton Laboratory,[29] we have measured infrared transmission spectra for microparticles levitated in the ultrasonic field. Four different types of particulate were studied: (i) carbonaceous soot, (ii) sand, and (iii, iv) two types of volcanic ash collected from Tenerife by the Earth Sciences Department at the Open University—one being a basalt (low percentage of silicate: $\approx 50\%$) and the other being a Rhyolite (high percentage of silicate, $\approx 75\%$). The volcanic ash samples were sieved into four band

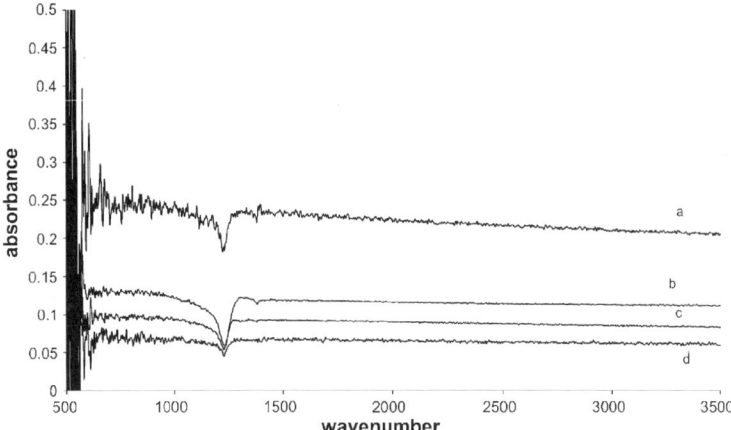

Fig. 4 IR absorbance of volcanic basalt sample as a function of grain size; a < 32 μm, b 32–45 μm, c 45–63 μm and d 63–90 μm. The distinct minimum is typical of a silicate material.

sizes: <32 μm, 32–45 μm, 45–63 μm and 63–90 μm. Small amounts of each sample were levitated using the ultrasonic trap and subsequently collected for size distribution analysis. Fig. 4 shows typical transmission spectra obtained from the basalt volcanic ash. A minimum characteristic of absorption from silicate material is clearly observed.

The key parameters that govern the scattering and absorption of radiation by any microparticle are:

(i) the wavelength of the incident radiation, λ;

(ii) the size of the particles;

(iii) the complex refractive index of a particle: $m = n + Ik$ where n is the real part of the refractive index, k is the imaginary part of the refractive index with both n and k depending upon the wavelength. The real part, n, is responsible for scattering and the imaginary part, k, is responsible for absorption. If k is equal to 0 at a given wavelength, the particle does not absorb radiation at this wavelength. It should be noted that since the refractive index of a particle is defined by its chemical composition, if a particle is made of a mixture of substances then an effective refractive index must be determined.

The information of most use in remote sensing technologies and climate prediction models is the complex refractive index, which can be found by measuring the transmission $T(\lambda)$ through an aerosol sample and hence evaluating the extinction coefficient;

$$T(\lambda) = \exp^{-\beta x} \quad (1)$$

where T is the transmission through the sample, β the extinction coefficient, x the path length within the sample and λ the wavelength.

$$\beta(\lambda) = \int_0^\infty \sigma_{\text{ext}}(r, m, \lambda) n(r) \mathrm{d}r$$

where σ_{ext} is the single particle spectral extinction cross section, $n(r)\mathrm{d}r$ the measured number of particles with a radius between r and $r + \mathrm{d}r$, and m the particle's complex refractive index. We are currently using the method recently developed by Thomas et al.[30] to derive the aerosol refractive index as a function of wavelength directly from measured extinction spectra. This method is somewhat iterative but provides the best method, to date, to determine complex refractive indices from direct measurements of extinction spectra. However, within this method there are several assumptions, the most critical of which is that the particles are near spherical in shape. This is a common approximation in all methods used to derive optical properties from spectral measurements and derives from the use of Mie scattering to evaluate n, but many atmospheric aerosols are fractal in nature (*e.g.*, carbonaceous material) and thus distinctly non-spherical. Light scattering models for such structures are, at present, not sufficiently robust to give accurate/consistent optical constants, thus direct measurements are needed. Eqn (1) can be rewritten

$$T(\lambda) = I/I_0 = \exp(-\sigma_s M_s x) \qquad (2)$$

where I_0 and I correspond to the incident and transmitted light intensities, x is the path length (assuming the aerosol sample to be homogeneous), M_s is the mass of aerosol per unit volume and σ_s is the mass-specific extinction coefficient. For non-spherical solid particulates, eqn (2) may be used to evaluate σ_s, measuring M_s by collecting the samples after levitation in the trap and weighing them. The volume in the trap is defined by the trapping field and, at least for larger particles, may be imaged directly using a helium neon laser. This method has been successfully adopted to evaluate optical extinction coefficients of combustion generated aerosols.[31] SEM images of the 'dust' prior and subsequent to trapping provide additional data on the morphology and structure of such particulates. Fig. 5 shows a SEM image of some of the volcanic particulates whose spectra were shown in Fig. 3. It can be clearly that seen that the assumption of a spherical surface is seriously in error. Therefore, it will be necessary to develop a modified version of Mie scattering codes to determine the refractive index of particulates that are, in general, non-spherical.

We have also been successful in trapping liquid droplets (see Fig. 2) and solid water ice samples in the trap by feeding water vapour from an atomiser into the trapping region, and cooling the reflector appropriately. This is our first step towards coating the dust samples with ice and water to investigate the changes in their optical properties as a function of the relative humidity in the chamber. Fig. 6 shows a preliminary IR spectrum of ice-covered carbonaceous soot particulates compared with an ice film grown on an IR transparent CaF_2 window at various temperatures. In the trapped ice/soot particulate, clear features pertaining to formation of a crystalline surface ice layer can be seen (identified by a shift in the peak to

Fig. 5 SEM images of volcanic particulates whose spectra are shown in Fig. 4.

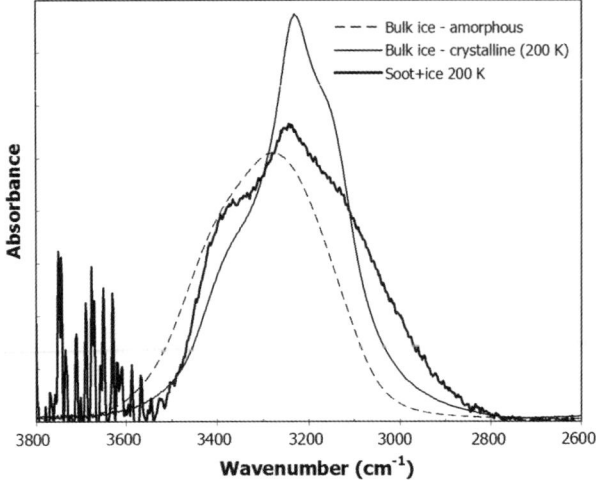

Fig. 6 Spectra of trapped ice-covered carbon particles in the ultrasonic trap (thick curve) compared with those of water ice at different temperatures deposited on a transmitting substrate. The fine structure at the higher wavenumbers is indicative of water vapour in the trap.

3200 cm^{-1}). This formation of a crystalline water ice on the surface of an aerosol has been observed on other hygroscopic particulates[32,33] and is expected to be prevalent in mixed organic/inorganic atmospheric aerosols (see below).

Future developments

The ultrasonic trap to store particulates may be used to study the spectroscopic properties of a wide variety of microparticles. Since the trap necessarily contains an ambient background gas, it is easy to probe the spectra as a function of Relative Humidity (RH); as water is ever-present in the terrestrial atmosphere, all aerosol/particulates will eventually contain an aqueous component. Once the RH exceeds a specific value known as the 'deliquescence point', water is condensed upon the aerosol/particulate surface, past the deliquescence point the aerosol takes up water and is diluted. Similarly, if an aerosol begins as a liquid and is dried, it continues to lose water until it reaches its 'efflorescence point'. The aerosol will then lose all of its water and return to a solid. Thus, the RH regulates the optical properties of any aerosol. By changing the RH in the trap, the extinction coefficient (and thence the refractive index) of these dust particles may be measured as a function of RH. The RH in the trap can be readily controlled by regulating the flow of background gas (helium or nitrogen) into the trap, particulates can be 'solvated' by passing moist air into the trap and dried by passing dry air, the uptake/removal of water being observed in the resultant IR spectrum (as in Fig. 5), hence, deliquescence and efflorescence effects can be probed and the hygroscopic properties of the aerosol/particulate explored.

The long residence time combined with the isolation of the particle(s) with respect to any disturbing surfaces also opens the possibility of studying heterogeneous chemistry and molecular synthesis on discrete aerosol/particulate surfaces. Reactive species seeded in the inert gas (helium or nitrogen) may be deposited upon suspended microparticles with the resultant chemical reactions explored by observation of the chemical products using IR and/or UV spectroscopy. As a first experiment, we will study a well known stratospheric reaction

$$HCl + ClONO_2 \text{ ----on ice} \rightarrow Cl_2(gas) + HNO_3(ice)$$

Introduction of HCl and ClONO$_2$ into the ultrasonic trap will allow us to observe the rate of uptake of such species onto the microparticle surface while simultaneously looking for formation of HNO$_3$ and the release of Cl$_2$ (readily recognised in the gas phase through IR and UV-Vis absorption in the trap).

The trap will also allow us to explore another phenomenon of particular relevance to astrochemistry: the method by which absorbed gases are returned to the gaseous phase—where they are detected by observational astronomy. Microparticle–microparticle collisions will be explored by directing a beam of microparticles charged in a Van der Graff accelerator upon a microparticle target in the trap.

Conclusions

We have designed and constructed a new apparatus that uses ultrasound to form an acoustic trap to levitate and trap microparticles. Once in the trap, the particles may be stored for long periods of time under isolated conditions. Using this device we have begun to study the optical properties of a variety of microparticles with the aim of deriving their refractive index and albedo.

Since the ultrasonic trap is both small and portable, it may be readily coupled to existing analytical devices used to detect and examine terrestrial aerosols. The trap may also be coupled to a Van der Graff accelerator[34] to explore particle–particle collisions of relevance to astrochemistry.[35] Hence, we expect to use the ultrasonic trap in the coming year to study the physical and chemical properties of microparticles.

Acknowledgements

This research was funded by the UK NERC research council for whom one of us (E. A. Drage) acknowledges receipt of a PhD. Studentship. Sarah Webb thanks the Open University for the support of a postgraduate studentship.

References

1　D. A. Williams, *Faraday Discuss.*, 1998, **109**, 1.
2　D. C. B. Whittet, *Dust in the Galactic Environment*, Cambridge University Press, Cambridge, 2003.
3　D. A. Williams and E. Herbst, *Surf. Sci.*, 2002, **500**, 823.
4　P. A. Gerakines, W. A. Schutte and P. Ehrenfreund, *Astron. Astrophys.*, 1996, **312**, 289.
5　R. I. Kaiser and K. Roessler, *Astrophys. J.*, 1998, **503**, 959.
6　M. P. Bernstein, S. A. Sandford, L. J. Allamandola, S. Chang and M. A. Scharberg, *Astrophys. J.*, 1995, **454**, 327.
7　C. Hollenstein, J. L. Dorier, J. Dutta, L. Sansonnens and A. A. Howling, *Plasma Sources Sci. Technol.*, 1994, **3**, 278.
8　A. A. Howling, L. Sansonnens, J.-L. Dorier and C. Hollenstein, *J. Phys. D*, 1993, **26**, 1003.
9　S. R. Solomon, R. R. Garcia, F. S. Rowland and D. J. Wuebbles, *Nature*, 1986, **321**, 755.
10　M. A. Tolbert, M. J. Rossi, R. Malhotra and D. M. Golden, *Science*, 1987, **238**, 1258.
11　J. Schreiner, *Science*, 1999, **283**, 968.
12　K. S. Carslaw, T. Peter and S. L. Clegg, *Rev. Geophys.*, 1997, **35**, 125.
13　K. S. Carslaw, M. Wirth, A. Tsias, B. P. Luo, A. Dörnbrack, M. Leutbecher, H. Volkert, W. Renger, J. T. Bachmeister and T. Peter, *J. Geophys. Res., [Atmos.]*, 1998, **103**, 5785.
14　S. Fuegistaler, *Atmos. Chem. Phys.*, 2002, **2**, 93.
15　S. Arnold, N. Wotherspoon and N. L. Goddard, *Rev. Sci. Instrum.*, 1999, **70**, 1473–1477.
16　L. M. Folan and S. Arnold, *Opt. Lett.*, 1988, **13**, 1–3.
17　S. E. Bauer, Y. Balkanski, M. Schulz, D. A. Hauglustaine and F. Dentener, *J. Geophys. Res., D*, 2004, 109.
18　Y. Rudich, I. Benjamin, R. Naaman, E. Thomas, S. Trakhtenberg and R. Ussyshkin, *J. Phys. Chem. A*, 2000, **104**, 5238–45.
19　D. Gerlich, *Hyperfine Interact.*, 2003, **146**, 293.
20　V. Garcés-Chávez, D. Roskey, M. D. Summers, H. Melville, D. McGloin, E. M. Wright and K. Dholakia, *Appl. Phys. Lett.*, 2004, **85**, 4001.
21　W. M. Lee, B. P. S. Ahluwalia, X. C. Yuan, W. C. Cheong and K. Dholakia, *J. Opt. A*, 2005, **7**, 1.

22 M. Hamilton and D. Blackstock, *Nonlinear acoustics*, Elsevier Press, 1998.
23 J. K. R. Weber, D. S. Hampton, D. R. Merkley, C. A. Rey, M. M. Zatarski and P. C. Nordine, *Rev. Sci. Instrum.*, 1994, **65**, 456.
24 E. H. Trinh, *Rev. Sci. Instrum.*, 2006, **56**, 2059.
25 S. M. Webb and N. J. Mason, *Eur. J. Phys.*, 2003, **25**, 101.
26 W. J. Xie and B. Wei, *Appl. Phys. Lett.*, 2001, **79**, 881.
27 S. Bauerecker and B. Neidhart, *Science*, 1998, **282**, 5397.
28 S. Bauerecker and B. Neidhart, *J. Chem. Phys.*, 2006, **109**, 3709.
29 http://www.msf.rl.ac.uk.
30 G. E. Thomas, S. F. Bass, R. G. Grainger and A. Lambert, *Appl. Opt.*, 2005, **44**, 1332.
31 J. Widemann, J. Duchez, J. C. Yang, J. M. Conny and G. W. Mulholland, *Aero. Sci.*, 2005, **36**, 283.
32 J. P. Devlin, C. Joyce and V. Buch, *J. Phys. Chem. A*, 2000, **104**, 1974.
33 R. Signorell, *Mol. Phys.*, 2003, **101**, 3385.
34 H. Sibata, K. Kobayshi, T. Iwai, Y. Hamabe, S. Sasaki, S. Hasegawa, H. yano, A. Fujiwara, T. Kawamura and K. Nogami, *Radiat. Phys. Chem.*, 2001, **60**, 277.
35 (a) A. G. G. M. Tielens, C. F. McKee, C. G. Seab and D. J. Hollenbach, *Astrophys. J.*, 1994, **431**, 321; (b) E. Meszaros, *Fundamentals of atmospheric aerosol chemistry*, Akademiai Kiado, Budapest, 1999.

Using dynamic light scattering to characterize mixed phase single particles levitated in a quasi-electrostatic balance

U. K. Krieger* and A. A. Zardini

Received 12th February 2007, Accepted 19th March 2007
First published as an Advance Article on the web 11th September 2007
DOI: 10.1039/b702148h

We use Dynamic Light Scattering (DLS) to characterize non-spherical, micrometre-sized, single aerosol particles levitated in an electrodynamic or in a quasi-electrostatic balance. These are either solid salt particles effloresced from an aqueous salt solution droplet upon drying, or mixed phase aerosol particles, *i.e.* aqueous solution droplets containing a single solid salt inclusion. We show that the shortest decay of the temporal intensity autocorrelation function measured in the far field scattering pattern can be quantitatively analyzed. We treat the scattering pattern as if arising from an equivalent sized Mie sphere, and we attribute the temporal intensity fluctuations to rotational Brownian motion of the whole particle. This analysis allows sizing of non-spherical particles. We have indications that the long tails of the autocorrelation functions are due to deviations of the scattering pattern from that of a Mie sphere, leading to spikes in the temporal evolution of the intensity because of the rotational Brownian motion. We also show that the diffusional motion of an inclusion within the aqueous solution of a host droplet is masked by rotational Brownian motion, prohibiting even a qualitative analysis.

1. Introduction

Within an aged atmospheric aerosol particle, a solid phase may coexist with a liquid over a wide range of ambient conditions. Aerosol particles often contain water insoluble components such as mineral dust or soot. Or, depending on composition, temperature, and relative humidity, a solid phase may form in an aged aqueous particle. To characterize size, morphology and location within the host droplet of such a solid inclusion is of great interest for atmospheric science, because these properties will influence the way the particles participate in heterogeneous chemistry and their scattering efficiency of solar radiation. Also, in single particle levitation experiments one could potentially use an inclusion of known size to probe the viscosity of a solution by analyzing its Brownian motion within the liquid host droplet, in contrast to bulk experiments that would allow to measure viscosities of oversaturated or supercooled liquids. Or, one could explore the effect of the close proximity of a wall on the Brownian diffusion of an inclusion in an ideal, namely spherical geometry.

Institute for Atmospheric and Climate Science, ETH Zurich, 8092, Zurich, Switzerland. E-mail: ulrich.krieger@env.ethz.ch

Up to now, techniques to fully characterize mixed phase aerosol particles as stated above are still missing, largely because investigating such a system without affecting it is not without difficulties. While optical microscopy might technically be able to characterize the size and location of an inclusion for particles of several tens of micrometres and inclusion of several micrometres, gravitation will have a major influence on the inclusion's location if the density of the inclusion and that of the droplet's liquid are not very similar, or the size of the inclusion is not sufficiently small. In this case, the gravitational potential forces the inclusion to be at either the top or bottom of the host droplet depending on the density difference.[1] The inclusion's Brownian motion will thus deviate and be substantially attenuated. However, naturally occurring aerosols have a size for which the influence of gravitation can be neglected.

Recently, we have shown that measuring light scattering intensity fluctuations can be used to estimate the area of a single, solid inclusion,[2] to study phase transitions in aerosol particles,[3] and to allow for the detection of changes in the morphology of complex aerosol particles.[4] In these studies, the measure for the intensity fluctuations was the root mean square deviation of the scattering intensity. Also, we measured photon-counting histograms to gain some limited information about the location of an inclusion within its host.[5]

Potentially, Dynamic Light Scattering (DLS) or Photon Correlation Spectroscopy[6] on single levitated aerosol particles with complex morphologies should yield information beyond what is possible by analyzing the root mean square deviation of intensity, because the temporal evolution of scattered light intensity is measured and analyzed by computing its autocorrelation function. In a traditional DLS experiment, the light scattered by a large number of particles is collected by a detector far from the sample volume, its intensity depending on the interference of light from all the scatterers. The random thermal motion of the particles shifts the relative phases of the scattered fields and yields temporal intensity fluctuations. A single spherical symmetric object in a homogeneous light field will scatter light with constant intensity, and hence will not produce temporal intensity fluctuations. Thus, applying the technique to a single, spherical object requires either a strongly focused laser beam, whose intensity profile changes over scales comparable to the size of the object,[7,8] or one has to use a space- and time-modulated illumination field to obtain data about its translational Brownian motion,[9,10] whereas rotational Brownian motion of a non-spherical symmetric object will induce temporal intensity fluctuations even in a homogeneous light field. Intensity fluctuations due to rotational Brownian motion have been used for indication of optical anisotropy and to size non-spherical sub-micrometre prolate-shaped hematite particles,[10] but quantitative analysis of rotational Brownian motion of objects in the Mie scattering size range has not been performed previously to our knowledge.

There has been one attempt by Bronk et al.[1] to use DLS for characterization of mixed phase aerosol particles. They studied aqueous solution droplets (ca. 15 μm in radius) containing, on average, ca. 150 spherical guest particles (0.25 μm in radius). With crossed polarizers at 90° scattering angles, they observed a "ring of fireflies" individually flashing on and off. The corresponding autocorrelation function of the scattered light intensity showed a complex decay, extending in time from roughly 10 ms to several seconds. They hypothesized that the long tails may be associated with diffusion of the guest particles with respect to intensity patterns of the optical field inside the host droplet, and attributed the shortest characteristic decay time to a translational diffusion of the guest particles.

Here, we extend their work and clarify the causes of the observed temporal light intensity fluctuations by studying a mixed phase aerosol droplet containing only one solid inclusion. The remainder of the paper is organized as follows. In the next section, we describe the theory of DLS as it applies to the system in question. In Section 3, we present its experimental realization, and in Section 4 we will show that the shortest characteristic decay time observed by Bronk et al. is due to rotational

Brownian motion of the whole particle and may be used to size non-spherical solid aerosol particles effloresced from aqueous salt droplets on drying. In Section 5, we will show the difficulties in interpreting the long tails observed in both the intensity autocorrelation measured in the far-field scattering pattern of mixed phase aerosol particles and in the autocorrelation function of the glare spot intensities. We draw some conclusions in Section 6.

2. Theoretical considerations

In this section, we follow the general derivation by Dahneke and Hutchins[9] (see also Berne and Pecora[6]) to obtain the scattered light intensity Autocorrelation Function (ACF) for an optically anisotropic particle conducting rotational Brownian motion in a light field of constant intensity. We extend their derivation to the case of an object large with respect to the wavelength and with its scattering pattern approximated by that of a Mie sphere. Since the incoming light has constant intensity over an area larger than the object size, and since the particle is kept at a fixed position by the balance, no variation of the scattered intensity occurs with particle translation, but only with particle rotation. The normalized ACF, $C_n(\tau)$, for a general optically anisotropic particle will be given by $C_n(\tau) = 1 + C_r(\tau)$, with $C_r(\tau) = \langle \gamma_A(t)\gamma_A(t-\tau)\rangle$ being the rotational ACF and $\gamma_A(t)$ being the relative fluctuation in scattering cross section with the particle rotation. Let us consider a rotationally symmetric particle. Here, $p(\theta, \phi, \tau; \theta_0, \phi_0)$ is the probability density that a particle having initial orientation $(\theta_0, \phi_0,)$ will obtain the orientation $(\theta(\tau),\phi(\tau))$ after the time interval τ, where θ and ϕ are the polar and azimuthal angles of the particle's polar axis with respect to the forward scattering direction. Without assuming any scattering law and without knowledge of the optical properties of the particle, $\gamma_A(t)$ may be written very generally as the weighted sum of a complete set of orthogonal functions:

$$\gamma_A(t) = \sum_{j=0}^{\infty} \sum_{k=-j}^{j} c_{jk} Y_{jk}(\theta(t), \phi(t)), \qquad (1)$$

where $Y_{jk}(\theta, \phi)$ are the complex spherical harmonics and c_{jk} are the weighting coefficients determined by the optical properties of the particle. The rotational ACF is given formally by:

$$C_r(\tau) = \frac{1}{(2\pi)^2} \int_0^{2\pi} d\phi_0 \int_0^{\pi} d\theta_0 \gamma_A(\theta_0, \phi_0) \int_0^{2\pi} d\phi \int_0^{\pi} d\theta p(\theta, \phi, \tau; \theta_0, \phi_0) \gamma_A(\theta, \phi), \qquad (2)$$

where the joint probability density, $p(\theta, \phi, \tau; \theta_0, \phi_0)$, can be derived from the rotational diffusion equation.[6] For a spherical symmetric particle with initial orientation θ_0, ϕ_0 rotationally diffusing to angles θ and ϕ within a time interval τ, the result is:

$$p(\theta, \phi, \tau; \theta_0, \phi_0) = \sum_{j=0}^{\infty} \sum_{k=-j}^{j} Y_{jk}(\theta_0, \phi_0) Y_{jk}(\theta, \phi) \exp\{-j(j+1)\sigma_{\text{re}}^2(\tau)/2\},$$

$\sigma_{\text{re}}^2(\tau)$ being the mean-square rotational displacement about an equatorial diameter. Evaluation of eqn (2) with $\gamma_A(t)$ and $p(\theta, \phi, \tau; \theta_0, \phi_0)$ leads to:

$$C_r(\tau) = \sum_{j=0}^{\infty} c_j \exp\left\{\frac{-j(j+1)}{2}\sigma_{\text{re}}^2(\tau)\right\}, \qquad (3)$$

where c_j are constants depending on the light scattering properties of the particle in question, properly normalized. Depending on the morphology of the particle, the light scattering pattern can be quite complex, requiring a significant number of weighting coefficients, c_j, in eqn (3). In contrast, for highly symmetric particles of small size, only a few c_j values are significant. For example, in Rayleigh scattering from a prolate spheroid or from a cylindrically symmetric particle, only c_2 is non-zero, $c_2 = 1$.

The size of our levitated particles are clearly beyond the Rayleigh limit and their light scattering patterns can be quite complex. However, in naturally observed aerosol particles the pattern often resembles that of a slightly perturbed sphere.[11] Further support to this assumption is given by experiments that show that the spatial frequency in the light scattering pattern of cubic sodium chloride particles correlates with their equivalent edge size.[12] If that holds also for the particles under consideration here, namely effloresced aqueous solution droplets and droplets containing an inclusion, we may further simplify eqn (3). We proceed in two steps: we start by assuming that our particle behaves like a spherical symmetric Rayleigh particle, which reduces the sum in eqn (3) to just the $j = 2$ term.[6] In contrast to the light scattering pattern of a Rayleigh particle, our particles—with size being about 10 μm in radius—show the typical ripple structure of particles with large size parameter x ($x = 2\pi a/\lambda$, with a being the radius of the particle and λ the wavelength of the incoming light).

Our second step is to correct for the ripple structure by approximating the interference pattern in the far field by the one obtained for a sphere in the geometrical optics limit, which yields for the angular increment between successive extrema in the intensity pattern at θ equal to 90°:[13]

$$\Delta\theta \simeq \frac{2\pi}{x\left(1 + \frac{\sqrt{2}}{2}\right)}. \tag{4}$$

This compares very favorable with exact Mie calculations for a spherical particle with a range of refractive indices expected for atmospheric aerosol particles, see also Davis and Periasamy.[14] Using this result we are able to calculate an "optical rotational degeneracy number", N_r, $N_r = 2\pi/\Delta\theta = x\left(1 + \frac{\sqrt{2}}{2}\right)$, which is equal to the number of periods in the scattered light intensity per 2π rotation of the particle.

By applying this optical degeneracy number, we overall approximate our light scattering problem as if a spherical particle exhibits hydrodynamically a rotational Brownian motion, and as if its scattering pattern shows the characteristic ripple structure of a Mie sphere independent of the choice of the axis of rotation. We imagine this intensity pattern in the laboratory frame to pass by the detector due to Brownian rotational motion of the particle, and hence inducing temporal fluctuations in intensity. Overall, these approximations lead to the following form of eqn (3):

$$C_r(\tau) = \exp\left\{-\frac{6}{2}\sigma_{re}^2(\tau)N_r^2\right\}. \tag{5}$$

In the "viscous" or "Einstein regime", where inertia effects are neglected, $\sigma_{re}^2 = 2D_{re}\tau$, with $D_{re} = k_B T/f_{re}$ being the rotational diffusion coefficient, and $f_{re} = 4\pi\eta a^3$ being the rotational friction coefficient for rotation about an equatorial axis (η is the viscosity of the gas phase the particle is levitated in).[15] At short time scales compared to the rotational relaxation time $1/\beta = I_m/f_{re}$, with I_m being the mass moment of inertia, friction and random forces become negligible ($I_m = \frac{8}{15}\pi\rho a^5$ for a spherical particle of density ρ). Following an analogous argument as in the case of translational Brownian motion,[16] the generalized expression for the mean-square rotational displacement σ_{re}^2 can be written as $\sigma_{re}^2 = 2D_{re}\tau(1 - [1 - \exp(-\beta t)]/(\beta t))$,[17] leading finally to the following light scattering ACF:

$$C_r(\tau) = \exp\left\{-\frac{6\pi k_B T}{\eta \lambda^2 a}\left(1+\frac{\sqrt{2}}{2}\right)^2 \tau \left(1 - \frac{1-\exp\left(-\frac{15\eta}{2\rho a^2}\tau\right)}{\frac{15\eta}{2\rho a^2}\tau}\right)\right\}. \quad (6)$$

Note that the characteristic time constant in the "Einstein regime" for the decay of the light scattering ACF, $\tau_{\text{scat}} = \frac{\eta \lambda^2 a}{6\pi k_B T\left(1+\frac{\sqrt{2}}{2}\right)^2}$, scales linearly in radius and quadratic with the wavelength. In Section 4 we will compare eqn (6) with measurements.

3. Experimental

The basic experimental setup has been described previously.[5] Briefly, an electrically charged particle (typically 5–25 μm in radius) is levitated in an electrodynamic balance, see a schematic of the setup in Fig. 1. The balance is hosted within a three wall glass chamber with a cooling agent flowing between the inner walls and an insulation vacuum between the outer walls. A constant flow (typically 20 sccm) of a N_2/H_2O mixture with a controlled H_2O partial pressure is pumped continuously through the chamber at a constant total pressure adjustable between 200 and 1000 mbar. The temperature can be varied between 330 K and 160 K with stability better than 100 mK and accuracy of ±0.5 K. Relative humidity (RH) in the chamber is set by adjusting the N_2/H_2O ratio, using automatic mass flow controllers. Three different lasers (HeNe laser, 633 nm, 3 mW; Ar^+ laser, 488 nm, 40 mW; tunable diode laser, 780 nm, 12 mW) can be used to illuminate the particle from below. All lasers are linearly polarized.

To size the particle, the two-dimensional angular scattering pattern is recorded with CCD sensor 2 by measuring the elastically scattered light from both lasers (HeNe and Ar^+) over observation angles ranging from 78° to 101°. If the particle is liquid, and therefore of spherical shape, the scattering pattern is regular, with the mean distance between fringes being a good measure of the radius of the particle, almost independent of its refractive index.[14]

In the present experiment we use mainly two methods to measure the temporal fluctuation of the scattered light of the particle.

First, the scattering intensity at 90° to the incident beam in the far field is detected within a very small conical detection angle (approximately 0.07° half angle) using a fibre coupled avalanche diode single photon counting module (Perkin Elmer, SPCM-AQR-14-FC). A filter wheel in front of the detector allows to select the wavelength seen by the detector to each of the incoming laser wavelengths. The detector is placed in the 90° scattering plane at an angle of about 45° relative to

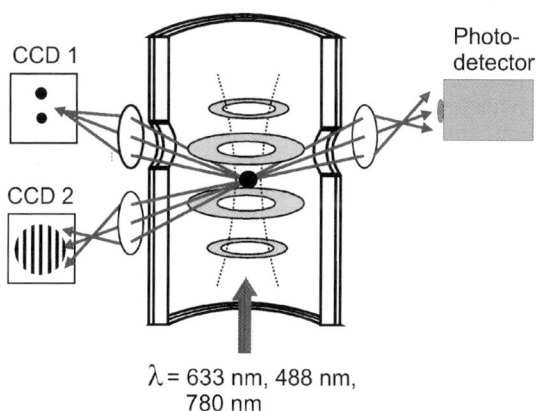

λ = 633 nm, 488 nm, 780 nm

Fig. 1 Schematic of the experimental setup.

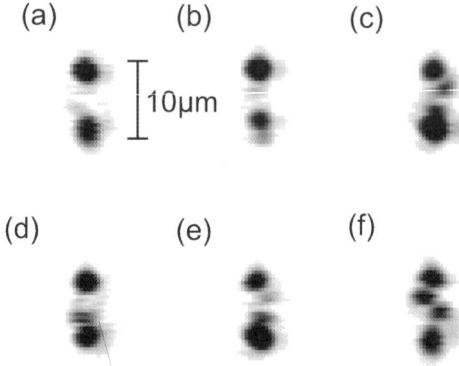

Fig. 2 Six frames taken at arbitrary times of a droplet containing an inclusion, illustrating that, at times (*e.g.* panel (a)), the glare spots appear almost undisturbed as if arising from a homogeneous spherical particle, whereas at other times additional features appear resulting from the presence of the inclusion. The visual impression is that described by Bronk *et al.*[1] as a "firefly flashing on and off".

the polarization vector of the incoming lasers and no polarizer is used in front of it. The output of this photon counting module is fed into a hardware correlator (Correlator.com, Flex99s480) running in multiple tau autocorrelation mode and covering a delay time from 480 ns to several hundreds of seconds.

Second, we observe the particle microscopically with a standard TV-type CCD camera (CCD 1 in Fig. 1) in the same geometry as the photon counting module but with a larger detection angle. For a homogeneous liquid particle we observe the two glare spots,[18] but if the droplet contains an inclusion the glare spots get disturbed, as shown in Fig. 2. A PC and a framegrabber are used to digitize the image (15 Hz sample rate), compute the integrated intensity of both glare spots and the integrated intensity of the region between the glare spots. Autocorrelation data are computed offline from these time series of intensities.

In a modification of the setup described previously, we are able to convert the electrodynamic balance to a Quasi-ElectroStatic Levitator[19] (QESL) by applying asymmetric DC potentials to the AC rings of the balance to stabilize the particle radially, and by using active feedback to stabilize the particle with respect to the gravitational force. By eliminating the need for the alternating gradient forces, which are intrinsic to the electrodynamic balance, the system in its quasi-electrostatic mode is shielded from unwanted noise, parametric instabilities and residual oscillational motion of the particle. In addition, any restoring force suppressing free rotational Brownian motion of the particle is not present.

Mixed particles (solid inclusion within a liquid host) were prepared by injecting an aqueous $NaCl/MgCl_2$ particle and choosing the NaCl to $MgCl_2$ mixing ratio and the ambient relative humidity within the chamber so that a small solid inclusion of NaCl is in thermodynamic equilibrium with an aqueous solution containing Na^+, Mg^{++} and Cl^- ions. Changing the temperature allows us then to adjust the viscosity of this aqueous solution, while affecting to a much lesser degree the viscosity of the nitrogen gas in which the particle is levitated.

4. Sizing a non-spherical particle using DLS

Numerous methods have been proposed for sizing single levitated particles. Light scattering measurements based on comparison with Mie theory constitute the most effective way to size small levitated spheres. For sizing non-spherical particles in an electrodynamic balance, either the springpoint method, sedimentation techniques, or methods analyzing the oscillation of the particle have been used. A detailed

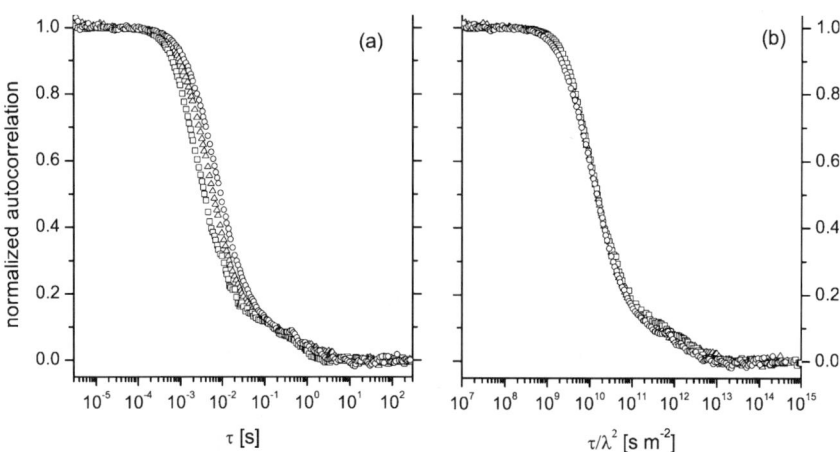

Fig. 3 Panel (a) shows the normalized intensity autocorrelation function measured at the three laser wavelengths (squares 488 nm, triangles 633 nm and circles 780 nm) of a solid NH_4NO_3 particle levitated in the quasi-electrostatic balance. Panel (b) shows the same data but with the time scaled by the square of the inverse wavelength.

comparison of these methods has been performed by Zheng et al.[20] In this section we want to evaluate the ability of dynamic light scattering, using the setup described above and applying the theory of Section 2, to size non-spherical solid ammonium nitrate particles that effloresced in the balance upon drying.

Fig. 3 shows the normalized temporal autocorrelation function of an ammonium nitrate particle levitated in the QESL measured at three different wavelengths over seven orders of magnitude in time. The ACF is plotted *versus* time in panel (a) and *versus* time divided by the square of the respective wavelength in panel (b). Clearly visible is a characteristic decay roughly in the 1 to 20 ms range depending on wavelength, and a longer tail extending to a few seconds (panel (a)). The three measurements at the different wavelengths collapse on one line when scaling the time with the inverse square of the wavelength, as predicted by eqn (6). Note, however, that at short correlation times (~ 0.5–1.5 ms) this scaling overcompensates, *i.e.* the data taken at the blue wavelength are now lagging behind those taken at the near infrared wavelength, whereas they appear reverse in the original data. We attribute this to the fact that in our derivation of the wavelength dependence, we account for it only in the form of the optical degeneracy number. At longer autocorrelation times this seems to be a valid approach, while at very short correlation times the almost sinusoidal form of the Mie interference pattern should be taken into account. Also, the long tail can not be explained by our approximation. We speculate that the cause lays in the fact that the scattering pattern of the solid particle is not that of a Mie sphere. This is clearly the case, as can be either seen when looking at the 2-dimensional scattering pattern,[11] or simply by looking at the intensity time series of the data taken at the blue wavelength of Fig. 3, as shown in Fig. 4. Here, the count rate averaged over 0.628 s time intervals is plotted *versus* time. The spikes seen even in the average counting rate can not be explained by a Mie type intensity interference pattern passing the detector. The modulation depth of a Mie pattern is rather modest if not in resonance, therefore we attribute the appearance of the spikes in Fig. 4 to enhanced scattering in certain directions due to the specific non-spherical morphology of the particle. Since the visible spikes appear on a time scale of seconds due to rotational Brownian motion, they will lead to a finite autocorrelation signal at long times.

A quantitative comparison of the measured ACFs with theory can be seen in Fig. 5. Here, the straight lines are a simultaneous fit of eqn (6) to the data at the three wavelengths for times up to 2 ms only. Note that there is one single free parameter in

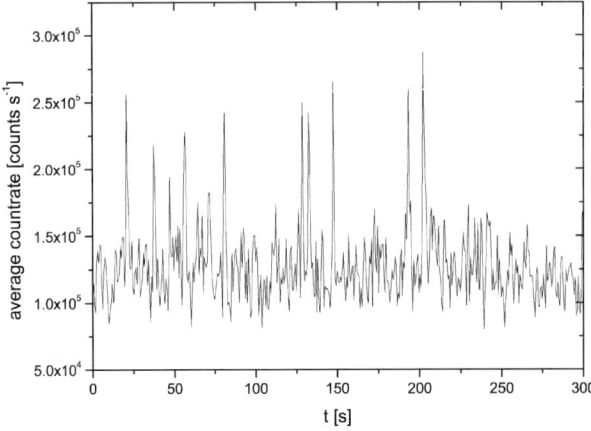

Fig. 4 Temporal evolution of scattered intensity averaged over 0.628 s intervals of the measurement shown in Fig. 3 as open squares (488 nm).

the fit procedure, namely the radius of the particle. The fit yields a radius of 19.8 μm (with the density of ammonium nitrate being 1730 kg m^{-3} and the viscosity of nitrogen being 1.8×10^{-5} N sm^{-2}). This compares favorably with the size of the deliquesced, liquid aqueous ammonium nitrate particle at high relative humidities, determined by comparing a wavelength dependent Mie pattern with theory.[21] If the fit repeated without taking into account the correction for inertia (dotted lines in panel (b) of Fig. 5) it yields a radius of 380 μm, much larger than the particle levitated. However, it is also apparent (see panel (b) of Fig. 5) that the fit with the inertia correction starts deviating strongly for correlation times beyond a millisecond. As discussed above, we attribute this to the deviation of the scattering pattern from that of a Mie sphere.

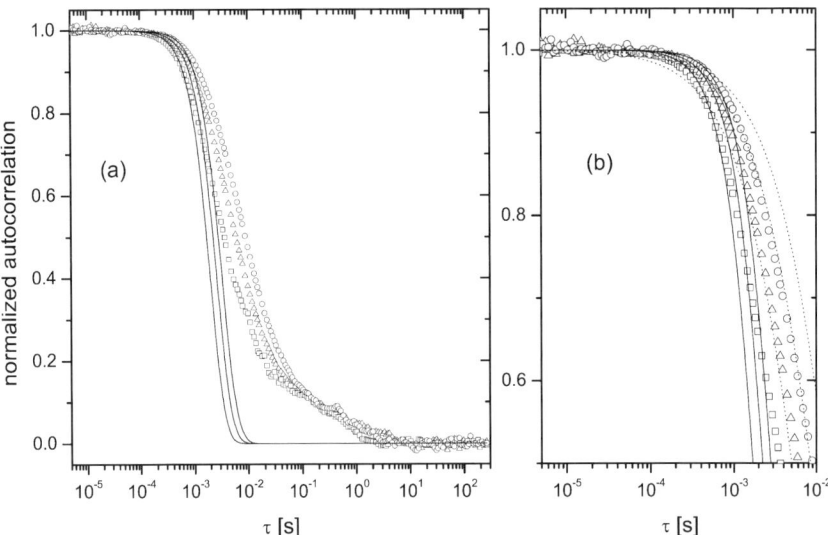

Fig. 5 Panel (a) shows the same data as Fig. 3, squares 488 nm, triangles 633 nm and circles 780 nm. The three lines, corresponding to the three different wavelengths, are a single fit to the data up to correlation times of 2 ms, using eqn (6). Panel (b) is an enlargement of panel (a) at the shortest decay times. In addition, the dotted lines show a fit to eqn (6) when neglecting the inertia correction, see text.

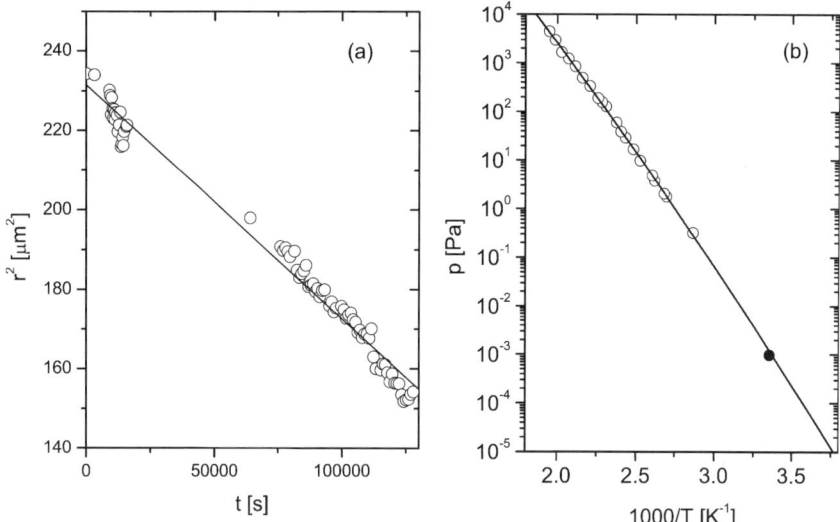

Fig. 6 Panel (a) shows the radius squared *versus* time obtained from measurements as shown in Fig. 5 of a single, solid NH$_4$NO$_3$ particle evaporating for more than a day. Using this data, the vapor pressure is calculated (solid circle in panel (b)) and compared to evaporation data of Brandner *et al.* obtained at higher temperatures[22] (opens circles). The line is an extrapolation of the data of Brandner *et al.* to room temperatures.

As a test of the ability of the technique to follow minor changes in the radius of a solid particle, we let the particle evaporate for more than a day and measured its size fitting the ACF, as shown in Fig. 5. Ammonium nitrate has a finite vapor pressure, releasing nitric acid and ammonia to the gas phase while evaporating.[22] The result is shown in Fig. 6, where we plot in panel (a) the square of the fitted radius *versus* time, the slope being directly proportional to the vapor pressure of ammonium nitrate[23] through:

$$p = \frac{1}{2}\frac{dr^2}{dt}\frac{RT}{M}\frac{\rho}{D},$$

where D is the gas diffusivity of the constituent species with respect to the ambient nitrogen, M the molecular weight and ρ the density of the particle, and R the gas constant. The vapor pressure at 298 K resulting from this measurement is 9.8×10^{-4} Pa. In Fig. 6(b), we compare this value with the data of Brandner *et al.*[22] taken at higher temperatures; the solid line shown in panel (b) of Fig. 6 is an extrapolation of their data to lower temperatures. The agreement between extrapolation and our measured vapor pressure at room temperature is excellent. The scatter of the data in Fig. 6(a) implies a precision of the radius measurement of about 3% for a 13 μm size particle. It is difficult from our measurement to estimate the accuracy, but based on the comparison with the vapor pressure data we establish an upper limit of 20% for a 13 μm size particle. In summary, we conclude that the approximations made in the derivation of the light scattering autocorrelation function of Section 2 are valid for the shortest correlation times and can be used to size solid particles of nearly spherical shape in levitation experiments.

5. Temporal autocorrelation of scattered light by mixed phase aerosol particles

As has been shown above, the shortest correlation times of a levitated non-spherical particle observed by light scattering with homogeneous illumination are due to rotational Brownian motion of the particle in the surrounding gas. In this section, we show how the ACF of a mixed phase particle compare to those of a solid particle. We prepare a mixed phase particle (aqueous NaCl/MgCl$_2$ particle) by adjusting relative humidity such that a solid inclusion of NaCl is in thermodynamic equilibrium with an aqueous solution of NaCl and MgCl$_2$, for details see Krieger et al.[5] In addition, we are able to change temperature from room temperature to 210 K. While no viscosity data exists for these aqueous solutions down to 210 K, by extrapolating existing data of MgCl$_2$ solutions to these temperatures, we estimate that the viscosity of the solution will change from about 2.7×10^{-3} N sm^{-2} at 293 K to about 22×10^{-3} N sm^{-2} at 210 K. By adjusting the relative humidity with temperature so that the total size of the particle remains approximately the same, the translational diffusion coefficient is decreased by almost an order of magnitude while keeping all other relevant parameters approximately constant. For a rough estimate of the characteristic time of the translational diffusion, we assume that the inclusion has to travel the equivalent mean distance of half of a wavelength of the incoming light, as the intensity distribution inside the droplet is assumed to be that of a standing wave with the wavelength of the incoming laser. Using Einstein's result[15] for the mean square displacement $\overline{\Delta x^2}$, a characteristic time of about 0.5 s is obtained; $\tau_{trans} = \frac{\lambda^2}{4} \frac{3\pi\eta a}{k_B T}$, with a being about 2 μm.

The results of the measurements of one single particle are shown in Fig. 7 and Fig. 8. Fig. 7 shows a set of ACFs measured as for those in Section 4, but using the electrodynamic balance instead of the quasi electrostatic levitator. The temperature was lowered from 290.1 K to 214.1 K. The ACFs show a long tail extending up to 100 s with a slope proportional to $\tau^{-0.2}$, but there is no discernible trend with temperature. However, there is a striking scatter between the individual correlation

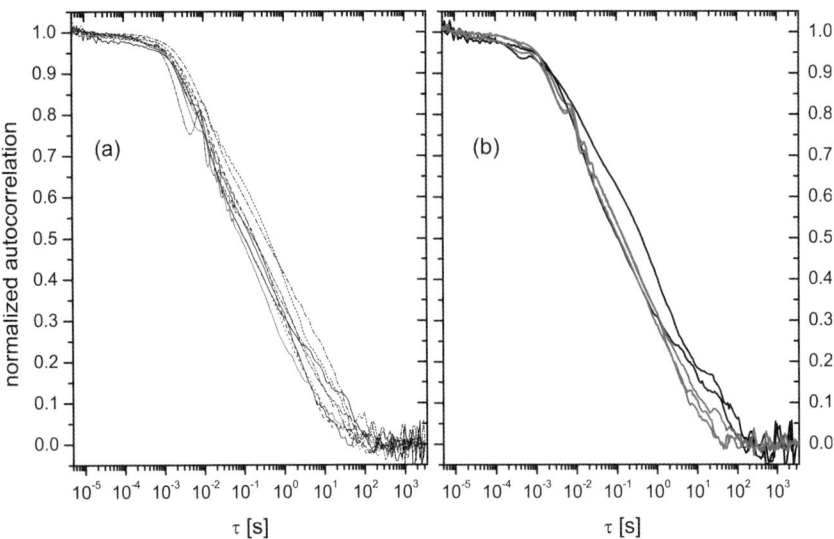

Fig. 7 Normalized autocorrelation functions measured at a wavelength of 633 nm of a mixed phase particle, solid NaCl inclusion in an aqueous solution droplet. Panel (a) shows autocorrelation functions measured at different temperatures: solid line 290.1 K, dashed line 280.2 K, dotted line 270.3 K, dash dot line 260.5 K, dash dot dot line 251.5 K, short dash line 242.8 K, short dot line 233.6 K, short dash dot line 224.3 K, grey solid line 214.1 K. Panel (b) shows two measurements at 290.1 K (black lines) and three measurements at 214.1 K (grey lines). Note, that because the data are taken with the particle in the electrodynamic balance and in contrast to the measurements shown earlier, where the particle was levitated in the quasi-electrostatic balance, a peak can be detected at the 90 Hz frequency of the AC-field and its lower harmonics.

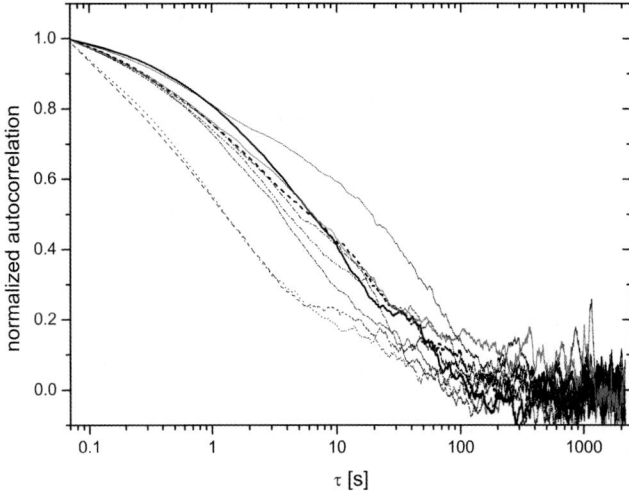

Fig. 8 Normalized autocorrelation functions of the intensity in microscopic image of the particle between glare spots, wavelength 633 nm. Data were simultaneously taken with those shown in Fig. 7. Solid line 290.1 K, dashed line 280.2 K, dotted line 270.3 K, dash dot line 260.5 K, dash dot dot line 251.5 K, short dash line 242.8 K, short dot line 233.6 K, short dash dot line 224.3 K, grey solid line 214.1 K.

times in the long tails. To illustrate this more clearly, we have plotted in panel (b) of Fig. 7 two ACFs taken at a temperature of 290.1 K together with three taken at 214.1 K. All three taken at the lower temperature and one taken at room temperature almost overlap, while the remaining one also taken at room temperature deviates substantially at times between 0.05 to 2 s. At correlation times of about 10 s up to 100 s, the ACFs at different temperatures seem to separate, but this separation is within the noise signal amplitude of our measurements.

Fig. 8 shows autocorrelation functions calculated from time series of the intensity measured in the microscopic image in the area between the two glare spots, as described in Section 3, taken simultaneously with the data shown in Fig. 7. Again, there is no discernible trend with temperature; most ACFs decay similarly, but three of them show a scatter decay time of about one order of magnitude.

While we can not explain the long tails of the ACFs and their scatter quantitatively, because of a quite similar behavior found in Section 4 for the ACFs of the solid particles, we suggest that these long tails are also due to spikes in the temporal fluctuations arising from specific positions of the inclusion within the sphere, see Fig. 2. However, a specific configuration of the inclusion within the sphere in the laboratory frame can be obtained either by diffusion of the inclusion within the sphere, or by a rotational diffusion of the whole particle. Our data suggest that the latter masks the diffusion of the inclusion within the sphere in light scattering data of single levitated mixed phase aerosol particles.

6. Conclusion and perspectives

This study demonstrates the ability of DLS to size solid particles effloresced from aqueous solution droplets upon drying. This could provide an easy tool to measure very low vapor pressures of solid materials, as we have shown for the example of ammonium nitrate evaporating at room temperature. Also, the technique may be used to complementary the more established methods for sizing levitated non-spherical particles. However, our measurements also show unambiguously that our approximate model for motion and light scattering of a non-spherical particle only explains the very early decay of the temporal autocorrelation function quantitatively. At these short correlation times, a more sophisticated model of the light scattering pattern might help to improve the description of the ACFs. At longer correlation times, probably both a less approximate description of light scattering and a less approximate description of the hydrodynamics of the rotational Brownian motion in a real fluid, taking into account the effect on the particle of time-dependent velocity fields set up in the fluid by the particle motion itself, is needed. In addition, our measurements do not show promise to characterize an inclusion in a mixed phase aerosol particle using DLS, because the rotational motion of the whole particle is effectively masking the translational motion of the inclusion within the host droplet.

Acknowledgements

We are very grateful to Thomas Peter and Thierry Corti for several valuable discussions, and to Edwin Hausammann for technical support.

References

1. B. V. Bronk, M. J. Smith and S. Arnold, *Opt. Lett.*, 1993, **18**, 93.
2. U. K. Krieger and C. Braun, *J. Quant. Spectrosc. Radiat. Transfer*, 2001, **70**, 545.
3. C. Braun and U. K. Krieger, *Opt. Express*, 2001, **8**, 314.
4. C. A. Colberg, U. K. Krieger and Th. Peter, *J. Phys. Chem. A*, 2004, **108**, 2700.
5. U. K. Krieger, T. Corti and G. Videen, *J. Quant. Spectrosc. Radiat. Transfer*, 2004, **89**, 191.
6. B. J. Berne and R. Pecora, *Dynamic Light Scattering with Applications to Chemistry, Biology, and Physics*, Wiley, New York, 1976.
7. R. Bar-Ziv, A. Meller, T. Tlusty, E. Moses, J. Stavans and S. A. Safran, *Phys. Rev. Lett.*, 1997, **78**, 154.
8. N. B. Viana, R. T. S. Freire and O. N. Mesquita, *Phys. Rev. E*, 2002, **65**(4), Art. No. 041921, Part 1.
9. B. E. Dahneke and D. K. Hutchins, *J. Chem. Phys.*, 1994, **100**, 7890.
10. D. K. Hutchins and B. E. Dahneke, *J. Chem. Phys.*, 1994, **100**, 7903.
11. K. B. Aptowicz, R. G. Pinnick, S. C. Hill, Y. L. Pan and R. K. Chang, *J. Geophys. Res., [Atmos.]*, 2006, **111**, DOI: 10.1029/2005JD006774.
12. B. Berge, K. Sudholz, B. Steiner, J. Rohmann and E. Rühl, *Phys. Chem. Chem. Phys.*, 1999, **1**, 5485.
13. W. J. Glantschnig and S.-H. Chen, *Appl. Opt.*, 1981, **20**, 2499.
14. E. J. Davis and R. Periasamy, *Langmuir*, 1985, **1**, 373.
15. A. Einstein, *Ann. Phys.*, 1906, **19**, 371.
16. G. E. Uhlenbeck and L. S. Ornstein, *Phys. Rev.*, 1930, **36**, 823.
17. J. Blum, S. Bruns, D. Rademacher, A. Voss, B. Willenberg and M. Krause, *Phys. Rev. Lett.*, 2006, **97**, 230601.
18. A. Ashkin and J. M. Dziedzic, *Appl. Opt.*, 1981, **20**, 1803.
19. S. Arnold, N. L. Goddard and N. Wotherspoon, *Rev. Sci. Instrum.*, 1999, **70**, 1473.
20. F. Zheng, M. L. Laucks and E. J. Davis, *J. Aerosol Sci.*, 2000, **31**, 1173.
21. A. A. Zardini, U. K. Krieger and C. Marcolli, *Opt. Express*, 2006, **14**, 6951.
22. J. D. Brandner, N. M. Junk, J. W. Lawrence and J. Robins, *J. Chem. Eng. Data*, 1962, **7**, 227.
23. N. A. Fuchs, *Evaporation and droplet growth in gaseous media*, Pergamon Press, London, 1959.

PAPER

Elastic light scattering from free sub-micron particles in the soft X-ray regime†

H. Bresch,[a] B. Wassermann,[a] B. Langer,[ab] C. Graf,[a] R. Flesch,[a] U. Becker,[c] B. Österreicher,[d] T. Leisner[de] and E. Rühl*[a]

Received 21st February 2007, Accepted 19th March 2007
First published as an Advance Article on the web 30th July 2007
DOI: 10.1039/b702630g

We report the first experimental results on angle-resolved elastic light scattering in the soft X-ray regime, where free sub-micron particles in the size regime between 150 and 250 nm are studied in the gas phase by using a continuous particle beam. Two different types of studies are reported: (i) Angle-resolved elastic light scattering experiments provide specific information on the scattering patterns in the regime of element-selective inner-shell excitation near the Si 2p-edge (80–150 eV). In addition to intense forward scattering, we observe distinct features in the angle-resolved scattering patterns. These are modelled by using Mie theory as well as a model that includes contributions from diffuse and specular reflection. The results are primarily attributed to scattering from soft X-rays in the surface layer. (ii) Spectroscopic experiments are reported, where the photon detector is placed at a given scattering angle while scanning the photon energy near the Si 2p-absorption edge. These results are also analyzed by a Mie model, yielding accurate information of the size distribution.

1. Introduction

Elastic light scattering from free particles is an approach that has been successfully used for many years to probe the size, shape, and index of refraction of microparticles.[1–4] Pioneering work goes back to Gustav Mie a century ago, who reported the assignment of light scattering patterns of colloidal metal particles.[5]

It is well-known that Mie scattering is accompanied by a typical intense forward-scattering lobe for $x \gg 1$, where x is the Mie size parameter ($x = \pi D/\lambda$). Here, D corresponds to the diameter (size) of the particle and λ is the wavelength of the incident radiation. Moreover, distinct variations in the scattered light intensity give a unique signature of the particle's optical properties. These unique features allow one in the UV and visible regime to derive, *via* Mie simulations, the particle size and the index of refraction with high accuracy.[2,6]

[a] *Physikalische und Theoretische Chemie, Institut für Chemie und Biochemie, Freie Universität Berlin, Takustr. 3, 14195, Berlin, Germany. E-mail: ruehl@chemie.fu-berlin.de; Fax: +49 30 8385 2717; Tel: +49 30 8385 2396*
[b] *Max-Born-Institut, Max-Born-Str. 2a, 12489, Berlin, Germany*
[c] *Fritz-Haber-Institut der MPG, Faradayweg 4-6, 14195, Berlin, Germany*
[d] *Fakultät für Mathematik und Naturwissenschaften, TU Ilmenau, Max-Planck-Ring 14, 98693, Ilmenau, Germany*
[e] *IMK, Forschungszentrum Karlsruhe, Postfach 3640, 76021, Karlsruhe, Germany*

† The HTML version of this article has been enhanced with colour images.

More recent work has focused on short wavelength radiation in Vacuum Ultra-Violet (VUV), as a promising way to study sub-micron particles.[7] It has been shown that sizes of free particles below $D \approx 300$ nm can only be probed by VUV-radiation, whereas ultraviolet and visible light are suitable to probe optical properties of microparticles by elastic light scattering. Note, that ensembles of nanoparticles in liquid dispersions can be properly sized by visible light. The difference compared to the present work is that light scattering from particles in solution relies on fluctuations of the refractive index, which are caused by fluctuations in density and concentration.[8] Thus, the scattered light signal may be influenced by other factors, such as interactions of the particles with the solvent. Note also, that size information on single nanoparticles in the gas phase *via* elastic light scattering cannot be derived in the visible regime. This is because with visible light, nanoparticles have Mie size parameters around or well below unity. In this regime, the scattering cross section drops rapidly with decreasing x and scales as R^6/λ^4, where R is the radius of the particle. Additionally, the angle-resolved scattering patterns become similar to that of a Hertz dipole in the regime of $x \leq 1$.

Angle-resolved light scattering patterns of free nanoparticles have been measured in a continuous particle beam approach, which is similar to the present experimental setup.[7,9,10] Short wavelength radiation in the VUV and soft X-ray regimes has the inherent advantage that the Mie size parameter of sub-micron particles becomes sufficiently large with increasing photon energy to observe properties of Mie scattering. However, besides intense forward scattering, previous work has shown little evidence of Mie resonances so far.[7,9,10] This is mostly due to changes in the index of refraction, particle size distributions in the beam, and a somewhat limited range of accessible scattering angles in previous measurements (*cf.* ref. 7, 9 and 10).

Evidence that elastic light scattering from single, size-selected particles has been observed, arising from an alternative approach for plotting the experimental results.[11] Angle-resolved light scattering can alternatively be presented by plotting the dimensionless parameter qR, where q is the scattering wave vector, instead of the scattering angle.[11–13] This approach yields, besides evidence for Mie resonances, slopes of −2 or −4, which have been assigned in terms of the phase shift, multiple scattering, and the illuminated portion of the particles. Indeed, previous work in the VUV regime clearly indicated that such even-numbered slopes occur.[9]

Besides free particles in the gas phase, numerous works have been performed in which deposited samples of particulate matter were studied in the soft X-ray regime.[14] X-Rays are known to be suitable to probe structured surfaces by scattering experiments.[15] Specifically, soft X-ray scattering from deposited nanoparticles of polymers has been studied recently, where the region of small values of q was investigated, yielding accurate size information.[16]

In general, it is expected that investigations of free nanoparticles in the gas phase are advantageous compared to studies on the corresponding adsorbed species, since there is no influence of any substrate. Moreover, multiple scattering between closely spaced particles (*cf.* ref. 17) is inhibited, which permits the application of Mie formalism for data analysis.

We report in this study first experimental results on angle-resolved elastic light scattering in the soft X-ray regime, where free sub-micron particles are studied in the continuous beam. This approach is complementary to single trapped nanoparticles, where the charging mechanisms were derived recently.[18] However, in trapped, single particle experiments, radiation damage may occur, which can change the particle properties.[19] The present approach overcomes this inherent drawback, since a constantly fresh sample of well defined, size-selected, sub-micron particles is studied. We use for the present work silica nanoparticles in the size regime between 150 nm and 250 nm as a simple and robust model system. Related to the present work are soft X-ray reflectivity measurements on amorphous SiO_2, which yield near-edge spectra and the optical constants in the Si 2p-regime.[20–24]

2. Experimental

Free variable size nanoparticles are prepared in a continuous beam, similar to previous work.[7,9,10] This approach makes use of the following components, which are described in more detail in the following: (i) sample preparation by generating and spraying a dispersion of nanoparticles into a controlled gas phase at ambient pressure by using an atomizer (TSI 3076); (ii) size selection is accomplished by an electrostatic classifier, yielding isolated nanoparticles of defined size (TSI 3080L); (iii) transfer of the particles into the high vacuum chamber and focusing of the continuous particle beam by an aerodynamic lens system, including a triple differential pumping system; (iv) interaction with a beam of monochromatic X-rays from the storage ring BESSY (Berlin, Germany); (v) angle-resolved detection of the scattered radiation.

Variable size silica nanoparticles are prepared by chemical syntheses,[9] which rely on the Stöber approach.[25,26] The particles are repeatedly cleaned by centrifugation and redispersion in purified ethanol. The particle sizes (and standard deviations) are determined by Transmission Electron Microscopy (TEM), yielding: (i) 147 nm ($\pm$4.8%), (ii) 188 nm ($\pm$4.2%), and (iii) 251 nm ($\pm$3.4%). Subsequently, the particle sizes (and relative widths of the size distributions at full width half maximum) are measured by the particle sizer: (i) 151 nm ($\pm$10.0%), (ii) 202 nm ($\pm$7.4%) and (iii) 250 nm ($\pm$6.0%). These sizes are determined by a calibrated Differential Mobility Analyzer (DMA), which is equipped by a condensation particle counter (particle sizer system TSI 3936L22). They are fully consistent with those from TEM, where the increased widths of the distributions in DMA-sizing are primarily due to the resolution of this device. The widths of the particle size distributions are expected to decrease with particle size,[9] as observed in TEM and DMA measurements. Considering the particle sizes and their size distributions, as determined by two independent approaches, we refer to these sizes by rounded values in the following discussion. These correspond to $\sim$150 nm, $\sim$200 nm, and $\sim$250 nm, respectively. We use these rounded values throughout this work.

The liquid dispersions containing the nanoparticles of well-defined size are diluted to $\sim$0.5 g/ L^{-1}. Recent work has shown that high particle concentrations in the liquid phase lead to the enhanced formation of aggregates after they are transferred into the gas phase by spraying the liquid samples into the atmosphere at ambient pressure.[9] The droplets are dried in a diffusion dryer (TSI 3062) and neutralized by a ^{85}Kr source (TSI 3077). This yields a beam of isolated silica nanoparticles of well-defined size. The particles are either singly or doubly charged, where the charge state is inferred from the particle sizer (TSI 3936L22), as shown in Fig. 1. The particle sizer consists of a differential mobility analyzer (TSI 3081) and a condensation particle counter (TSI 3022A). As a result of the control of particle size in the gas phase, the particle beam has been used in this study without primary size selection. This provides a substantial increase in target density. Typical particle densities of $\geq 6 \times 10^6$ particles per cm^3 are obtained from this particle preparation, which is sufficient for angle-resolved light scattering experiments. The particles are focused by an aerodynamic lens system. It is mounted behind a primary aperture of 180 μm, so that the pressure at its entrance is reduced from ambient pressure to $\sim$10 mbar. The lens system has a total length of 300 mm. It consists of six apertures, which are mounted at equal distances of 50 mm with orifices decreasing in downstream direction from 5.3 mm to 3.9 mm. We note that this design is similar to that published earlier in ref. 27. The present lens system yields an optimum transmission range between 70 nm and 300 nm, where the maximum of particles is detected at sizes of 200 nm at an air flow of 0.37 L min^{-1}. The pressure at the exit of the aerodynamic lens is typically of the order of 0.1 mbar. The focused aerosol beam is transferred into the scattering chamber *via* a triple differential pumping stage.

The scattering chamber is kept under high vacuum during operation. The base pressure in the scattering chamber is $\sim$10^{-8} mbar, it increases to $\sim$3 $\times$ 10^{-7} mbar

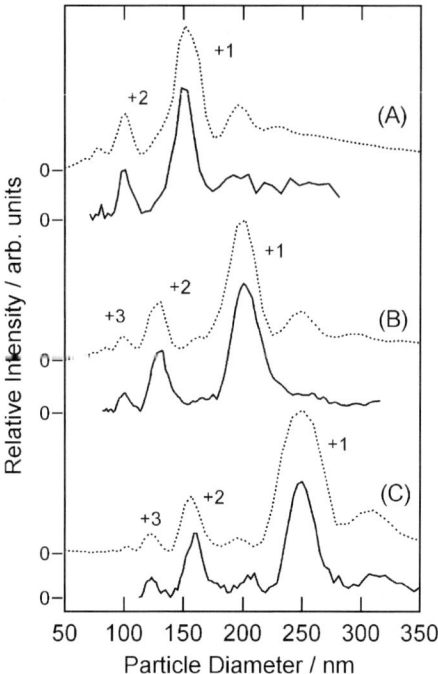

Fig. 1 Size distribution of variable size silica nanoparticles: (A) 150 nm; (B) 200, and (C) 250 nm. The size distributions are either probed by monochromatic synchrotron radiation ($E = 85$ eV, $\lambda = 14.59$ nm, 17° scattering angle) in a nanoparticle beam (full curves) or by a condensation particle counter (dotted curves).

during the experiments. The particle beam has in the scattering chamber a diameter of ≤500 μm. It is crossed in the middle of this chamber by monochromatic synchrotron radiation from the UE49/2-PGM1 and UE52-SGM beam lines at the storage ring facility BESSY (Berlin, Germany). The typical energy resolving power ($E/\Delta E$) of these undulator beam lines is approximately 10^4, where the photon energy range between 80 eV and 150 eV is covered. This is suitable to excite the valence continuum below the Si 2p-edge as well as the Si 2p-continuum. The size of the monochromatic synchrotron radiation in the scattering center depends on the size of the exit slit of the monochromator. Typical values are 17 μm × 60 μm for large scattering angles >6° and 17 × 6 μm for 4°–6° scattering angles, where the slits are narrowed in order to avoid saturation of the detector due to strong forward scattering and the primary beam of synchrotron radiation.

The scattered radiation from free silica particles in a high vacuum surroundings is detected by a CsI-sensitized microchannel plate detector, which is mounted behind a variable size slit. We have used for the present experiments a slit width of 1.3 mm, where the distance between the scattering center and the detector is 65 mm. This yields an angular resolution of 1.15°. In this study, the detector is moved in the polarization plane of the linearly-polarized synchrotron radiation by using a stepping motor in the range of scattering angles between 4° and 100°, where the precision of the angle is ±0.1°. The absolute scattering angle is calibrated by the particle beam impinging on the detector, providing a distinct signal at 90° scattering angle (Fig. 2), which is similar to previous work.[9] Note, that the detector can be mechanically adjusted to values between forward scattering, corresponding to 0°, and a 150° scattering angle. This detector can be replaced by a total electron yield detector, which allows us to measure near-edge spectra of free nanoparticles.

The experimental setup is used for the following experiments:

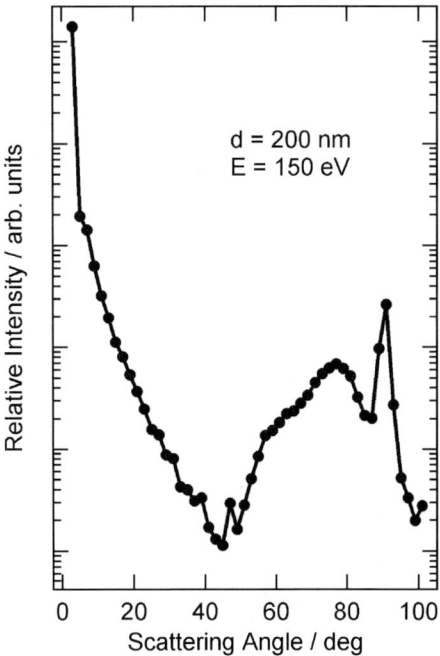

Fig. 2 Angle-resolved scattered light intensity in a logarithmic scale of 200 nm sized silica nanoparticles probed at 150.0 eV photon energy. The sharp maximum at 90° is a result of the particle beam hitting the detector. See text for further details.

(i) Angle-resolved elastic light scattering provides specific information on the light scattering patterns in the regime of element-selective inner-shell excitation.

(ii) Spectroscopic experiments have been performed, where the photon detector is placed at a given scattering angle while scanning the photon energy near core level absorption edges.

The present work is focused on the Si 2p-regime, which is well-known from studies on macroscopic samples.[20,23,24]

3. Results and discussion

Fig. 1 shows a comparison of the particle size distribution recorded at different detection schemes: (i) use of a condensation particle counter (Fig. 1, dashed curves) and (ii) scattered light intensity of $E = 85$ eV photons ($\lambda = 14.59$ nm), as recorded after aerodynamic focusing, where the photon detector is placed at 17° scattering angle (Fig. 1, full curves). This photon energy is located below the Si 2p-edge. The scattering angle is well within the cone of forward scattering, as can be seen from Fig. 2, indicating that the scattered light intensity is preferentially due to elastic light scattering from particles. This is similar to previous work using VUV radiation.[9] Note that the dynamic range of the scattered light intensity covers about five orders of magnitude, as shown in Fig. 2. This is about three orders of magnitude higher than in previous work.[9] The main difference compared to ref. 9 is that smaller scattering angles are accessible in the present setup. These contribute to a massive increase in scattered light intensity at low scattering angles. Furthermore, the Mie size parameter x is substantially increased by a factor of ~ 8 in the present work, which is a result of shorter wavelength excitation (8.2 nm $\leq \lambda \leq$ 14.6 nm) compared to recent results.[9] Specifically, in the case of particle sizes under investigation in this work, we derive $32 \leq x \leq 54$. Note that x would be < 1.5 in the visible regime, where most light scattering experiments are performed.

The size distributions shown in Fig. 1 probe particles as well as their aggregates. The present results indicate that the method of particle preparation is a sensitive parameter. The results obtained from the use of the condensation particle counter show that singly, doubly, and with minor intensity triply charged particles occur (dashed curves in Fig. 1). Besides isolated particles, one also observes aggregates of nanoparticles, which occur at larger sizes than the singly charged ones. The relative intensity of these aggregates is of the order of 20%. This is substantially less than observed in previous work.[9] Aggregates of nanoparticles are formed if the particles already stick together in the liquid dispersion used to produce aerosol particles, or more than one particle is contained in a liquid aerosol droplet. The latter is easily avoided by diluting the suspension that contains the particles. The present results clearly indicate that the fraction of aggregates is substantially decreased if the scattered light intensity is used for probing the particle size. Evidently, this finding is a result of the transmission function of the aerodynamic lens, which is slightly different in design compared to earlier work.[9] The present suppression of aggregates is advantageous for the analysis of light scattering data, since the formation of aggregates is known to change the light scattering patterns.[9,17] Thus, such corrections are not needed in the present work, so that we may safely assume that we are preferentially studying isolated silica nanoparticles at low charge states.

The raw data of the scattered light intensity are shown in Fig. 2, where highest intensity is observed at low scattering angles. Alternatively, the scattered light intensity can be plotted as a function of the scattering wave vector q, as shown in Fig. 3. The scattering wave vector is given by $q = 4\pi\lambda^{-1} \sin(\theta/2)$, where λ is the wavelength of the radiation and θ is the scattering angle. This presentation of the experimental results is similar to earlier work of Sorensen and Fischbach,[11] who have used the dimensionless product qR.

Fig. 3 presents the experimental results along with fits of the q-dependences of the normalized scattered light intensity of differently sized SiO_2 particles recorded at 107.2 eV and 150.0 eV photon energy, respectively. The lower photon energy is located in the Si 2p-near-edge regime, whereas the higher one corresponds to the Si 2p-continuum. The dashed curves result from calculated results, where the transmitted part of scattered radiation is calculated in the framework of Mie theory by using the program package MiePlot.[2,6,9] This requires to use proper values of the refraction and absorption coefficients, as obtained from reflectivity measurements on thin SiO_2 films.[24] A comparison of the experimental results with a simulation that considers exclusively Mie scattering (see dotted curve at the top of each plot shown in Fig. 3) indicates that there are distinct differences as a function of q. We conclude from this, that Mie-scattering cannot be exclusively used to model the experimental results over the entire range of q. This is specifically evidenced by a distinct minimum, which occurs near $q \approx 0.43$ nm^{-1} at $E = 107.2$ eV and near $q \approx 0.55$ nm^{-1} at 150.0 eV in the experimental curves, as shown in Fig. 3.

Mie theory predicts that the distinct minimum should occur at a scattering angle of 90°, which corresponds to $q \approx 0.77$ nm^{-1} at 107.2 eV and $q \approx 1.07$ nm^{-1} at 150 eV, respectively (cf. dotted curves in Fig. 3). This is unlike the experimental results. At a first glance one might assign this discrepancy to a strong increase in the phase shift parameter ρ, leading to a violation of the Rayleigh–Debye–Gans (RDG) theory, which requires that $\rho < 1$ and $|n - 1| < 1$, respectively.[2] The phase shift parameter ρ is expressed by $\rho = 2k_0R|n - 1|$, (see ref. 11 for details), where $k_0 = 2\pi/\lambda$ is the wave vector of the incident wave, R is the particle radius, and n is the refractive index of the sphere. In the present case, using e.g., $E = 150$ eV, $R = 125$ nm, and $n \approx 0.98$, we derive the phase shift value of $\rho \approx 7.6$, which brings us to the intermediate regime between the applicability of RDG theory and anomalous diffraction.

Thus, one possibility to rationalize the deviation of the first, strong Mie minimum in the experimental scattered light intensity as a function of q, shown in Fig. 3, could be found in an unexpected strong change in refractive index n in the surface region.

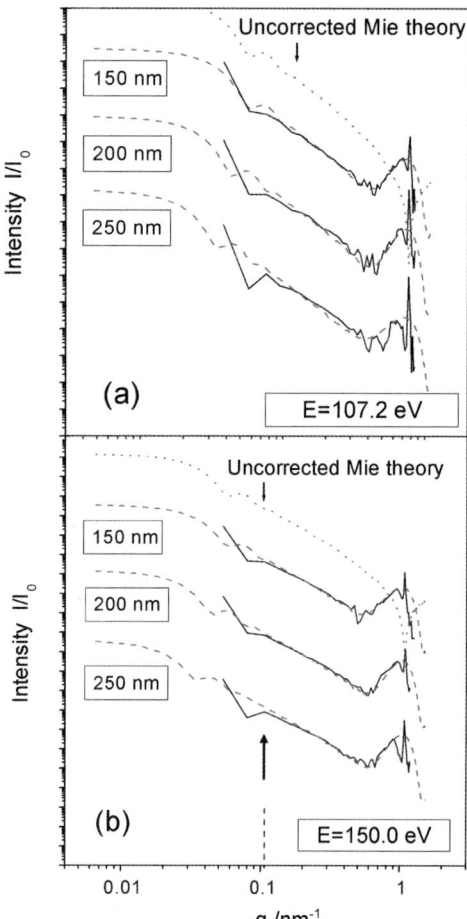

Fig. 3 Dependence of the normalized scattered light intensity from variable size silica nanoparticles at different photon energies E as a function of the scattering wave vector q: (a) $E = 107.2$ eV, (b) $E = 150.0$ eV. The experimental data (solid lines) are compared to model results, as indicated by dashed curves (for further details: see text and eqn (3)). The results are compared to a Mie model for a perfectly smooth 250 nm SiO_2 particle (dotted curve at the top of each plot). The position corresponding to the total external reflection angle is indicated by an upward arrow. The curves are vertically displaced in order to visualize changes in shape.

A possible origin of such a change might be found either as a result of surface contamination or a strong structural change in the surface layer. However, attempts to fit the q-dependence of Mie scattering to the simple model of a coated sphere, by considering either surface contaminations or a rough surface of coated spheres, failed. This is because one obtains from such fits unrealistically high values for the index of refraction, which range between 1.7 and 2.5. Therefore, we discount this assignment. Further, additional fits to the experimental results using pure Mie theory by varying the index of refraction of the spheres, which are different from published values,[24] also failed. They give similarly large values of $n \gg 1$, which is unlike the published values ranging between 0.97 and 0.99 in the energy regime under investigation.[24]

A meaningful and correct fit of the experimental results is only achieved by considering strong reflectivity attenuation of the incoming beam from a rough surface of SiO_2, as will be outlined below. As a result, the expression for the detected

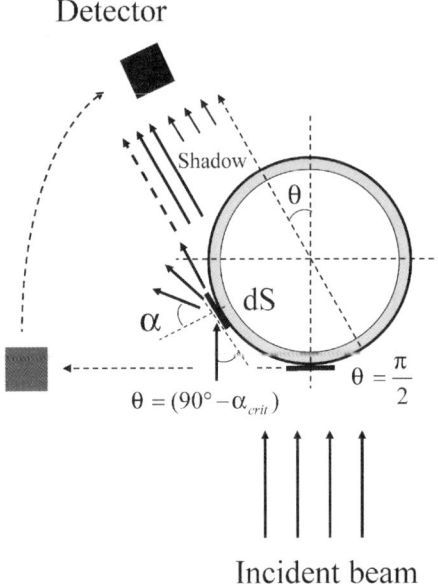

Fig. 4 Schematic presentation of the scattering geometry. The edge of the diffusely scattered beam reaching the detector at the scattering angle θ is indicated by a coarsely dashed arrow.

scattered light intensity contains two parts, one coming from the Mie scattering from an ideal sphere and the other one is due to changes in reflectivity caused by a rough and graded surface, yielding $I_{res} = I_{scat}^{ideal} + \Delta I_{refl}$. The scattered part for perfect spheres I_{scat}^{ideal} results from Mie calculations using the tabulated values of the highly dispersive extinction coefficient, corresponding to n and k (see ref. 24) and by taking into account the experimental particle size distribution, which is derived from the present experiments (see below). The change in the reflective part is modelled by a combination of diffuse R_{diff} and specular reflectivity R_{spec}. The relative intensity of reflected radiation $I = R_{diff}/(R_{diff} + R_{spec})$ is corrected for the attenuation due to the experimentally estimated surface roughness σ. We use for σ typical experimental values, ranging between 1 and 2 nm,[24] which is added to the ~ 20 nm thickness of a radial density gradient layer. Such a layer may, at least in part, result from residual solvent on the particles. This appears to be plausibe, as the particles are transferred within 5 to 10 s from the liquid phase into the high vacuum chamber, where the scattering process occurs. Therefore, it is likely that some solvent remains on the surface. Such a change in reflected light is treated in the framework of geometrical optics, since anomalous diffraction can still be ignored at $\rho = 7.6$.[28] The consideration of diffuse reflection is applicable to the present case, since particle diameters in the range of 150–250 nm are much larger than the wavelength, which is of the order of 8–15 nm.

According to van de Hulst,[28] the scattering patterns of small spheres with elements obeying the Lambert law of diffuse reflection can be calculated by using the theory of Schoenberg.[29] We show in Fig. 4 the geometry of diffuse reflection, which is discussed below in greater detail.

The experimentally detected phase shift observed in Fig. 3 is related to the phase shift across the spherical particle in Mie theory ($\rho = 2k_0 R |n - 1|$),[11] as mentioned above. We have included the anisotropy of the scattering angle θ in the Gaussian power spectral density function of the surface roughness by using the correlation function $g(k)$:

$$g(k) \sim \pi\sigma^2 (2R)^2 \exp(-[2k_0 R|n-1|\cos\theta]^2) = \sigma^2 \exp(-[\rho\cos\theta]^2) \quad (1)$$

Here, $2R|n-1|\cos\theta$ is the optical path difference through the sphere along θ (cf. ref. 28 and 30) and the correlation length is taken to be $2R$, corresponding to the particle diameter. Note that eqn (1) represents an empirical approximation of the exact solution for calculating the reflectivity of a rough surface using the values mentioned above. This also requires the use of results from ref. 30, where the integration over the illuminated surface of a spherical SiO_2 particle has to be taken into account.

In the schematic presentation shown in Fig. 4, a surface element dS reflects the flux $\sim \cos(\alpha)dS/\pi$ per unit solid angle in the direction of a given angle α relative to the surface normal. By integrating over all angles up to the shadow and using the attenuation weight given by eqn (1), we estimate the contribution of the diffusely scattered light:

$$R_{\text{diff}} \sim \pi\sigma^2 (2R)^2 \exp(-[\rho\cos\theta]^2)[\sin\theta + (\pi-\theta)\cos\theta] \qquad (2)$$

It is clear from Fig. 4 that for surface elements dS, where the photon beam is incident at glancing angles $\theta \leq (90° - \alpha_{\text{crit}})$ and α_{crit} is the angle of Total External Reflection (TER), the entire intensity will be diffusely reflected and some part of it will reach the photon detector. Hence, surface elements coming out of the shadow by a further decrease in scattering angle will not change the distribution between the reflected and transmitted light.

The contribution of specular reflectivity R_{spec} is considered for a comparison by the Fresnel formula for parallel polarized light[31]

$$R_{\text{spec}} = \left| \frac{\varepsilon\cos\alpha - (\varepsilon - \sin^2\alpha)^{1/2}}{\varepsilon\cos\alpha + (\varepsilon - \sin^2\alpha)^{1/2}} \right|^2 \qquad (3)$$

where the notation of ref. 24 is used with $\varepsilon = 1 - \varepsilon_1 + i\varepsilon_2 = (n + ik)^2$. ε_1 and ε_2 correspond to the real and imaginary parts of the dielectric constant, respectively. Further, n and k represent the real and imaginary parts of refractive index, and α is the angle relative to the normal. In general, it is not possible to distinguish quantitatively between the contributions from specular and diffuse reflection, simply by fitting the experimental curves. However, one important observation in the region of small values of q is deduced from the curve fits, as shown in Fig. 3: there is a clear step visible near $q \approx 0.1$ nm^{-1}. Specifically, this feature occurs at $q = 0.12$ nm^{-1} in the case of 250 nm particles studied at $E = 150$ eV (see arrow in Fig. 3(b)). This feature is quite close to the calculated value of $\alpha_{\text{crit}} = 80.8°$ in total external reflection using the realistic value of $n_{SiO_2} = 0.987$. Thus, we assign this distinct step in the experimental results to total external reflection. Note, that the expression $\varepsilon - \sin^2\alpha$ in the numerator of R_{spec} in eqn (3) is mainly governed by absorption at $\alpha = \alpha_{\text{crit}}$. This correlates well with the step in the experimental curves, as indicated by the arrow in Fig. 3(b). One can also see from Fig. 3 that all experimental curves are successfully reproduced by the model outlined above reaching down to small glancing angles, which correspond to small values of q. These are close to the critical angle α_{crit}. Interestingly, there is no such feature modelled for the pure Mie scattering part in this regime of q. However, there is a similar feature that occurs as a low amplitude wiggle in the Mie scattering at somewhat smaller q (see Fig. 3). The absence of the experimentally observed shoulder in Fig. 3 in the theoretical curve at low q values ($q \sim 0.1$ nm^{-1}) is evidently related to TER, according to Fresnel formalism shown in eqn (3). The occurrence of this feature at lower values of q compared to the experimental results is likely a result of limitations of the present model, which may be improved by using the approach outlined in ref. 30.

The expected values for $2R|n-1|$ for the given photon energies and particle sizes are ranging between 1.7 and 3.5. Consistently, the values extracted from the fit to the experimental data shown in Fig. 3 are observed in the value range between 1.9 and 3.0. This agreement indicates that the only adjustable parameter in our model is the fraction of the incoherently reflected light, which is due to the surface roughness and

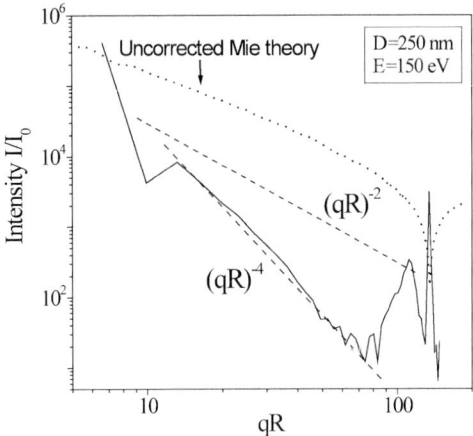

Fig. 5 Scattered light intensity from 250 nm SiO_2 particles at 150.0 eV photon energy as a function of the dimensionless parameter qR (full curve). Slopes corresponding to $(qR)^{-2}$ and $(qR)^{-4}$ are indicated by dashed lines. The dotted line is obtained from a Mie model of a perfect sphere.

radial density gradient of residual solvent, as discussed above. Note, that the exact modelling of the theory outlined in ref. 30 would even make this single fit parameter superfluous. Similarly, there are small but detectable differences in particle size-dependent phase shifts. These are found to be smaller than expected from the definition of the phase shift ρ. Evidently, the deviations from RDG theory, which are solely a function of qR, are not too strong. At the same time the size of the particles does not affect their roughness, indicating that the roughness remains a constant quantity.

The position of the broad minimum mentioned above evidently depends on the particle radius. Fig. 3(a) and (b) indicate that the broad minimum is well modelled by the present approach and the position of this minimum is shifted by a factor of ∼1.5 to larger values of q when changing the particle size from $D = 150$ nm to 250 nm. This difference in energy-dependent shift of the minimum is at a first glance somewhat unexpected, since q scales with the photon energy E and will lead to a constant value of the minimum in the scattering curves as a function of q. A plausible reason for this discrepancy with the experimental results is found in the energy dependence of the index of refraction. Specifically, one derives from ref. 24 that $|n - 1| = 0.011$ at $E = 107.2$ eV and $|n - 1| = 0.014$ at $E = 150.0$ eV, respectively. The position of the minimum scales with the product $|n - 1|\cos \theta$, which indicates that this feature should be found at larger values of q, when the photon energy is increased. This is in full agreement with the experimental results shown in Fig. 3.

Another presentation of the experimental results, that has been suggested before,[11] is to plot the scattered light intensity over the dimensionless parameter qR. This yields power laws, as shown in Fig. 5, where the slopes are indicative for the properties of the scattering process. One expects to obtain from this approach for systems, which slightly deviate from RDG theory, slopes of $(qR)^{-2}$.[11] These have to be followed by a $(qR)^{-4}$ Porod law at higher qR values.[11] We show in Fig. 5, as one typical example, such a plot, which is recorded at $E = 150$ eV using 250 nm silica particles. We show also for comparison in Fig. 5 results from Mie simulations of a sphere (dotted curve in Fig. 5), yielding a $(qR)^{-2}$ dependence. The same slope can also be rationalized by connecting neighboring Mie maxima, as indicated by the dashed line in Fig. 5. There is, however, a distinct $(qR)^{-4}$ dependence in the present results shown in Fig. 5. This is assigned to the occurrence of a gradient of a smoothly decreasing scattering intensity of ~30 nm thickness in the surface region of the particles, similar to the discussion given in ref. 22).

Expulsion of the electromagnetic field to the boundary of spherical particles is another important aspect to be considered.[32] In the photon energy range of this study (85–150 eV), in which the incoming photon wavelength λ is by about one order of magnitude smaller than the particle size, the electromagnetic field is predominantly localized in the region close to the surface. The thickness of this surface layer accounts for ~20% of the particle radius.[32] This implies that contributions from the particle core to the elastically scattered light intensity can be neglected. As a result, the relative importance of the surface contribution and the sensitivity to surface properties of free nanoparticles probed by the reflected light is significantly enhanced in the soft X-ray regime. This supports our conclusion that the $(qR)^{-4}$ regime shown in Fig. 5 comes from the density gradient of the surface layer, which is most likely dominated by residual solvent.

The Si 2p near-edge regime of various SiO_2 polymorphs as well as amorphous SiO_2 have been studied before.[23] The features occurring in the near-edge regime (NEXAFS) of 150 nm silica nanoparticles are shown in Fig. 6, which are in agreement with earlier results.[20,23,24] Details on the spectral assignments of the near-edge features can be found in these previous works. Additional experiments indicate that the near-edge spectra of differently-sized silica particles do not show any particle size dependence. This is quite expected, since the local surrounding near the absorbing Si sites is probed, which evidently does not change in the regime of nanoparticles. Fig. 6 also shows the scattered light intensity in the Si 2p-regime recorded at different scattering angles. The photon energy scale and the spectral shape of both reflection and absorption appears to be in general agreement with previous results.[20,23]

Simulations of the energy scans of the scattered light intensity are also shown in Fig. 6 (dashed lines), where a pure Mie model[2,6] is used. The applicability of this approach is justified for the regime of large scattering angles $\theta \geq 18°$. This is because the Mie model yields reasonable agreement with the experimental results at sufficiently large q-values, which are above the distinct step, that is discussed above in terms of TER, and below the broad minimum (cf. Fig. 3). The fine structure in the 100–125 eV regime, corresponding to the highly dispersive part of the complex refraction index, taken from ref. 24, is well reproduced by the present model results on variable-size nanoparticles. However, the results shown in Fig. 6 indicate that there are significant deviations from the Mie model, especially below 90 eV. These deviations from the Mie simulations to experimental spectra can be either attributed to diffuse scattering from a surface layer of the remaining solvent or to the lack of further corrections given in eqn (2), as well as previous work.[30]

Fitting the experimental energy scans in the Si 2p-regime to Mie theory is found to be sensitive to the SiO_2 particle size distribution in the nanoparticle beam (cf. dashed curves in Fig. 6). This is similar to recent results recorded in the valence regime, where it is found that Mie oscillations smear out with increasing the size distribution.[9] Fig. 7 clearly shows how sensitive the size distribution changes the photon energy dependencies, by using different log-normal size distributions for the $D = 250$ nm sample. This represents an arbitrarily chosen example, where the scattered light is detected at $\theta = 23°$. Best agreement is observed for these conditions at 4.6% size distribution (Table 1), whereas monodisperse particles (0% size distribution) also show distinct oscillations in the pre-edge regime. These are evidently not observed in the experimental results. Such oscillations smear out with increasing size distribution and the narrow features near 105 eV start to sharpen up, as is observed in the experimental spectrum (cf. Fig. 6). Increasing the width of the size distribution to larger values beyond the best fit to the experimental results indicates that there are less significant changes than those observed for narrow size distributions (<4%). There are, however, characteristic changes in amplitudes of the spectral features, as is evident from a comparison between a 4% and an 8%

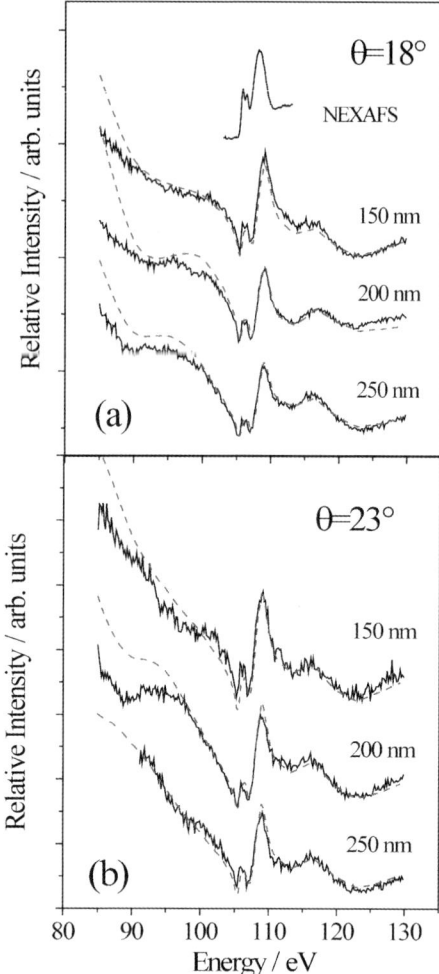

Fig. 6 Dependence of the normalized scattered light intensity from variable size silica particles on the incident photon energy E in the Si 2p-regime (full curves): (a) 18° scattering angle, (b) 23° scattering angle. The dashed curves correspond to the modelled spectra using a pure Mie model including the size distributions compiled in Table 1 along with the optical constants taken from ref. 24. The near-edge spectrum (NEXAFS) at the top of the Figure is obtained from total electron yield measurements of 150 nm silica particles.

distribution, indicating that an 8% size distribution is too wide. A comparison with previous work on amorphous SiO_2 [24] indicates that the clearly split feature near 105 eV, as shown in Fig. 6, also occurs. It is similar in shape as observed from the size distributions, whereas the monodisperse particles indicate no such splitting, as observed from Mie simulations (cf. Fig. 7, bottom trace). Clearly, this similarity in spectral shape indicates that these are not unique. The major difference between the particle size distributions shown in Fig. 6 and the amorphous films (cf. ref. 20) is the significantly different scattering angle: a small scattering angle of ~0.6° from the solid yields a similar spectral signature to the present particle distributions at considerably larger scattering angles.

The resulting log-normal distributions of the particle sizes that are used to fit the results shown in Fig. 6 are compiled in Table 1. These are consistent with the TEM and DMA results discussed above, indicating that size distributions can be

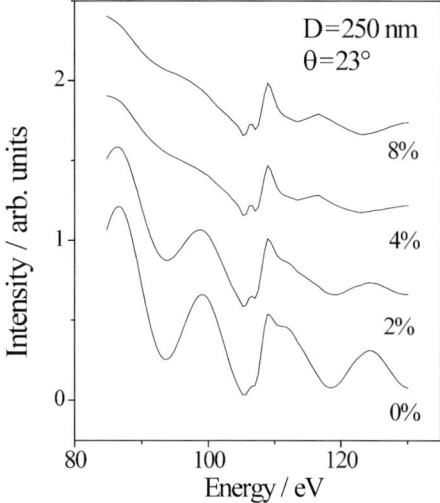

Fig. 7 Mie simulations for a SiO$_2$ sphere (D = 250 nm at 23° scattering angle θ) in the Si 2p-regime of variable size log-normal distributions, as indicated. The size distribution of 0% corresponds to monodisperse particles.

Table 1 Log-normal particle size distribution as obtained from a fit to the Mie model from the energy resolved experiments shown in Fig. 6

Nominal diameter D/nm	Center of size distribution/nm	Width of size distribution (in %)
150	147	7.1
200	192	5.0
250	240	4.6

sensitively and reliably derived from the scattered light intensity in the near-edge regime. The data shown in Table 1 also indicate that the relative widths of the distributions decrease as a function of particle size, which is consistent with earlier work[9] and results from electron microscopy.[26]

4. Conclusions

It is shown that properties of isolated nanoparticles are sensitively probed in the soft X-ray regime. This is accomplished by studying a free nanoparticle beam containing variable size silica particles in the Si 2p-regime (85–150 eV). Angle-resolved light scattering provides detailed information on the scattering mechanisms, which contribute to the inner-shell excitation regime. The present results indicate that Mie scattering cannot fully explain the angle-resolved scattered light intensity. A simple model is developed, which considers the surface roughness, total external reflection, and specular reflection. It allows us to fit the experimental results and to rationalize distinct deviations from Mie theory. The present results clearly show evidence for the surface sensitivity of the present approach.

Energy scans near the Si 2p-edge at a fixed scattering angle show size-effects of the scattered light intensity. This provides a sensitive measure for both the size as well as the size distribution of free nanoparticles. The results are consistent with other approaches of particle size determination.

The present results indicate that inner-shell excitation has the potential to provide a sensitive probe for size effects in nanoscopic matter. Specifically, it is anticipated that surface properties of free variable size and surface functionalized nanoparticles can be selectively probed by element-specific core level excitation and elastic light scattering, which is the subject of ongoing work.

Acknowledgements

Financial support by the BMBF (contract 05 KS4WW1/7) and the Fonds der Chemischen Industrie is gratefully acknowledged.

References

1. M. Kerker, *The Scattering of Light and Other Electromagnetic Radiation*, Academic Press, New York, 1969.
2. C. F. Bohren and D. R. Huffman, *Absorption and Scattering of Light by Small Particles*, Wiley, New York, 1983.
3. R. Finsy, *Adv. Colloid Interface Sci.*, 1994, **52**, 79.
4. *Aerosol Measurement*, ed. P. A. Baron and K. Willeke, Wiley, New York, 2001.
5. G. Mie, *Ann. Phys. (Leipzig)*, 1908, **25**, 377.
6. P. Laven, *Appl. Opt.*, 2003, **42**, 436.
7. J. Shu, K. R. Wilson, A. N. Arrowsmith, M. Ahmed and S. R. Leone, *Nano Lett.*, 2005, **5**, 1009.
8. M. B. Huglin, *Light Scattering from Polymer Solutions*, Academic Press, London, 1972.
9. J. N. Shu, K. R. Wilson, M. Ahmed, S. R. Leone, C. Graf and E. Rühl, *J. Chem. Phys.*, 2006, **124**, 034707.
10. J. N. Shu, K. R. Wilson, M. Ahmed and S. R. Leone, *Rev. Sci. Instrum.*, 2006, **77**, 043106.
11. C. M. Sorensen and D. J. Fischbach, *Opt. Commun.*, 2000, **173**, 145.
12. M. Xu and R. R. Alfano, *Opt. Commun.*, 2003, **226**, 1.
13. J. C. Jonsson, G. B. Smith and G. A. Niklasson, *Opt. Commun.*, 2004, **240**, 9.
14. (a) K. S. Harnard, R. Roth, J. Rockenberger, T. van Buuren and A. P. Alivisatos, *Phys. Rev. Lett.*, 1999, **83**, 3474; (b) A. S. Ethiraj, N. Hebalkar, S. K. Kulkarni, R. Pasricha, C. Dem, M. Schmitt, W. Kiefer, L. Weinhardt, S. Joshi, R. Fink, C. Heske, C. Kumpf and E. Umbach, *J. Chem. Phys.*, 2003, **118**, 8945.
15. U. Pietsch, *Phys. Rev. B*, 2002, **66**, 155430.
16. T. Araki, H. Ade, J. M. Stubbs, D. C. Sundberg, G. E. Mitchell and A. L. D. Kilcoyne, *Appl. Phys. Lett.*, 2006, **89**, 124106.
17. (a) M. Quinten and U. Kreibig, *Surf. Sci.*, 1986, **172**, 557; (b) M. Quinten and U. Kreibig, *Appl. Opt.*, 1993, **32**, 6173.
18. M. Grimm, B. Langer, S. Schlemmer, T. Lischke, U. Becker, W. Widdra, D. Gerlich, R. Flesch and E. Rühl, *Phys. Rev. Lett.*, 2006, **96**, 066801.
19. H. Döllefeld, C. McGinley, S. Almousalami, T. Möller, H. Weller and A. Eychmüller, *J. Chem. Phys.*, 2002, **117**, 8953.
20. O. A. Ershov, D. A. Goganov and A. P. Lukirskii, *Sov. Phys. Sol. State*, 1966, **7**, 1903.
21. A. V. Vinogradov, N. N. Zorev, I. V. Kozhevnikov, S. I. Sagitov and A. G. Tur'ynskii, *Sov. Phys. JETP*, 1988, **67**, 1631.
22. S. K. Sinha, E. B. Sirota, S. Garoff and H. B. Stanley, *Phys. Rev. B*, 1988, **38**, 2297.
23. D. Li, G. M. Bancroft, M. Kasrai, M. E. Fleet, R. A. Secco, X. H. Feng, K. H. Tan and B. X. Yang, *Am. Mineral.*, 1994, **79**, 622.
24. E. Filatova, V. Lukjanov, R. Barchewitz, J. M. André, M. Idir and Ph. Stemmler, *J. Phys.: Condens. Matter*, 1999, **11**, 3355.
25. W. Stöber, A. Fink and E. J. Bohn, *J. Colloid Interface Sci.*, 1968, **26**, 62.
26. A. van Blaaderen, J. van Geest and A. Vrij, *J. Colloid Interface Sci.*, 1992, **154**, 481.
27. (a) P. Liu, P. J. Ziemann, D. B. Kittelson and P. H. McMurry, *Aerosol Sci. Technol.*, 1995, **22**, 293; (b) P. Liu, P. J. Ziemann, D. B. Kittelson and P. H. McMurry, *Aerosol Sci. Technol.*, 1995, **22**, 315.
28. H. C. van de Hulst, *Light Scattering by Small Particles*, Dover, New York, 1981.
29. E. Schoenberg, *Grundlagen der Photometrie*, in *Theoretische Astrophysik*, Springer, Berlin, 1929, p. 64.
30. J. M. Elson, J. P. Rahn and J. M. Bennet, *Appl. Opt.*, 1983, **22**, 3207.
31. F. Stern, in *Solid State Physics*, ed. F. Seitz and D. Turnbull, Academic, London, 1963, vol. 15, p. 320.
32. D. Q. Chowdhury, P. W. Barber and S. C. Hill, *Appl. Opt.*, 1992, **31**, 3518.

General Discussion

Dr Leach opened the discussion of Dr Bartlett's paper by presenting his work: Parametric Resonance of Optically Trapped Aerosols: Complementary work to that of Dr P. Bartlett.

We have observed parametric resonance in optically trapped aerosols. We trap a 5 μm water droplet in air and modulate the intensity of the trapping laser at twice the natural resonant frequency of the droplet. The resonance is excited as we are in an under-damped regime, in contrast to the solid particle in water case, which is heavily over-damped.

Professor Arnold addressed Dr Leach: The problem of the motion of a particle in air in a parametric trap was worked out in the early 1990s. There is no reference to this work in this paper.

Dr Leach responded: This work is cited in our paper.[1]

1 R. Di Leonardo, G. Ruocco, J. Leach, M. J. Padgett, A. J. Wright, J. M. Girkin, D. R. Burnham and D. McGloin, *Phys. Rev. Lett.*, 2007, **99**, 010601.

Professor Goree remarked: You have modeled motion in a trap as a spring. Any physical spring will be nonlinear. A signature of the nonlinearity is harmonic generation when the spring is driven by an external periodic driving force like yours. Is there any detectable harmonic generation in your experimental power spectrum?

Dr Bartlett responded: The forces on a particle held in an optical trap are closely approximated by a harmonic interaction provided the particle is displaced by a small distance compared with its radius. In our experiments the periodic displacement produced by the electric field is typically $\leq$ 20 nm so a harmonic description should be very accurate. The power spectra in Fig. 2 of our paper reveals a single sharp peak at the driving frequency and no higher-order harmonics, consistent with this argument.

Professor Arnold stated: It may be useful to connect up the torque on your liquid crystal drop with its fundamental origin; the spin of the photon.

Dr Bartlett replied: Yes, the torque on the particle (see eqn (18) in our paper) is directly related to the amount of spin angular momentum lost by the circularly-polarized light beam as it passes through the particle.

Professor Rühl addressed Dr Bartlett and Dr Leach:

(1) Comment on the measurement of the charge on a single microparticle: It may be likely as a result of the liquid environment that a fraction of an elementary charge is determined on a particle. If a single particle is surrounded by ultra high vacuum, then only integer values of the elementary charge occur (*cf.* Grimm *et al.*[1]).

(2) A question to both presenters: Is the amplitude of the trapped particles in the regime of Brownian motion different if there is liquid or gaseous surroundings of the particles?

1 M. Grimm, B. Langer, S. Schlemmer, T. Lischke, U. Becker, W. Widdra, D. Gerlich, R. Flesch, and E. Rühl, *Phys. Rev. Lett.*, 2006, **96**, 066801.

Dr Leach responded: The amplitude of the oscillations that we measure in the water drop in air case are of the order of 10s of nm. This is the same as in the solid sphere in water case.

Dr Bartlett responded: Certainly the charge on a single microparticle in a liquid environment will fluctuate as ions thermally adsorb or desorb from the surface of the particle. Our measurements are on long timescales compared to the characteristic time for this process (which probably occurs on times of ms for our system) so we expect to measure a statistically-averaged charge which, of course, need not be integral.

In answer to your second point, the amplitude of the thermal fluctuations of the trapped particle depend only on the temperature and the stiffness of the trapping potential.

Professor Bain remarked: You mentioned the controversy generated by a PRL in 2005 over a method for measuring the force constant of a trap from the amplitude of the response of an over-damped particle to an oscillating force. Valentine *et al.* showed in 1996[1] that the phase shift between the particle motion and the driving force was a sensitive method for measuring the force constant of an optical trap. Mellor *et al.* used a similar approach and extended the analysis to two coupled particles.[2]

1 M. T. Valentine, L. E. Dewatt, H.-D. Ou-Yang, *J. Phys.: Condens. Matter*, 1996, **8**, 9477.
2 C. D. Mellor, M. A. Sharp, C. D. Bain, A. D. Ward, *J. Appl. Phys.*, 2005, **97**, 103114.

Dr Reid asked: What determines the saturation level of charge observed on PMMA microspheres in dry dodecane, as seen in Fig. 3?

Dr Bartlett replied: Recent experiments have shown that the saturation charge is determined by the difference between the number of positive and negatively-charged surfactant micelles which are absorbed on the surface of the particle.

Dr Ward commented: If I were to apply this technique to measuring the charges on individual particles that are optically trapped in aqueous solution I would expect a significantly higher surface charge density. Is there an upper limit on the surface charge that can be measured in this way? Would you expect the charge resolution of 0.25 e to be valid in these conditions?

Dr Bartlett replied: There is no limit on the maximum charge that is measurable. All that is important is that the displacement of the particle remains within the linear force regime of the trap. For a highly charged particle however the electric field would need to be reduced below the values used here. The smallest charge that can be detected by our technique is fixed by the point where the peak in the spectral density at the driving frequency becomes indistinguishable from the Brownian background. A straightforward analysis gives the minimum charge detectable as

$$e|Z| = \frac{\sqrt{12 k_B T \eta a \Delta \omega}}{E}$$

where α is the radius, η is the viscosity, E is the applied field and $\Delta \omega$ is the minimum width of a frequency channel. Reducing the applied field would accordingly reduce the charge resolution.

Dr Ward commented: I would expect the laser power oscillations that you describe to result in axial (*i.e.* z-direction) movement of the trapped droplet. Have you looked at the analagous measurement in the x–y plane by oscillating the laser position without power variation? The results should be comparable unless there is an effect from the variation of the trapping efficiency as the laser power is modulated.

Dr Leach replied: As the laser is modulated, the optical trap is effectively turned on and off. When the trap is off, air currents, gravity and thermal fluctuations cause the water droplet to move away from its central position in all 3 dimensions, x, y and z. When the trap is on, the particle is brought back to the central position. Repeating this process at the correct frequency excites the resonance that we observe. Note that we measure the positions of the particle in the x–y plane in our experiment. We have not investigated the effect of moving the position of the trap with respect to the particle.

Professor Arnold stated:
(1) Arthur Ashkin measured the charge on a particle in air using optical levitation with higher accuracy than is demonstrated in this paper.
(2) Although Ashkin's work (the invention of optical tweezers) is essential to this paper, there are no references to Ashkin in the paper.

Dr Bartlett responded: Many thanks for your comments. I am not familiar with this early work of Ashkin and I would be very interested to hear more. Finally, thank you for pointing out that we forgot to refer to Arthur Ashkin's seminal contributions in this area. Obviously we (and many others) rely heavily upon his efforts. I have modified the draft paper accordingly.

Professor Arnold remarked: Since the equilibrium change is not reached for a long time, it should be possible to measure the change by forcibly ionizing the particle in the trap, followed by rebalancing. This technique is known as 'electron stepping'.[1]

1 S. Arnold, *J. Aerosol Sci.*, 1979, **10**, 49–53.

Dr Bartlett responded: This is a very interesting suggestion.

Professor Bain said: A number of authors have argued that the bare hydrocarbon–water interface is charged and have used this evidence to explain the stability of oil-in-water emulsions in the absence of surfactants. For example, Beattie and Djerdjev report a charge of -4.6 μC cm^{-2} at the hexadecane–water interface and postulate that this charge arises from adsorption of hydroxide ions.[1] Could your technique be used to measure directly the charge on oil-in-water, or water-in-oil emulsion droplets?

1 J. K. Beattie and A. M. Djerdjev, *Angew. Chem., Int. Ed. Engl.*, 2004, **43**, 3568.

Dr Bartlett replied: In our experiments, we have found that all nonpolar suspensions have some degree of charge, although in some cases the degree of charging can be very small. In which case I'm not surprised that electrostatic effects have been invoked to account for the stability of oil-in-water emulsions. It would certainly be fascinating to try and measure the charge in these systems using our optical electrophoresis technique.

Mr Taylor communicated: When you are trapping liquid crystal (LC) droplets, can you be confident that the director of the LC droplets is perpendicular to the beam axis? If it is not, this will affect the degree of birefringence of the droplets. In the extreme case, if the director was parallel to the beam axis the LC would exhibit no birefringence and thus would not rotate at all in the laser field.

There might be a sufficient force on the droplet to ensure the desired alignment, arising from the energy of a dielectric particle in an electric field (which gives rise to the familiar gradient force in optical tweezers) energy [is proportional to] [minus] [integral over volume] (relative permittivity × intensity). Since the extraordinary refractive index is greater than the ordinary refractive index in most LCs, the minimum-energy configuration should be with the director perpendicular to the beam axis as assumed in the paper. I don't know whether that force would be sufficient to stabilize the orientation against Brownian motion, though.

Dr Bartlett communicated in reply: The free energy associated with holding the director so that it lies in the plane of the electromagnetic field is equal to

$$\tfrac{1}{2}\varepsilon_0 \Delta\varepsilon \int I(r) \mathrm{d}V$$

[term 2 of eqn (1) in Škarabot et al.[1]] when the director and electromagnetic fields coincide. Since the particle radius is of the order of microns, the intensity of the focused optical trap is very high, typically 10^{13} W m^{-2}. The energy to rotate a micron radius particle is accordingly of the order of 10^{-14} J, or $\sim 10^6 \times k_B T$. This alignment energy is large enough to ensure that the director remains essentially parallel to the electric field of the beam despite Brownian motion. Experimentally we have observed that particles which are not initially aligned with the laser beam (so that we see the maltese cross at the dipole end through crossed polarizers) quickly rotate in the presence of the laser field. After optical trapping we see no further rotation out of plane.

1 M. Škarabot, M. Ravnik, D. Babič, N. Osterman, I. Poberaj, S. Žumer and I. Muševič, A. Nych, U. Ognysta, and V. Nazarenko, *Phys. Rev. E*, 2006, **73**, 021705.

Dr Corbett communicated: The author states that the PALS (phase analysis light scattering) technique did not successfully measure the 2.9 e charge. What happened with the 50 e charge experiment?

Dr Bartlett communicated in reply: We did not repeat all of our measurements of Fig. 3 using PALS. PALS was certainly not able to detect the very low particle charges of $-2.9\ e$ which we report. However I would expect that the PALS technique would have sufficient sensitivity to detect charges of order $-50\ e$.

Dr Corbett communicated: Have known standard charged particles been measured to validate the technique?

Dr Bartlett communicated in reply: Yes. Our optical electrophoresis technique has been checked against a commercial PALS instrument (Brookhaven ZetaPlus) using a sample of highly charged particles. We found agreement within experimental error.

Dr Corbett communicated: What is the lowest particle size that they can detect in their experimental set-up?

Dr Bartlett communicated in reply: We have not explored how small a particle we can detect. However the size of the smallest particle will be limited by the requirement that the particle be stably confined within the optical trap. Other workers have shown that particles as small as a few 10's of nm may be optically trapped so our approach should work, in principle, down to these limits.

Dr Mitchem opened the discussion of Dr McGloin's paper: Why would you expect larger droplets to be trapped for a given power using green light compared to 1064 nm?

Dr McGloin answered: The smaller spot size that can be produced with shorter wavelength light gives rise to larger gradient forces, so I believe that slightly large particles could be trapped in optical tweezers with green rather than IR laser light.

Professor Kaye asked: How large can particles in air be and still be controlled and moved by a trap?

Dr McGloin replied: We've trapped particles with diameters above 10 microns. We haven't explicitly looked for an upper limit. In liquid based optical tweezers tens of microns is usually the limit and I would expect the same to be true here. with the dual beam trap we could maybe go higher and push towards 100 microns (although this is supposition and I have no experimental evidence to back this up).

Dr Reid commented: Much of the work that has been done on aerosol tweezing has concentrated on particles larger than 1 μm in size. What prospect is there for trapping particles below 1 μm to the nm size?

Dr McGloin answered: In our case we have concentrated on these larger particles due to the fact that our nebulizer produces particles with an average size > 1 μm in diameter. I see no reason why we shouldn't be able to trap smaller particles if we were able to produce them and reliably load them into the trap volume. Trapping very small particles will be more difficult due to issues with viewing the particles and how the particles react to airflows and other ambient effects but these should be surmountable.

Dr Reid commented: We find that the ability to trap aerosol particles is degraded by how well the coverslip wets, the accuracy of the optical alignment and variations in relative humidity. Under optimal conditions, trapping is facile and we feel that coupling additional techniques to aid trapping would unnecessarily complicate the technique.

Dr McGloin responded: I agree with the comment in part. However my argument would be that if I want to load 20 particles all with diameters of 3 μm into the trap in succession that this would not be so easy. It is the reliable loading of particles with specified properties that is desirable for quick and reliable quantitative studies and this is a goal I would like to see pursued.

Dr Ward asked: In the counter-propagating fibre trap could you clarify if there is a discrete point of focus where the droplet is levitated. Is it possible to put microlenses onto the fibre ends to enhance the trapping? What is the effective working distance of the fibre trap?

Dr McGloin responded: The dual beam fibre trap works by radiation pressure holding the particle at the point where the force from each beam balances. So there is not a discrete point of focus. One could put microlenses on the fibres which would enhance the available gradient force but might make the treatment of the radiation pressure forces more complicated and would likely make alignment more difficult.

There have been demonstrations of single fibre gradient force traps but here the trapping is usually much closer to the fibre tip and this might prove a problem when trapping droplets.

The effective working distance of the fibres is from 10 s to 100 s microns (the separation of the fibre ends) but we are still working on properly quantifying this.

Professor Kaye asked: When the SLM (spatial light modulator) is reconfigured to move the beam focus to a new position, presumably there is momentarily no trapping force on the particle? What is the response time of the SLM and how does this limit the rate of movement of a trapped particle?

Dr McGloin responded: The SLM we use (Holoeye LC-R 2500) has a specified refresh rate of ~72 Hz although in practice we would normally run it around normal video frame rates (~25 Hz). So there is a time when the hologram is 'off' and not displaying a specified pattern. We find that the diffusion of the droplets in this time is not sufficient for them to fall out of the trap. For more complicated or accurate motion than we demonstrate this refresh time may become an issue but we have not found it to be a limiting factor in any of our experiments to date.

Dr Hudson asked: Does the field of the trapping laser influence the evolving geometry of a pair of liquid droplets during a coalescence vent?

Dr McGloin replied: It has some influence, in the sense that if the droplets get too close one can jump from one trap into the other (a process which is laser power dependent). We have done some work on this, in collaboration with Jonathan Reid's group.[1] Essentially the finding here is that the laser has little influence on the droplet coagulation unless the sum of the droplet radii is very small ($\ll 7\mu m$). However we find that very often the droplets seem to get closer together than the sum of their radii before coagulation and the exact reason for this is not wholly understood, but is probably due to one of the aerosols moving under the other before coagulation.

1 J. Buajarern, L. Mitchem, A. D. Ward, N. H. Nahler, D. McGloin and J. P. Reid, *J. Chem. Phys.*, 2006, **125**, 114506.

Professor Kaye asked: Is there any possible method of guiding a particle into the optical trap (from an ambient atmosphere or aerosol) rather than waiting for one to fall in by chance?

Dr McGloin replied: Yes, there are various ways. We have shown in the paper that Bessel beams could be used to guide particles and this might be one option to demonstrate guiding into a trap. We are also considering options such as hollow core photonic crystal fibres and one could also employ some form of aerodynamic focussing to improve trap loading rates. My view is that it is important for basic studies to be able to load reliably particles of known composition and size into a trap and we are looking at how this might best be achieved.

Dr Mitchem asked: Which of your aerosol trapping techniques do you think would be most successful for sampling aerosol particles in the field (including Bessel beam, and single and multimode fibre traps)?

Dr McGloin responded: Of the techniques presented in the paper I believe the dual beam fibre trap is the best option for field based work. This is because the forces available in the dual beam trap are larger than in the other optical techniques and so this type of trap has the best chance of catching a aerosol from a field based sample. However the preliminary data we have on the dual beam trap suggests that we can trap only a limited size range of particles (Fig. 11 in the paper) and we are working to see why this should be. My guess is that it is to do with the way in which we are introducing the aerosols into the trap rather than anything more fundamental.

Professor Jaenicke stated: Design an experiment with an ice crystal and a water droplet close together (vary the distance, temperature could be around -10 °C). Observe, if the ice crystal grows at the cost of the water droplet.

Dr McGloin responded: This sounds very achievable, assuming we can trap ice particles. This type of experiment is something I would like to try and looking at cooling aerosols in the trap is on our list of things to try.

Professor Kaye asked: To what extent can non-spherical particles (such as ice crystals) be trapped and manipulated?

Dr McGloin answered: I believe that non-spherical particles can be trapped but their motion within the trap may be rather erratic. To date we haven't done any work on non-spherical particles but this is something we will look at as part of instrument development. A paper was published last year demonstrating a dual beam trap that could trap ice crystals[1] but trapping times were limited and the laser powers required were very high. I believe our systems are more robust, require much lower powers and are slightly more flexible.

1   Kazuki Taji, Maki Tachikawa, and Kazushige Nagashima, *Appl. Phys. Lett.*, 2006, **88** 141111.

Dr Murray commented: Could this technique be used to measure the rate of contact nucleation of ice in supercooled water droplets? This might involve trapping a supercooled droplet and bringing a solid (also trapped) particle (possibly soot or mineral dust) into contact with the droplet in a controlled manner. You could potentially measure the nucleation rate which would be very useful in establishing how important this mode of ice formation is in the Earth's atmosphere.

Dr McGloin answered: This is certainly a possibility. Assuming we can trap particles like soot or mineral dust I don't see why this couldn't be tried out. The trap geometry might have to be a bit involved and in order to supercool the droplets the environmental control would have to be more precise than at present. Studying cooling processes and simple nucleation are certainly on the list of things to try.

Professor Chang communicated: In the poster (Summers, Burnham and McGloin: New developments in the optical trapping of aerosol) you present a method to optically levitate solid particles. Could you give an indication of how long you can hold these silica particles, what size range of particles you tried and what are the key parameters to achieve stable levitation?

Dr McGloin communicated in reply: This is very much a work in progress but we have been able to trap 0.96 and 3 micron diameter particles. The smaller particles have only been captured in very small numbers (about 1 or 2 to date). The 3 micron particles trapping is rather more robust. We use ~1.5 mW of power at 1064 nm. We hope to parametrise the experiment shortly and will publish the results in due course but details are lacking at present.

I should also note that our modelling indicates that smaller particles will be more difficult to trap than larger ones but we haven't verified this experimentally as yet.

Professor Kaye asked: To what extent can you operate the trap at reduced pressure to more realistically simulate the environment that may be experienced by astronomical dust particles?

Dr McGloin answered: This should certainly be possible. Ashkin showed optical levitation in a vacuum in the 1970s[1] and so while you need to take care trapping at lower pressures can be done. The damping of the air actually aids the trapping which is one issue and carrying out experiments at low pressure on liquid droplets might present some challenges. But assuming we can tweeze solid particles routinely in air, lower pressure would be more of an engineering challenge than a scientific one.

1   A. Ashkin and J. M. Dziedzic, *Appl. Phys. Lett.*, 1976, **28**, 333.

Professor Goree opened the discussion of Dr Reid's paper: Does the method of measuring relative humidity that you have reported offer an improved precision or other advantage as compared to other methods of measurement?

Dr Reid answered: The typical capacitance probes used early in this work have an associated accuracy of ±2%, a slow response time of 50 s and a maximum relative humidity of ∼90%. A dew point hygrometer provides a more accurate method for determining relative humidity (RH) to higher RH and can determine the dew point (and hence RH) with an accuracy of 0.2 °C, an accuracy of better than ±1%. The ability to make measurements of RH close to 100% with an accuracy of better than ±0.1% makes this an extremely accurate and responsive probe for the RH with a time resolution of better than 1 s. Further, it is an extremely local probe, allowing small gradients in RH to be measured over micron length scales. You could also rightly say that it is complex to use and not too versatile!

Professor Jaenicke asked: To what extent is the probe—a droplet—influencing the humidity field of the target—also a droplet?

Dr Reid answered: At all times the droplet should exist in a state of equilibrium with the surrounding gas, with a vapour pressure that is equal to the surrounding relative humidity. Thus, the droplet should not influence the surrounding humidity field of the target. Further, the droplets are maintained at a separation of many diameters, to avoid droplet interference. In future work, we intend to examine the possibility of coupling between the droplets in greater detail.

Dr Krieger asked: Can you use your double droplet arrangement to characterize the effect of absorptive heating in your optical trap?

Dr Reid replied: This is indeed possible. Specifically, we can vary the trapping laser power forming each trap independently. Thus, by retaining one trap at constant power and varying the power of the second trap alone, the absorptive heating of the droplet in the second trap can be varied while maintaining a control droplet at constant temperature in the first trap. In preliminary measurements we have performed similar measurements on a single trapped droplet without making a control measurement. In this work, however, absorptive heating was investigated for a droplet doped with a blue absorbing dye. Initially when trapped with green light, the dye doped droplet does not absorb. However, with additional illumination from a blue diode laser operating at 410 nm, the absorption by the dye can be turned on and the response in droplet size investigated. Droplet heating leads to an increase in vapour pressure above the surrounding relative humidity. The droplet evaporates to compensate by increasing the salt concentration, lowering the vapour pressure. A new equilibrium point is achieved with an elevated droplet temperature. When the blue laser is turned off, the droplet returns to its initial equilibrium size.

It is also important to note that we generally use green light to trap the aerosol droplets to minimise droplet heating. An invariance of droplet size recorded with varying laser power confirms that droplet heating is avoided.[1]

1   Rebecca J. Hopkins, Laura Mitchem, Andrew D. Ward and Jonathan P. Reid, *Phys. Chem. Chem. Phys.*, 2004, **6**, 4924.

Professor Chang asked: Kirk Fuller from NASA in Alabama, USA made some predictions about new peaks being formed and the original peaks disappearing when two spheres are close to touching each other. When you brought 2 spheres, each acting as a separate resonator, close to each other, did you see any interaction in the form of new peaks in your Raman spectra?

Dr Reid responded: We have seen some signs of optical interaction, particularly with light coupling between whispering gallery modes in the two neighbouring droplets. At this stage I cannot comment more fully, but this is something that we intend to examine more closely.

Dr Davies commented: What is the spectroscopic evidence, if any, for the organic being at or near the surface? Is there a direct spectroscopic signature from the surfactant?

Dr Reid answered: Glutaric acid is a water soluble organic. At the concentrations it is present in the droplet, it will exhibit a surface excess but will also be present in the bulk of the droplet. Thus, there is no discrete boundary in refractive index between an aqueous core and an organic shell. This precludes modelling the droplet as a core–shell structure. The presence of a surfactant monolayer is not easy to discern with this form of spectroscopic measurement, although indirect evidence could come from the different time responses of the two droplets to changes in surrounding relative humidity.

Professor Arnold asked: Since Raman is spontaneous, involving dipoles, it should be possible to use photoselection at the surface to distinguish between molecular transition dipoles oriented parallel *vs.* perpendicular to the surface. Thereby one can determine the orientation of surfactant molecules at the surface. This has been done for fluorescence emission.[1]

1 S. Arnold, N. L. Goddard and S. C. Hill, *J. Chem. Phys.*, 1999, **111**, 10407–10411.

Dr Reid responded: This is indeed something we have considered and would like to pursue.

Professor Bain asked: The effect of organic monolayers on the transport of water through interfaces has been widely studied in the context of reducing evaporation from reservoirs in hot climates (see, for example, G. T. Barnes[1]). Evaporative loss is normally governed by diffusion and convection in the gas phase and the rate of permeation through an organic layer has to be greatly reduced before interfacial transport becomes rate determining. Effective barriers to evaporation in reservoir applications typically contain long-chain fatty alcohols that form densely packed crystalline monolayers on the surface of water. To what extent are aerosol droplets in the same mass transfer regime as planar water surfaces? Is the evaporation rate limited by transport in the gas phase, or can small surface-active molecules significantly reduce the evaporation rate (other than by depressing the equilibrium vapour pressure through colligative effects in the bulk)?

1 G. T. Barnes, *Colloids Surf. A*, 1997, **126**, 149.

Dr Reid answered: Gas phase diffusion is often the limiting process in governing the rate of evaporation or growth, although surface processes become more significant as the particle under consideration diminishes in size. To characterise this, it is normal to calculate the Knudsen number, the ratio of the mean free path of the gas phase molecules to the radius of the particle. For large particles at high pressures, the surrounding gas phase can be considered to be continuous fluid, while for small particles at low pressures, the particle is small enough that it 'looks like' another gas molecule, with gas molecules moving in large steps between collisions around it. For droplets 4 microns in radius, the Knudsen number is 0.003 for water vapour diffusing in nitrogen at 100 kPa. Under these conditions within the continuum regime, gas diffusion limits any surface process. As the pressure is decreased to 10 kPa, the Knudsen number increases to 0.03 and to 0.3 at 1 kPa. Under these conditions, the timescales for gas diffusion and mass transfer across the interface become more comparable and the nature of the interfacial transport must be considered. Clearly, for sub-micron particles, the pressure at which interfacial processes limit the mass transfer increases above that for the large particle considered above. Further, if the probability of an incoming molecule undergoing mass accommodation is significantly less than 1, the interfacial transport can become limiting at higher pressures. If a picture that allows decoupling of the timescales for the different processes that are occurring is assumed, when the mass accommodation coefficient is 0.02, surface accommodation rather than gas diffusion becomes the rate limiting process at pressures less than 35 kPa.

Mr Tuckermann communicated: Monolayers of long-chain amphiphiles (*e.g.* stearic acid or octadecanol) are able to reduce the evaporation rate of water. This has been recently investigated on acoustically levitated water drops[1] with typical diameters of approx. 0.5–1 mm. Surface layers of organics with a shorter or snapped-off carbon chain or a large hydrophilic head group are not able to reduce the evaporation rate efficiently. I would expect similar results on micro droplets trapped in your optical tweezer.

1 R. Tuckermann, S. Bauerecker and H. K. Cammenga, *J. Colloid Interface Sci.*, 2007, **310**, 559–569.

Mr Hunt asked: In what way is doping of the droplet of interest monitored when trapping the RH probe droplet? By monitoring a size change in the droplet of interest it is impossible to determine the initial size of the probe droplet.

Dr Reid replied: This is primarily a question of time and size resolution. We can directly measure discrete changes in size through coagulation and separate these steps from a slower time dependence arising from RH changes. This then allows a direct analysis of the changing composition and changing water content from the time dependence of the size. Further, in new measurements we are able to simultaneously monitor the size of the droplet of interest and the control, allowing direct knowledge of both droplets simultaneously.

Professor Davidovits addressed Dr Reid, Professor Bain and Dr Krieger: Has anyone observed capillary wave effects in controlled droplet experiments?

Professor Arnold replied: I believe that Richard Chang's group has demonstrated capillary wave effects as droplets are emitted from Vibrating Orifice Aerosol Generator (VOAG).

Dr Krieger answered: While I am not aware of any work on thermally activated capillary waves in controlled droplet experiments, there is the work by T. Leisner's group[1] who studied shape oscillations close to the Rayleigh limit, both theoretically and experimentally.

1 D. Duft, H. Lebius, B. A. Huber, C. Guet, and T. Leisner, *Phys. Rev. Lett.*, 2002, **89**, 084503.

Professor Bain responded: The capillary wave broadening for the free water surface is only about 0.4 nm. The amplitude of capillary waves scales as the inverse root of the surface tension. Except near critical points, where the surface tension tends to zero and the capillary wave amplitude diverges, the capillary wave broadening of aerosol droplets is negligible on the micron length scales of controlled droplet experiments.

Dr Reid replied: Not to my knowledge.

Professor Chang communicated: Did you try to protect/isolate the glutaric acid/sodium chloride droplet in the first optical trap during nebulisation and trapping of the sodium chloride droplet in the second trap to avoid dosing the initial glutaric acid/sodium chloride droplet with further sodium chloride?

Dr Reid communicated in reply: This is difficult to achieve with a physical barrier as the length scales are so short and the aerosol flow so diffuse. However, we are in the process of studying the loading of holographic arrays of traps, through which many droplets can be simultaneously manipulated. We anticipate that using the optical scattering force to our advantage, we may be able to create certain geometries of traps that would permit more controlled loading of the optical traps with droplets of specific composition.

Dr McGloin opened the discussion of the Professor Mason's paper: In thinking about loading optical traps using acoustic traps, one of the problems I encountered was the difficulty in trapping small (<10 μm) diameter particles in such traps. Clearly your data demonstrates this is possible—so what are the technical challenges in trapping these small particles?

Professor Mason responded: The main problem is coagulation (as I mentioned in the text) both in the trap and while injecting them into the trap.

Professor Rühl asked: An ultrasonic trap appears to be ideal to trap ensembles of aerosol particles in a defined atmosphere. Can this approach be used to model in the laboratory multiple scattering of sunlight, so that reliable values of an albedo can be derived?

Professor Mason replied: I have not thought this through in detail but multiple scattering can certainly be observed and the main aim of the present work is to determine refractive indices and albedo of particles in the trap.

Dr Rudić said: Please can you say something about the sensitivity of the presented technique in general and compared to other techniques (for example cavity ring down spectroscopy) which equally rely on the Beer–Lambert law? Additionally what is the influence on measurements of the particles position with respect to the incident radiation (laser beam) intensity profile?

Professor Mason replied: At present this is a single pass experiment the incident light beam passing through the trap containing the material, so the sensitivity is the same as any laboratory single pass device. It might be possible to set this up as a multipass system which would improve the sensitivity.

Dr Reid asked: Can you comment on the spatial variability in gas phase temperature and composition within the trapping region? Can this be modelled or measured to help understand/interpret any dynamical measurements you perform?

Professor Mason replied: The temperature in the trap is lower than in the surrounding gas, this is why water vapour freezes out in the trap to form liquid/ice. This is discussed in Bauerecker and Neidhart.[1]

1 S. Bauerecker and B. Neidhart, Formation and growth of ice particles in stationary ultrasonic fields, *J. Chem. Phys. B*, 1998, **109**, 3709–3712.

Ms Kahan said: Regarding the chemical reactions you intend to monitor on trapped ice particles, do you know whether they will be sensitive to atmospherically relevant reagent partial pressures?

Professor Mason answered: These experiments have yet to be performed so I can not be definitive. The pressure in the trap is atmospheric (nitrogen) but we can levitate particles in an air mixture with partial pressures of trace gases. What we don't know is the sticking coefficients of each gas under these conditions and hence on particles in the trap.

Dr Parker said:
(1) How do the infrared spectra of the individual particles compare to the bulk? Do factors such as size, shape or whether single or polycrystalline modify the spectra to complicate identification?
(2) For individual inorganic particles, would infrared microscopy be a simpler method?

Professor Mason replied:
(1) One of the major aims of the research programme is to answer this question and determine if the spectra of ices on such particles is the same as that in a bulk ice sample. We do expect the size and shape to influence the spectra as shown by Signorell.
(2) The two methods may be complimentary.

Dr McGloin commented: Can you shape acoustic fields in a similar manner to optical beam shaping?

Professor Mason responded: Yes. This is the science of acoustics! By shaping reflecting plates you can change the size and shape of the nodes set up.

Professor Goree commented: Would it be possible to cause low-speed collisions of two dust particles in your ultrasonic trap by trapping them in two different nodes and using transducers with two different frequencies so that the nodes move?

Professor Mason replied: In principle this may be possible. We have observed particles in one node fall out into another but not in a way that we can at present control. It is probably due to instabilities in the sonic field such that the trap is not quite symmetrical. The idea of using more than one transducer to set up multiple traps (as in an optical lattice) is an interesting one but may be limited by the size of the transducers—which at present are quite large.

Dr Mitchem remarked: How cleanly can the different nodes in the ultrasonic trap be loaded?
Is it possible to load one of the nodes with one reagent, and another with a different compound and bring the nodes together so the compounds react?

Professor Mason answered: The traps are loaded by injection of the particles into the nodal region. So yes you can load each trap with different particles but transfer of material from one trap to the other is still problematic (as discussed with Professor Goree) it is easier to mix inside each trap.

Professor Chang communicated: In your poster (Tuckermann: Evaporation and crystallization of acoustically levitated salt/water drops) you give a diameter range for droplets levitated in acoustical traps of 0.05 to 5 mm. Did you also try to levitate smaller (sub-micron) droplets and do you expect specific difficulties for smaller droplets in addition to the problem of coagulation close to the resonant frequency of the transducer discussed by Mason et al.? What is your estimate of the errors in the relative humidity, temperature and pressure measurements compared to the actual relative humidity, temperature and pressure experienced by the levitated droplet?

Mr Tuckermann communicated in reply: There is a theoretical lower threshold for acoustic levitation given by

$$d \geq \sqrt{\frac{\eta}{\rho_\infty 2\pi f}}$$

with the particle diameter d, the dynamic viscosity of the gaseous environment η, its density ρ_∞ and the ultrasound frequency f used.[1] Using a 27-kHz levitator and assuming $\eta/\rho_\infty = 13.3 \times 10^{-6}$ m^2 s^{-1} the minimum diameter of suspendable particles is about 22 μm at 1 bar and 16 μm at 2 bar.

In my own studies I am using a 20-kHz and a 58-kHz acoustic levitator, whose lower theoretical thresholds are 26 μm and 15 μm, respectively. Nevertheless, I have never been successful in stable levitation of particles smaller than approx. 50 μm in diameter. In general, particle coagulation in acoustic fields (and not only in stationary ultrasonic fields used for acoustic levitation) depends on the frequency of the acoustic field and the viscosity of the gaseous environment:

$$f_{opt} = \frac{9\eta}{\rho_p 4\pi r_2 r_1}$$

with the particle radii r_1 and r_2, particle density ρ_p, the dynamic viscosity of the environment η and the optimum frequency f_{opt} of the acoustic field for particle coagulation.[2] Assuming $d/2 = r_1 = r_2$, $\eta_{air} = 16.1 \times 10^{-6}$ Pa s, $\rho_p = 103$ kg m^{-3} and $f_{opt} = 27$ kHz, particle diameter d results in 1.3 μm. Furthermore, as dynamic viscosity η is independent of the pressure of the environmental gas, the optimum frequency f_{opt} does not depend on it, too. But besides the particle collision rate there are two more aspects which have to be taken into account by investigating particle agglomeration in stationary ultrasonic fields: (i) For wet particles (droplets) it can be assumed that nearly for each particle collision agglomeration takes place, i.e. a larger secondary particle is formed. That is not the case for dry particles. On water–ice particles a quasi-liquid layer exist down to temperatures of about −25 °C at normal atmospheric pressure. (ii) The levitation forces in a stationary ultrasonic field always force larger particles/droplets ($d > 10$–20 μm) into the pressure nodes, that additionally increases the collision rate.

For example, I have been able to form macroscopic particles/droplets of 1–5 mm in diameter in the stationary ultrasonic field of a 20-kHz levitator from ice aerosol ($T \geq -25$ °C), water fog or cigarette smoke, whose primary particles have been in the range of a few micrometers and have been wet or covered by a quasi-liquid layer. Furthermore, as shown by a Spanish research group strong ultrasonic fields can be used for technical cleaning of industrial exhaust flumes by acoustic agglomeration of sub-micron particles to particles of a few micrometers.[3]

In conclusion, from my experiences I do not expect (i) the possibility of stable acoustic levitation of particles/droplets smaller than approx. 50 μm. (ii) For wet aerosols with particle diameters of 1-5 μm I expect strong secondary particle formation of macroscopic particles in stationary ultrasonic fields.

Regarding the second question, both, the temperature offset within the stationary ultrasonic field in respect to the gaseous environment as well as the temperature variation inside a stationary ultrasonic field between the pressure antinodes and the pressure nodes (whereas the temperature is lower within the pressure nodes), are in the range of $\Delta T \approx 1$–2 °C at sound pressure levels of about 160-170 dB normally used for acoustic levitation. The temperature variation within the pressure nodes is only about a tenth of °C.

I have never tried to measure RH variation inside a stationary ultrasonic field. But because temperature variation within the stationary ultrasonic field I also would expect a small variation of RH. (Note, in ambient air temperature variation of 1 °C can result in a change of RH of about 5 %!) But as conventional RH measurements have usually an error of 1–2%, it will be difficult to measure RH variation within a stationary ultrasonic field correctly. Sound pressure levels of approx. 160 dB result in pressure variation of about 1 kPa. (Note, assuming adiabatic expansion/compression of ambient air at 293 K and 1013 hPa the variation of pressure within stationary ultrasonic fields leads to a temperature variation of approx. 0.8 °C, which fits well to the estimation/measurements given above.)

1 L. D. Landau and E. M. Lifschitz, *Lehrbuch der Theoretischen Physik VI*, Akademie-Verlag, Berlin, 1966.
2 E. P. Mednikov, *Acoustic coagulation and precipitation of aerosols*, Consultants Bureau, New York, 1965.
3 J. A. Gallego, E. Riera, G. Rodríguez, T. L. Hoffmann, J. C. Galvez, L. Elvira, F. Vazquez, F. Montoya, J.J. Rodríguez, F. J. Goméz, M. Martin, A Pilot Scale Acoustic System for Fine Particle Removal from Coal Combustion Fumes, from the 1st World Congress on Ultrasonics 95, Berlin, 3–7 September 1995, *Proceedings of the World Congress on Ultrasonics 1995*, ed. J. Herbertz, IEEE, Berlin, 1995, 737–740.

Dr Pavlů communicated: The connection of an ultrasonic trap with a microparticle Van der Graff accelerator would cause trouble with system compatibility—while the particle accelerator probably needs vacuum conditions, the ultrasonic trap will not work under too low pressures. How would you propose to overcome this trouble (differential pumping?) and what is the lowest possible operational pressure for the trap?

Professor Mason communicated in reply: You are correct there has to be a large pressure drop along the vacuum system from the trap to the Van der Graff where the dust is produced. The dust is at high speed and can easily move through the gas in the ultrasonic trap. As for the lowest operating pressure we have not really investigated that at present, it will depend on your transducer power and you will need sufficient pressure to sustain a sonic field. I would say a limit is in the order of 1/10 ths of atmosphere but this is a guess.

Dr Hudson opened the discussion of the paper by Krieger: The paper by Dr Krieger and co-workers has highlighted the difficulty in monitoring the translational motion of an inclusion within a single aerosol droplet from light-scattering measurements. We are also interested in resolving the dynamics of mixed-phase droplets, and some preliminary data from our experiments will be described.

We are studying liquid droplets with radii in the range of 2 to 5 μm using optical tweezing of single particles in an aerosol. The motion of an inclusion formed from the presence of an organic phase within the aqueous solution of a host droplet is revealed by the addition of a fluorescent probe that partitions to the organic component in the aerosol. In these measurements, an optical design that achieves a sensitive discrimination between fluorescence and scattered light from the droplet is needed to identify the small signals generated by highly dilute concentrations of the fluorescent probe.

A schematic of the experimental apparatus is illustrated in Fig. 1(a); the optical tweezers (single-beam, gradient-force) are formed in an inverted microscope by passing a continuous-wave laser (532 nm) through the back aperture of an objective ($\times 100$, 1.25 N.A.). A small chamber is located on the microscope to provide a controlled environment for the stable trapping of liquid droplets. The aerosol is generated using a commercial medical nebuliser and directed into the chamber across the path of the laser. A small quantity of quantum dots (Molecular Probes Inc.) is added to organic liquids in the nebuliser to give solutions with a concentration of 1 nM. The nanoparticles have a diameter of approx. 20 nm with a hydrophobic coating and are used in preference to an organic dye because the rate of photobleaching is much slower. The nanoparticles are excited by the same laser wavelength used for the optical tweezers and the emission (a narrow band, centered at 605 nm) from trapped droplets is observed through the microscope objective. The fluorescence is isolated from scattered light by a dichroic mirror and bandpass filter (see Fig. 1(a)), and an image of the spatial profile of the intensity is measured using an electron-multiplied charge-coupled device (PhotonMax 512B, Princeton Instruments).

An initial test was performed using an aerosol of decane containing single-phase droplets, and the optical tweezers were optimized for trapping particles with an average radius of 2 µm. Fluorescence images of the liquid droplets were observed following the addition of quantum dots (1 nM) to the aerosol. If the evaporation of solvents in the apparatus is assumed to be insignificant following the generation of an aerosol, then the number of quantum dots present in a 2 µm-radius droplet can be estimated as ∼20 particles. This assumption is realistic as the sizes of both organic and aqueous droplets captured in the optical tweezers are observed to be stable for a significant time (5–10 min). The measured intensity of fluorescence from decane droplets indicates that detection of a single quantum dot in a liquid droplet would be possible with a small improvement in the isolation of the signal from background and scattered light.

The experimental technique used to form a decane inclusion in the aqueous solution of a host droplet is outlined in Fig. 1(a). The optical tweezers are now optimized for trapping aqueous droplets with radii in the range of 3 to 5 µm. The host droplet (containing 20 g mol^{-1} of NaCl salt) is initially captured by the optical tweezers and then it is bombarded with an aerosol containing the decane liquid (with 1 nM quantum dots). The quantum dots are confined to the organic phase in the droplet and an example of the fluorescence image obtained from a mixed-phase droplet is shown in Fig. 1(b). In this case, an inclusion with a radius of approx. 1 µm

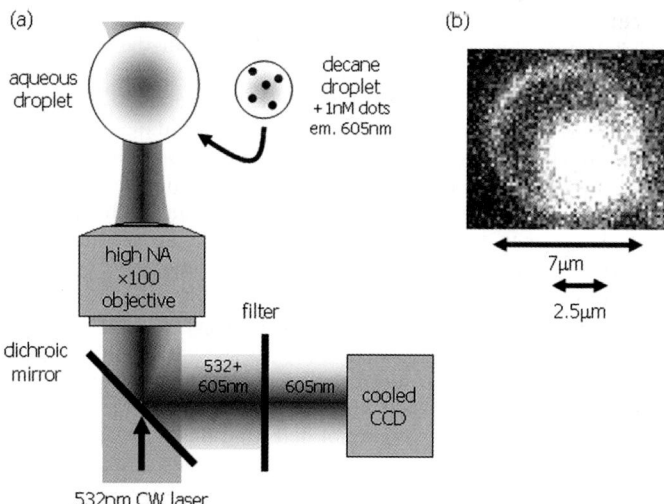

Fig. 1 (a) Apparatus for the formation and observation of an organic inclusion in an aqueous droplet (b) fluorescence image showing the organic phase illuminated by quantum dots. The concentration of the fluorescent probe was 1 nM implying that less than 10 particles are present in the volume of liquid.

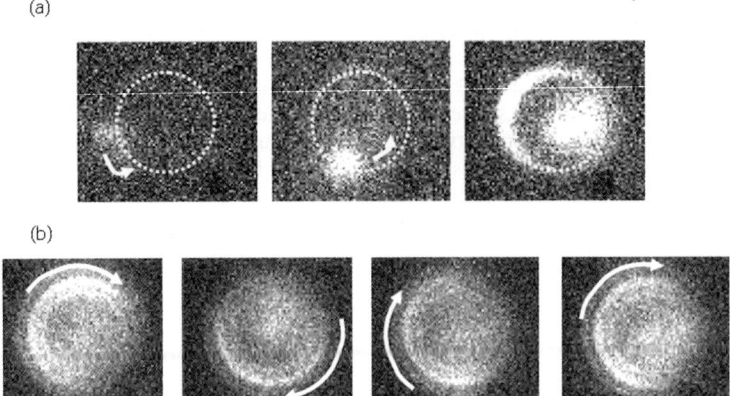

Fig. 2 Sequences of frames showing the location of the organic component in different mixed-phase droplets: (a) the coagulation of a decane droplet with an aqueous droplet and formation of a large inclusion, (b) the translational motion of a very small inclusion observed through the presence of a strong reflection at a position on the circumference of the host droplet.

was formed in a host droplet with a radius of approx. 3.5 µm. The size of the inclusion implies that the fluorescence image was generated from the presence of < 10 nanoparticles inside the liquid volume.

Sequences of frames obtained to show the movement of decane inclusions inside host droplets are shown in Fig. 2. In these examples, an integration time of 10 ms and a video capture rate of 20 fps was used. In Fig 2(a), the collision between a decane droplet in an aerosol and the aqueous particle held in the optical trap is shown in the 1st frame, followed by uptake of the organic liquid on the water surface in the 2nd frame and finally the formation of an inclusion inside the host droplet in the 3rd frame. In Fig. 2(b), it is believed that a particularly small inclusion of the organic phase is moving inside the host droplet and, in each separate frame, the location of the inclusion lies across from the region on the circumference of the host droplet where a strong reflection of light is detected. Although the images were obtained using an acquisition rate of 20 fps (corresponding to a 50 ms interval between each frame), the intensity of the fluorescence in the recorded images is sufficient to suggest that a much better time resolution for the motion of the inclusion would be possible with the same experimental setup. The measurements shown in Fig. 2 are the result of a preliminary study to validate the apparatus, and a more detailed investigation is under development. The new studies will use a shorter integration time and faster acquisition rate for data collection. In addition, the apparatus will be modified to use a laser for optical tweezing with a longer operating wavelength. This laser is not capable of exciting the quantum dots and, instead, a suitable wavelength of light will be supplied (on demand) by a second laser attenuated to a much reduced power density. This will enable the measurement of fluorescence from the liquid droplet for much longer periods of time without photobleaching of the probe.

Our research has a number of objectives beyond the measurement of the translational motion of an inclusion in a mixed-phase droplet. The results shown above have demonstrated the possibility of detecting the fluorescence from a single molecule in the liquid volume. This is the first demonstration of a continuous measurement, with this level of sensitivity, from an individual droplet in the size range of 10 to 100 femtolitres (note that single-molecule emission has already been recorded in streams of liquid droplets using photon-counting methods; for example see ref. 1). We intend to exploit the novelty of this approach to develop a new experimental tool for single-molecule fluorescence spectroscopy.

I am very grateful to Ms Kerry Knox and Dr Jonathan Reid for support in the experimental measurements, and John McCaffrey from Princeton Instruments for the loan of an electron multiplied CCD.

1 M. D. Barnes, N. Lermer, C.-Y. Kung, W. B. Whitten, and J. M. Ramsey, *Opt. Lett.*, 1997, **22**, 1265.

Dr Krieger replied: These are very interesting experiments. We did some experiments with these types of quantum dots as well, using photon correlation spectroscopy at the emission wavelength of the quantum dots. Our thought was that because of their very small size (15–20 nm in diameter) the characteristic timescale of the intensity autocorrelation function of their translational Brownian motion within the host droplet becomes shorter than the characteristic timescale of the intensity autocorrelation function caused by the rotational Brownian motion of the host droplet. But the characteristic timescale of the intensity autocorrelation function of the rotational Brownian motion scales only linear with radius, if the scattering pattern of the host particle is taken into account (see eqn (6) of our paper). Taking into account also the different viscosities of air and liquid, this leads us to the conclusion that the radius of the inclusion has to be smaller than about 10^{-4} times the radius of the host droplet to prohibit the masking of the Brownian motion of the inclusion by the rotational Brownian motion of the host droplet. A direct imaging approach such as yours helps, because the characteristic time of the rotational Brownian motion scales with the third power of the radius since the phase function does not have to be considered when imaging the particle. However, as we show in Fig. 8 of our paper, one still has to consider the rotational Brownian motion at timescales larger than typically 1/100 th of a second for these sizes of host droplets.

Therefore, I believe that what you observe in Fig. 2(b) is actually the rotational motion of the host droplet and not the translational movement of the inclusion within the host droplet.

Professor Arnold addressed Dr Hudson and Dr Krieger: How can you infer the position of the particle from the images you show? The spherical enclosure is expected to distort the measured position from the actual position?

Dr Krieger answered: We absolutely agree: the spherical enclosure will distort the measured position. However, the resolution of our glare spot images shown in Fig. 2 of our paper is in any case too low to track the position of the inclusion. They are meant only to illustrate that some frames show the glare spot image of an almost undisturbed sphere, while others clearly indicate the presence of an inclusion and sometimes even distort the internal optical field in a way that you would believe naively that two inclusions are present within the host droplet (see Fig. 2f).

So, instead of trying to calculate the positional autocorrelation function from tracking the inclusion, we simply calculated the intensity autocorrelation related to the inclusions movement. This was done as described in section 3 of our paper: A PC and a frame grabber were used to digitize the glare spot image (15 Hz sample rate), compute the integrated intensity of both glare spots and the integrated intensity of the region between the glare spots. Autocorrelation data (and cross correlation data) were computed offline from these time series of intensities. Fig. 8 of the paper shows for example the temporal autocorrelation of the intensity in the region between the glare spots. However, it is evident from the non-existing trend with temperature that here also the rotational movement of the host droplet masks any translational movement of the inclusion! We agree with you that even if there had not been such masking a quantitative interpretation of these autocorrelation functions would have been a very challenging task because of the complicated internal field patterns of the mixed phase particle.

Dr Hudson answered: There can be little doubt about the location of the organic component within the host droplets shown previously in Fig. 1b and Fig. 2a. Light scattering from different surfaces is certainly responsible for the absence of precise definition for the interface between the organic and aqueous phases. Nevertheless, the data demonstrates that an accurate characterisation of mixed-phase droplets (with a significant proportion of each component) can be made.

Admittedly, the images represent a projection of the (spherical) host droplet on a 2D plane and it might seem difficult to estimate the depth of the organic inclusion in the host droplet. However, our apparatus has the capability of recording two simultaneous images of the droplet from parallel and perpendicular orientations with respect to the optical tweezers. This was not used for the purpose of the pilot experiments described above but, in the future, it will be done to characterise mixed-phase droplets in greater detail.

Professor Arnold's comment has more relevance to the example in Fig. 2b. In this case, the position of a small organic component is challenging to locate. However, the image will have a unique intensity distribution for every possible location of the inclusion and we are developing a model (including the scattering and reflection of light from the host droplet) to predict the position of an inclusion from the measured fluorescence image of the host droplet.

Professor Chang said: Is there a possibility of introducing a fluorescent dye in the form of a localized capsule that can be broken down on demand and thereby act as a marker for internal circulations which have been shown to exist in moving droplets by Michael Winters at UTRC?

Dr Krieger responded: Let me first note that the droplets in our experiments are typically only a few micrometers in radius and suspended in a very gentle flow (about 20 cm^3 min^{-1}), which means that you do not expect an internal circulation to occur. However, for droplets in the 100 μm range using a fluorescent marker to visualize the internal flow pattern may be feasible. We have used quantum dots (their size is approximately 15–20 nm in diameter) which can be chemically targeted to blend in only in an organic phase to investigate mixed phase organic/aqueous particles with respect to their internal morphology. Since targeted quantum dots are now commercially available they are attractive alternatives to the more complicated scheme of releasing a fluorescent dye by braking a capsule on demand.

Mr Taylor said: You introduce quantum dots into your organic droplet to better visualize the position of the organic droplet within the host droplet. Dr Krieger observed that in their case they believed the motion of the droplets to be altered by the force from the field of the laser (specifically by the standing wave set up within the host droplet).

The intention of the quantum dots was to improve the scattering characteristics of the organic droplet: it seems that this should also significantly alter the force exerted by the probe laser on the organic droplet. The droplet now contains strongly-scattering inclusions which will alter its own scattering characteristics. Will the altered force have a significant effect on the dynamics of the system that you are trying to study, or is the force negligible in your case?

Dr Hudson answered: The motion of an inclusion inside a liquid droplet will very likely be influenced by the laser tweezers as suggested by Mr Taylor. The optical force exerted on the inclusion will derive from the gradient field of the laser inside the host droplet (cavity modes or standing waves are unlikely to play a major role).

Dr Krieger remarked that an inclusion might be trapped close to an internal glare spot within the host droplet. However, this effect was not seen in their experiments because the rotational Brownian motion of the host droplet masks any such effect because it leads to a rotational movement of the inclusion relative to the internal field of the host droplet.

In our experiments, the influence of the gradient field of the laser on the location of an inclusion has not been investigated in detail. The internal field in the host droplet will have been modified significantly by the presence of an inclusion inside the aqueous solution but it is unlikely that it will be affected by the quantum dots dissolved in the organic phase. The concentration of quantum dots used in the experiment was 1 nM, giving rise to ~ 10 particles inside the organic phase of the droplet (note that the equipment is also capable of single-particle sensitivities).

Professor Ray remarked: Is it possible to probe the motion of a small inclusion particle (say 1/100 th of the host particle) from inelastic scattering signal by illuminating the host particle through a resonating wavelength light? Since we can predict the internal field, we should, in principle, be able to calculate the position of the inclusion particle from the relative fluorescence intensity!

Dr Krieger responded: This is a very interesting proposition. I do not see any reason why this could not be done in principle. There will be a certain ambiguity in even an ideal experiment without noise, because there will be locations within the host droplet with equal intensity. In practice such an experiment may be quite difficult because it involves two different wavelengths, namely the excitation wavelength and the fluorescence wavelength to be considered. There has been some work in this direction, *e.g.* U. Waggon *et al.*[1] It is not clear to me whether it is experimentally feasible with a levitated droplet in air.

1 U. Woggon, R. Wannemacher, M. V. Artemyev, B. Möller, N. LeThomas, V. Anikeyev and O. Schöps, *Appl. Phys. B*, 2003, **77**, 469–484.

Professor Ray asked: How is the Brownian motion affected by the confinement of the host droplet?

Dr Krieger replied: That is one of the questions which motivated the research of the paper in the first place. Clearly, there will be an influence of the confinement; a spherical droplet could be a perfect model system to study this influence in a well defined geometry. The problem of Brownian motion in confinement remains the subject of discussion, *e.g.* Benesch *et al.*,[1] Lancon *et al.*[2]

1 T. Benesch, S. Yiacoumi, C. Tsouris, *Phys. Rev. E*, 2003, **68**, 021401.
2 P. Lancon, G. Batrouni, L. Lobry, N. Ostrowsky, *Europhys. Lett.*, 2001, **54**, 28.

Professor Rühl remarked: The experimental approach appears to be suitable to derive vapour pressures of semi-volatile compounds. Can this be extended to derive vapour pressures of low volatile species?

Dr Krieger answered: The lowest vapour pressure we can measure may be roughly estimated by looking at the data shown in Fig. 6 of our paper. The scatter of the square of the radius derived from the autocorrelation data is about 10 μm^2 out of 200 μm^2. Thus, it may be possible to measure a rate of 1 μm^2 per 100 000 s with sufficient accuracy which would correspond to a vapour pressure of 2.5×10^{-5} Pa.

Professor Signorell opened the discussion of Professor Rühl's by communicating:
With reference to Fig. 1: The X-ray scattering method seems to me much more complicated and expensive compared with the standard DMA/CPC method. I do not see any advantage of this scattering method for the size characterization of *non-volatile* aerosols (such as silica particles). A really interesting application of the X-ray

scattering method could be the determination of size distributions of *volatile* aerosols. This is not possible with DMA/CPC. Can the scattering method be applied to volatile substances or are there problems with fragmentation, ionization *etc.*?

Professor Rühl communicated in reply: Certainly, the experimental approach is not only applicable to measure a particle size distribution. This would be by far too elaborate and there are indeed other ways to measure the particle size reliably. However, Fig. 1 shows several important features: We can probe the particles that arrive at the scattering region (full curves in Fig. 1). There are distinct differences compared to the size distribution measured by the condensation nuclei counter (dotted curves in Fig. 1). It is evident that the difference between both detection schemes is due to the transmission of the aerodynamic lens system, which evidently prevents aggregates of particles to penetrate into the scattering region. As a result, Fig. 1 serves to characterize the sample in the interaction region. Otherwise it can be easily objected that we do not know the size distribution of the target under investigation.

The aim of the present approach is not just to characterize the particle size, rather than to study elastic light scattering properties of free nanoscopic particles. This yields information on optical processes occurring in the soft X-ray regime. There are several advantages compared to other approaches. This includes: (i) element-selective detection of nanoscopic matter and; (ii) high surface sensitivity. The present results are just a first attempt to use tunable soft X-rays for probing properties of free nanoparticles in a dense, continuous beam. It is already evident that this approach will turn out to be quite useful for solving numerous scientific problems, where also different detections schemes, such as electron, ions, fluorescence, *etc.* can be used. We have already started with this work, where the issue of volatile/non-volatile particles can be addressed in future work.

Professor Signorell communicated: With reference to Fig. 3: You explain the deviation of the first strong Mie minimum by strong reflectivity attenuation of the incoming beam from a rough surface of SiO_2. Why do you exclude multiple scattering from the particle ensemble in your explanation? Measurements on coated particles (modified surface) could be used to test your explanation.

Professor Rühl communicated in reply: The density of particles in the beam is too small for multiple scattering among different particles to play a significant role. We have outlined this issue in a related publication, where a similarly designed nanoparticle beam was used (*cf.* Shu *et al.*[1]). It is certainly possible to verify this conclusion by using coated particles of different reflection properties. However, the estimated particle density in the scattering region is so low that it will be difficult to prove the significant importance of multiple scattering processes in this way. Further, the particle density cannot be significantly enhanced in the present setup. Moreover, we have performed additional sensitivity studies, which indicate that the surface roughness is indeed a crucial parameter to explain the position of the first strong Mie minimum.

1 J. N. Shu, K. R. Wilson, M. Ahmed, S. R. Leone, C. Graf, and E. Rühl, *J. Chem. Phys.*, 2006, **124**, 034707.

Dr Corbett communicated: What does the author think is a reasonable lower size limit on this technique?

Professor Rühl communicated in reply: Currently, this approach is limited to *ca.* 40 nm particles. This is a result of the aerodynamic lens that is used. Its transmission is the crucial factor besides the particle density of the beam. However, it appears to be quite possible to extend the size range to even smaller particles in the size regime well below 20 nm. This size regime has been explored by other groups more recently (cf. e.g. Wang and McMurry[1]). A combination of such lenses along with experiments using synchrotron radiation remains to be done. A possible aim of such work can be to investigate properties of free quantum dots in the gas phase.

1 X. Wang and P. H. McMurry, *Int. J. Mass Spectrom.*, 2006, **258**, 30.

Dr Corbett communicated: Have you investigated the ensemble problem rather than single or trains of particles?

Professor Rühl communicated in reply: We have performed in the past the following complimentary experiments on nanoparticles in the gas phase using vacuum-UV-radiation and soft X-rays:

(i) Particle beam experiments, as presented in this work, and related studies in collaboration (see for example: Shu et al.[1]

(ii) Studies on single particles using an electrodynamic trap: see Grimm et al.[2]

The particle density in the present nanoparticle beam approach is too low to experience any ensemble problems, such as multiple scattering processes between particles. This issue is outlined in greater detail in ref. 1.

1 J. Shu, K. R. Wilson, M. Ahmed, S. R. Leone, C. Graf and E. Rühl, *J. Chem. Phys.*, 2006, **124**, 034707.
2 M. Grimm, B. Langer, S. Schlemmer, T. Lischke, U. Becker, W. Widdra, D. Gerlich R. Flesch and E. Rühl, *Phys. Rev. Lett.*, 2006, **96**, 066801.

Dr Reid said: The model of the scattering presented in Fig. 4 is dependant on surface roughness and on a radial density gradient layer. How unique are the fits of the model to the experimental data? How sensitive are the model predictions to these two variables?

Professor Rühl replied: As mentioned in our paper the presented model of a rough and graded surface layer is the only physically meaningful way to assign the present results. Alternatively, we have tried the model coated spheres and strong changes in refractive index of the entire sphere. This gave results that appeared to be unreasonable. At the present stage, our model does not allow us to distinguish between the surface roughness and a graded surface layer. Further calculations and model improvements are on the way to resolve this question.

Dr Reid commented: How accurately can the surface roughness be assessed?

Professor Rühl answered: As pointed out in the paper the surface roughness that is included in our model is of the order of 2 nm. This is in accordance with results from electron microscopy. The accuracy of this value is estimated to be in the same order of magnitude.

Professor Mason remarked: I recall work in France where they looked at X-ray spectra of dust (carbon) in flames. How does this relate to your work?

Professor Rühl answered: We are aware of recent work of J. B. A. Mitchell and co-workers from Rennes (France). They published work *e.g.* on small angle scattering of nanoparticles that are formed in flames (see *e.g.* Mitchell et al.[1]), indicating that the size of nanoparticles can also be derived from this approach. This is complimentary to our work, where they used hard X-rays for the experiments. In contrast, the present work makes use of tunable soft X-rays, where we derive more specific information on surface properties of the particles. The Mitchell group also published a few years ago results that are related to X-ray absorption from nanoparticles (cf. Mitchell et al.[2]) as well as results from neutron scattering.

1   J. B. A. Mitchell, J. Courbe, A. I. Florescu Mitchell, S. di Stasio, T. Weiss, *J. Appl. Phys.*, 2006, **100**, 124918.
2   J. B. A. Mitchell, J. L. Le Garrec, D. Travers, B. R. Rowe, R. J. Randler, A. Plech, M. Wulff, *Nucl. Instrum. Methods, Sect. B*, 2003, **207**, 227.

Dr Huthwelker remarked: Would it be possible to study aerosols under non-vacuum conditions? Here I'm thinking about examples of studies on salt particles during and after their deliquescence. What would be the maximum gas phase pressure in the region where the X-ray beam hits the target?

Professor Rühl replied: In general, it may even be possible to work at ambient pressure. In the soft X-ray regime this requires the use of a short path length of radiation along with efficient differential pumping, which is used to maintain the ultra high vacuum of the beamline and the soft X-ray monochromator. Preferentially, a carrier gas of low absorption cross section should be used. Most suitable would be helium for such experiments. Alternatively, thin windows can be used, which seal off the beam line from the experiment. Another option is to use hard X-rays that penetrate easily samples at ambient pressure.

The spectroscopy and dynamics of microparticles

Paul Davidovits

Received 18th July 2007, Accepted 18th July 2007
First published as an Advance Article on the web 14th August 2007
DOI: 10.1039/b711018a

The 137th Faraday Discussion covered a wide range of subjects divided into the four categories of Spectroscopic Techniques, Dusty Plasmas and X-Ray Characterization, Atmospheric Aerosols, and Particle Manipulation. These divisions organized the thinking into specific areas of research and allowed one to see interconnections between the two central foci of physical chemistry; techniques and applications. Physical chemists excel at developing and mastering a wide range of new techniques and applying them to a variety of tasks as the need arises. At times specific tasks present themselves and in response new techniques are developed. The presentations provided examples of both such interplays. In these remarks the presentations are summarized, common features are highlighted, and possible directions for future research are suggested.

Firstly, I would like to thank Prof. Richard K. Chang for his lucid and very useful Introduction summarizing past, present and future applications of spectroscopy applied to the study of microparticles. His presentation highlighted the great progress that has been made in this field during the past twenty years and the exciting work that remains yet to be done.

The year 2007 marks the 100th anniversary of *Faraday Discuss.* and I think, our Discussion continues in the spirit of Michael Faraday. Faraday's scientific interests ranged widely. He moved undaunted in the world of science from chemistry to electrochemistry, to electromagnetism, to optics and to technology. Of course, science has changed since Faraday's time. No individual can now span all of physical chemistry. But our Discussion did encompass the full range of what is now at the center of physical chemistry and here I also include the closely related fields included in *Faraday Discuss.* that is, chemical physics and biophysical chemistry.

The subjects we discussed ranged from the extraterrestrial to the down-to-earth. I mean this literally. In the category of the extraterrestrial we learned from Ruth Signorell about aerosols related to the atmosphere of Io and Europa, satellites of Jupiter and about simulated interplanetary charged dust particles studied in the experiments of Jiří Pavlů. While in the down-to-earth category, Tom Choularton described field studies sampling aerosol particles in clouds that contacted ground at Jungfraujoch in the Alps.

Division of the presentations into the four categories of Spectroscopic Techniques, Dusty Plasmas and X-ray Characterization, Atmospheric Aerosols, and Particle Manipulation and Characterization was very useful. These divisions organized our thinking into specific areas of research and allowed us to see interconnections

---

Chemistry Department, Boston College, Chestnut Hill, MA 02467, USA

between the two central foci of physical chemistry; techniques and applications (or tasks). Physical chemists excel at developing and mastering a wide range of new techniques and applying them to a variety of tasks as the need arises. At times specific tasks present themselves and in response new techniques are developed. Our presentations provided examples of both such interplays.

One common feature of all the presentations is contained in the overall title of our Discussion. That is "Microparticles". Although the particle sizes we encountered in the presentations range over a factor of 10^5, from about 5 nm to a few hundred μm, in the context of the specific studies the particles are small. That is, their surface to volume ratio is large. In some of the studies this attribute is used to levitate and manipulate the particles. In other studies such as the whispering gallery mode experiments, a certain particle size-range is necessary to manifest the specific phenomena. Finally, in several studies particle size is mandated by nature, as is the case for atmospheric aerosols, interstellar particles and undesirable contaminants in plasmas.

I will now survey the Discussion from the perspective of the categories listing the presentations. I will certainly not be able to do justice to any of the specific works. For this and also for any misrepresentations I apologize in advance.

Table 1 lists the Spectroscopic Techniques presentations. The table lists the presenting author, the technique and the application. Included in Table 1 are three additional spectroscopic techniques used in projects discussed under the other categories.

We display here seven very different spectroscopic techniques each suited to a specific task. This is an example of the ability of physical chemists to use their large repertory of techniques to shed light on a variety of problems. Signorell *et al.* showed that infrared spectroscopy applied to aerosols in the size range ∼50 nm, can differentiate between a homogeneous composition and a core shell structure. The experiments of Ray *et al.* used Raman scattering from larger particles to reveal similar morphological distinctions. These techniques may become useful in the study of atmospheric aerosols where the surface composition often determines chemical reactivity and the potential of the particle to promote cloud condensation.

Wolff *et al.* and Arnold *et al.* applied spectroscopic techniques to detecting bacterial and viral material, respectively. Biosensing is a relatively new application of spectroscopy that is likely to expand in the near future. The field will be driven primarily by the need to provide techniques for prevention of bio-terrorism. Biosensing will also likely provide new methods of medical diagnosis.

In Table 2 we list presentations in the Dusty Plasmas and X-ray Characterization category. The work of Stoffels *et al.* and Kersten *et al.* are aimed at monitoring of particles in plasmas and observing plasma properties of importance in

Table 1 Presentations in the category of spectroscopic techniques

Authors	Spectroscopic technique	Application
Wolf *et al.*	Femtosecond pump–probe spectroscopy	Detection of biological microparticles
Signorell *et al.*	Infrared spectroscopy	Characterization of aerosol internal composition
Arnold *et al.*	Whispering gallery mode spectroscopy	Viral biosensing
Ray *et al.*	Morphology-dependent Raman scattering from suspended particles	Characterization of aerosol internal composition
	Microwave spectroscopy	
	X-Ray spectroscopy	
	Cavity ring down spectroscopy	

Table 2 Presentations in the category of dusty plasmas and X-ray characterization

Authors	Experimental technique	Application
Stoffels et al.	Microwave spectroscopy	Charging kinetics nanoparticles in plasmas
Shiratani et al.	Nanoparticle transport via AM rf discharge	Producing nanocomposite materials
Pavlů et al.	Monitoring oscillation frequency of trapped charged particles	Measuring dynamics of interstellar-like dust particles
Kersten et al.	Electrostatic probes and whispering gallery mode spectroscopy	Plasma diagnostic
Rühl et al.	Elastic scattering of X-rays	Measuring size of nanoparticles

manufacturing processes such as plasma etching and deposition. The studies of Shiratani et al. might well develop into a useful method for producing nano-materials. The work of Ruhl et al. is an elegant X-ray spectroscopic technique for measuring the size of nanoparticles.

Table 3 lists the presentations in the Atmospheric Aerosols category. The effect of atmospheric aerosols on climate may be as large as that of greenhouse gases but much more uncertain. (See for example *IPCC WGI Assessment Report*, 2007.[1]) The high uncertainties are due to the complex composition of the aerosol and the currently inadequate representation of the aerosol interactions with climate.

Aerosols affect climate through direct and indirect interactions. Through the indirect effect, hydrophilic aerosols may serve as cloud condensation nuclei (CCN) affecting cloud cover and hence the radiation balance. Through direct interactions most aerosols scatter light and produce cooling but aerosols containing black carbon (BC) or perhaps other substances absorb incoming light heating the atmosphere. It has been estimated that the direct radiative effect of BC is the second-most important contributor to global warming after absorption by CO_2. The estimates however are highly uncertain.

An important goal of atmospheric aerosol studies is to reduce the uncertainties in the data used to evaluate the climate forcing effects of aerosol. The measurements of Chan et al., Treuel et al. and to some extent also the field experiments of Choularton

Table 3 Presentations in the category of atmospheric aerosols

Authors	Experimental technique	Application
Choularton et al.	Field measurement of aerosol composition using an AMS	Determine CCN activity of transported, aged aerosols
Grothe et al.	X-Ray diffraction and ESEM	Measure kinetics of NAD formation
Jaenicke	Calculation of aerosol residence time as a function of size	Determine time resolution required for accurate aerosol measurements
Chan et al.	Raman spectroscopy of electrodynamically balanced single particles	Understanding hygroscopic growth of inorganic and organic-coated particles
Treuel et al.	Microscopy and Mie scattering of electro-dynamically balanced single particles	Studies of deliquescence of mixed composition droplets
Rudich et al.	Index of refraction measurements with cavity ring-down aerosol spectrometer	Determine heating of the atmosphere due to light absorption of humic-like substances in aerosols

CCN – Cloud condensation nuclei; ESEM – Environmental scanning electron microscopy; AMS – Aerosol mass spectrometer; NAD – Nitric acid dihydrate.

Table 4 Presentations in the category particle manipulation and characterization

Authors	Experimental technique	Application
Barlett et al.	Driven, optically trapped, charged microparticles in a liquid	Study Brownian fluctuations and optical anisotropy
McGloin et al.	Optical trapping and guiding airborne particles, monitored with holography	Study single particle phenomena e.g. coagulation
Mason et al.	Ultrasound trapping of particles in air	Study physical, optical, and chemical properties of single particles
Krieger et al.	Light scattering from electro-dynamically suspended particles	Sizing non-spherical particles
King et al.	Optical trapping of particles and Raman spectroscopy	Studying oxidation of atmospherically relevant organic particles
Reid et al.	Monitor Raman scattering from simultaneous optically trapped particles of different composition.	Accurate measurement of difference in properties due to composition

et al. are directed toward studying the hygroscopic properties of aerosol particles. Experiments measuring hygroscopicity of aerosols relate to both direct and indirect interactions. An aerosol that is hygroscopic will grow in size when it takes up water and therefore its light scattering efficiency will increase. Likewise a hygroscopic aerosol will display a higher CCN activity.

Choularton et al. used an Aerosol Mass Spectrometer built by Aerodyne Research Inc., to perform their field measurements. This instrument is an example of a technique developed by physical chemists in response to a task that needed to be performed. As was pointed out in the presentation of Ruprecht Jaenicke, ensemble properties of aerosol particles at a given location may change on a relatively short time scale. Aerosol mass spectrometers were developed to measure aerosol size and composition in real time. The newest versions of this instrument provide data on a particle-by-particle basis.

The cavity ring down measurements of Rudich et al. address direct effects of aerosols. They measured the light absorption by humic-like substances in the visible and ultraviolet region of the spectrum and concluded that the effect may be significant. The studies of Groethe et al. relate to the formation of polar stratospheric clouds that remain an important component of atmospheric chemistry.

Presentations in the Particle Manipulation and Characterization category are summarized in Table 4. The growing awareness of the importance of small particles in the environment and in various technological applications has motivated the development of techniques for the study of single isolated particles. Three techniques are commonly used to trap and suspend particles for experimental studies; optical, acoustical and electrodynamic. Table 4 shows applications of all three techniques.

These techniques can be used to study a variety of phenomena and new applications continue to be developed. However, as pointed out by several of the authors, the techniques are still in the early stages of development. Barlett et al., Mason et al. and Krieger et al. applied trapping techniques to study basic properties of microparticles. The work of McGloin et al. is at this point mainly concerned with improving our understanding of particle trapping phenomena. King et al. and Reid et al. are moving their trapping experiments toward atmospheric applications. These presentations demonstrate the remarkable potential of the techniques to study a wide range of important single particle phenomena.

In the "Concluding remarks" speakers often highlight possible future directions for the field. We discussed such a wide range of topics touching on so many fields that it is not possible here to consider all the exciting possibilities that the presentations evoke. Nor is this necessary. The studies presented are so vibrant that in most cases the direction of future work is already outlined in the presentations themselves. Consider two examples from the first and fourth category of presentations. Jean-Pierre Wolff showed that using pump–pump excitation techniques it is possible to distinguish between bacteria and organic urban aerosol. He also outlined a possible solution for the next challenge that is, distinguishing one species of bacteria from another.

Jonathan Reid in the fourth category of presentations showed that with optical tweezers two droplets of different composition can be positioned close to each other and using Raman spectroscopy their behavior can be studied. At present this technique is confined to studying thermodynamic properties. Reid suggested a future application; the extension of the method to study interfacial kinetics. This possibility touches on some of my research. Experimental and modeling studies show that surfaces of some aqueous solutions are enriched in the solute species. The kinetics of the species at the interface is often quite different from the kinetics of the species in the gas or liquid phases. (For a review see Davidovits et al.[2]) There are compelling reasons for expecting surface species to be more reactive than the corresponding gas phase or fully solvated molecules, with important ramifications in a number of fields. While surface reactions have been observed for several species, the measurements are not yet accurate enough to yield a clear picture of the process. By comparing the reactivity of two droplets of same size but different composition or two droplets of the same composition but different size, the technique described by Reid may provide the accurate and sensitive tool needed to elucidate these surface processes.

The presentations we heard these past three days point to many other important and interesting studies to be performed. I think each of us could list several favourites. I will suggest one more possible subject for future research in an area that has not been discussed at this meeting.

On December 29th 1959 at the annual meeting of the American Physical Society at Caltech Richard Feynman presented a talk titled "There's Plenty of Room at the Bottom". In the now classic often-quoted presentation he said:[3]

"It is a staggeringly small world that is below. In the year 2000, when they look back at this age, they will wonder why it was not until the year 1960 that anybody began seriously to move in this direction."

In reality, serious attention was not paid to this world of the very small until perhaps the mid 1980s. However, at this point the nano-world is certainly at the center of research attention. Last year the US federal government funded nano-technology research at the one billion dollar level. Investment by other countries in nano-research is likewise high. Increasingly nanotechnology is also at the center of public attention. The July 2007 issue of Consumer Reports[4] featured an article titled "Nanotechnology—Untold promise, unknown risk." The article describes several upcoming applications of nanotechnology and points out associated risks. Cosmetics is one area where nanoparticles are already widely used. A number of sunscreens contain nanoparticles of zinc oxide or titanium dioxide. These particles that absorb ultraviolet radiation are white on the micron size scale. They become transparent and therefore cosmetically more appealing when the particle size is reduced to a few hundred nanometers. In another application, some manufacturers have added C_{60} fullerenes to "anti-aging" creams because these particles can act as antioxidants.

As is pointed out in the article, normally inert materials can become toxic and damaging when they are nano-sized. A recent IRSST report[5] reviews the current knowledge about the health effects of nanoparticles. Some of the effects of nanoparticles on tissue hinted at by the available experimental evidence are alarmingly dangerous.

The use of nanoparticles in clothing, in appliances, in building materials and in medical applications of various types is expected to increase rapidly within the next decade. It is important that we understand how such particles interact with the environment. The first step in any such interactions is the penetration of the particles into the environmental systems. For example, the health risks posed by the cosmetic use of nanoparticles depend in part on how easily such particles penetrate through the skin into the tissue. At this point very little is known about these transport processes. Understanding such transport processes is important not only in the evaluation of the health and environmental effects, but also for the applications themselves such as for example targeted delivery of drugs.

The treatment of such transport processes is in the experimental and modeling domain of physical chemistry. I suggest that in the near future, modeling and experimental kinetics studies will begin to elucidate the mechanisms of nanoparticle transport through biological barriers such as for example across the blood–brain barrier, across the air-blood barrier in the lungs, across skin (including damaged skin) and across plant tissue.

Acknowledgements

I want to thank Drs Jonathan Reid, Paul Davies, Christian George, John Goree, Paul Kaye and David McGloin for organizing this very interesting and important Faraday Discussion and for inviting me to present the Concluding Remarks. I also thank Ms. Amanda Middleton for administering this meeting and resolving so promptly and efficiently a host of issues I raised.

References

1. IPCC (Intergovernmental Panel on Climate Change) WGI Fourth Assessment Report 2007.
2. P. Davidovits, C. E. Kolb, L. R. Williams, J. T. Jayne and D. R. Worsnop, *Chem. Rev.*, 2006, **106**, 1323.
3. R. P. Feynman, December 29th 1959 Annual meeting of the American Physical Society at the California Institute of Technology.
4. Consumer Reports, July 2007, p. 40.
5. IRSST Report on Health Effects of Nanoparticles, August 2006, On the web at http://www.irsst.qc.ca.

Poster titles

Array formation in evanescent wave optical traps, **L. Y. Wong, C. D. Mellor, T. A. Fennerty, M. R. Cargill and C. D. Bain**, *University of Durham, UK*

The crystallisation of aqueous droplets relevant to the Earth's atmosphere, **Benjamin J. Murray and Allan K. Bertram**, *University of Leeds, UK*

Uranium oxide particulate contamination surrounding a former processing facility, **Nicholas S. Lloyd, Randall R. Parrish, Tim S. Brewer, Simon R. Chenery and Sarah V. Hainsworth**, *British Geological Survey, UK*

Morphology and chemical composition of micro atmospheric particles (PM_{10}) in Glasgow, Scotland, UK, by scanning electron microscopy (SEM) and energy dispersive X-ray fluorescence (EDX), **Mark Gibson, Iain Beverland, David Bache, Mike Jackson, Andrew Hursthouse, Margaret Corrigan, Louise Rowley, Colin Clark, Araceli Sanchez Jimenez, David Fleming and Judy Guernsey**, *Dalhousie University, Canada*

Selective manipulation of actinide microparticles, **M. Soames**, *AWE, Aldermaston, UK*

Evaporation and crystallization of acoustically levitated salt/water drops, **R. Tuckermann**, *Universität Göttingen, Germany*

New techniques in the optical manipulation of aerosols, **Michael Summers, Daniel Rudd, Daniel Burnham and David McGloin**, *University of St Andrews, UK*

Laser Raman tweezers study of reactions of ozone with proxies for atmospheric aerosol, **Christian Pfrang, Martin D. King, Andy D. Ward and Oliver Hunt**, *Royal Holloway University of London, UK*

Cavity enhanced spectroscopy on microparticles in plasmas, **G. Thieme, R. Basner, J. Ehlbeck, J. Röpcke, H. Kersten, J. P. Reid and P. B. Davies**, *Institut für Niedertemperatur—Plasmaphysik (INP), Germany*

Manipulation of liquid aerosols with holographic optical traps, **J. R. Butler, L. Mitchem, J. Buajarern, D. Burnham, A. D. Ward and J. P. Reid**, *University of Bristol, UK*

Examining the presence of organic compounds in aqueous aerosol droplets by spectroscopic techniques, **Adele M. C. Laurain, Laura Mitchem, Jariya Buajarern and Jonathan P. Reid**, *University of Bristol, UK*

Rapid measurement of aerosol optical extinction using optical feedback cavity ring-down spectroscopy, **Timothy Butler, Daniel Mellon and Andrew Orr-Ewing**, *University of Bristol, UK*

New methods for the characterisation of optically-tweezed aerosol droplets: fluorescence and axial displacement, **K. J. Knox, K. L. Hanford, A. J. Hudson, L. Mitchem and J. P. Reid**, *University of Bristol, UK*

Understanding the influence of organic compounds on the hygroscopicity of aerosol particles, **Kate L. Hanford, Laura Mitchem and Jonathan P. Reid**, *University of Bristol, UK*

Surface characteristics of black carbon particles in relation to their toxicity, **Steve Smith, Phil Barnes, Howard Williams, Ken Donaldson, Andy Brown and Rik Brydson**, *King's College London, UK*

Characterising the optical properties of non-absorbing aerosol, **Rachael E. H. Miles, Svemir Rudić, Thomas A. Brenda, Andrew J. Orr-Ewing and Jonathan P. Reid**, *University of Bristol, UK*

Observation of a transition in the water–nanoparticle formation process at 167 K, **S. Bauerecker, A. Wargenau, M. Schultze, T. Kessler, R. Tuckermann and J. Reichardt**, *Universität Göttingen, Germany*

Optical trapping of liquid aerosol particles using counter-propagating beams, **D. Rudd, C. López-Mariscal and D. McGloin**, *University of St Andrews, UK*

Micro spectroscopy on high vapour pressure condensed matter under controlled gas and temperature environment: planned experiments, technical developments, and first results at the POLLUX Beamline, **Thomas Huthwelker, G. Tvetkov, M. Birrer, S. Sjogren, J. Raabe, E. Weingartner, M. Heuuberger-Vernooij and Markus Ammann**, *Paul Scherrer Institute, Switzerland*

Dynamics of optically trapped liquid aerosol droplets, **D. R. Burnham, P. J. Reece and D. M. McGloin**, *University of St Andrews, UK*

Classification of small ice crystal shapes using Fourier analysis of azimuthal scattering patterns, **Z. Ulanowski, C. Stopford, E. Hesse, P. H. Kaye, E. Hirst and M. Schnaiter**, *University of Hertfordshire, UK*

Phase separation during hygroscopic cycles in polyethylene glycol 400/ammonium sulfate system, **Gabriela Ciobanu, Claudia Marcolli, Ulrich K. Krieger, Uwe Weers and Thomas Peter**, *ETH Zurich, Switzerland*

Laser-based studies of reactions of important atmospheric radicals with organic compounds in the aqueous phase, **T. Schaefer, D. Hoffmann, C. Weller, K. Parajuli, P. Barzaghi and H. Herrmann**, *Leibniz Institute for Tropospheric Research, Germany*

Analysis and scaling of measured nucleation rates and comparison with classical theory, **Jussi Malila, Antti-Pekka Hyvärinen, Yrjö Viisanen and Ari Laaksonen**, *University of Kuopio, Finland*

The characterisation of multiphase organic/inorganic/aqueous aerosol droplets, **Jariya Buajarern, Laura Mitchem and Jonathan P. Reid**, *University of Bristol, UK*

Parametric resonance of optically trapped aerosols, **Jonathon Leach, Miles Padgett, Roberto Di Leonardo, David McGloin, Dan Burnham, John Girkin and Amanda Wright**, *University of Glasgow, Scotland*

Later-particle hydrodynamic couplings in optical tweezers, **S. Keen, J. Leach, G. M. Gibson, M. J. Padgett, R. Di Leonardo, S. Saunter and G. Love**, *University of Glasgow, Scotland*

Size effects in molecular ice nanoparticles: A study of the umbrella vibrational mode in ammonia particles, **George Firanescu**, **David Luckhaus and Ruth Signorell**, *University of British Columbia, Canada*

The Skinner prize for the best poster was awarded jointly to Rachael Mills from the University of Bristol, UK for her poster on characterising the optical properties of non-absorbing aerosols and Daniel Burnham from the University of St Andrews, UK for his poster on the dynamics of optically trapped liquid aerosol droplets.

List of participants

Miss J. Al Ajmi, *Public Authority of Applied Education and Training, Kuwait*
Dr M. Ammann, *Paul Scherrer Institute, Switzerland*
Professor S. Arnold, *Polytechnic University, New York, USA*
Professor C. Bain, *Durham University, UK*
Dr P. Bartlett, *University of Bristol, UK*
Dr K. Bower, *University of Manchester, UK*
Mr D. Burnham, *University of St Andrews, UK*
Mr J. Butler, *University of Bristol, UK*
Mr T. Butler, *University of Bristol, UK*
Professor C. Chan, *Hong Kong University of Science and Technology, Hong Kong*
Professor R. Chang, *Yale University, USA*
Professor T. Choularton, *University of Manchester, UK*
Ms G. Ciobanu, *ETH Zurich, Switzerland*
Professor D. Clary, *University of Oxford, UK*
Dr J. Corbett, *Malvern Instruments Ltd., UK*
Dr T. Cox, *University of Cambridge, UK*
Dr H. Crichton, *Royal Society of Chemistry, UK*
Professor P. Davidovits, *Boston College, USA*
Dr P. Davies, *University of Cambridge, UK*
Dr C. Delval, *Copenhagen Centre for Atmospheric Research, Denmark*
Ms R. Doherty, *Royal Society of Chemistry, UK*
M. Elliot, *Elliot Scientific Ltd., UK*
Ms G. Firanescu, *University of British Columbia, Canada*
Mr D. Gazzard, *Environment Agency, UK*
Dr M. Gibson, *Dalhousie University, Canada*
Professor J. Goree, *University of Iowa, USA*
Mr A. Gravell, *Environment Agency, UK*
Dr P. Griffiths, *University of Cambridge, UK*
Dr H. Grothe, *Vienna Institute of Technology, Austria*
Miss K. Hanford, *University of Bristol, UK*
Mr C. Homer, *University of Bristol, UK*
Dr R. Hopkins, *DSTL, UK*
Dr A. Hudson, *University of Bristol, UK*
Mr O. Hunt, *Rutherford Appleton Laboratory, UK*
Dr T. Huthwelker, *Paul Scherrer Institute, Switzerland*
Professor R. Jaenicke, *Universität Mainz, Germany*
Ms T. Kahan, *University of Toronto, UK*
Professor P. Kaye, *University of Hertfordshire, UK*
Mr S. Keen, *University of Glasgow, UK*
Professor H. Kersten, *University of Kiel, Germany*
Dr M. King, *Royal Holloway University of London, UK*
Dr U. Krieger, *ETH-Zurich, Switzerland*
Miss A. Laurain, *University of Bristol, UK*
Dr J. Leach, *University of Glasgow, UK*
Mr N. Lloyd, *University of Leicester, UK*
Dr H. Lunn, *Royal Society of Chemistry, UK*
J. MacLeod, *Photonics Solutions Plc, UK*
Mr J. Malila, *University of Kuopio, Finland*
Professor W. Martin, *University of Hertfordshire, UK*
Professor N. Mason, *Open University, UK*
Dr D. McGloin, *University of St Andrews, UK*
Mr D. Mellon, *University of Bristol, UK*

Mrs H. Meresman, *University of Bristol, UK*
Miss A. Middleton, *Royal Society of Chemistry, UK*
Miss R. Miles, *University of Bristol, UK*
Dr L. Mitchem, *University of Bristol, UK*
Dr B. Murray, *University of Leeds, UK*
Miss R. Needham, *Royal Society of Chemistry, UK*
Dr S. Parker, *Rutherford Appleton Laboratory, UK*
Dr J. Pavlů, *Charles University, Czech Republic*
Dr C. Pfrang, *Royal Holloway University of London, UK*
Dr F. Pope, *University of Cambridge, UK*
Professor A. Ray, *University of Kentucky, USA*
Dr J. Reid, *University of Bristol, UK*
Mr D. Rudd, *University of St Andrews, UK*
Dr S. Rudić, *University of Bristol, UK*
Professor Y. Rudich, *Weizmann Institute of Science, Israel*
Professor E. Rühl, *Freie Universität Berlin, Germany*
Mr T. Schaefer, *Leibniz-Institute for Troposphere Research, Germany*
Dr D. Shallcross, *University of Bristol, UK*
Professor M. Shiratani, *Kyushu University, Japan*
Professor R. Signorell, *University of British Columbia, Canada*
Dr S. Smith, *King's College London, UK*
Dr M. Soames, *AWE Plc, UK*
Professor W. Stoffels, *T U Eindhoven, The Netherlands*
Mr C. Stopford, *University of Hertfordshire, UK*
Mr M. Summers, *University of St Andrews, UK*
Mr J. Taylor, *Durham University, UK*
Ms G. Thieme, *INP Greifswald, Germany*
Mr L. Treuel, *University of Duisberg-Essen, Germany*
Mr R. Tuckermann, *Universität Göttingen, Germany*
F. Turner, *University of Manchester, UK*
Dr A. Ward, *STFC Rutherford Appleton Laboratory, UK*
Professor J. Wolf, *University of Geneva, Switzerland*
Miss L. Wong, *University of Durham, UK*
Mr A. Woodward, *University of Bristol, UK*
Professor Y. Zhang, *Beijing Institute of Technology, China*

Index of contributors*

Abo Riziq, A., **279**
Alfarra, M. R., **205**
Ammann, M., 193, 297
Anand, S., **335**
Arnold, S., **65**, 99, 403
Bain, C., 297, 403
Baltensperger, U., **205**
Bartlett, P., **319**, 403
Basner, R., **157**
Becker, U., **389**
Bonacina, L., **37**
Bonnet, C., **37**
Boutou, V., **37**
Bower, K., **205**, 403
Bresch, H., **389**
Burnham, D. R., **335**
Butler, J. R., 297, **351**, 403
Butler, T., 297
Čermák, I., **139**
Chan, C. K., **245**
Chang, R. K., **9**, 99, 193, 297, 403
Choularton, T., **205**, 297
Coe, H., **205**
Connolly, P., **205**
Corbett, J., 403
Courvoisier, F., **37**
Cox, T., 297
Cozic, J., **205**
Crawford, I., **205**
Crosier, J., **205**
Davidovits, P., 99, 193, 403, **425**
Davies, P., 99, 403
Dawes, A., **367**
Devarakonda, V., **85**
Dewar, N., **335**
Dinar, E., **279**
Drage, E. A., **367**
Eaimkhong, S., **319**
Erlick, C., **279**
Extermann, J., **37**
Flesch, R., **389**
Flynn, M., **205**
Gallagher, M. W., **205**
Gao, Z., **85**
Goree, J., 99, 193, 403
Graf, C., **389**
Grothe, H., **223**, 297
Guyon, L., **37**
Gysel, M., **205**
Hanford, K. L., **351**
Hayes, G., **367**
Homer, C., 297

Hopkins, R., 297
Hudson, A., 99, 403
Hughes, B. R., **173**
Hunt, O., 99, 297, 403
Huthwelker, T., 99, 297, 403
Iwashita, S., **127**
Jaenicke, R., 99, 193, **235**, 297, 403
Jetzki, M., **51**
Kahan, T., 193, 297, 403
Kaye, P., 99, 403
Keng, D., **65**
Kersten, H., 99, **157**, 193
King, M. D., **173**, 193, 297
Kiss, G., **279**
Koga, K., **127**
Kolchenko, V., **65**
Kreiger, U. K., 99, 193, 297, **377**, 403
Langer, B., **389**
Leach, J., 403
Lee, A. K. Y., **245**
Leisner, Th., **265**, **389**
Ling, T. Y., **245**
Malila, J., 297
Mason, N. J., 99, 193, 297, **367**, 403
McGloin, D., 193, **335**, 403
McPheat, R., **367**
Mellon, D., 297
Miles, R., 297
Mitchem, L., 297, **351**, 403
Murray, B., 297, 403
Nunomura, S., **127**
Němeček, Z., **139**
Ortega, K., **223**
Österreicher, B., **389**
Pan, Y.-P., **9**
Parker, S., 403
Pavlů, J., 99, **139**, 193, 403
Pfrang, C., **173**, 297
Rabitz, H., **37**
Ramjit, R., **65**
Ray, A. K., **85**, 99, 193, 297, 403
Reid, J., 99, 193, 297, **351**, 403
Remy, J., **115**
Richterovà, I., **139**
Roberts, G. S., **319**
Roth, M., **37**
Rudd, D., **335**
Rudić, S., 297, 403
Rudich, Y., 99, 193, **279**, 297
Rühl, E., 99, 193, 297, **389**, 403
Šafránková, J., **139**
Schulze, S., **265**

Shiratani, M., 99, **127**, 193
Signorell, R., **51**, 99, 193, 297, 403
Sjogren, S., **205**
Sorokin, M., **115**
Splinder, C., **279**
Stoffels, W. W., 99, **115**, 193
Summers, M. D., **335**
Targino, A., **205**
Taylor, J., 99
Teraoka, I., **65**
Thieme, G., **157**
Thompson, K. C., **173**
Thuiller, B., **37**

Tizek, H., **223**
Treuel, L., **265**, 297, **351**
Tuckermann, R., 99, 403
Verheggen, B., **205**
Ward, A. D., **173**, 297, 403
Wassermann, B., **389**
Webb, S. M., **367**
Weingartner, E., **205**
Wiese, R., **157**
Wood, T. A., **319**
Wolf, J.-P., **37**, 99
Zardini, A. A., **377**
Zellner, R., **265**

* The page numbers in **bold** type indicate papers submitted for discussions.